Innovations in Biotechnology for a Sustainable Future

Naga Raju Maddela • Luz Cecilia García
Editors

Innovations in Biotechnology for a Sustainable Future

Springer

Editors
Naga Raju Maddela
Facultad de Ciencias de la Salud and
Instituto de Investigación
Universidad Técnica de Manabí
Portoviejo, Ecuador

Luz Cecilia García
Instituto de Investigación and Facultad de
Ingeniería Agronómica
Universidad Técnica de Manabí
Portoviejo, Ecuador

ISBN 978-3-030-80107-6 ISBN 978-3-030-80108-3 (eBook)
https://doi.org/10.1007/978-3-030-80108-3

This Springer imprint is published by the registered company Springer Nature Switzerland AG
The registered company address is: Gewerbestrasse 11, 6330 Cham, Switzerland

Foreword

The book *Innovations in Biotechnology for a Sustainable Future* collects a good part of the biotechnological advances presented at the IV International Scientific Convention organized by the Technical University of Manabí, Ecuador (CCIUTM 2020). This book presents a series of independent chapters where different topics of great relevance and impact are addressed. Its creation is a reflection of our passion for scientific research and dissemination, and we hope that it will constitute a significant contribution to biotechnological knowledge, and especially that it challenges readers to verify-contrast innovative findings and results on different topics. In total, technical-scientific results are presented from more than 60 researchers from different countries, in which multidisciplinary and transdisciplinary studies are developed in fields such as phytochemistry, bioremediation, microbiology, and recovery of plant residues.

We live in the era of biotechnological revolution, and all the countries of the world are adopting measures to overcome their problems in the sectors: pharmaceutical, agricultural, food, environmental, and energy. At a general level, human society relates its future to the progress of biotechnology; it is expected that through different biotechnological procedures we can, among other things, reverse environmental damage caused by anthropogenic actions, sustainably take advantage of agro-industrial waste, increase the quality and production of food to meet the growing planetary demand, and find the cure for cancer and many other diseases. Currently, this transdisciplinary science has been leading in the creation of vaccines to combat the global pandemic due to COVID-19, which we have not yet been able to fully overcome today, but thanks to the collaboration of world scientists working towards the same goal, we hope that, sooner than later, this planetary health crisis is part of the past in the contemporary history of humanity.

However, the motivation to be part of this era must be supported by a great sense of responsibility and social commitment, paying special attention to scientific, social, political, and ethical issues. Governments, legislators, and society in general will increasingly depend on the knowledge and biotechnological heritage developed by researchers worldwide, which is why it is increasingly important to create and publish books like this one whose main objective is to socialize the advances and provide newer and more impressive results with the scientific community, students, teachers, and in general with those interested in expanding knowledge and experiences related to various topics of this vast discipline: Biotechnology.

Universidad Técnica de Manabí,
Portoviejo, Ecuador
3 May 2021

Vicente Véliz Briones

Preface

The book entitled *Innovations in Biotechnology for a Sustainable Future* typically aims to establish advances made by the allied fields of biotechnology to date. Advancement of science and technology for the welfare of humankind is gradually reaching the zenith. The contribution of biotechnology in this aspect is immense. This book in its present form has been designed in such a manner so that any enthusiast can know about recent achievements that have taken place in the environmental- and agricultural biotechnology fields. It can serve as a "handbook" dealing with modern technologies that evolved recently.

The book focuses on topics that comprise industrial, agricultural, environmental, and medical fields related to biotechnological aspects and covers studies that exhibit correlation between biological world and human impact over it in a nutshell. With increasing world population, there is a huge demand for food. Along with irregularities and constantly changing environmental conditions, the agricultural sector is faced with problems like plant diseases, soil toxicity, and degradation of soil characteristics. Biotechnology has been constantly developing technologies for dealing with improved crop production by means of genetic alterations. Another exploding problem arising presently is the huge generation of agro-industrial wastes. Biotechnology has vehemently formulated sustainable techniques to handle and convert agro-industrial wastes into alternate energy or other environment friendly products. Apart from it, a constructive study has been presented regarding the major issues of "*Agricultural Biotechnology*." Ranging from microscale to macroscale studies, it covers a huge domain of "*Environmental Biotechnology*." This book further gives immense importance to studies related to the fields that deal with the mitigation of environmental degradation components. The development of bio-based technologies in controlling and restoring soil and water characteristics has been provided in detail. Additionally, a separate section "*Medical Biotechnology*" has been included, which addresses various facets in the light of contemporary developments in antimicrobial therapeutic agents.

Overall, the book portrays a very clear idea about the emerging modern technologies and also directs young minds in the same path. This book has been designed to serve as a kind of information hub about modern sciences of biotechnology and its applied fields. It will also serve as a ready reference for practicing students and researchers in biotechnology, environmental engineering, chemical engineering, and other allied fields likewise. In order to reflect the proposed title of this book, 5 parts have been designed comprising 20 chapters: *Part I: Biotechnology Overview; Part II: Industrial Biotechnology; Part III: Agricultural Biotechnology; Part IV: Environmental Biotechnology*; and *Part V: Medicinal Biotechnology*. Part I consists of two chapters, where the first chapter provides a general description and the purpose of this book. The second chapter provides information about the current trends in biotechnology and its advances. Part II consists of five chapters, which focus on microbial proteases, essential oils in food safety, use of green waste for the production of unicellular biomass, probiotics, and bioplastics. Part III consists of three chapters, which focus on the impact of phosphorous in agricultural soils, impact of environmental factors on plant secondary metabolite production, and arbuscular mycorrhizal fungi. Part IV consists of six chapters, which were designed to present latest insights on microbial reductive dehalogenation, microbial utilization of nitroaromatics, microbe-metal interactions, microbial remediation of pharmaceutical and personal care products, biodetoxification of heavy metals in marine system, and biocoagulants. Part V consists of four chapters, which emphasize on the drug resistance mechanism in *Staphylococcus aureus*, plant secondary metabolites with anticancer properties, and medical and pharmaceutical applications of *Cassia grandis* L. fruit extractions. Towards the end of this book, i.e., Chapter "Challenges and Future Prospects of Biotechnology" have been discussed in detail. The chapters were contributed by 65 academicians/scientists of 12 different countries (Argentina, Brazil, Chile, Colombia, Ecuador, Honduras, India, Iran, Mexico, Nigeria, Spain, USA) across the world.

Portoviejo, Ecuador — Naga Raju Maddela
Portoviejo, Ecuador — Luz Cecilia García

Acknowledgments

We greatly acknowledge the support of the contributors to the book for their valuable contribution and timely responses during this project. Without their enthusiasm and support, this volume would not be ready in the scheduled time; therefore we really appreciate their cooperation and collaboration. We also thank anonymous reviewers for their constructive criticism, which helped us in improving the quality of this book by inviting experts to contribute additional chapters. We greatly acknowledge the Springer Editorial and Production team for their valuable support; without their guidelines, this project would not be finished in such a very short time. It is our honor to work with them, honestly. Finally, yet importantly, we are very much thankful to the colleagues at Universidad Técnica de Manabí (Ecuador) for their unconditional support and for the provision of valuable suggestions at the time of book proposal and final book preparation.

Naga Raju Maddela, Ph.D.,
Universidad Técnica de Manabí,
Portoviejo, Ecuador

Luz Cecilia García, Ph.D.,
Universidad Técnica de Manabí,
Portoviejo, Ecuador

Contents

Contributors

O. P. Abioye Department of Microbiology, Federal University of Technology, Minna, Nigeria

Cristóbal N. Aguilar Bioprocesses and Bioproducts Research Group, Food Research Department, School of Chemistry, Autonomous University of Coahuila, Saltillo, Coahuila, Mexico

S. A. Aransiola Bioresources Development Centre, National Biotechnology Development Agency, Ogbomoso, Nigeria

Swapna B. Department of Botany, Vikrama Simhapuri University PG Centre, Kavali, Andhra Pradesh, India

Ricardo J. Baquerizo-Crespo Departamento de Procesos Químicos, Facultad de Ciencias Matemáticas Físicas y Químicas, Universidad Técnica de Manabí, Portoviejo, Ecuador

Isabel Borrás-Linares Functional Food Research and Development Centre (CIDAF), Granada, Spain

Carlos Alfredo Cedeño-Palacios Departamento de Procesos Químicos, Facultad de Ciencias Matemáticas, Físicas y Químicas, Universidad Técnica de Manabí, Portoviejo, Ecuador

Gabriela S. Cedeño-Solórzano Departamento Producción, Empresa Purificadora Aqua Heredia, Aquaher S. A., Rocafuerte, Ecuador

M. Subhosh Chandra Department of Microbiology, Yogi Vemana University, Kadapa, Andhra Pradesh, India

Mónica L. Chávez-González Bioprocesses and Bioproducts Research Group, Food Research Department, School of Chemistry, Autonomous University of Coahuila, Saltillo, Coahuila, Mexico

Nanobioscience Group, School of Chemistry, Autonomous University of Coahuila, Saltillo, Coahuila, Mexico

María Hipatia Delgado-Demera Departamento de Ciencias Veterinarias, Facultad de Ciencias Veterinarias, Universidad Técnica de Manabí, Portoviejo, Ecuador

Alex Alberto Dueñas-Rivadeneira Departamento de Procesos Agroindustriales, Facultad de Ciencias Zootécnicas, Universidad Técnica de Manabí, Chone, Manabí, Ecuador

Kevin Jhon Fernández-Andrade Programa de Posgrado en Ingeniería Química, Instituto de Posgrado, Universidad Técnica de Manabí, Portoviejo, Ecuador
Departamento de Investigación y Desarrollo, Empresa Purificadora Aqua Heredia, Aquaher S. A., Rocafuerte, Ecuador

Andrea Guadalupe Flores-Valdes Bioprocesses and Bioproducts Research Group, Food Research Department, School of Chemistry, Autonomous University of Coahuila, Saltillo, Coahuila, Mexico
Nanobioscience Group, School of Chemistry, Autonomous University of Coahuila, Saltillo, Coahuila, Mexico

Luz Cecilia García Instituto de Investigación Científica, Universidad Técnica de Manabí, Portoviejo, Ecuador
Facultad de Ingeniería Agronómica, Universidad Técnica de Manabí, Portoviejo, Ecuador

Archana Giri Centre for Biotechnology, Institute of Science and Technology, Jawaharlal Nehru Technological University, Hyderabad, India

Yunet Gómez-Salcedo Departamento de Procesos Químicos, Facultad de Ciencias Matemáticas Físicas y Químicas, Universidad Técnica de Manabí, Portoviejo, Ecuador

Raquel Guerrero-Chuez Universidad Técnica Estatal de Quevedo, Quevedo, Ecuador

Luis Guillermo Hernádez-Montiel Centro de Investigaciones Biológicas del Noroeste, S.C., Calle Instituto Politécnico Nacional No. 195, La Paz, Baja California Sur, México

A. A. Ikhumetse Department of Microbiology, Federal University of Technology, Minna, Nigeria

Anna Ilyina Nanobioscience Group, School of Chemistry, Autonomous University of Coahuila, Saltillo, Coahuila, Mexico

Seyed Mehdi Jazayeri Faculty of Biology, University-College of Science, University of Tehran, Tehran, Iran
Departamento de Biología, Facultad de Ciencias, Universidad Nacional de Bogotá, Bogotá, Colombia

Srinivasan Kameswaran Department of Botany, Vikrama Simhapuri University PG Centre, Kavali, Andhra Pradesh, India

Priyanka Kandula Centre for Biotechnology, Institute of Science and Technology, Jawaharlal Nehru Technological University, Hyderabad, India

B. Vijay Kumar Department of Food Technology, Vikrama Simhapuri University, Nellore, Andhra Pradesh, India

K. Naveen Kumar Department of Chemistry, Vikrama Simhapuri University, Nellore, Andhra Pradesh, India

Liliana Lara-Capistrán Centro de Investigaciones Atmosféricas y de Ecología, perteneciente a la Universidad Popular Autónoma Veracruzana (UPAV), Veracruz, México
Facultad de Ciencias Agrícolas, Universidad Veracruzana, Veracruz, México

Caroline Lissette Loor-Moreira Programa de Posgrado en Ingeniería Química, Instituto de Posgrado, Universidad Técnica de Manabí, Portoviejo, Ecuador

Virginia Monserrate López-Zambrano Maestría en Agroindustria, Instituto de Posgrado, Universidad Técnica de Manabí, Portoviejo, Ecuador

Naga Raju Maddela Departamento de Ciencias Biológicas, Facultad de Ciencias de la Salud, Universidad Técnica de Manabí, Portoviejo, Ecuador
Grupo de Investigación en Biodiversidad y Ecología de Ecosistemas Acuáticos, Departamento de Acuicultura Pesca y Recursos Naturales Renovables, Facultad de Ciencias Veterinarias, Universidad Técnica de Manabí, Bahía de Caráquez, Manabí, Ecuador

A. Madhavi Department of Microbiology, Sri Krishnadevaraya University, Anantapuramu, India

Silvina Maldonado Facultad de Ingeniería Industrial, Universidad Nacional de Jujuy, San Salvador de Jujuy, Argentina

Gema M. Manzaba-Salazar Departamento de Procesos Químicos, Facultad de Ciencias Matemáticas Físicas y Químicas, Universidad Técnica de Manabí, Portoviejo, Ecuador

Jhunior Marcía-Fuentes Technology Sciences Faculty, Universidad Nacional de Agricultura, Catacamas, Honduras

José L. Martínez-Hernández Nanobioscience Group, School of Chemistry, Autonomous University of Coahuila, Saltillo, Coahuila, Mexico

G. Jaffer Mohiddin Ciencia De la vida y Agricultura, Sede Santo Domingo, Universidad de las Fuerzas Armadas – ESPE, Sangolquí, Ecuador

Carlos Alberto Molina Hidrovo Instituto Nacional de Investigaciones Agropecuarias, Guayaquil, Ecuador

Anjaneyulu Musini Centre for Biotechnology, Institute of Science and Technology, Jawaharlal Nehru Technological University, Hyderabad, India

Byron Oviedo-Bayas Universidad Técnica Estatal de Quevedo, Quevedo, Ecuador

Dante Pinochet Tejos Institute of Agricultural Engineering and Soils, Faculty of Agricultural Sciences, University Austral of Chile, Valdivia, Chile
Centro de investigación de Suelos Volcánicos (CISVO), Universidad Austral de Chile, Valdivia, Chile

Gopi Krishna Pitchika Department of Zoology, Vikrama Simhapuri University PG Centre, Kavali, Andhra Pradesh, India

Carla Quilumbango Grijalva School of Biological Sciences and Engineering, Yachay Tech University, Urcuquí, Ecuador

M. Ramakrishna Department of Botany, Vikrama Simhapuri University PG Centre, Kavali, Andhra Pradesh, India

Bellamkonda Ramesh Department of Food Technology, Vikrama Simhapuri University, Nellore, Andhra Pradesh, India

V. Rangaswamy Department of Microbiology, Sri Krishnadevaraya University, Anantapuramu, India

Pabbati Ranjit Centre for Biotechnology, Institute of Science & Technology, Jawaharlal Nehru Technological University Hyderabad, Hyderabad, Telangana, India

G. Vidya Sagar Reddy Department of Biotechnology, Vikrama Simhapuri University, Nellore, Andhra Pradesh, India

Kondakindi Venkateswar Reddy Centre for Biotechnology, Institute of Science & Technology, Jawaharlal Nehru Technological University Hyderabad, Hyderabad, Telangana, India

Juan José Reyes-Pérez Universidad Técnica Estatal de Quevedo, Facultad de Ciencias Agropecuarias, Quevedo, Los Ríos, Ecuador

María Antonieta Riera Facultad de Ciencias Matemáticas, Físicas y Químicas, Universidad Técnica de Manabí, Portoviejo, Ecuador
Doctorado en Ingeniería Industrial, Universidad Nacional de Cuyo, Mendoza, Argentina

Joan Manuel Rodríguez-Díaz Departamento de Procesos Químicos, Facultad de Ciencias Matemáticas Físicas y Químicas, Universidad Técnica de Manabí, Portoviejo, Ecuador
Programa de Pós-graduação em Engenharia Química, Universidade Federal da Paraíba, João Pessoa, Brazil

Jesús Lozano Sánchez Functional Food Research and Development Centre (CIDAF), Granada, Spain
Department of Food Science and Nutrition, University of Granada, Granada, Spain

Michelle Sánchez García School of Biological Sciences and Engineering, Yachay Tech University, Urcuquí, Ecuador

Ricardo Santos-Aleman School of Nutrition and Food Sciences, Louisiana State University, Baton Rouge, LA, USA

Verónica Segovia Motesdeoca Universidad Técnica Estatal de Quevedo, Quevedo, Ecuador

Somavarapu Silpa Department of Food Technology, Vikrama Simhapuri University, Nellore, Andhra Pradesh, India

M. Srinivasulu Department of Biotechnology, Yogi Vemana University, Kadapa, Andhra Pradesh, India

Yenny Torres-Navarrete Universidad Técnica Estatal de Quevedo, Quevedo, Ecuador

Chittari Amaravathi Sneha Latha Varma Centre for Biotechnology, Institute of Science & Technology, Jawaharlal Nehru Technological University Hyderabad, Hyderabad, Telangana, India

Gregorio Vásconez Montúfar Universidad Técnica Estatal de Quevedo, Quevedo, Ecuador
Graduate School, Faculty of Agricultural Sciences, University Austral of Chile, Valdivia, Chile

Ch. Venkatrayulu Department of Food Technology, Vikrama Simhapuri University, Nellore, Andhra Pradesh, India
Department of Marine Biology, Vikrama Simhapuri University, Nellore, Andhra Pradesh, India

M. O. Victor-Ekwebelem Department of Biology/Microbiology/Biotechnology, Alex Ekwueme Federal University, Abakaliki, Ebonyi, Nigeria

Ch. Vijaya Department of Marine Biology, Vikrama Simhapuri University, Nellore, Andhra Pradesh, India

Ronald Oswaldo Villamar-Torres Universidad Técnica Estatal de Quevedo, Facultad de Ciencias Agropecuarias, Quevedo, Los Ríos, Ecuador
Instituto Superior Tecnológico "Ciudad de Valencia"–Tecnología en Producción Agrícola y Tecnología en Procesamiento de Alimentos, Quevedo, Los Ríos, Ecuador

P. Suresh Yadav Department of Microbiology, Yogi Vemana University, Kadapa, India

José Daniel Zambrano-Veliz Carrera de Industrias Agropecuarias, Facultad de Ciencias Zootécnicas, Universidad Técnica de Manabí, Chone, Ecuador

Ramón Zulueta-Rodríguez Facultad de Ciencias Agrícolas, Universidad Veracruzana, Veracruz, México

About the Editors

Naga Raju Maddela received his M.Sc. (1996–1998) and Ph.D. (2012) in Microbiology from Sri Krishnadevaraya University, Anantapuramu, India. During his doctoral program in the area of Environmental Microbiology, he investigated the effects of industrial effluents/insecticides on soil microorganisms and their biological activities and worked as a Faculty in Microbiology for 17 years, teaching undergraduate and postgraduate students. He received "Prometeo Investigator Fellowship" (2013–2015) from Secretaría de Educación Superior, Ciencia, Tecnología e Innovación (SENESCYT), Ecuador, and "Postdoctoral Fellowship" (2016–2018) from Sun Yat-sen University, China. He also received external funding from "China Postdoctoral Science Foundation" in 2017, internal funding from "Universidad Técnica de Manabí" in 2020, worked in the area of Environmental Biotechnology, participated in 20 national/international conferences, and presented research data in China, Cuba, Ecuador, India, and Singapore. Currently, he is working as a full-time Professor at the Facultad de Ciencias de la Salud, Universidad Técnica de Manabí, Portoviejo, Ecuador. He has published 5 books (Springer), 10 chapters (InTech Open/Springer), and 45 research papers.

Luz Cecilia García received her Ph.D. in Forest Sciences from the Universidad Austral de Chile (UACH), Valdivia, Chile, in 2014. Since 2016, she is a full-time Professor at the Faculty of Agronomic Engineering, Universidad Técnica de Manabí (UTM), Portoviejo, Ecuador. She is also working as a Director of Institute of Investigation at UTM since 2017. Her line of research is "Conservation and improvement of plant genetic resources." She directs a program of "Ecophysiology, conservation and genetic improvement of Theobroma cacao" in which higher genotypes are being developed with better production, disease tolerance, drought tolerance, and excellent organoleptic characteristics. To her credit, there are more than 20 scientific publications and she has participated as a speaker in several national and international conferences. She has received some distinctions and awards, such as best graduate of the doctoral program (UACH), scholarships from MECESUP (Programa de Mejoramiento de la Calidad y la Equidad de la Educación Superior, Govt. of Chile, Chile) and SENESCYT (Secretaría de Educación Superior, Ciencia, Tecnología e Innovación, Govt. of Ecuador, Ecuador) for her doctoral study, awards as a leading researcher at the Universidad Técnica Estatal de Quevedo, Ecuador (2014–2016) and at the UTM (2016–present), best teacher of the Faculty of Agronomic Engineering 2019, and II Place in National MAGAP (Ministerio de Agricultura, Ganadería, Acuacultura y Pesca, Govt. of Ecuador, Ecuador) Agri-Economic Contest, GIZ (Gesellschaft für Internationale Zusammenarbeit).

Part I
Biotechnology Overview

Biotechnology: An Editorial Overview

Naga Raju Maddela and Luz Cecilia García

1 Historical Background

"Biotechnology" is considered as a broad area of biology, which depends on the living systems to develop or make products. Biotechnology has expanded to different fields in the late-twentieth and early twenty-first centuries by the successful implementations of genomics, recombinant gene technologies, immune techniques, emerging therapeutic approaches, and diagnostics. There is a closely similar or overlapping field to biotechnology is bioengineering, however, bioengineering mainly emphasizes higher system approaches. Therefore, bioengineering is the application of engineering and natural science to tissues, cells and molecules; and such studies are likely to improve the functions of plants and animals. The term biotechnology has been defined in different angles as shown below (SLH, 2010; Verma et al., 2011; Wikipedia, 2021):

> *Production of products from raw materials with the aid of living organisms—Biotechnology*—Karl Ereky, 1919.

> *Application of biological organism, systems, or processes by various industries to learning about the science of life and the improvement of the value of materials and organism such as pharmaceuticals, crops, and livestock*—American Chemical Society.

N. R. Maddela (✉)
Departamento de Ciencias Biológicas, Facultad de Ciencias de la Salud, Universidad Técnica de Manabí, Portoviejo, Ecuador

Grupo de Investigación en Biodiversidad y Ecología de Ecosistemas Acuáticos, Departamento de Acuicultura Pesca y Recursos Naturales Renovables, Facultad de Ciencias Veterinarias, Universidad Técnica de Manabí, Bahía de Caráquez, Manabí, Ecuador
e-mail: raju.maddela@utm.edu.ec

L. C. García
Instituto de Investigación Científica, Universidad Técnica de Manabí, Portoviejo, Ecuador

Facultad de Ingeniería Agronómica, Universidad Técnica de Manabí, Portoviejo, Ecuador

N. R. Maddela, L. C. García (eds.), *Innovations in Biotechnology for a Sustainable Future*, https://doi.org/10.1007/978-3-030-80108-3_1

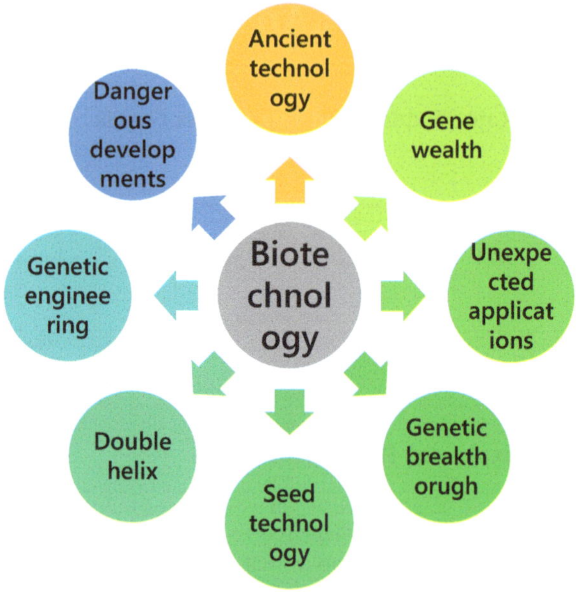

Fig. 1 Advances in biotechnology

> *Biotechnology is the integration of natural science and organism, cells, parts, therefore, and molecular analogues for products and services*—European Federation of Biotechnology.

> *The use of living things to make products*—American Association for the Advancement of Science (AAAS).

> *Any technological application that uses biological systems, living organisms, or derivatives thereof, to make or modify products or processes for specific use*—Convention on Biological Diversity.

> *The application of science and technology to living organisms, as well as parts, products and models therefore, to alter living or non-living materials for the production of knowledge, goods and services*—Organization for Economic Cooperation and Development (OECD).

According to the growth in human population and their needs, biotechnology incorporates a wide range emerging and innovative techniques and procedures for the modification of living systems, these include but not limited to nano techniques, genetic engineering, cell and tissue culture procedures. According to the developments and achievements in biotechnology with time (Fig. 1), biotechnology can be classified into three stages: (1) Ancient Biotechnology, (2) Classical Biotechnology, and (3) Modern Biotechnology (Verma et al., 2011). *Ancient Biotechnology* belongs to the period before the year of 1800; main breakthroughs in this era were based on the basic observations about nature which could help in the improvisation of human life during this period. One of the historical developments in the Ancient Biotechnology period was fermentation technology. The *Classical Biotechnology* period belongs to the year 1800 to mid of twentieth century. The main achievements in the period were; discovery of laws of inheritance by Gregor John Mendel in 1865,

Table 1 Major inventions in Medical Biotechnology the period 2000–2020 (Timeline, 2021)

Year	Achievement
2001	Human genome sequence draft created by Ceiera Genomics and Human Genome project.
2002	Compete genome of rice has been decoded, and it becomes the first crop to have its genome decoded.
2003	Location and sequence of human genes on all 46 chromosomes has been completed.
2008	First Medical Experiment Module (Kibo) has been launched by Japanese astronomers; Kibo will be used on the ISS (International Space Station).
2009	Modified SAN heart genes were used by Cedars-Sinai Heart Institute to create the first viral pacemaker in experimental animal i.e., guinea pigs.
2012	A nervous system-controlled bionic leg was successfully used to climb the Chicago Willis Tower.
2019	A new method of genetic engineering i.e., Prime Editing has been described which is superior than CRISPR.
2020	3D bioprint graphene oxide with a protein has been developed.
	Development of synthetic red blood cells
	Human thymus was produced by using stem cells and a bioengineered scaffold.
2021	A tenfold effective super-bug targeting formicamycin antibiotic has been produced by using CRISPR/Cas9 genome.

theory of evolution by Charles Darwinin 1858, nucleus in cells by Robert Brown in 1831, DNA as a genetic material and its role of DNA in the transfer of genetic information by Fredrich Miescher in 1869, first ever solid medium for culturing microorganisms by Robert Koch in 1881, principle of genetics in inheritance and theory of gene by T.H. Morgan in 1910, world's first antibiotic i.e., penicillin by Alexander Flemming in 1928, etc. The *Modern Biotechnology* era belongs to the period after mid-twentieth century. Hallmarks of this period include the discovery of double helix model of DNA by JD Watson and FHC Crick in 1953, concept of operon by Jacob and Monad in 1961, synthesis of DNA in test tube by Karl Mullis in 1983, animal cloning by Ian Wilmut in 1996, etc. Likewise, research related to modern biotechnology has led to many breakthroughs in medicine such as transgenics, monoclonal antibodies, vaccines, antibiotics, human genome projects etc. (Table 1).

2 Scope and Branches of Biotechnology

Biotechnology has a marked impact on different aspects of human welfare including food processing, protecting the environment and human health. The principal scope of biotechnology is as follows:

- Sustainable food production by using available land to meet the demand of a growing population.
- Search for disease-resistant and high-yield crop varieties.

- Introducing harmless biofertilizers and subsequently minimize the use of harmful chemical fertilizers.
- Integrated pest management by introducing biocides in agriculture.
- Preservation of germplasm of higher (plants and animals) and lower organisms (microorganisms).
- Production of sustainable pharmaceutical products for the treatment of life-threatening diseases in humans and animals.
- Production of biofuels in order to discourage deforestation which is a common practice fuel wood.
- Achieving sustainability in food safety by means of microbiological applications.
- Highlight the environmental applications of microorganisms such as extraction of minerals for poor quality ores, remediation of contaminated sites, production of microbial based nanomaterials, etc.

Keeping in view of above scope, principally biotechnology has the following branches—Industrial Biotechnology, Agricultural Biotechnology, Environmental Biotechnology, Medical Biotechnology. *Industrial biotechnology* deals with the activities related to production of different organics (such as acetic acid, citric acid, acetone, glycerin, antibiotics etc.) that have significant importance in the medicine and other areas (e.g., food industry). *Agricultural biotechnology* emphasizes on the development of transgenic plants with a better resistance to the biotic and abiotic stress factors, development of haploids, rescue of embryos, multiplication of clones, cryopreservation, mitigation of plant diseases by using biological agents (e.g., virus, bacteria, fungi, etc.). *Environmental biotechnology* deals with various aspects such as detoxification of waste and industrial effluents, treatment of water, and wastewater, etc. (Kuppusamy et al., 2020b; Maddela et al., 2021). Whereas the areas of *Medical biotechnology* including but not limited to diagnosis of diseases, large-scale production of drugs, antibiotics, hormones, and vaccines.

3 Outlines of the Volume: Parts, Chapters

This volume consists of five parts—(I) Biotechnology overview, (II) Industrial biotechnology, (III) Agricultural biotechnology, (IV) Environmental biotechnology, and (V) Medical biotechnology. Two chapters have been included under the part of "Biotechnology Overview." Chapter "Biotechnology: An Editorial Overview" is an "Editorial Overview" where we intended to briefly discuss the purpose of this volume and its salient features. Chapter "Biotechnology of Twenty-First Century" emphasizes on different topics such as major advances in biotechnology between 2000 and 2020, benefits due to advances in biotechnology, global research in biotechnology, and expansion of biotechnology in the private and public sector. Additionally, this chapter has tried to focus on different branches of biotechnology. There is in-depth information on the components and importance of environmental biotechnology, importance of biotechnological applications in bioremediation and phytoremediation, mitigation of pollution in different components of the environment (air, water, and soil). There is also focus on the importance of biotechnology in plant growth and yield, different agricultural engineering techniques (such as breeding techniques, genetic engineering approaches, organic farming practices), and global food security. In this chapters, information related to Industrial biotechnology has also been provided, where an in-depth review of literature has been done related to industrial products and food/dairy industry. Toward the end of this chapter, there is information on biotechnological advances in pharmaceutics, vaccinology, and emergence of novel methodologies. Finally, challenges and knowledge gaps for the future developments in biotechnology have been discussed. It is important to note that a very recent literature review has raised a doubt that is organic produce is free from the environmental contaminants (Ramakrishnan et al., 2021), which implies that the available biotechnological practices in organic farming should be revalidated.

Part II of this volume belongs to Industrial Biotechnology, which has been designed by incorporating five chapters (chapters "Enzymes from Microorganisms", "Biotechnological Applications of Essential Oils: Post-harvest and Food Preservation", "Use of Waste from the Citrus Industry for the Production of Unicellular Biomass", "Organic Waste: A Cheaper Source for Probiotics Production", "Agro-Industrial Waste as an Option for the Sustainable Development of Bioplastic") with recent advances in the respective domain. Chapter "Enzymes from Microorganisms" has paid special attention toward proteases, such as exopeptidases, endopeptidases, cysteine proteases, distribution and sources of proteases, microbial proteases, and applications of proteases in different industries (e.g., detergent, leather, food and feed, silk degumming, photographic, pharmaceutical, and biofuels). Nevertheless, this chapter highlights the microbial proteases over plant and animal counterparts. It is noteworthy that proteases have significant importance in the therapeutic uses (Xue et al., 2021), plant diseases control (Wang et al., 2020b), industrial applications (Barzkar, 2020), etc. Chapter "Biotechnological Applications of Essential Oils: Postharvest and Food Preservation," was intended to provide information on the following parameters—food preservation, essential oils as food preservatives, food biotechnology, conventional and emerging methods of essential oil extraction, essential oils and their biotechnological applications in postharvest, physicochemical parameters of essential oils, essential oils as secondary metabolites, and antimicrobial activities of essential oils. Overall, this chapter compiles information on the biotechnological applications of essential oils to reduce the proliferation of microorganisms that cause foodborne diseases and that decrease the shelf life during storage of the fruits. Nowadays, there is a significant attraction toward plant essential oils for a food safety (Bhavaniramya et al., 2019; Chen et al., 2021a; Zhu et al., 2021). Chapter "Use of Waste from the Citrus Industry for the Production of Unicellular Biomass" deals with the production of unicellular biomass from the waste of the citrus industry. The main contents of this chapter include treatment of lignocellulosic materials, chemical composition of citrus peel powder, growth kinetics and fermentation by *Candida utilis*, and optimization. Overall, this chapter provides useful insights in understanding the viability of using orange peel residues as a substrate for the production of unicellular biomass of *C. utilis*, thus this chapter highlights the reuse of green waste as sustainable substrate in fermentation technology. It should be remembered that biomass production is a sustainable way to achieve bioeconomy (Antar et al., 2021), hence, there is a much attention on the reuse of waste for biomass production (Wang et al., 2020a; Makaroglou et al., 2021; Shahid et al., 2021). Chapter "Organic Waste: A Cheaper Source for Probiotics Production" emphasizes on the production of probiotics by using organic waste as an economical substrate. The chapter starts with a brief overview on organic wastes and probiotics, then this chapter emphasizes on the different organic wastes (e.g., agricultural wastes, waste from vegetable processing, dairy wastewaters, fish processing wastes, fermented silages, waste from meat processing, municipal wastes) as substrate for probiotic production. Likewise, this chapter deals with various sources of organic wastes which can be used as cheaper sources for production of different kinds of probiotics; this could have significant importance in the sustainable production of probiotics (Ashayerizadeh et al., 2017; Vodnar et al., 2019). Chapter "Agro-Industrial Waste as an Option for the Sustainable Development of Bioplastic" is related to the production of bioplastics from agro-industrial wastes. The main contents of this chapter include consumer society versus sustainable production, valorization of agro-industrial waste, biorefineries and transformation processes, and the futurity of bioplastics. On the whole, this chapter addresses the use of agro-industrial waste as second-generation raw material, to obtain bioplastics through sustainable processes that have characteristics similar to traditional plastics, capable of meeting the various needs of use existing in the market. In the recent time, there is a much attention toward bioplastic production of different waste substrates (Tsang et al., 2019; Jõgi & Bhat, 2020; Khatami et al., 2021).

Part III of this volume belongs to the Agricultural Biotechnology, which includes three chapters (chapters "Flow and Distribution of Phosphorus in Soils from a Geochemical and Agronomic Approach", "Environmental Factors Enhance Production of Plant Secondary Metabolites Toward More Tolerance and Human Health: Cocoa and Coffee Two Model Species", "Diversity and Ecology of Arbuscular Mycorrhization Fungi"). Chapter "Flow and Distribution of Phosphorus in Soils from a Geochemical and Agronomic Approach" deals with the flow and distribution of phosphorus in soils as studied by geochemical and agronomic approaches. The focused areas of this chapter include residual effect of phosphorus in soils, phosphorus shapes (organic and inorganic) in soils, fractionation of phosphorus contained in soils, and considerations for studies of phosphorus fertility of soils. Overall, this chapter concludes that agronomic availability of phosphorus, which is estimated through routine laboratory methods, is the result of the distribution and subsequent balance of phosphorus added between the fractions that make up phosphorus in soils. Levels of phosphorus in soil have significant influence on crop yield (Mian et al., 2021; Waani et al., 2021), therefore it is necessary to characterize the flow and distribution of phosphorus in agricultural soils. Chapter "Environmental Factors Enhance Production of Plant Secondary Metabolites Toward More Tolerance and Human Health: Cocoa and Coffee Two Model Species" provides insights on the impact of environmental factors in enhancing the production of plant (cocoa and coffee) secondary metabolites toward tolerance and human health. The contents of this chapter include stress and plant responses, secondary metabolites in plants, production of secondary metabolites under environmental cues and stress, and multifunctionalities of secondary metabolites. Finally, there is a special emphasis on the cocoa and coffee as model species for potential secondary metabolites production and adaptations. Overall, this chapter concludes that cacao and coffee are two plant models having known SMs with pharmaceutical/medicinal/nutritional values that make them tolerant to adverse conditions and positively produced upon plant exposure to stress. Characterization of plant secondary metabolites is one of the active research areas in the field of medicine (Kongkham et al., 2020; Mahajan et al., 2020; Ogbe et al., 2020). Chapter "Diversity and Ecology of Arbuscular Mycorrhization Fungi" deals with the following topics—arbuscular mycorrhizal fungi (AMF), systematics (taxonomy) of AMF, and diversity and ecology of AMF. Overall, this chapter emphasizes on some groups of taxonomists of AMF and some morphological characteristics such as a group of walls, shapes, color, etc., of this group of fungi., as well as its diversity and ecology of this symbiosis in natural ecosystems and agroecosystems. AMF is one of the key elements in the soil fertility and plant growth by minimizing the stress (biotic and abiotic), hence research lines in the area of AMF are always in trending (Riaz et al., 2021; Shen & Zhu, 2021; Zhao et al., 2021).

Part IV of this volume consists of six chapters (chapters "Microbial Reductive Dehalogenation and Its Role in Bioremediation", "Microbial Capacities for Utilization of Nitroaromatics", "Microbial Interaction with Metals and Metalloids", "Microbial Remediation of Pharmaceuticals and Personal Care Products", "Biodetoxification of Toxic Heavy Metals by Marine Metal Resistant Bacteria: A Novel Approach for Bioremediation of the Polluted Saline Environment", "Generalities of the Coagulation-Flocculation Process: A Perspective on Biocoagulants") which are related to Environmental Biotechnology. Chapter "Microbial Reductive Dehalogenation and Its Role in Bioremediation" deals with microbial reductive dehalogenation and its role in bioremediation. This chapter presents the following topics—mechanism of reductive dehalogenation, aerobic growth on halogenated aliphatic and aromatic compounds, halorespiration, factors influencing the dichlorination of polychlorinated biphenyls (PCBs), molecular aspects of dehalogenase gene, genomic diversity of dehalorespiration, PCB-dehalogenating bacteria and consortia, and acquisition and distribution of dehalogenase genes. Overall, this chapter focuses on the types of halo-organic compounds that have become a significant, potent, theoretically significant soil pollution, to be bioremediated. Microbial reductive dehalogenation is one of the principal mechanisms in the restoration of organohalide-contaminated sites (Nijenhuis & Kuntze, 2016; Lu et al., 2021). It is important to note that the remediation of contaminated media is always challenging, and contaminants can only be removed at field level with proper lab- and plot-scale level experiments (Maddela et al., 2015a, b; Kuppusamy et al., 2020a). Chapter "Microbial Capacities for Utilization of Nitroaromatics" is intended to provide latest insights on anaerobic and aerobic biodegradation of nitroaromatics, degradation pathways at molecular level, challenges in the biodegradation, a special emphasis on the biodegradation of selective nitroaromatics such as nitrobenzoate, nitrobenzaldehyde, trinitrotoluene, and chloronitrobenzene, microorganisms degrading nitroaromatics, and their genetics. It is noteworthy that nitroaromatics are important environmental contaminants that are released and have toxic effects on ecosystems; and microbial utilization capacities for nitroaromatics under aerobic conditions is presented in this chapter. Microbial removal of nitroaromatics is one if the active research areas in the bioremediation (Tiwari et al., 2020; Luo et al., 2021). Chapter "Microbial Interaction with Metals and Metalloids" presents latest insights on microbial interactions with metals and metalloids. Topics that are covered in this chapter include types and levels of microbial interactions with metals and metalloids, natural occurrences of interactions, metal–microbes interactions, bioremediation by adsorption, biosorption, natural occurrences of metal–microbe interaction, metal mobilization and immobilization, metal tolerance and resistance in microbes, molecular insights on microbial interactions with the selected metals, and biological remediation of selected metals. The main conclusions of this chapter are—heavy metals are an essential and important trace element, but as these heavy metals increase in concentration due to natural or industrial activities, they become toxic to many microbes; on the other hand, microbes have adapted to tolerate minerals or can even use them for grow; hence this interaction between microbes and minerals on environmental matrices is an essential part of the Earth's biogeochemical cycle. Microorganisms present in the crude oil-polluted soils have great potential to absorb heavy metals (Maddela et al. 2015a), and understanding over the interactions between microorganisms and metals will help in the development of an efficient strategies for the remediation of metals-contaminated sites (Yu et al., 2020). Also, there is a great threat to the ecosystem and human health due to the presence of heavy metals in the soil (Maddela et al., 2020b). Chapter "Microbial Remediation of Pharmaceuticals and Personal Care Products" deals with the microbial remediation of pharmaceuticals and personal care products (PPCPs). In this chapter, there is a special attention on biosorption of PPCPs, role of bacteria, fungi and plants in the remediation of PPCPs-contaminated sites, biodegradation of pharmaceutical compounds, pure and mixed culture studies, and toxic effects of PPCPs. The main conclusion of this chapter is that PPCPs have adverse toxic effects on ecosystems, as well as human health, therefore, it is essential to remediate the PPCPs-contaminated sites by using novel microorganisms. There is a great concern about the toxicity of PPCPs, hence there is a continuous search for the development of emerging bioremediation techniques (Kang et al., 2021). Chapter "Biodetoxification of Toxic Heavy Metals by Marine Metal Resistant Bacteria: A Novel Approach for Bioremediation of the Polluted Saline Environment" is about biodetoxification of toxic heavy metals by marine metal resistant bacteria; and in this direction, this chapter mainly highlights the sources, toxic effects and microbial detoxification of selected heavy metals in the marine ecosystem, The principal conclusions of this chapter are: heavy metals are generally toxic to the body at very low level; the main mechanism of heavy metal toxicity include the generation of free radicals to cause oxidative stress, damage of biological molecules such as enzymes, proteins, lipids, and nucleic acids, damage of

DNA which is key to carcinogenesis as well as neurotoxicity; microbes have various mechanisms of metal sequestration that hold greater metal biosorption capacities; and several microorganisms like bacteria, fungi, and algae have been used to clean up heavy metal contaminated environments. Several microbial systems have been identified for the effective removal of heavy metals in the marine system (Poo et al., 2018; Chen et al., 2021b; Djinni & Djoudi, 2021). Chapter "Generalities of the Coagulation-Flocculation Process: A Perspective on Biocoagulants," is intended to provide in depth insights on colloidal systems, and fundamentals and kinetic aspects of coagulation-flocculation. Overall, the rationale of this chapter is—water treatment is a necessity for social and industrial development; coagulation-flocculation is a fundamental process for the reduction of colloidal particles present in the water to be treated. The use of synthetic coagulants in effluent or wastewater treatment leads to a high production of nonbiodegradable sludge and water containing trace elements that are harmful to ecosystems. Therefore, biocoagulants are a very efficient alternative that produces a low volume of sludge and has no harmful effects on flora or fauna. Now a days, biocoagulants are widely used in the restoration of contaminated media (Frantz et al., 2020; Miyashiro et al., 2021).

Finally, four chapters (chapters "Drug Resistance Mechanism in *Staphylococcus aureus*", "Anticancer Secondary Metabolites Found in Native Ecuadorian Plant Species *Uncaria tomentosa* DC. (Rubiaceae), *Croton lechleri* Müll. Arg. (Euphorbiaceae), and *Equisetum giganteum* L. (Equisetaceae)", "The Carao (*Cassia grandis* L.): Its Potential Usage in Pharmacological, Nutritional, and Medicinal Applications", "Challenges and Future Prospects of Biotechnology") have been included under Part V, and this part is about Medical Biotechnology. Chapter "Drug Resistance Mechanism in *Staphylococcus aureus*" focuses on risk groups, epidemiology, genetic mobile components and *S. aureus* genome, plasmids encode antibiotic resistance, action of antibiotics and mechanism, kinetic mechanism of resistance of *S. aureus* to penicillin, methicillin, Biofilms and antibiotic resistance, and quorum sensing. Overall, this chapter provides in depth insights over the drug resistance mechanism of MRSA (Methicillin-resistant *Staphylococcus aureus*) at the molecular level is of great importance for the treatment of *S. aureus* infections. MRSA is one of the potential bacterial pathogens which is difficult to control, hence, it is always hot-topic in the area of medical microbiology/biotechnology (Hemeg, 2021; Yeager et al., 2021). Furthermore, now a days, quorum sensing and quorum quenching related research is giving much importance (Maddela et al., 2019, 2020a; Maddela & Meng, 2020) as these strategies offer several advantages in the mitigation of biofilm-mediated problems. Chapter "Anticancer Secondary Metabolites Found in Native Ecuadorian Plant Species *Uncaria tomentosa* DC. (Rubiaceae), *Croton lechleri* Müll. Arg. (Euphorbiaceae), and *Equisetum giganteum* L. (Equisetaceae)" is about anticancer secondary metabolites of native Ecuadorian plant species, and the contents of this chapter include detailed description on secondary metabolites (such as alkaloids, terpenoids, phenols), native plant species (*Uncaria tomentosa* DC, *Croton lechleri* Mull. Arg, *Equiseum giganteum* L.) that yield secondary metabolites with anticancer properties. Overall, this review is useful to have a better understanding of the different characteristics, diversity, and concentration of secondary metabolites present in these plants, its biological activity as a therapeutic agent and potential use for medical purposes against diseases such as cancer. There are several plant secondary metabolites have been emerged with anticancer properties (Alzandi et al., 2021; Ramakrishna et al., 2021). Chapter "The Carao (*Cassia grandis* L.): Its Potential Usage in Pharmacological, Nutritional, and Medicinal Applications" is about the potential usage of *Cassia grandis* L. in pharmacological, nutritional, and medicinal applications. This chapter focuses on bioactive compounds and its properties, extraction and characterization techniques of phenolic compounds, antioxidant activities, and proximal analysis. Likewise, this chapter summarizes its chemical composition and describes its potential nutritional, pharmacological, and medicinal applications. Due to its proximal, mineral, and bioactive compounds content, the *Cassia grandis* L. fruit is considered a potential functional and nutraceutical food, which can be used as an active ingredient for the fortification and enrichment of foods in people with special diets. Very recently, several studies have been focused on *Cassia grandis* fruit extract for its pharmacological and medicinal implications (Prada et al., 2018; Lafourcade Prada et al., 2020). Chapter "Challenges and Future Prospects of Biotechnology" is considered as a concluding chapter of this volume, and it mainly highlights the challenges and future directions of Biotechnology for a sustainable future.

4 Contributors

As this volume has been designed to publish the selected papers of IV Convención Científica Internacional de la Universidad Técnica de Manabí (CCIUTM 2020) Ecuador, most of the contributors are the participants of this event. Overall, the contributors of all 20 chapters are subject experts in their concerned chapters. Professionally, contributors are academicians and scientists and are geographically belonging to different regions. Overall, 67 contributors of 12 countries (Argentina, Brazil, Chile, Colombia, Ecuador, Honduras, India, Iran, Mexico, Nigeria, Spain, USA) have been involved in this volume. We strongly believe that this volume could be a single source of information that provides latest insights several emerging topics of in the domain of industrial-, agricultural-, environmental-, and medical biotechnology.

References

Alzandi, A. A., Taher, E. A., Al-Sagheer, N. A., Al-Khulaidi, A. W., Azizi, M., & Naguib, D. M. (2021). Phytochemical components, antioxidant and anticancer activity of 18 major medicinal plants in Albaha region, Saudi Arabia. *Biocatalysis and Agricultural Biotechnology, 34*, 102020.

Antar, M., Lyu, D., Nazari, M., Shah, A., Zhou, X., & Smith, D. L. (2021). Biomass for a sustainable bioeconomy: An overview of world biomass production and utilization. *Renewable and Sustainable Energy Reviews, 139*, 110691.

Ashayerizadeh, O., Dastar, B., Samadi, F., Khomeiri, M., Yamchi, A., & Zerehdaran, S. (2017). Study on the chemical and microbial composition and probiotic characteristics of dominant lactic acid bacteria in fermented poultry slaughterhouse waste. *Waste Management, 65*, 178–185.

Barzkar, N. (2020). Marine microbial alkaline protease: An efficient and essential tool for various industrial applications. *International Journal of Biological Macromolecules, 161*, 1216–1229.

Bhavaniramya, S., Vishnupriya, S., Al-Aboody, M. S., Vijayakumar, R., & Baskaran, D. (2019). Role of essential oils in food safety: Antimicrobial and antioxidant applications. *Grain & Oil Science and Technology, 2*, 49–55.

Chen, K., Zhang, M., Bhandari, B., & Mujumdar, A. S. (2021a). Edible flower essential oils: A review of chemical compositions, bioactivities, safety and applications in food preservation. *Food Research International, 139*, 109809.

Chen, Q., Li, Y., Liu, M., Zhu, B., Mu, J., & Chen, Z. (2021b). Removal of Pb and Hg from marine intertidal sediment by using rhamnolipid biosurfactant produced by a Pseudomonas aeruginosa strain. *Environmental Technology & Innovation, 22*, 101456.

Djinni, I., & Djoudi, W. (2021). Streptomyces sp. WR1L1S8 a potent endophytic marine strain for heavy metal resistance and copper removal enhanced by RSM modeling. *Acta Ecologica Sinica*. https://doi.org/10.1016/j.chnaes.2021.04.004 (In press, available online 17 April 2021).

Frantz, T. S., de Farias, B. S., Leite, V. R. M., Kessler, F., Cadaval, T. R. S. A., Jr., & de Almeida Pinto, L. A. (2020). Preparation of new biocoagulants by shrimp waste and its application in coagulation-flocculation processes. *Journal of Cleaner Production, 269*, 122397.

Hemeg, H. A. (2021). Determination of phylogenetic relationships among methicillin-resistant Staphylococcus aureus recovered from infected humans and companion animals. *Saudi Journal of Biological Sciences, 28*, 2098–2101.

Jõgi, K., & Bhat, R. (2020). Valorization of food processing wastes and by-products for bioplastic production. *Sustainable Chemistry and Pharmacy, 18*, 100326.

Kang, B. R., Kim, S. Y., Kang, M., & Lee, T. K. (2021). Removal of pharmaceuticals and personal care products using native fungal enzymes extracted during the ligninolytic process. *Environmental Research, 195*, 110878.

Khatami, K., Perez-Zabaleta, M., Owusu-Agyeman, I., & Cetecioglu, Z. (2021). Waste to bioplastics: How close are we to sustainable polyhydroxyalkanoates production? *Waste Management, 119*, 374–388.

Kongkham, B., Prabakaran, D., & Puttaswamy, H. (2020). Opportunities and challenges in managing antibiotic resistance in bacteria using plant secondary metabolites. *Fitoterapia, 147*, 104762.

Kuppusamy, S., Maddela, N. R., Megharaj, M., & Venkateswarlu, K. (2020a). *Case studies on remediation of sites contaminated with total petroleum hydrocarbons. Total petroleum hydrocarbons* (pp. 225–256). Springer.

Kuppusamy, S., Maddela, N. R., Megharaj, M., & Venkateswarlu, K. (2020b). *Total petroleum hydrocarbons*. Springer.

Lafourcade Prada, A., Achod, L. D. R., Keita, H., Carvalho, J. C. T., de Souza, T. P., & Rodríguez Amado, J. R. (2020). Development, pharmacological and toxicological evaluation of a new tablet formulation based on Cassia grandis fruit extract. *Sustainable Chemistry and Pharmacy, 16*, 100244.

Lu, Q., Liu, J., He, H., Liang, Z., Qiu, R., & Wang, S. (2021). Waste activated sludge stimulates in situ microbial reductive dehalogenation of organohalide-contaminated soil. *Journal of Hazardous Materials, 411*, 125189.

Luo, J., Xu, Y., Wang, J., Zhang, L., Jiang, X., & Shen, J. (2021). Coupled biodegradation of p-nitrophenol and p-aminophenol in bioelectrochemical system: Mechanism and microbial functional diversity. *Journal of Environmental Sciences, 108*, 134–144.

Maddela, N.R., Reyes, J.J.M., Viafara, D. and Gooty, J.M. (2015a). Biosorption of copper (II) by the microorganisms isolated from the crude oil contaminated soil. *Soil and Sediment Contamination: An International Journal, 24*, 898–908.

Maddela, N., Masabanda, M., & Leiva-Mora, M. (2015b). Novel diesel-oil degrading bacteria and fungi from Ecuadorian Amazon rainforest. *Water Science and Technology, 71*, 1554–1561.

Maddela, N. R., & Meng, F. (2020). Discrepant roles of a quorum quenching bacterium (Rhodococcus sp. BH4) in growing dual-species biofilms. *Science of the Total Environment, 713*, 136402.

Maddela, N. R., Burgos, R., Kadiyala, V., Carrion, A. R., & Bangeppagari, M. (2016a). Removal of petroleum hydrocarbons from crude oil in solid and slurry phase by mixed soil microorganisms isolated from Ecuadorian oil fields. *International Biodeterioration & Biodegradation, 108*, 85–90.

Maddela, N. R., Scalvenzi, L., & Kadiyala, V. (2016b). Microbial degradation of total petroleum hydrocarbons in crude oil: A field-scale study at the low-land rainforest of Ecuador. *Environmental Technology, 38*(20), 2543–2550.

Maddela, N. R., Sheng, B., Yuan, S., Zhou, Z., Villamar-Torres, R., & Meng, F. (2019). Roles of quorum sensing in biological wastewater treatment: A critical review. *Chemosphere, 221*, 616–629.

Maddela, N. R., Cruzatty, L. C. G., Leal-Alvarado, D. A., Olaya, J. C., Chakraborty, S., & Mukherjee, A. (2020a). *Quorum quenching for sustainable environment: Biology, mechanisms, and applications. Microbial technology for health and environment* (pp. 73–112). Springer.

Maddela, N. R., Kakarla, D., García, L. C., Chakraborty, S., Venkateswarlu, K., & Megharaj, M. (2020b). Cocoa-laden cadmium threatens human health and cacao economy: A critical view. *Science of the Total Environment, 720*, 137645.

Maddela, N. R., Cruzatty, L. C. G., & Chakraborty, S. (2021). Advances in the domain of environmental biotechnology.

Mahajan, M., Kuiry, R., & Pal, P. K. (2020). Understanding the consequence of environmental stress for accumulation of secondary metabolites in medicinal and aromatic plants. *Journal of Applied Research on Medicinal and Aromatic Plants, 18*, 100255.

Makaroglou, G., Marakas, H., Fodelianakis, S., Axaopoulou, V. A., Koumi, I., Kalogerakis, N., & Gikas, P. (2021). Optimization of biomass production from Stichococcous sp. biofilms coupled to wastewater treatment. *Biochemical Engineering Journal, 169*, 107964.

Mian, I. A., Ahmad, B., Khan, S., Khan, B., Dawar, K., Tariq, M., Mussarat, M., Muhammad, M. W., Ali, S., Bibi, H., Muhammad, F., & Khan, K. (2021). Improving wheat productivity and soil quality through integrated phosphorous management with residual effect of biochar. *Journal of Saudi Chemical Society, 25*, 101175.

Miyashiro, C. S., Mateus, G. A. P., dos Santos, T. R. T., Paludo, M. P., Bergamasco, R., & Fagundes-Klen, M. R. (2021). Synthesis and performance evaluation of a magnetic biocoagulant in the removal of reactive black 5 dye in aqueous medium. *Materials Science and Engineering: C, 119*, 111523.

Nijenhuis, I., & Kuntze, K. (2016). Anaerobic microbial dehalogenation of organohalides—State of the art and remediation strategies. *Current Opinion in Biotechnology, 38*, 33–38.

Ogbe, A. A., Finnie, J. F., & Van Staden, J. (2020). The role of endophytes in secondary metabolites accumulation in medicinal plants under abiotic stress. *South African Journal of Botany, 134*, 126–134.

Poo, K.-M., Son, E.-B., Chang, J.-S., Ren, X., Choi, Y.-J., & Chae, K.-J. (2018). Biochars derived from wasted marine macro-algae (Saccharina japonica and Sargassum fusiforme) and their potential for heavy metal removal in aqueous solution. *Journal of Environmental Management, 206*, 364–372.

Prada, A. L., Amado, J. R. R., Keita, H., Zapata, E. P., Carvalho, H., Lima, E. S., de Sousa, T. P., & Carvalho, J. C. T. (2018). Cassia grandis fruit extract reduces the blood glucose level in alloxan-induced diabetic rats. *Biomedicine & Pharmacotherapy, 103*, 421–428.

Ramakrishna, W., Kumari, A., Rahman, N., & Mandave, P. (2021). Anticancer activities of plant secondary metabolites: Rice callus suspension culture as a new paradigm. *Rice Science, 28*, 13–30.

Ramakrishnan, B., Maddela, N. R., Venkateswarlu, K., & Megharaj, M. (2021). Organic farming: Does it contribute to contaminant-free produce and ensure food safety? *Science of the Total Environment, 769*, 145079.

Riaz, M., Kamran, M., Fang, Y., Wang, Q., Cao, H., Yang, G., Deng, L., Wang, Y., Zhou, Y., Anastopoulos, I., & Wang, X. (2021). Arbuscular mycorrhizal fungi-induced mitigation of heavy metal phytotoxicity in metal contaminated soils: A critical review. *Journal of Hazardous Materials, 402*, 123919.

Shahid, A., Usman, M., Atta, Z., Musharraf, S. G., Malik, S., Elkamel, A., Shahid, M., Abdulhamid Alkhattabi, N., Gull, M., & Mehmood, M. A. (2021). Impact of wastewater cultivation on pollutant removal, biomass production, metabolite biosynthesis, and carbon dioxide fixation of newly isolated cyanobacteria in a multiproduct biorefinery paradigm. *Bioresource Technology, 333*, 125194.

Shen, Y., & Zhu, B. (2021). Arbuscular mycorrhizal fungi reduce soil nitrous oxide emission. *Geoderma, 402*, 115179.

SLH. (2010, February 1). *Science Learning Hub—Definitaions of biotechnology*. https://www.sciencelearn.org.nz/resources/1202-definitions-of-biotechnology

Timeline. (2021, April 30). *Timeline of biotechnology, Wikipedia*. https://en.wikipedia.org/wiki/Timeline_of_biotechnology#21st_century

Tiwari, J., Gandhi, D., Sivanesan, S., Naoghare, P., & Bafana, A. (2020). Remediation of different nitroaromatic pollutants by a promising agent of Cupriavidus sp. strain a3. *Ecotoxicology and Environmental Safety, 205*, 111138.

Tsang, Y. F., Kumar, V., Samadar, P., Yang, Y., Lee, J., Ok, Y. S., Song, H., Kim, K.-H., Kwon, E. E., & Jeon, Y. J. (2019). Production of bioplastic through food waste valorization. *Environment International, 127*, 625–644.

Verma, A. S., Agrahari, S., Rastogi, S., & Singh, A. (2011). Biotechnology in the realm of history. *Journal of Pharmacy & Bioallied Sciences, 3*, 321–323.

Vodnar, D. C., Călinoiu, L. F., Mitrea, L., Precup, G., Bindea, M., Păcurar, A. M., Szabo, K., & Ştefănescu, B. E. (2019). 15—A new generation of probiotic functional beverages using bioactive compounds from agro-industrial waste. In A. M. Grumezescu & A. M. Holban (Eds.), *Functional and medicinal beverages* (pp. 483–528). Academic Press.

Waani, S. P. T., Irum, S., Gul, I., Yaqoob, K., Khalid, M. U., Ali, M. A., Manzoor, U., Noor, T., Ali, S., Rizwan, M., & Arshad, M. (2021). TiO2 nanoparticles dose, application method and phosphorous levels influence genotoxicity in Rice (Oryza sativa L.), soil enzymatic activities and plant growth. *Ecotoxicology and Environmental Safety, 213*, 111977.

Wang, F., Chen, J., Zhang, C., & Gao, B. (2020a). Resourceful treatment of cane sugar industry wastewater by Tribonema minus towards the production of valuable biomass. *Bioresource Technology, 316*, 123902.

Wang, Y., Wang, Y., & Wang, Y. (2020b). Apoplastic proteases: Powerful weapons against pathogen infection in plants. *Plant Communications, 1*, 100085.

Wikipedia. (2021, April 22). *Biotechnology*. https://en.wikipedia.org/wiki/Biotechnology

Xue, R.-Y., Liu, C., Xiao, Q.-T., Sun, S., Zou, Q.-M., & Li, H.-B. (2021). HtrA family proteases of bacterial pathogens: Pros and cons for their therapeutic use. *Clinical Microbiology and Infection, 27*, 559–564.

Yeager, S. D., Oliver, J. E., Shorman, M. A., Wright, L. R., & Veve, M. P. (2021). Comparison of linezolid step-down therapy to standard parenteral therapy in methicillin-resistant Staphylococcus aureus bloodstream infections. *International Journal of Antimicrobial Agents, 57*, 106329.
Yu, G., Wang, G., Li, J., Chi, T., Wang, S., Peng, H., Chen, H., Du, C., Jiang, C., Liu, Y., Zhou, L., & Wu, H. (2020). Enhanced Cd2+ and Zn2+ removal from heavy metal wastewater in constructed wetlands with resistant microorganisms. *Bioresource Technology, 316*, 123898.
Zhao, Z., Chen, L., & Xiao, Y. (2021). The combined use of arbuscular mycorrhizal fungi, biochar and nitrogen fertilizer is most beneficial to cultivate Cichorium intybus L. in Cd-contaminated soil. *Ecotoxicology and Environmental Safety, 217*, 112154.
Zhu, Y., Li, C., Cui, H., & Lin, L. (2021). Encapsulation strategies to enhance the antibacterial properties of essential oils in food system. *Food Control, 123*, 107856.

Biotechnology of Twenty-First Century

**Pabbati Ranjit, Chittari Amaravathi Sneha Latha Varma,
Naga Raju Maddela, and Kondakindi Venkateswar Reddy**

1 Introduction

Biotechnology can be defined as the "implementation of engineering and biological science theory to produce new products from biologically derived raw materials" or, in other words, it can also be explained as "the manipulation of living organisms or their products to alter or enhance human health and the environment of our planet" (Verma et al., 2011). The word biotechnology was first coined by KarolyEreky in 1919 in a book entitled *Biotechnology of Fat, Meat and Milk Production in Large-Scale Agricultural Farm* (Ereky, 1919).

1.1 Biotechnology: Major Advances Between 2000s and 2020

The following are some of the significant events in the modern era of biotechnology:

2000	Synthesis and amplification of DNA in a test tube by Har gobind Khorana and Kary mullis, respectively (Verma et al., 2011). Completion of rough copy of human genome by Celeria Genomics and

(continued)

P. Ranjit · C. A. S. L. Varma · K. V. Reddy (✉)
Centre for Biotechnology, Institute of Science & Technology, Jawaharlal Nehru Technological University Hyderabad, Hyderabad, Telangana, India

N. R. Maddela
Departamento de Ciencias Biológicas, Facultad la Ciencias de la Salud, Universidad Técnica de Manabí, Portoviejo, Ecuador

Grupo de Investigación en Biodiversidad y Ecología de Ecosistemas Acuáticos, Departamento de Acuicultura Pesca y Recursos Naturales Renovables, Facultad de Ciencias Veterinarias, Universidad Técnica de Manabí, Portoviejo, Ecuador

N. R. Maddela, L. C. García (eds.), *Innovations in Biotechnology for a Sustainable Future*, https://doi.org/10.1007/978-3-030-80108-3_2

	Human Genome Project (Verma et al., 2011). Kenya's first biotech crop, a virus-resistant sweet potato, was field-tested (Colwell, 2020). Sir Ian Wilmut cloned an adult sheep and named it "Dolly" (Bhatia & Goli, 2018).
2001	The complete human **DNA** sequence was published in the Science and Nature journals (Bhatia & Goli, 2018). Gleevec® (imatinib), the first drug- gene-targeted for patients with leukaemia chronic myeloid approved by the Food and Drug Administration (FDA) (Colwell, 2020).
2002	The genome of rice is decoded for the first time (Timeline of biotechnology, Wikipedia contributors, 2021). First time cloning of an endangered banteng species (Colwell, 2020). Approval of the first transgenic rootworm-resistant corn by the Environmental Protection Agency (EPA) (Colwell, 2020). Completion of the period of high-throughput shotgun sequencing of major genomes which include rat, chimpanzee, dog and hundreds of animals (Bhatia & Goli, 2018).
2003	Successful completion of the human genome sequencing by Celera and the National Institute of Health (NIH) (Bhatia & Goli, 2018). Gendicine (Shenzhen SiBionoGenTech, China), which expresses the p53 gene as a treatment for squamous cell neck and head cancer, receives the world's first regulatory approval (Colwell, 2020). The first genetically modified pet animal, TK-1 (GloFish), was marketed in Taiwan (Colwell, 2020).
2004	FDA approval of Avastin®, the first antiangiogenic medication for cancer therapy (Colwell, 2020). Approval of DNA microarray analysis device by FDA that aids in the selection of medications for various ailments which is a big move forward in the field of personalised medicine (Bhatia & Goli, 2018). Biotech crops are endorsed by the UN Food and Agriculture Organization which can benefit consumers and poor farmers in developing countries (Colwell, 2020).
2005	The Act called Energy Policy was signed and enacted into law, allowing for multiple bioethanol production incentives (Colwell, 2020).
2006	FDA approval of Gardasil®, first vaccine recombinant developed against papillomavirus (HPV) for human (Colwell, 2020). The 3D structure of the AIDS-causing human immunodeficiency virus (HIV) was deduced (Bhatia & Goli, 2018). Dow Agro Sciences earns the first regulatory approval for a plant-based vaccine from the USDA (Colwell, 2020). Stelarc, an artist, had an ear produced in a lab and transplanted onto his arm (Colwell, 2020). Launch of a 10,000-patient study for 10-year by NIH that employs a DNA test to determine direct care and breast cancer relapse (Colwell, 2020).
2007	Approval of H5N1 vaccine by FDA, making it the first avian flu vaccine to be approved (Colwell, 2020). Researchers have established how to develop embryonic stem cells from human skin cells (Bhatia & Goli, 2018).
2008	Japanese chemists build the first DNA molecule that is almost completely made up of synthetic components which may be useful in the field of gene therapy (Bhatia & Goli, 2018).

(continued)

2009	FDA approval of the first genetically modified animal to produce recombinant human antithrombin (Colwell, 2020). Three new genes linked to Alzheimer's disease have been discovered, allowing for new diagnosis and therapy (Bhatia & Goli, 2018). The first FDA-approved clinical trial for involving in embryonic stem cells is launched by Geron (Bhatia & Goli, 2018). Cedars-Sinai Heart Institute produces the first viral pacemaker in guinea pigs, now known as iSANs, using modified Sinoatrial node (SAN) heart genes (Wikipedia contributors, 2021).
2010	Dr. J. C. Venter reveals the finishing of "synthetic life" by incorporating a self-replicating synthetic genome into a recipient bacterial cell (Bhatia & Goli, 2018). Development of "lung on a chip" technology by Harvard researchers (Colwell, 2020). Researchers developed malaria-resistant mosquitos (Colwell, 2020). FDA approval of a personalised new prostate cancer drug that enhances a patient's immune cells' ability to identify and attack cancer cells (Bhatia & Goli, 2018). FDA approval of an osteoporosis drug, first medicine based on genomic research (Bhatia & Goli, 2018). ReNeuron has begun a clinical trial to treat stroke patients with a genetically modified neural stem cell line (Colwell, 2020). Neural stem has begun a clinical trial to treat patients with ALS (Lou Gehrig's disease) employing human embryonic stem cells (Colwell, 2020).
2011	Stem cell-derived trachea was grafted into a human recipient (Bhatia & Goli, 2018). Progressions in 3D printing technologies have enabled "skin-printing" (Bhatia & Goli, 2018). FDA approval for employing first cord blood therapy in the transplantation of hematopoietic stem cells (Bhatia & Goli, 2018).
2012	Synthesis of the polymer, Xeno nucleic acid (XNA) by the molecular biologists Vitor Pinheiro and Philipp Holliger. XNA can be exploited unlike DNA. Complete genome of the foetus was successfully sequenced using only the fragments of DNA present in the mothers blood. Zac Vawter, 31, climbs the Chicago Willis Tower with the help of a bionic leg powered by his nervous system (Wikipedia contributors, 2021).
2013 (Colwell, 2020)	Development of the CRISPR system for editing genes. Generation of functional liver tissue of humans using reprogrammed skin cells.
2014 (Colwell, 2020)	Developments in research discovered that a young mouse blood would restore the muscles and brain of an older mouse. Researchers discovered a way to transform stem cells in human into functioning pancreatic cells. Researchers developed new (deoxy ribonucleic acid) DNA bases testing in lab, extending genetic code for life's and allowing for the development of new types microbes. Woman delivered a baby after undergoing a womb transplant for the first time. Creation of an artificial and highly operational yeast chromosome. The remarkable advance, which took 7 years to achieve, could eventually goes to custom-built species (including humans).

(continued)

2015 (Colwell, 2020)	Singapore's Institute of Nanotechnology and Bioengineering developed small strands of peptides which assemble themselves as a fibrous gel in case of water is applied, allowing them to be used like a healing nanogel. CRISPR and 2015: Using the CRISPR gene-editing technology, scientists made a series of breakthroughs. In a controversial move, Chinese developers reported changing the DNA of a nonviable embryo in human. Harvard University scientists introduced DNA a long-extinct into the living cells woolly mammoth for a modern elephant in petri dish. Researchers have also used CRISPR as theoretically alter pig organs into human transplantation to eliminate malaria by mosquitos. Swedish researcher designed a blood test which can diagnose cancer in its early stages using only one blood drop. Researchers discover a new type of antibiotic for the first time in after 30 years, which could shows the way for a next new generation of antibiotics and help in minimise drug resistance. Teixobactin antibiotic that can be used to give treatment in a variety of infections caused by bacteria, including septicaemia and tuberculosis. Stanford University researchers unveiled a mechanism for forcing malignant leukaemia into harmless immune cells to turn known as macrophages.
2016 (Colwell, 2020)	The mosquito-borne disease Zika, which was first detected in 1947 in Uganda country, which blows onto the world wide when it started rapidly spreading around Latin America. Scientists have separated a human antibody which "significantly minimises" Zika virus. CRISPR, ground breaking DNA-editing technology which aims to reduce diseases and fix disasters cause environmentally, took a big forward step in this year when a group of Chinese researchers used it for the first time to treat a human patient. GK-PID, an ancient molecule discovered by scientists, are the reason for organisms single-celled began to involve into organisms multicellular about 8 billion years ago. Bioengineers produced a fully "heart on a chip" 3D-printed for the first time.
2017 (Colwell, 2020)	The first step toward epigenetically enhanced cotton has been taken. The genome sequencing of a green alga offers a model for developing renewable energy and bioproducts. Disease-resistant rice that doesn't sacrifice yield was developed. For the first time stem cells of blood were grown in a lab. Scientists in Sahlgrenska Academy, which is part of the Gothenburg University in Sweden, used a 3D-bioprinter to print cartilage tissue.
2019	16 April 2019—For the first time, scientists described how they used CRISPR technology to alter human genes for treating cancer patients who had failed to respond to standard therapies (Fingas, 2019; Staff, 2019). 21 October 2019—In a new report, researchers define "prime editing," a new method of genetic engineering that outperforms previous methods such as CRISPR (Anzalone et al., 2019; Gallagher, 2019; NPR, 2019).
2020	27 January—Demonstration of designer nanoparticle "Trojan horse" that causes blood cells to eat away at portions of atherosclerotic plaque, which causes heart attacks which is the world's leading cause of death (Michigan State University, 2020; New Atlas, 2020a; Flores et al., 2020; ScienceDaily, WHO). 9 March—Discovery of CRISPR-Cas12b, a promising third CRISPR editing method for plant genome engineering, in addition to Cas9 and Cas12a, phys.org).

(continued)

	16 March—Development of a new type of CRISPR-Cas13d screening platform for designing successful guide RNA for target RNA. This technology was made accessible through an interactive website and free and open source software, along with a guide on how to create guide RNAs to target particular genes such as SARS-CoV2 RNA genome (Wessels et al., 2020, phys.org). 10 April—Wireless regulation of secretion of adrenal hormone in genetically unmodified rats using injectable magnetic nanoparticles (MNPs) (Rosenfeld et al., 2020, phys.org 16 May 2020). 8 May—Scientists claim to have created artificial chloroplasts (Barras, 2020b, phys.org 12 June 2020; Miller et al., 2020; New Atlas, 2020b). 10 November—Microorganisms could be used to mine useful elements from basalt rocks through bioleaching in space, according to scientists who conducted an experiment on the International Space Station with different gravity environments (Cockell et al., 2020; Crane, 2020). 18 November—For the first time, researchers announce that CRISPR/Cas9 was successfully used to treat cancer in a living animal using a lipid nanoparticle delivery system (Rosenblum et al., 2020; Tel Aviv University, 2020). 25 November—Development of symbiotic algal-bacterial multicellular spheroid microbial reactors that can produce oxygen and hydrogen through photosynthesis (Xu et al., 2020, phys.org 9 December 2020). 30 November—In tests of the biennial CASP evaluation with AlphaFold2, an artificial intelligence company shows how an AI algorithm-based method for protein folding, one of biology's most difficult problems, achieves a 90% precision in protein structure prediction (BBC News, 2020; DeepMind, 2020). 2 December—The Government of Singapore grants the world's first regulatory approval for a cultivated meat product (Shanker, 2019). 11 December—Scientists announce that they have used stem cells and a bioengineered scaffold to reconstruct a human thymus (Francis Crick Institute, 2020; Campinoti et al., 2020).
2021	12 January—CRISPR/Cas9 genome editing has resulted in a tenfold rise in superbugs that target formicamycin antibiotics, according to researchers (Devine et al., 2021; EurekAlert, 2021).

1.2 Benefits Due to Advances of Biotechnology

There are a huge number of benefits attributed to the innovative advances made in the field of biotechnology. Complete human DNA sequence published enabled researchers all over the world to begin researching new therapies for diseases with genetic roots, such as heart disease, Alzheimer's disease, cancer, etc. (Bhatia & Goli, 2018). Sequencing of genomes of crops like rice can guide in the development of resistant crops on the other hand sequencing of genomes of hundreds of animals can help in bringing back the endangered species to life like in the case of banteng. Discovery of genes related to different diseases may aid in the complete cure of the diseases. Another major achievement is the creation of functional trachea, liver, pancreas using stem cells. Now-a-days, stem cell banks are maintained just like blood banks which can be used in future in case of incidents where new organs or skin are needed to be created. Discovery of CRISPR-cas system can be considered as a boon to cancer patients as it can alter human genes. Not only human genes, this technology can also be employed in plant genome engineering.

1.3 Global Research in Biotechnology/Biotechnology in Private/Public Sector

From the above discoveries/advances made in the start of twenty-first century, it can be understood that global research in biotechnology focussed majorly in the medical science field in order to develop treatments for various diseases which humans are suffering and also in the agricultural field to develop genetically modified crops. In the growth of the biotechnology industry, public-private collaboration is crucial. In India, there are several government agencies, including the Science and Technology Department (DST), the Scientific and Industrial Research Department (DSIR), the Scientific and Industrial Research Council (CSIR), the Agricultural Research Indian Council (ICAR), the Medical Research Indian Council (ICMR), the Atomic Energy Department (DAE) and the Grant Commission University (UGC) which are focused on the development of biotechnology in India. In addition, there are research institutions such as Immunology National Institute (NII), New Delhi; Centre for DNA Diagnostics and Fingerprinting (CDFD), Hyderabad, etc. which work under the supervision of Department of Biotechnology (DBT) (Konde, 2008). In the private sector, industries, colleges and other institutes play a major role in developing biotechnology.

Biotechnology is being used in a wide range of fields, including bioremediation, forensics and agriculture, where fingerprinting DNA is widely used. Similarly techniques like PCR, immunoassays, and recombinant DNA are commonly used in both industry and medicine. The first reason that biology is known now as a potential science and biotechnology as a leading industry is because of genetic manipulation (Colwell, 2020). Biotechnology has applications in a number of fields, ranging from agriculture to medicine. Based on applications, biotechnology is divided into several branches, each of which is referred to by a different name, which is highlighted by different colours to distinguish the biotechnological area in which it is used (Fig. 1). Four major branches of biotechnology are Industrial biotechnology (White biotechnology), Medical biotechnology (also called Red biotechnology), Environmental biotechnology (Grey biotechnology) and Agricultural biotechnology (Green biotechnology). This chapter gives a brief introduction to the four major branches of biotechnology.

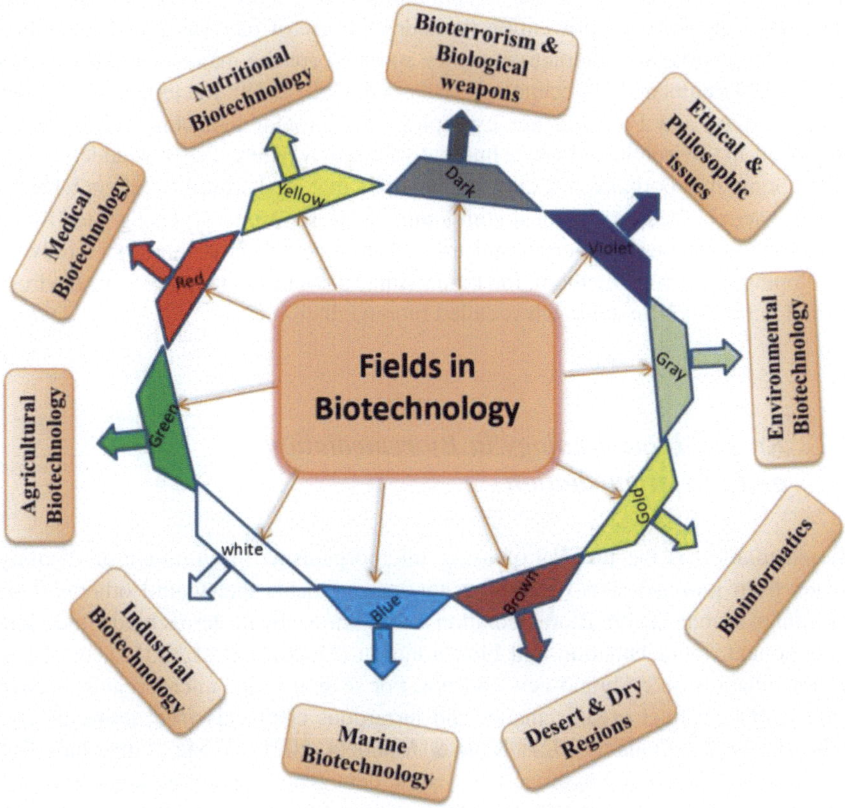

Fig. 1 Biotechnology classification. (Modified after Indira Padhy et al., 2020)

2 Environmental Biotechnology

Environmental biotechnology can be defined as the body of science and engineering expertise combined together which is concerned with the use of microbes and their products in the surveillance, treatment and prevention of environmental contamination (Ivanov & Hung, 2010).

2.1 Components and Importance of Environmental Biotechnology

Environmental monitoring/Biomonitoring of environment (employs biosensors to diagnose environmental issues) and treatment process, biotreatment of solid, liquid, and gaseous wastes, bioremediation/biodegradation of contaminated environment (degradation of organic molecules or contaminants in the environment by employing microbes) and pollution prevention (includes the use of renewable resources, biodegradable goods, and alternative energy sources) are the major concerns/components of environmental biotechnology (Bhatia & Goli, 2018). Bacteria and Archaea, Fungi, Algae, and Protozoa are examples of microbial biotechnological agents employed in environmental biotechnology (Ivanov & Hung, 2010; Maddela et al., 2016).Some microorganisms consume materials that are harmful to others whereas certain bacteria feed on chemical compounds of waste products. Using such living organisms particularly microorganisms, environmental biotechnology study is developing successful methods for minimising, preventing and reversing environmental harm. Such an approach is called bioremediation.

2.2 *Role of Biotechnology in Bioremediation and Phytoremediation*

Bioremediation is the process of using microorganisms to eliminate or detoxify toxins from polluted sites of soils, water, or sediments that would otherwise be harmful to human health. Bioremediation is also known by the terms biodegradation, biotreatment, bioreclamation and biorestoration (Maddela et al., 2017a, b, 2019). Bioremediation isn't a brand-new concept. For several years, microorganisms have been used to extract organic matter and hazardous chemicals from domestic and industrial waste (Godani, 2021; Kaur & Maddela, 2021). Toxic compounds like organics, metals, oil and hydrocarbons, dyes, detergents, etc. are broken down into less toxic and less complex metabolites such as inorganic minerals, H_2O, CO_2 (aerobic) or CH_4 (anaerobic) (Alexander, 1999). Bioremediation is a more effective and budget friendly cleaning method than other cleaning techniques such as chemical or physical techniques (Kamaludeen et al., 2003). For bioremediation, natural microorganisms are used; these natural microorganisms may be indigenous or non-indigenous (introduced). The chemical structure of the pollutant is taken into account when selecting microorganisms for bioremediation (Prescott et al., 2002). Occasionally, naturally occurring microbial species are insufficiently active or suitable for bioremediation of pollutants resistant to microbial assault (Dejonghe et al., 2000). Here comes the role of environmental biotechnology in developing Genetically Modified Microorganisms. Genetic engineering, a branch of molecular biology, builds novel strains with desired traits where the properties of naturally occurring microbes are modified in order to construct novel pathways, alter the existing regulatory mechanisms, alter and assemble various degradative enzymes extracted from different microorganisms into a single microorganism for degradation of pollutants and enhance the genetic stability of catabolic activities of microbes (Timmis & Pieper, 1999; Chen et al., 1999).

On the other hand, phytoremediation employs plants in place of microbes. The use of fast-growing, high-biomass plants capable of absorption and accumulation of large quantities of toxic metals in their aboveground harvestable sections is the most significant prerequisite for phytoremediation. Bioengineering of non-accumulators with high biomass is important for successful phytoremediation because many metal hyper accumulators are slow growing and have low biomass (Buhari et al., 2016). Biotechnology allows for the transfer of hyper accumulator phenotypes into fast-growing, high-biomass plants, which could be very useful in Phytoremediation (Rupali & Dibyengi, 2004). Biotechnology/Genetic engineering develops methods to boost plants' ability to withstand various contaminants and the efficacy of phytoremediation.

2.3 Pollution Control: Air, Water and Soil

The introduction of hazardous and toxic substances called pollutants into the environment is referred to as pollution. Volcanic ash, for example, is a natural pollutant. Human activity, such as garbage or industry runoff also causes pollution. Pollutants have a negative impact on the quality of the air, water, and land. Air, water, and land/soil pollution are the three primary kinds of pollution. There has been a substantial rise in the levels of environmental pollution over the last two decades as a result of direct or indirect human activities (because all kinds of human activity produce wastes). Industries, anthropogenic sources (man-made activities primarily in urban areas), biogenic sources, and other sources of emissions are currently the primary sources of environmental pollution (Saranya et al., 2020). Environmental pollution is often linked to the global industrial explosion, which is designed to meet the needs of the world's growing population (Okpokwasili, 2007). Therefore, it can be concluded human activities are the major reason for the massive environmental pollution during these days which can be resolved using environmental biotechnology. Environmental biotechnology is primarily used in wastewater treatment (water pollution) (Maddela et al., 2019), soil treatment to eliminate contaminants (land/soil pollution) (Maddela et al. 2015a, b; 2017a, b), and gaseous pollutant (air pollution) removal using microbiological catabolic operation.

Environmental, or outdoor, air quality has been the primary subject of air pollution control in developed countries. This entails the regulation of a limited number of unique "criteria" pollutants linked to urban smog and chronic public health issues. Fine particulates and gases (carbon monoxide, sulphur dioxide, nitrogen dioxide, ozone) and lead are among the criteria pollutants (Nathanson, 2019). The major air pollution control technologies are incinerators, gravitational settling chambers, electrostatic precipitators, cyclone separators, selective catalytic reduction systems, cloth filters, biofilters, biotrickling filters, bioscrubbers and membrane bioreactors (Kalender, 2019). Biofilters, biotrickling filters, bioscrubbers and membrane reactors come under the biological waste gas purification technologies which employ microbial communities to remove the criteria pollutants from the air. There are different approaches to treat water pollution which include aerobic, anaerobic and physicochemical processes in fixed-bed filters and in bioreactors. In all these

Table 1 List of microbes responsible for chemical compound degradation (Godani, 2021)

Compound	Organisms
Petroleum hydrocarbons	*Acinetobacter, Mycobacter, Pseudomonas, yeasts, Cladosporium, Scolecobasidium*
Pesticides (Aldrin, Dieldrin, parathion, malathion)	*Xylaria xylestrix*
Hydrocarbons, phenols, organophosphates, polychlorinated biphenyls and polycyclic aromatics.	*Pseudomonas*
Nitrate, nitrite, phosphate and heavy metals	*Phormidium laminosum*
Paraquat	*Lipomyces* sp.
Formaldehyde	*Candida* sp.
Benzaldehyde to benzyl alcohol	*Rhodotorula* sp.
Tannins	*Aspergillus Niger, Chaetomium cupreum*
Recalcitrant, pentachlorophenol	*Phanerochaete chrysosporium*
Volatile organic chemicals (VOCs)	*Nocardia* sp., *Xanthomonas* sp.

methods, the waste water containing industrial effluents and other contaminants and microbes are held in suspension (Godani, 2021). In order to treat soil pollution, a mixture of either naturally occurring or GEM'S (Genetically Engineered microbes) are added to the contaminated sites. The microbes responsible for degrading chemical compounds are given the below Table 1 (Godani, 2021).

3 Agricultural Biotechnology

Agricultural biotechnology, or agritech, is the application of modern scientific techniques based on our knowledge of DNA to boost crops and livestock in ways that traditional breeding alone cannot. Modern molecular plant breeding techniques including marker-assisted selection (MAS) can help with this.

3.1 Plant Growth and Yield

The world's growing population has posed a serious threat to food security. By 2050, population growth, especially in developing economies, would necessitate a 70% increase in food production, making substantial increases in agricultural productivity a priority over the next several decades. In this regard, biotechnology has focused its attention on developing technologies that can boost crop yields (Freddy et al., 2020).

The potential role of developmental features in growing crop yield was demonstrated in the "green revolution," where semi-dwarf rice and wheat varieties were bred to achieve unparalleled yield increases.

Reduction in plant height, an example is, by altering signalling and biosynthesis of gibberellic acid (GA) (Spielmeyer et al., 2002; Peng et al., 1999), reduced lodging and increased the tillers number. Given that the primary site of photosynthesis is a leaf, it's reasonable for assuming that plants may be genetically modified for the production of leaves in ideal size and shape for more effective harvesting, which results in faster higher yield and growth (Horton, 2000).

Another function that controls a plant's overall output is its vasculature, which not only provides strength but also gives as a conduit for the transport of minerals, water, and photosynthesis (Sack & Scoffoni, 2013; Brodribb et al., 2007). As a result, manipulations in genetics is that they change this developmental traits in favourable manner may be a large step forward in terms of crop yield (Jazayeri et al., 2020). Domestication impact on crop plant and also architecture of leaf adds to the argument that manipulating these developmental traits will boost crop production (Meyer & Purugganan, 2013).

Developmental traits are engineered with the goal of increasing efficient photosynthesis and therefore yield necessitates a detailed understanding of genetic basis. The molecular basis and genetics of the developmental procedure regulating these traits has made significant progress, especially in the model plant thaliana Arabidopsis. Anyhow, only a few crop plant examples that have grasped this concept exist. Despite this, it can be concluded that basic knowledge gained from the entire plant developmental studies can be applied to increase crop yield.

3.2 Agricultural Engineering (Breeding Techniques/Genetic Engineering/Organic Farming)

Techniques of breeding plant are methods allowing the development of new plant varieties with desired traits, by modifying the DNA of the seeds and plant cells. There are three major procedures for manipulating plant chromosome combination in general. To begin, plants from a given population can be selected for desired traits and used for further breeding and cultivation, a process known as pure line selection. Second, desired traits from different plant lines may be combined to create plants that display both traits at the same time, a process known as hybridisation. Third, crop improvement can be aided by polyploidy (an increase in the number of chromosome sets). Finally, natural or artificially induced mutations may introduce new genetic variability (Pandey et al., 2011).

The process of transferring individual genes between species or altering genes in an organism to eliminate or add a desired trait or feature is known as genetic engineering. Genetically modified crops or species are created by genetic engineering. GMOs or genetically modified crops are used to make biotech foods. Figure 2 shown global area distribution of the genetically modified crops (Beck et al., 2016). Restriction Fragment Length Polymorphism is the most effective and commonly used process of this kind in plant breeding (RFLP). Restriction endonucleases are

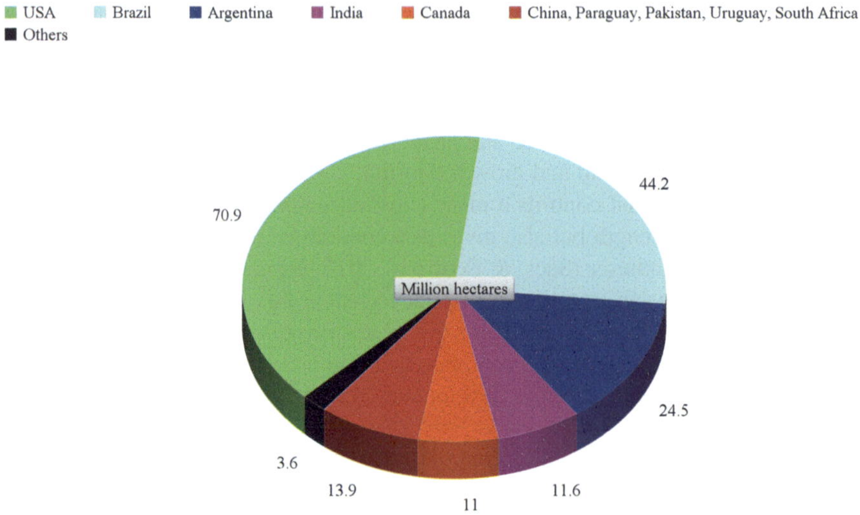

Fig. 2 Global area of genetically modified crops. (Modified after Beck et al., 2016

used in RFLP. Enzymes that identify and cut unique nucleotide sequences in DNA are known as restriction enzymes. Gene transfer allows any organism (other plants, bacteria, fungi, animals, viruses) to introduce useful traits coded for by unique genes into the genome of any plant. Transgenes are normally incorporated into a plant cell's nuclear genome (Ghosh et al., 2011).

Northbourne coined the word "organic" in his book *Look to the Land*, published in 1940. Organic farming is a form of farming that prevents or limits the use of synthetic fertilisers, pesticides, growth regulators, and feed additives in livestock (Ramakrishnan et al., 2020). Organic farming's fundamental goals are natural, social, and economic sustainability (Stockdale et al., 2001). Protecting long-term soil fertility by maintaining organic matter levels, fostering soil biological activity, careful mechanical intervention, biological nitrogen fixation, nitrogen self-sufficiency through the use of legumes and effective recycling of organic materials such as livestock wastes, weed and crop residues are the significant characteristics for organic farming. To reduce the distance between (Nitrogen, Phosphorous, Potassium) NPK removal from the soil and NPK addition, a strong focus is put on maintaining soil fertility by returning all wastes to soil primarily through compost (Chhonkar, 2002). Many countries are now forced to use pesticides and fertilisers to increase farm productivity in order to meet their ever-increasing food demands as a result of rising population pressure. Long-term and excessive chemical use, on the other hand, has resulted in human and soil health risks, as well as environmental contamination. As a result, farmers in developing countries are encouraged to turn their current farms to organic farms.

3.3 Global Food Security

The background of agriculture food security is expected to feed a growing population globally, which is expected to exceed 7500 million people by 2020, with 6300 million of them living in developing countries. There is a high-level concern over food insecurity due to trophic transfer of environmental pollutants (Maddela et al., 2020). Despite the fact that the population growth rate is slowing, the rise in people absolute numbers to feed could soon exceed the carrying capacity of agricultural lands, given current technology (FAO, 1999). Biotechnologies, for example, offer a responsible way to boost agricultural productivity now and in the future if properly oriented.

The issues of poverty reduction, food security and environmental protection in the developing world are all very important to the biotechnology revolution. However, for many, it poses serious ethical, intellectual property, and biosafety concerns (Johnson, 1999). Protests against the spread of agrobiotechnology have been widespread. Scientists are concerned that "novel" products would obliterate transforming agricultural trends, agricultural diversity into unrecognisable and uncontrollable types. Civil society organisations have staged several marches based on ethical or environmental concerns.

Fears of a new period of competitive disadvantage and increased dependency in the developing world have been posed by the domination of a highly concentrated private sector (British Medical Association, 1999). Patenting and intellectual property rights are also hot topics. Supporters of patenting argue that in order for the private sector to mobilise and spend substantial amounts of money in agrobiotechnology research and development, it must be able to protect and recoup its investment (Biotechnology Industry Organization, 2021).On the other hand, there is concern that patenting could lead to knowledge monopolisation, limited access to germplasm, research process restrictions, research focus selectivity, and increased marginalisation of the major world's population (Rural Advancement Foundation International, 2021). These issues can't and shouldn't be overlooked. To ensure that the effect of agrobiotechnology is both benign and positive, effective regulatory frameworks and protections must be implemented globally. In order to promote food security and assist the vulnerable, every tool of agricultural transformation should be used.

In the past last decade, cereals harvested per person were 370 kg, up from 275 kg in the 1950s; an increase of more than 33% per capita (FAO, 1995). Despite the fact that there are twice as many people on the planet today as there were 40 years ago, and there are 1.5 billion hungry people. Despite this remarkable increase in crop productivity, still more progress is needed to feed 2 billion people additionally by the early twenty-first century (Anderson, 1996), to generate income, combat food insecurity and jobs.

However, it is unclear if any of the current benefits agricultural biotechnology development would reach consumers and poorer farmers without significant public sector involvement. Farmers with limited financial resources are unlikely to get easy passage to agricultural inputs like improved pesticides, seeds, fertilisers and irrigation. Given that women are over 70% of those living in poverty in the developing countries, with the many of them are staying in rural areas (IDRC-UNCTAD, 1998), science alone is unlikely to provide a "scientific remedy" for eradicating poverty. Lack of access to land, low buying power and other productive resources, fragile ecosystems, political powerlessness and isolation from markets are just some of the systems, causes and socioeconomic structures that contribute to rural people's poverty. Biotechnology must be integrated as a supplement to agricultural technology, resulting in better and more widely available seeds and more sustainable production methods.

4 Industrial Biotechnology

Industrial biotechnology is the industrial application of biotechnology for the manufacture and processing of chemical products, materials, and fuels in a sustainable manner.

4.1 Industrial Products

Biotechnological processing employs enzymes and microorganisms to create goods for a variety of industries, including pharmaceutical, chemical, human, pulp and paper, energy, fabrics, textiles, and polymers, all of which rely on renewable raw materials.

Many of these sectors are more effective and environmentally sustainable as a result of the use of biotechnology to replace conventional technologies, leading to industrial sustainability in a number of ways. This paradigm shift affects a variety of fields, including the most well known, such as pharmaceuticals and agriculture, as well as the manufacture of biopolymers and bioplastics (Wikipedia, 2018).

The following are the seven most popular biotechnology applications in industry:

1. Fermentation Product Improvement.
2. Synthetic Fuels Generated by Microbes.
3. Bioleaching or Microbial Mining.
4. Single Cell Protein Production and Microbial Biomass.
5. Enzyme and Human Protein Production.
6. Secondary Metabolites Produced by Cultured Plant Cells.
7. Molecular Farming for Healthcare Products.

Modern biotechnology has the ability to provide a wide range of useful products to help us avoid diseases, diagnose and treat illnesses, and improve our overall health. Medicine, pharmacology, bioremediation, nutrition, food processing, energy production and forensics are all fields where biotech products are used. Biological medicines such as antibodies, vaccinations, recombinant proteins, antibiotics (e.g. enzymes, hormones, blood products, growth factors etc.), packages, diagnostic tests, cell therapy products, and gene therapy are several of the biotech-based health products. Haemophilia, diabetes, other illnesses and cancer are currently treated or diagnosed with biotech-based health products. In addition, genetic modification techniques have resulted in the development of many improved crops in terms of improved nutritional quality and material, disease and insect resistance, and crops that can produce antibodies, edible vaccines, pharmaceutically essential compounds. Food biotechnology is a relatively new branch of molecular biology that is rapidly expanding. Biotechnology is widely used in the food industry. It aids in the recovery of texture, food edibility and storage; it helps to prevent the attack of virus-like bacteriophage on food, especially dairy; it aids in the destruction of unwanted microorganisms in food that cause toxicity; and it aids in the prevention of mycotoxins and other toxins and anti-nutritional elements present in food. It also has the potential to play a significant role in protein engineering. Pathogens, toxins, and anti-nutritional elements in food can all be classified using this technology.

4.2 Food/Dairy Industry

Dairy products are widely regarded as natural, organic foods (Ramchandran & Shah, 2009) Biotechnology has the potential to play a huge role in improving the country's food and nutritional welfare. Biotechnology has been used in the processing of dairy products for centuries (cultured milk products, cheeses, and refined milk byproducts) by using starter cultures or enzymes for milk clotting, cheese ripening acceleration, fat, protein, or lactose hydrolyzate production, and antimicrobial purposes.

Modern biotechnology advances have opened up new and exciting possibilities in dairying, putting milk and milk products within reach of the poor and catering to the needs of a wide segment of the population. The dairy industry, in particular, will benefit greatly from biotechnological interventions that boost not only the overall quality and protection of processed dairy foods, but also their commercial value for both domestic and international consumption. Because the dairy industry's primary responsibility is to provide customers with high-quality, nutritious, and affordable dairy foods, biotechnological activity at various stages of milk production and processing has become anticipated. Biotechnology has previously significant offerings in dairy industry.

Possible applications are

Dairy Production

- Recombinant vaccines.
- Recombinant bovine.

DNA fingerprinting

- Animal cloning.
- Gene forming and transgenic.
- Embryo transmit technology.

Dairy Processing

- Dairy enzymes/proteins.
- Food grade bio-preservatives.
- Probiotics.
- Dairy waste organisation and pollution control.
- Functional foods and nutraceuticals.

5 Medical Biotechnology

Medical biotechnology is a branch of medicine that studies and then manufactures pharmaceutical and diagnostic products using living cells and cell materials. These items aid in the treatment and prevention of diseases. Medical biotechnology is a relatively new and rapidly growing area in which biotechnology concepts are applied to drug production (Maddela et al., 2021a). Biotechnology has a variety of effects on the medical industry and has the potential to alter its characteristics (Shan et al., 2018).

5.1 *Pharmaceutical and Vaccinology*

Biotechnology concepts such as recombinant DNA technology are used by pharmaceutical firms who have marketed bioformulations to design more successful protein-based drugs.

For treating the symptoms of an illness or disease, traditional pharmaceutical formulations are relatively simple molecules produced primarily by trial and error. Biopharmaceuticals, on the other hand, are complex biological molecules, widely known as proteins, that are used to cure diseases by removing the underlying mechanisms (Tables 2 and 3).

Table 2 Disease and examples of biopharmaceuticals used in treatment (Almeida et al., 2011)

Disease	Active substance
Multiple sclerosis	Interferon β
Hepatitis	Interferon α
Haemophilia	C factor VIII and factor IX
Renal cancer	Interleukin
Anaemia	Erythropoietin
Diabetes	Human insulin

Table 3 Marketed biotechnology products by different pharma and biotech companies (Evens & Kaitin, 2015)

Company	Name of the product	Uses
Sanofi	Gardasil	Prevention of certain strains of human papillomavirus (HPV)
Roche	Actemra	Moderate to severe treatment for active rheumatoid arthritis (RA)
Novartis	Eyelea	Treatment of patients with wet age-related macular degeneration (AMD)
Pfizer	Enbrel	Treatment of the symptoms of rheumatoid arthritis, ankylosing spondylitis, plaque psoriasis and psoriatic arthritis
Merck KGaA	Erbitux	Treatment for people with neck and head cancers and certain advanced colorectal
Novo Nordisk	Procrit, Remicade	Treatment for anaemia (low red blood cell count) for people with long-term kidney disease
GlaxoSmithKline	Pediarix	A vaccine used to immunise children against diphtheria, pertussis and tetanus
AbbVie	Humira	Treatment of rheumatoid arthritis
Eli Lilly	Erbitux, Forteo	Treatment of cancer of the colon
Astra Zeneca	Synagis	Treatment for respiratory syncytial virus (RSV) in children; disease prevention of serious lung caused

However, in some cases, such as type 1 diabetes mellitus, where insulin is used to treat only the symptoms of the condition rather than the underlying causes, this is not the case. Pharmaceutical biotechnology is basically the use of living cells to create complex larger molecules like those found in the human body. Living cells such as yeast cells, animal or plant cells and bacteria cells are being used. The large molecules are normally injected into the patient's body, unlike the smaller molecules that are given to them through tablets (Nehal et al., 2011).

The majority of medication therapies, such as nucleic acid products, antibodies and vaccines, that are commonly used for molecular diagnostics today are the result of biotechnology formulations. The first biotechnologically produced drug is Insulin, which is one of the most popular examples. Aside from that, biotechnology offers specialised medical facilities and equipment for both prevention and diagnosis (Vijayakuma & Sasikala, 2012).

A vaccine is a biological preparation that is used to develop artificial active immunity to a specific disease. The main goal is to boost immunity; the antigen is referred to as a vaccine (Afzal et al., 2016).

Vaccination is a medical procedure that involves eliciting an immune response that decreases the risk of developing a specific disease. Vaccines are one of the most active achievements against a variety of infectious diseases and mortality in the twenty-first century, outperforming any other medical advance (Plotkin, 2001). Recent advances in molecular biology have resulted in two vaccine approaches:

DNA vaccines and therapeutic vaccines (Poland et al., 2002). Cancer, allergic disorders, and autoimmunity may all be treated or prevented with DNA vaccines (Wahren & Liu, 2014).For example therapeutic vaccine against HIV that will induce virus specific cytotoxic T lymphocytes against HIV and activate T cells to destroy latently infected cells.

The United States Food and Drug Administration has licensed more than 170 biotechnology-related drugs and vaccines, with 113 currently on the market. Another 350 biotechnology medicines are in the final stages of growth, with a combined goal of over 200 diseases. Medicines to treat pneumococcal diseases in children, diabetes, cancer and haemophilia were among those approved in 2000 (BIO). In the future, DNA technology is expected to revolutionise vaccine production. DNA vaccines have only recently begun to be tested, but they are expected to potentially replace other vaccine manufacturing methods (Human Genome Project Information, 2021).

5.2 Novel Methodologies

The rapid growth of both agricultural and medical biotechnology may have recently provided the perception of technological separation due to specialisation. Microencapsulation, immobilised bacterial cells and enzymes, genetic modification of microorganisms, liposome processing, creation of novel vaccines from plants, biocomputational methods and epigenomics of mammalian cells and organisms for disease modelling and bioinformatics are some of the novel techniques used in Medical Biotechnology. In the treatment of thrombosis, immobilised enzymes (Plasmin and Heparin) are used to achieve a balance between coagulation and degradation of coagulated blood (fibrinolysis) (Baianu et al., 2004). In the recent past, there is a much emphasis on the quorum quenching mechanism, which is a sustainable tool for the control of bacterial biofilms in the field of medicine and industry (Maddela & Meng, 2020).

6 Challenges in Biotechnology

6.1 Challenges/Gaps for Further Developments

1. Setting biotechnology goals: National officials are often forced to make decisions about priorities, user feedback, with limited financial resources and scientific knowledge.
2. Developing suitable and affordable technologies: Technologies that complement current farming systems and native crops, are affordable, and are healthy for humans and the environment must be developed.
3. Involving citizens: To assess needs and resolve issues, public participatory processes are needed.

4. Investing in science and local capacity: In developing countries, investment is needed to establish and improve national scientific expertise.
5. Creating long-term alliances: Partnerships may help to stimulate research in resource-poor countries, but it's crucial to understand each partner's priorities.
6. Participating in global discussions and negotiations about biosafety, biodiversity, trade and intellectual property rights: Scientists and lawyers from developed countries must engage in negotiations and discussions about biodiversity, biosafety, trade and intellectual property rights to ensure that agreements can be enforced in ways that help their countries achieve their objectives.
7. Anticipating directions and future needs: Researchers and policymakers must predict shifts in agricultural production and market demand.

Given how quickly the future unfolds, current biotechnology research has a greater potential to serve as the foundation for future innovation to address society's major challenges, such as ensuring food security for an ever-increasing population, providing sustainable healthcare, resource conservation, precision agriculture, climate change, and meeting energy shortages.

6.2 *Future Directions of Research/Suggestions*

Major areas Biotechnology with tremendous future opportunities which can impact all include:

(a) Drug and pharmaceutical development;
(b) Medical device and diagnostics;
(c) Noninvasive sensor in agriculture;
(d) Rapid testing of pathological conditions;
(e) Fast testing for food adulteration;
(f) Precision Agriculture and Biofortification;
(g) Machine learning and Artificial Intelligence in Biotechnology.

Over the last two decades, rapid advances in biotechnology and information technology have occurred in lockstep. Several biotechnology studies that would have taken years to complete are now simpler and cheaper thanks to the vast amount of data available and modern testing technologies developed in recent years (Maddela et al., 2021b). The large amounts of data produced and stored in biotechnology research, mostly extracted via omics-technologies, open up a slew of new possibilities for researchers and companies that provide products and services in this field.

Machine learning (ML) and Artificial intelligence (AI) technologies to explore, process and analyse broad data sets are new emerging fields in biotechnology at the moment, and thus AI and ML are exciting areas essential in advancing biotechnology's benefits. Biotechnology will be able to address complex social problems with its own systemic testing by adapting AI and ML. For example, in the health sector, electronic health record systems are increasingly integrating so that health-related data can be accessed internationally, allowing for a comprehensive approach.

In the field of biotechnology, rapid technological innovation has resulted in remarkable discoveries and innovations over the last three decades. The Human DNA Project, for e.g. provided a comprehensive map of human genome. From the fertilised egg cell to death, the human being's growth is driven by inherited instructions encoded in DNA. As a consequence, decoding the human DNA sequence possesses tremendous strength. DNA sequencing technology has advanced so quickly that in 2014, a single human genome that cost $100 million to sequence in 2001 cost just $5000 (National Human Genome Research Institute, 2015).

Parallel to this, assembling artificial DNA and technology for synthesising has improved to the point that whole genomes and genes of bacteria and viruses can now be recreated (Andrew, 2002, Eckard et al., 2009, Gibson et al., 2010; Monya, 2011). Recent advances in DNA engineering technology have made modifications in genetics of bacteria, plants, insects and animals, more feasible including human gene therapy (Jennifer and Emmanuelle, 2014; Cossins, 2013). Herbicide-resistant soybeans, corn and cotton are examples that have lowered prices, increase national yield, enhanced personal protection and reduced the amount of pesticides used (National Academies of Sciences, 2010).

"Pharmacogenomics or Personalised medicine," is the field in which the genetics for individual patient suggest the possible reaction of the disease to the biologic treatment, would be a key initiative in the future. Prescribers would be able to customise individual care regimens to include medications with a high probability of providing a beneficial clinical outcome while eliminating therapies that can cause severe side effects. Trastuzumab's effectiveness in aggressive metastatic breast cancer, for example, is dependent on the presence of the HER2/neuoncogene, which is found in approximately 1/4% of patients and is the only breast cancer patients group that can respond to trastuzumab. According to a report involving nearly 30 firms, personalised medicine was present in 12–50% of product pipelines, with a median of about 25% expected for 2015 (Tufts Center for the Study of Drug Development, 2010).

In the coming decade, product production will continue to grow. According to a recent study, 88 fully human monoclonal antibodies, as well as a comparable number of chimeric and humanised molecules, were in clinical development in 2010. The two most common disease types were cancer and immune disorders (Nelson et al., 2010). Novel formulations are being tested to improve product delivery by increasing effectiveness while reducing toxicity.

Over 900 biotech-related molecules are currently being studied in clinical trials around the world (PhRMA, 2013). With 431 and 276 molecules, monoclonal antibodies and vaccines were the two most common product categories. Even if we assume a 10% clinical success rate for biologic candidates, we can expect 100 new molecules to hit the market in the next 5–10 years.

The regulatory landscape will continue to change. Biosimilar regulations were recently finalised by the FDA, and the first biosimilar products have hit the market in early 2015. Ten of the top-selling biotech drugs have already lost their patents, and several more will lose their patents in the next 5–6 years, presenting huge opportunities for biosimilar developers. However, because of the biotechnology expertise needed, the complex and sophisticated biologic testing required, and the extremely high cost and complexity of development, obtaining FDA approval for bio-similars is likely to be difficult.

7 Conclusions

Biotechnology research has made a significant contribution and will continue to do so in the future to fulfil human needs.

An everchanging biological and economic climate, technical and scientific growth, and need-of-the-moment biotechnology advances have characterised the last few years. This is especially true in light of the recent COVID-19 pandemic and other emerging threats, as biotechnology is at the heart of finding solutions to current and future problems.

Identifying the most important biological, social, economic, and technical patterns will aid scientists in determining future biotechnology research directions.

The study of recent progress in the field's long-term effects will open up new research areas that will serve as a springboard for new innovation addressing future needs. It necessitates ongoing cooperation with stakeholders and the identification of both routine and new emerging problems in order to improve biotechnology innovation production.

Furthermore, many ethical and regulatory questions about biotechnology-based products, such as patents on living organisms, have been raised. Biotechnology is currently used to manufacture antibiotics, carbohydrates, hormones, monoclonal antibodies and vaccines, among other medicinal products. These items are used to treat and prevent a variety of diseases that affect a large number of people.

Many promising biotechnology-based approaches are currently being established for the advancement of medicine and the treatment of various diseases. Biotechnology-based treatments such as gene therapy, pharmacogenomics, and stem cell therapy, for example, have the ability to significantly improve the treatment process in a variety of ways.

Biotechnology has had a huge effect on health care over the last few years. If our understanding of the pathophysiology of many currently incurable diseases improves, this trend will continue in the near future governments all over the world are advancing policies to encourage biotech innovation, and market strategies are evolving to handle the costly, time-consuming, and dangerous phase of product growth. As a result, a steady stream of new drugs will be created, leading to major advances in patient care.

References

Afzal, H., Zahid, K., Ali, Q., Sarwar, K., Shakoor, S., Nasir, U., & Nasir, I. A. (2016). Role of biotechnology in improving human health. *Journal of Molecular Biomarkers & Diagnosis, 8* (309), 2.

Alexander, M. (1999). *Biodegradation and bioremediation* (2nd ed.). Academic Press.

Almeida, H., Amaral, M. H., & Lobão, P. (2011). Drugs obtained by biotechnology processing. *Brazilian Journal of Pharmaceutical Sciences, 47*(2), 199–207.

Anderson, J. (1996). Feeding a hungrier world. *Phytopathology News, 30*(6), 90–91.

Andrew P., "Traces of terror: the science; scientists create a live polio virus."(New York Times, July 12, 2002). https://www.nytimes.com/2002/07/12/us/traces-of-terror-the-science-scientists-create-a-live-polio-virus.html. Accessed 3 Feb 2021

Anzalone, A. V., Randolph, P. B., Davis, J. R., Sousa, A. A., Koblan, L. W., Levy, J. M., et al. (2019). Search-and-replace genome editing without double-strand breaks or donor DNA. *Nature, 576*(7785), 149–157.

Baianu, I. C., Lozano, P. R., Prisecaru, V. I., & Lin, H. C. (2004). Applications of novel techniques to health foods, medical and agricultural biotechnology. *Other Quantitative Biology (q-bio.OT).* https://arxiv.org/abs/q-bio/0406047

Barras, C. (2020a, May 7). Cyber-spinach turns sunlight into sugar. *Nature.* https://doi.org/10.1038/d41586-020-01396-4. PMID: 32393873

Barras. (2020b). *Researchers develop an artificial chloroplast.* phys.org

BBC News. (2020, November 30). *One of biology's biggest mysteries 'largely solved' by AI.*

Beck, A., Haerlin, B., & Richter, L. (2016). *Agriculture at a crossroads: Finding and recommendations for future farming*. The Foundation on Future Farming.

Bhatia, S., & Goli, D. (2018). *Introduction to pharmaceutical biotechnology, Volume 1: Basic techniques and concepts*. IOP Science.

Biotechnology Industry Organization. (2021). http://www.bio.org

British Medical Association. (1999, May 17). Press release. Action Network–Asia Pacific (www.poptel.org.uk/panap) and Union of Concerned Scientists (www.ucsusa.org).

Brodribb, T. J., Feild, T. S., & Jordan, G. J. (2007). Leaf maximum photosynthetic rate and venation are linked by hydraulics. *Plant Physiology, 144*, 1890–1898. https://doi.org/10.1104/pp.107.101352

Buhari, M., Sulaiman, B., Vyas, N. L., Badaru, S., & Harisu, U. (2016). Role of biotechnology in phytoremediation. *Journal of Bioremediation & Biodegradation, 7*(2), 330.

Campinoti, S., Gjinovci, A., Ragazzini, R., Zanieri, L., Ariza-McNaughton, L., Catucci, M., & Bonfanti, P. (2020). Reconstitution of a functional human thymus by postnatal stromal progenitor cells and natural whole-organ scaffolds. *Nature Communications, 11*(1), 1–16.

Chen, W., Brühlmann, F., Richins, R. D., & Mulchandani, A. (1999). Engineering of improved microbes and enzymes for bioremediation. *Current Opinion in Biotechnology, 10*(2), 137–141.

Chhonkar, P. K. (2002, December). Organic farming myth and reality. In *Proceedings of the FAI Seminar on Fertilizer and Agriculture Meeting the Challenges*, New Delhi, India.

Cockell, C. S., Santomartino, R., Finster, K., Waajen, A. C., Eades, L. J., Moeller, R., et al. (2020). Space station biomining experiment demonstrates rare earth element extraction in microgravity and Mars gravity. *Nature Communications, 11*(1), 1–11.

Colwell, B. (2020). Biotechnology timeline: Humans have manipulated genes since the 'dawn of civilization'. *Genetic Literacy Project.* https://geneticliteracyproject.org/2020/09/08/biotechnology-timeline-humans-manipulating-genes-since-dawn-civilization/

Cossins, D. (2013). Gene therapy coming of age? *The Scientist.*

Crane, L. (2020). Asteroid-munching microbes could mine materials from space rocks. *New Scientist.*

DeepMind. (2020, November 30). AI cracks 50-year-old problem of protein folding. *The Guardian 30 November 2020. Retrieved 30 November 2020.*

Dejonghe, W., Goris, J., El Fantroussi, S., Höfte, M., De Vos, P., Verstraete, W., & Top, E. M. (2000). Effect of dissemination of 2,4-dichlorophenoxyacetic acid (2,4-D) degradation plasmids on 2,4-D degradation and on bacterial community structure in two different soil horizons. *Applied and Environmental Microbiology, 66*(8), 3297–3304.

Devine, R., McDonald, H. P., Qin, Z., Arnold, C. J., Noble, K., Chandra, G., & Hutchings, M. I. (2021). Re-wiring the regulation of the formicamycin biosynthetic gene cluster to enable the development of promising antibacterial compounds. *Cell Chemical Biology, 28*(4), 515–523.e5.

Eckard, W., Steffen, M., Terrence, M. T., & Jeffery, K. T. (2009). Synthetic viruses: a new opportunity to understand and prevent viral disease. *Nature Biotechnology, 27*(12), 1163–1172.

Ereky, K. (1919). *Biotechnologie der Fleisch-, Fett-, und Milcherzeugung im landwirtschaftlichen Grossbetriebe: für naturwissenschaftlich gebildete Landwirte verfasst*. P. Parey.

EurekAlert. (2021, January 12). Gene-editing produces tenfold increase in superbug slaying antibiotics. *EurekAlert*!

Evens, R., & Kaitin, K. (2015). The evolution of biotechnology and its impact on health care. *Health Affairs, 34*(2), 210–219.

FAO. (1999, January 25–29) *Committee on Agriculture: Biotechnology*. COAG. http://www.fao.org/3/X1551e/X1551e.htm

FAO (Food and Agriculture Organisation of the United Nations). (1995). *Dimensions of need: An atlas of food and agriculture*. FAO, Rome, Italy. http://www.fao.org/3/u8480e/U8480E00.htm. Accessed 6 Feb 2021.

Fingas, J. (2019, April 16). CRISPR gene editing has been used on humans in the US. *Engadget*.

Flores, A. M., Hosseini-Nassab, N., Jarr, K. U., Ye, J., Zhu, X., Wirka, R., et al. (2020). Pro-efferocytic nanoparticles are specifically taken up by lesional macrophages and prevent atherosclerosis. *Nature Nanotechnology, 15*(2), 154–161.

Francis Crick Institute. (2020, December 11). *Scientists build whole functioning thymus from human cells*.

Freddy, Z. G., Diva, S. A., Claudemir, Z., Eduardo, H. H., João, S. Y., Andre, P. B., Luiz, A. R. A., Luz, G. C., Maddela, N. R., & Maria, F. G. (2020). Co-inoculation of cyanobacteria with Azospirillum brasilense on the development of maize hybrids. *Rhizosphere, 15*, 100224.

Gallagher, J (2019, October 21). Prime editing: DNA tool could correct 89% of genetic defects. *BBC News*.

Ghosh, S., Ghosh, P., Basu, K., Das, S. K., & Daefler, S. (2011). A discrete event based stochastic simulation platform for 'in silico' study of molecular-level cellular dynamics. *Journal of Biotechnology & Biomaterials, S6*, 001.

Gibson, D. G., Glass, J. I., Lartigue, C., Noskov, V. N., Chuang, R. Y., Algire, M. A., et al. (2010). Creation of a bacterial cell controlled by a chemically synthesized genome. *Science, 329*(5987), 52–56.

Godani, K. (2021). Environmental Biotechnology: Meaning, Applications and Other Details. Biology Discussion. https://www.biologydiscussion.com/biotechnology/environmental-biotechnology/environmental-biotechnology-meaning-applications-and-other-details/8528. Accessed 3 Feb 2021.

Horton, P. (2000). Prospects for crop improvement through the genetic manipulation of photosynthesis: Morphological and biochemical aspects of light capture. *Journal of Experimental Botany, 51*, 475–485.

Human Genome Project Information. (2021). http://www.ornl.gov/hgmis. Accessed 6 Feb 2021

IDRC-UNCTAD. (1998). *Technology for small scale economic activities to address the basic needs of low income populations* (pp. 248–258). FAO. https://www.idrc.ca/sites/default/files/openebooks/800-7/index

Ivanov, V., & Hung, Y. T. (2010). Applications of environmental biotechnology. In *Environmental biotechnology* (pp. 1–17). Humana Press.

Jazayeri, S. M., Villamar-Torres, R. O., Zambrano-Vega, C., Cruzatty, L. C. G., Oviedo-Bayas, B., Santos, M. A., Maddela, N. R., Ghafoor, A. M. H. A. S., & Viot, C. (2020). Transcription factors and molecular markers revealed asymmetric contributions between allotetraploid upland cotton Gossypium hirsutum and its two diploid ancestors. *Bragantia, 76*, 1.

Jennifer, A. D., & Emmanuelle, C. (2014). The new frontier of genome engineering with CRISPR-Cas9. *Science, 346*(6213), 1258096.

Johnson, B. (1999, March 26). Personal communication (from a presentation at BioVision, Lyon, France). http://www.bibalex.org/cssp/publications/BioVision2006.pdf

Kalender, S. S. (2019). Air pollution prevention technologies. In C. Hussain (Ed.), *Handbook of environmental materials management*. Springer. https://doi.org/10.1007/978-3-319-73645-7_109

Kamaludeen, S. P. B., Arunkumar, K. R., & Ramasamy, K. (2003). Bioremediation of chromium contaminated environments. *Indian Journal of Experimental Biology, 41*(9), 972–985.

Kaur, J., & Maddela, N. R. (2021). Microbial bioremediation: A cutting-edge technology for xenobiotic removal. In N. R. Maddela, L. C. García Cruzatty, & S. Chakraborty (Eds.), *Advances in the domain of environmental biotechnology. Environmental and microbial biotechnology*. Springer. https://doi.org/10.1007/978-981-15-8999-7_16

Konde, V. (2008). Biotechnology in India: Public–private partnerships. *Journal of Commercial Biotechnology, 14*(1), 43–55.

Maddela, N. R., & Meng, F. (2020). Discrepant roles of a quorum quenching bacterium (Rhodococcus sp. BH4) in growing dual-species biofilms. *Science of the Total Environment, 713*, 136402.

Maddela, N. R., Scalvenzi, L., Pérez, M., Montero, C., & Gooty, J. M. (2015a). Efficiency of indigenous filamentous fungi for biodegradation of petroleum hydrocarbons in medium and soil: Laboratory study from Ecuador. *Bulletin of Environmental Contamination and Toxicology, 95*(3), 385–394.

Maddela, N. R., Reyes, J. J. M., Viafara, D., & Gooty, J. M. (2015b). Biosorption of copper (II) by microorganisms isolated from crude oil contaminated soil. *Soil and Sediment Contamination: An International Journal, 24*(8), 898–908.

Maddela, N. R., Burgos, R., Kadiyala, V., Banganegiri, M., & Carrión, A. R. (2016). Removal of crude oil from soil by using novel microorganisms of Ecuador soils: Solid and slurry phase methods. *International Biodeterioration and Biodegradation, 108*, 85–90.

Maddela, N. R., Rodriguez, L., Sanaguano, S. H., Morán, R. E. B., Venkateswarlu, K., & Scalvenzi, L. (2017a). Biodegradation of diesel, crude oil and spent lubricating oil by soil isolates of Bacillus spp. *Bulletin of Environmental Contamination and Toxicology, 98*, 698–705.

Maddela, N. R., Scalvenzi, L., & Venkateswarlu, K. (2017b). Microbial degradation of total petroleum hydrocarbons in crude oil: A field-scale study at the low-land rainforest of Ecuador. *Environmental Technology, 38*, 2543–2550.

Maddela, N. R., Sheng, B., Shasha, Y., Zhou, Z., Villamar-Torres, R., & Fangang, M. (2019). Roles of quorum sensing in biological wastewater treatment: A critical review. *Chemosphere, 221*, 616–629.

Maddela, N. R., Kakarla, D., Garcia, L. C., Chakraborty, S., Venkateswarlu, K., & Megharaj, M. (2020). Cocoa-laden cadmium threatens human health and cacao economy: A critical view. *Science of the Total Environment, 720*, 137645.

Maddela, N. R., Chakraborty, S., & Prasad, R. (Eds.). (2021a). *Nanotechnology for the advances in medical microbiology*. Springer Nature Singapore Pte Ltd. https://doi.org/10.1007/978-981-15-9916-3; ISBN: 978-981-15-9916-3, (pp X, 467).

Maddela, N. R., Garcia, L. C., & Chakraborty, S. (Eds.). (2021b). *Advances in the domain of environmental biotechnology*. Springer Nature Singapore Pte Ltd. https://doi.org/10.1007/978-981-15-8999-7; ISBN: 978-981-15-8999-7, (pp XVIII, 717).

Meyer, R. S., & Purugganan, M. D. (2013). Evolution of crop species: Genetics of domestication and diversification. *Nature Reviews. Genetics, 14*, 840–852. https://doi.org/10.1038/nrg3605

Michigan State University. (2020, January 27). *Nanoparticle chomps away plaques that cause heart attacks*.

Miller, T. E., Beneyton, T., Schwander, T., Diehl, C., Girault, M., McLean, R., & Erb, T. J. (2020). Light-powered CO2 fixation in a chloroplast mimic with natural and synthetic parts. *Science, 368*(6491), 649–654.

Ming, M., Ren, Q., Pan, C., He, Y., Zhang, Y., Liu, S., & Qi, Y. (2020). CRISPR–Cas12b enables efficient plant genome engineering. *Nature Plants, 6*(3), 202–208.

Monya, B. (2011). The next step for the synthetic genome. *Nature, 473*(7347), 403–408.

Nathanson, J. A. (2019, December 24). Air pollution control. *Encyclopedia Britannica*. https://www.britannica.com/technology/air-pollution-control

National Academies of Sciences. (2010). The impact of genetically engineered crops on farm sustainability in the United States. In *The National Research Council Committee on the impact of biotechnology on farm-level economics and sustainability*. The National Academies Press. https://www.nap.edu/catalog/12804/the-impact-of-genetically-engineered-crops-on-farm-sustainability-in-the-united-states

National Human Genome Research Institute. (2015). https://www.nih.gov/about-nih/what-we-do/nih-almanac/national-human-genome-research-institute-nhgri

Nehal, M. M., Patel, P. M., & Patel, N. M. (2011). A review on regulatory aspects of biotechnology derived product. *International Journal of Research in Ayurveda and Pharmacy, 2*(5), 1495–1500.

Nelson, A. L., Dhimolea, E., & Reichert, J. M. (2010). Development trends for human monoclonal antibody therapeutics. *Nature Reviews. Drug Discovery, 9*(10), 767–774.

New Atlas. (2020a, January 28). *Nanoparticle helps eat away deadly arterial plaque.*

New Atlas. (2020b, May 11). New technique makes thousands of semi-synthetic photosynthesis cells. *New Atlas*.

NPR. (2019). *Scientists create new, more powerful technique to edit gene.s*

Okpokwasili, G. C. (2007, November). Biotechnology and clean environment. In *Proceedings of the 20th Annual Conference of the Biotechnology Society of Nigeria (BSN), 14th–17th.*

Padhy, I., et al. (2020). Role of biotechnology in pharmaceutical research: A comprehensive review. *Indo American Journal of Pharmaceutical Sciences, 7*(5), 472–486.

Pandey, S., Negi, Y. K., Marla, S. S., & Arora, S. (2011). Comparative *insilico* analysis of ascorbate peroxidase protein sequences from different plant species. *Journal of Bioengineering and Biomedical Sciences, 1*, 103.

Peng, J., Richards, D. E., Hartley, N. M., Murphy, G. P., Devos, K. M., Flintham, J. E., Beales, J., Fish, L. J., Worland, A. J., Pelica, F., et al. (1999). 'Green revolution' genes encode mutant gibberellin response modulators. *Nature, 400*, 256–261. https://doi.org/10.1038/22307

Pharmaceutical Research and Manufacturers of America. (2013). *Medicines in development: Biologics*. PhRMA.

Plotkin, S. A. (2001). Vaccines in the 21st century. *Infectious Disease Clinics of North America, 15*, 30727.

Poland, G. A., Murray, D., & Bonilla-Guerrero, R. (2002). New vaccine development. *British Medical Journal, 324*(7349), 1315–1319.

Prescott, L. M., Harley, J. P., & Klein, D. A. (2002). *Microbiology* (5th ed.). McGraw-Hill.

Ramakrishnan, B., Maddela, N. R., Venkateswarlu, K., & Megharaj, M. (2020). Organic farming: Does it contribute to contaminant-free produce and ensure food safety? *The Science of the Total Environment, 769*, 145079.

Ramchandran, L., & Shah, N. P. (2009). Effect of EPS on the proteolytic and ACE inhibitor activities and textural and rheological properties of low-fat yogurt during refrigerated storage. *Journal of Dairy Science, 92*, 895–906.

Rosenblum, D., Gutkin, A., Kedmi, R., Ramishetti, S., Veiga, N., Jacobi, A. M., et al. (2020). CRISPR-Cas9 genome editing using targeted lipid nanoparticles for cancer therapy. *Science Advances, 6*(47), eabc9450.

Rosenfeld, D., Senko, A. W., Moon, J., Yick, I., Varnavides, G., Gregureć, D., et al. (2020). Transgene-free remote magnetothermal regulation of adrenal hormones. *Science Advances, 6* (15), eaaz3734.

Rupali, D., & Dibyengi, S. (2004). Biotechnology in phytoremediation of metal-contaminated soils. *Proceedings of the Indian National Science Academy, B701*, 99–108.

Rural Advancement Foundation International. (2021). https://www.rafiusa.org/. Accessed 10 Feb 2021.

Sack, L., & Scoffoni, C. (2013). Leaf venation: Structure, function, development, evolution, ecology and applications in the past, present and future. *The New Phytologist, 198*, 983–1000. https://doi.org/10.1111/nph.12253

Saranya, K., Maddela, N. R., Mallavarapu, M., & Kadiyala, V. (2020). *Total petroleum hydrocarbons—Environmental fate, toxicity, and remediation*. Springer International Publishing AG. https://doi.org/10.1007/978-3-030-24035-6, ISBN 978-3-030-24035-6. (pp. VIII 262).

Shan, M., Khan, Y. A., & Nazar, H. (2018). Application of biotechnology in medicinal field. *Single Cell Biol, 7*, 174.

Shanker, D. (October 22, 2019). *These $50 Chicken Nuggets Were Grown in a Lab*. Bloomberg. Retrieved February 27, 2020

Spielmeyer, W., Ellis, M. H., & Chandler, P. M. (2002). Semidwarf (sd-1), "green revolution" rice, contains a defective gibberellin 20-oxidase gene. *Proceedings of the National Academy of Sciences of the United States of America, 99*, 9043–9048. https://doi.org/10.1073/pnas.132266399

Staff. (2019, April 17). CRISPR has been used to treat US cancer patients for the first time. *MIT Technology Review*.

Stockdale, E. A., Lampkin, N. H., Hovi, M., et al. (2001). Agronomic and environmental implications of organic farming systems. *Advances in Agronomy, 70*, 261–327.

Tel Aviv University. (2020, November 18). *TAU breakthrough may increase life expectancy in brain and ovarian cancers*.

Timmis, K. N., & Pieper, D. H. (1999). Bacteria designed for bioremediation. *Trends in Biotechnology, 17*(5), 201–204.

Tufts Center for the Study of Drug Development. (2010). *Personalized medicine is playing a growing role in development pipelines*. The Center. https://www.policymed.com/2010/12/tufts-study-shows-drug-development-for-personalize-medicine-on-the-rise.html

Verma, A. S., Agrahari, S., Rastogi, S., & Singh, A. (2011). Biotechnology in the realm of history. *Journal of Pharmacy & Bioallied Sciences, 3*(3), 321.

Vijayakuma, S., & Sasikala, M. (2012). Application of biotechnology: A current review. *International Journal of Pharmacy, 2*, 59–66.

Wahren, B., & Liu, M. A. (2014). DNA vaccines: Recent developments and the future. *Vaccine, 2* (4), 785–796.

Wessels, H. H., Méndez-Mancilla, A., Guo, X., Legut, M., Daniloski, Z., & Sanjana, N. E. (2020). Massively parallel Cas13 screens reveal principles for guide RNA design. *Nature Biotechnology, 38*(6), 722–727.

Wikipedia. (2018). *Biotechnology*. Wikimedia Foundation. https://en.wikipedia.org/wiki/Biotechnology

Wikipedia contributors. (2021). Timeline of biotechnology. *Wikipedia, The Free Encyclopedia*. https://en.wikipedia.org/w/index.php?title=Timeline_of_biotechnology&oldid=1009136337

Xu, Z., Wang, S., Zhao, C., Li, S., Liu, X., Wang, L., & Mann, S. (2020). Photosynthetic hydrogen production by droplet-based microbial micro-reactors under aerobic conditions. *Nature Communications, 11*(1), 1–10.

Part II
Industrial Biotechnology

Enzymes from Microorganisms

Silpa Somavarapu, Bellamkonda Ramesh, G. Vidya Sagar Reddy, Srinivasan Kameswaran, M. Subhosh Chandra, Ch. Venkatrayulu, and B. Vijay Kumar

1 Introduction

Enzymes form a distinct class of molecules. They are the biological catalysts, which fasten the biochemical reaction by decreasing their activation energy.

IUB, which stands for International Union of Biochemistry depending on their mechanism of action, categorizes enzymes into six distinct classes namely

E.C 1 Oxidoreductases
E.C 2 Transferases
E.C 3 Hydrolases
E.C 4 Lyases
E.C 5 Isomerases
E.C 6 Ligases

During catalysis, enzymes catalyze the reaction i.e., conversion of substrate into products without themselves getting involved in the reaction and are required in very minute concentrations (Die Kinetik der Invertinwirkung Taylor et al., 2015) (Fig. 1).

The potential catalytic activity of enzymes is expressed as a constant k_{cat} (catalytic rate constant) also called as the turnover number (Fig. 2). The turnover number

S. Somavarapu · B. Ramesh (✉) · C. Venkatrayulu · B. V. Kumar
Department of Food Technology, Vikrama Simhapuri University, Nellore, Andhra Pradesh, India

G. V. S. Reddy
Department of Biotechnology, Vikrama Simhapuri University, Nellore, Andhra Pradesh, India

S. Kameswaran
Department of Botany, Vikrama Simhapuri University PG Centre, Kavali, Andhra Pradesh, India

M. S. Chandra
Department of Microbiology, Yogi Vemana University, Kadapa, Andhra Pradesh, India

N. R. Maddela, L. C. García (eds.), *Innovations in Biotechnology for a Sustainable Future*, https://doi.org/10.1007/978-3-030-80108-3_3

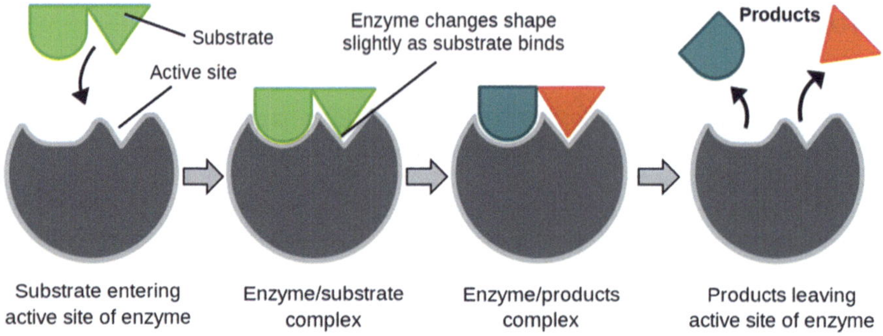

Fig. 1 Enzymatic reaction

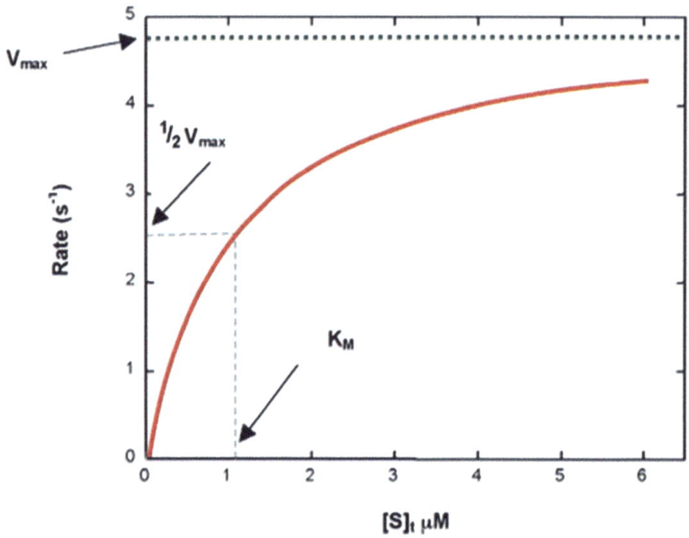

Fig. 2 Enzyme kinetics

is defined as the substrate converted to product by single enzyme unit in unit time (Johnson & Goody, 2011; Changeux, 2013).

Enymes are highly specific. They exhibit group specificity, absolute specificity and broad specificity. This specific nature of enzymes makes it significant in biosensors and biochemicals analysis (Adrio & Demain, 2014).

2 Enzymes Form Distinctive Class of Protease

For about a century, the enzymes that play the central role in the degradation of proteins by hydrolyzing peptide bonds have been known as "proteases" and the term protease is therefore equivalent to "peptide hydrolase." They hold first place in the world market of enzymes, estimated at ~US$3 billion (Leary et al., 2009). A little over 50 years ago, the German physiological chemists realized that there was an important distinction to be made between the hydrolases which act directly on proteins, and those act on peptides which are intermediates in proteolysis. The proteases which act on proteins were called proteinases, and those acting on peptides were called peptidases (Grassman & Dyckerhoff, 1928; Waldschmidt-Leitz, 1931). Later, the preference of proteases for action on intact proteins or long polypeptide chains, and that of "peptidases" for action on small peptides, became clear. These depended on the acceptability of the amino and carboxyl termini of polypeptides in the specificity sites of the enzymes. Thus, the end groups were not required and typically were not tolerated, in the specificity sites of the proteases, so that the enzymes acted well on long chains, away from the ends, but acted less well on the peptide products. In contrast, the specificity of the "peptidases" required at least one terminus to occupy a specificity site, so that these enzymes had little action on intact proteins, in which few of the peptide bonds are close to a terminus.

Proteolysis is an irreversible process of polypeptide cleavage with important physiological roles in a number of cellular processes where it is essential to confine the cleavage of peptides in space and time (Majerle & Jerala, 2003). These enzymes can also hydrolyze peptides when a pteroyl moiety or acyl groups replaces the peptidyl group.

In the course of discussion of the mechanism of proteolysis (Linderstrom-Lang, 1949), the understanding of the general characteristics of the action of the two subdivisions of proteases was embodied in the alternative names endopeptidases for those acting in the interior of polypeptide chains and exopeptidases for those acting at the termini (Bergmann & Ross, 1936). These names are, of course, analogous to those used for hydrolases acting on other polymers such as polysaccharides and polynucleotides.

Most proteases are unmistakably endopeptidases or exopeptidases, but a few have marked activities of both types. In the physiological context, the specificities of the endopeptidases and exopeptidases is that the former are responsible for the early stages of protein breakdown, and the latter enzymes take over at an intermediate stage and complete the generation of free amino acids. In biochemical pathways, the endopeptidases catalyzed initial stages that are typically the rate-limiting ones. Once the degradation of a protein molecule has started, it proceeds rapidly, and there is little or no accumulation of partial degradation products. The way in which the proteases are classified is summarized in Fig. 3.

Note: The numbers given in parentheses indicate the divisions in which the enzymes have been placed in the enzyme nomenclature given by International Union of Biochemistry and Molecular Biology.

2.1 Exopeptidases

The exopeptidases are those proteases, which act only near the terminal ends of polypeptide chains (Table 1). As every polypeptide chain has an amino terminal and

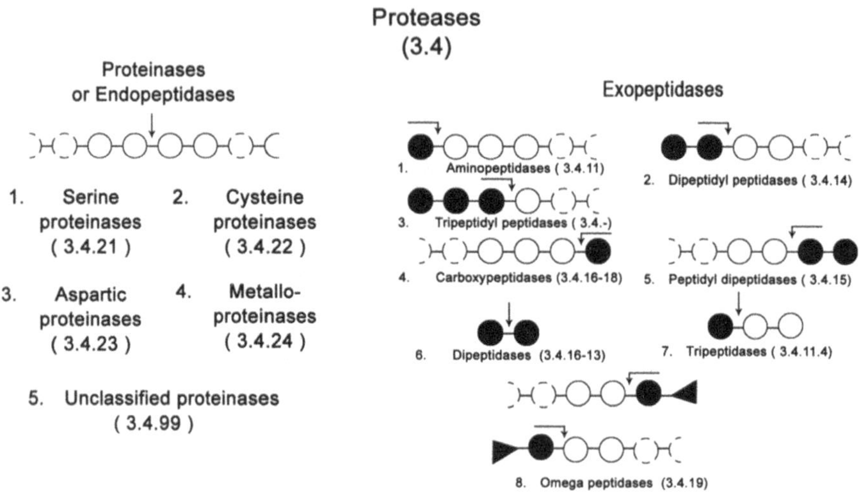

Fig. 3 Classification of proteases

Table 1 Exopeptidases

S. no	Exopeptidase	Source
1	Aminopeptidase I	*Escherichia coli* (De Marco & Dick, 1978)
2	Aminopeptidase	*Bacillus licheniformis*
3	Aminopeptidase II	*B. stearothermophilus* (Stoll et al., 1978)
4	Serine carboxypeptidases	*Penicillium* spp.
		Saccharomyces spp., *Aspergillus* spp.
5	Metallocarboxypeptidases	*Saccharomyces* spp. (Felix & Brouillet, 1966)
		Pseudomonas spp.

carboxy terminal ends, based on their site of action these proteases are classified as amino and carboxy peptidases, respectively.

2.2 *Endopeptidases*

Endopeptidases are the proteases which act at the peptide bonds that occur in the inner regions of the polypeptide chain (Table 2). Based on their mechanism of action these endopeptidases are further divided into four subgroups viz.,

1. Serine proteases
2. Aspartic proteases
3. Cysteine proteases and
4. Metalloproteases

Table 2 Endopeptidases

S. no	Endopeptidase	Source
1.	*Serine proteases*	
	A. Serine Alkaline proteases	*Arthrobacter, Streptomyces, Flavobacterium* spp., *S. cerevisiae, Conidiobolus* spp., *Aspergillus, Neurospora* spp. (Lindberg et al., 1981)
	A.B. Subtilisins	*Bacillus licheniformis*
		Bacillus amyloliquefaciens
2	*Aspartic proteases*	
	A. Pepsin-like enzymes	*Aspergillus, Penicillium, Rhizopus*, and *Neurospora*
	B. Rennin-like enzymes	*Endothia* and *Mucor* spp. (Sielecki et al., 1991)
3	*Cysteine proteases*	
	A. Clostripain	*Clostridium histolyticum*
	B. Streptopain	*Streptococcus* spp.
4	*Metallo proteases*	
	A. Thermolysin	*B. stearothermophilus (*Weaver et al., 1977)
	B. Collagenase	*Clostridium histolyticum*

2.3 Cysteine Proteases

Cysteine proteases (EC 3.4.2.2) are found in bacteria, (Morihara, 1974), eukaryotic microorganisms, (North, 1982), plants, and animals. Cysteine proteases are represented by 70 families belonging to 12 different classes. However, there may be more distinct super families of cysteine proteases. The most studied cysteine protease is papain, from the latex of *Carica papaya*, and it is now clear that the papain superfamily is a large one (Barrett et al., 1984). Amino acid sequence data show that cysteine proteases of other higher plants are members of the papain superfamily, as are cathepsins B, H, and L of mammalian lysosomes. Many protozoa contain cysteine proteases that have molecular weight of about 25,000 Da and other properties consistent with their being related to papain. It is conceivable that the first of the cysteine proteases of the papain superfamily functioned in the digestive vacuoles of a protozoan. The enzymes have continued in evolution to tend to be associated with the intracellular vacuoles of plant cells and of animal cells.

Two bacterial cysteine proteases that currently seem to represent independent superfamilies not occurring in higher plants and animals are clostripian from *Clostridium histolyticum* (Siffert et al., 1976), and the protease from a *Streptococcus* species (Tai et al., 1976). Clostripian is highly specific for substrates with an arginyl residue at P1, like some serine proteases (Siffert et al., 1976), and is calcium activated. The amino acid sequence around the essential cysteine is unlike that of other known cysteine proteases (Gilles et al., 1983). The Streptococcal protease has an inactive zymogen that is activated by limited proteolysis (Yonaha et al., 1982).

Zymogens of serine, aspartic, and metalloproteases are common in higher animals, but are uncommon among the cysteine proteases and in primitive organisms.

The identification of cysteine protease activity is not usually difficult, since the activity is generally stimulated by low M_r thiol compounds, as well as being inhibited by thiol blocking reagents such as iodoacetate. The reactivity of the essential thiol with iodoacetate is generally much greater than that of a low M_r thiol compound, so that iodoacetate will readily inactivate the enzyme, even in the presence of an excess of the thiol activator and at low pH. The covalent reactions with iodoacetamide and *N*-ethylmaleimide tend to be slower than that with iodoacetate and the mercurial reagents also are less satisfactory, reacting reversibly. In addition to the low M_r inhibitors, the cysteine proteases are sensitive to protein inhibitors. The papain-like enzymes are mostly inhibited by a group of proteins called cystatins, of which chicken egg white cystatin is not too difficult to obtain as a reagent (Anastasi et al., 1983). Cystatins do not inhibit the calpains, but these enzymes are sensitive to another protein, calpastatin (Murachi, 1983).

The three potential roles of cysteine proteases in plant defense include invader perception, downstream signaling pathway activation, and defense response (Vander Hoorn & Jones, 2004). Cysteine proteases are found to be useful in proteolysis during senescence, under drought and during programmed cell death. They are the critical components of growth, cell differentiation, signaling, and host invasion of various human and livestock pathogens (pathogenic parasites causing malaria, Chag's disease and schistosomiasis) as well as allergens (Lecaille et al., 2002, Mc Kerrow et al., 2006). Bromeliaceae cysteine proteases had potential industrial or biotechnological applications as they could be easily modified by protein engineering. They are used as antitumour agents to stimulate monocytic cytotoxicity, in wound healing and to reduce inflammation by altering migration and activation of lymphocytes. They are anti-edematous, anti-thrombic, and anti-inflammatory with less side effects. The best known cysteine proteases are caspase like proteins, vacuolar processing enzymes, papain-like peptides and cathepsin type proteases.

3 Distribution of Proteases

Proteases are distributed widely in different parts of the biological sources. In occurrence of proteases, plant kingdom occupies the highest rank (43.85%) followed by bacteria (18.09%), fungi (15.08%), animals (11.15%), algae (7.42%), and viruses (4.41%). Isolation of proteases from biological sources contributes 27 to 67% either animal, microbial or plant origin. Cysteine protease occurs abundantly in plants (34.92%). Microbes secrete serine and aspartic proteases in large quantities of 13.21% and 8.81%, respectively. The distribution of protease enzymes in biological sources are summarized in Figs. 4 and 5.

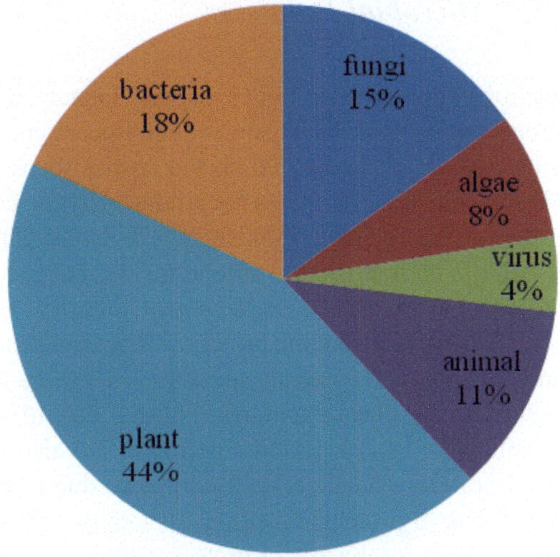

Fig. 4 Distribution of proteases from biological sources

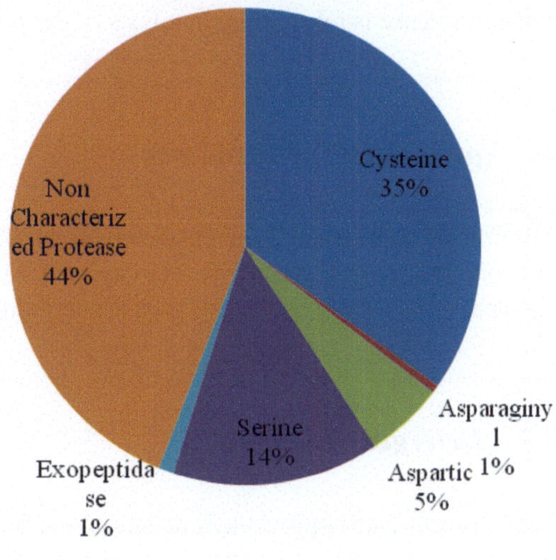

Fig. 5 Distribution of plant proteases

3.1 *Sources of Enzymes*

Enzymes from microbial sources have been proved to be excellent eco-friendly biocatalysts. Thermophilic, mesophilic, and extremophilic bacteria are excellent sources of thermostable enzymes. Microbes are an attractive source of proteases as they have short generation time and they require limited space for their cultivation and they are susceptible to genetic manipulation. The inability of the plant and animal proteases to meet current world demands has led to an increased interest in microbial proteases.

3.2 Microbial Proteases

Microbial proteases account for approximately 40% of the total worldwide enzymes. Proteases from microbial sources surpass enzymes from plant and animal sources as they contain all the desired characteristics for varied biotechnological applications. Most commercial proteases are produced by the genus *Bacillus*. These enzymes have pH range from pH 5 to 8 and are thermolabile. Bacterial alkaline proteases are active at pH 10 and temperature around 60 °C. They show broad substrate specificity and are suitable for use in the detergent industry. Fungi *Aspergillus oryzae* produces acid, neutral, and alkaline proteases. They are active over a wide pH range (pH 4–11) and exhibit broad substrate specificity. They are thermo labile. They are particularly useful in the cheese making industry due to their narrow pH and temperature specificities. Fungal alkaline proteases are also used in food protein modification. Viral proteases have gained importance due to their functional involvement in the processing of proteins of viruses that cause certain fatal diseases such as AIDS and cancer. Retroviral aspartyl proteases that are required for viral assembly and replication are homodimers and are expressed as a part of the polyprotein precursor. The mature protease is released by autolysis of the precursor (Kuo & Shafer, 1994).

4 Applications of Proteases

All proteolytic enzymes have characteristic properties with regards to temperature, pH, ion requirement, specificity, activity, and stability. These biochemical parameters determine the application of protease in industry and other fields.

4.1 Detergent Industry

Development and improvement of household and industrial detergents are greatly contributed by proteases. The enzymes used in detergent industry are proteases, lipases, amylases, cellulases, etc. Among these, protease plays a major application as detergent additive due to their ability to hydrolyze and remove proteinaceous stains such as blood, gravy, egg, and milk at high pH conditions.

4.2 Leather Industry

In leather industry, processing of leather involves three main steps known as soaking, dehairing, and bating. The conventional methods of leather processing which involves the usage of chemical reagents (sodium sulfide, sodium hydroxide, hydrogen sulfide, etc.) release a large amount of hazardous pollutants into the surrounding environment. However, application of proteases in place of chemical treatments has been identified as an environmental friendly alternative for leather processing. Keratinolytic activity of novel proteases has the potential to replace sodium sulfide in the dehairing process (Arunachalam & Sarita, 2009).

4.3 Food and Feed Industry

Proteases are used widely in the preparation of protein hydrolysates to be used as additives to food and feed to improve their nutritional value (Sumantha et al., 2006). These are also used in brewing, cheese elaboration, and bread manufacturing (Pande et al., 2006). Proteases also play a role in protein storage, mobilization, senescence, programmed cell death, hormone signaling and defense and are regulated by various types of environmental stresses (Grudkowska & Zagdanska, 2004). These have a role as meat tenderizers and as plant milk clotting enzymes for novel dairy products (Fadyloglu, 2001). Chymosin is preferred in cheese making industry due to its high specificity for casein (Aguilar et al., 2008). Measuring hydrolytic activity on synthetic substrates is a simple way to know the cleavage specificity of these enzymes, which provide important information for biotechnological applications, as with the production of bioactive peptides from food proteins (Silva & Malcata, 2005a, b).

They generate less bitterness in hydrolyzed food proteins, hence valuable in the food industry. Neutrase, a neutral protease, being insensitive to the natural plant proteinase inhibitors proves useful in brewing industry. Their low thermotolerance is advantageous for controlling their reactivity during the production of food hydrolysates with a low degree of hydrolysis. Some of the neutral proteases belong to the metalloprotease type and require divalent metal ions for their activity, while others are serine proteinases, which are not affected by chelating agents (Fernandes, 2010; Vashist et al., 2011).

Saccharomyces cerevisiae, *Aspergillus oryzae*, *Bacillus licheniformis*, *Bacillus stearothermophilus* are regarded as GRAS (Generally Regarded as Safe) by US Food and Drug Administration have a significant role in food industry in the production of α-amylase (Novoigt, 2008; Abdulaal, 2018; Abu et al., 2017; Acer et al., 2016).

4.4 Silk Degumming

Threads of raw silk must be degummed to remove sericin protein that covers the silk fiber. Traditionally, degumming is performed in an alkaline solution containing soap. This is a harsh treatment because the fiber itself is attacked. Use of alkaline proteases to remove the sericin without attacking the fiber is a better method. Freddi et al. (2003) have attempted degumming of crepe fabric, a very difficult fabric substrate with alkaline and neutral proteases.

4.5 Photographic Industry

The photographic films contain 1.5–2.0% silver by weight in their gelatin layers. The conventional method used for silver recovery is the burning of films which causes the problem of environmental pollution and in addition, the polyester based film cannot be recycled by this method. Thus, the use of enzymatic hydrolysis of gelatin layers not only extracts silver from the films but also polyester based films can be recycled (Gupta et al., 2002).

4.6 Pharmaceutical Industry

Proteases are also used in pharmaceutical industry in developing therapeutic agents (Walsh, 2002). Proteases from the plant extracts have been used as traditional medicine in treating cancer as antitumorals (Guimaraes-Ferreira et al., 2007; Otsuki et al., 2010), for digestive disorders and for immune modulation problems (Otsuki et al., 2010) Several plant latex proteases are known to interfere in homeostasis as procoagulant suggesting its unique substrate preference over other proteases (Richter et al., 2002; Rajesh et al., 2007). Some peptides are hidden and inactive in the original peptides, but when liberated they can have diverse biomedical applications, as antihypertensive or antioxidant agents (Perpetuo et al., 2003). Plant latex is a natural source of pharmaceuticals and pesticides (Upadhyay, 2011).

4.7 Wide Applications in Agriculture, Pharmaceutical, Biofuel Industries

They have important roles in the production of sweetening agents and the modification of antibiotics, they are used in washing powders and various cleaning products, and they play a key role in analytical devices and assays that have clinical, forensic and environmental applications (Maddela et al., 2021). Elected microbial-sourced HemG PPO enzyme variants present an opportunity for building new herbicide tolerance biotechnology traits. These traits provide tolerance to PPO-inhibiting herbicides and, therefore, could provide additional tools for farmers to employ in their weed management systems (Heap, 2019; Glenn et al., 2017; Dayan et al., 2010). Soil enzymes and microbial elemental stoichiometry as bioindicators of soil quality in diverse cropping systems and nutrient management practices of Indian Vertisols (Mangalassery et al., 2019).

References

Abdulaal, W. H. (2018). Purification and characterization of α-amylase from *Trichoderma psedokoningii*. *BMC Biochemistry, 19*(4), 1–6. https://doi.org/10.1186/s12858-018-0094-8

Abu, M. L., Nooh, H. M., Oslon, S. N., & Salleh, A. B. (2017). Optimization of physical conditions for the production of thermostable T1 lipase in *Pichia guilliermundii* strain SO using response surface methodology. *BMC Biotechnology, 17*(78), 1–10. https://doi.org/10.1186/s12896-017-0397-7

Acer, O., Bekler, F. M., Pirincchioglu, H., Guven, R. G., & Guven, K. (2016). Purification and characterization of thermostable and detergent stable α-amylase from *Anoxybacillus* species AH1. *Food Technology and Biotechnology, 5*(1), 70–77. https://doi.org/10.17113/ftb54.01.16-4122

Adrio, J. L., & Demain, A. L. (2014). Microbial enzymes: Tools for biotechnological processes. *Biomolecules, 4*, 117–139.

Aguilar, C. N., Sanchez, G. G., Barragan, P. A. R., Herrera, R. R., Hernandez, J. L. M., & Esquivel, C. C. (2008). Perspectives of solid state fermentation for production of food enzymes. *American Journal of Biochemistry and Biotechnology, 4*(4), 354–366. https://doi.org/10.3844/ajbbsp.2008.354.366

Anastasi, A., Brown, M. A., Kembhavi, A. A., Nicklin, M. J. H., Sayers, C. A., Sunder, D. C., & Barrett, A. J. (1983). Cystatin, a protein inhibitor of cysteine proteinases: Improved purification from egg white, characterization, and detection in chicken serum. *Biochemistry Journal, 211*, 129–138. https://doi.org/10.1042/bj2110129

Arunachalam, C., & Sarita, K. (2009). Protease enzyme: An eco-friendly alternative for leather industry. *Indian Journal of Science and Technology, 2*(12), 29–32. https://doi.org/10.17485/ijst/2009/v2i12.10

Barrett, A. J., Nicklin, M. J. H., & RawlingsND. (1984). The papain superfamily of cysteine proteinases and their protein inhibitors. *Symposia Biologica Hungarica, 25*, 203. https://doi.org/10.2174/138920312804871102

Bergmann, M., & Ross, W. F. (1936). On proteolytic enzymes X. The enzymes of papain and their activation. *Journal of Biological Chemistry, 114*, 717–726. https://doi.org/10.1016/0006-291X(69)90776-1

Changeux, J. P. (2013). 50 years of allosteric interactions: The twists and turns of the models. *Nature Reviews. Molecular Cell Biology, 14*, 819–829. https://doi.org/10.1038/nrm3695

Dayan, F. E., Daga, P. R., Duke, S. O., Lee, R. M., Tranel, P. J., & Doerksen, R. J. (2010). Biochemical and structural consequences of a glycine deletion in the *α*-8 helix of protoporphyrinogen oxidase. *Biochimica et Biophysica Acta, 1804*, 1548–1556. https://doi.org/10.1016/j.bbapap.2010.04.004

De Marco, A. C., & Dick, A. J. (1978). Aminopeptidase I activities in several microorganisms. *Canadian Journal of Biochemistry, 56*, 66–71. https://doi.org/10.1139/o78-010

Fadyloglu, S. (2001). Immobilization and characterisation of ficin. *Nahrun, 45*, 143–146. https://doi.org/10.1002/1521-3803(20010401)

Felix, F., & Brouillet, N. (1966). Purification and properties of two peptidases from brewer's yeast. *Biochimica et Biophysica Acta, 122*, 127–144. https://doi.org/10.1016/0926-6593(66)90096-8

Fernandes, P. (2010). Enzymes in food processing: A condensed overview on strategies for better biocatalysts. *Enzyme Research, 2010*, 862537. https://doi.org/10.4061/2010/862537

Freddi, G., Mossotti, R., & Innocenti, R. (2003). Degumming of silk fabric with several proteases. *Journal of Biotechnology, 106*, 101–112. https://doi.org/10.1016/j.jbiotec.2003.09.006

Gilles, A. M., DeWolf, A., & Kell, B. (1983). Amino-acid sequences of the active-site sulfhydryl peptide and other thiol peptides from the cysteine proteinase alpha-clostripain. *European Journal of Biochemistry, 130*, 473–479. https://doi.org/10.1111/j.1432-1033.1983.tb07174.x

Glenn, K. C., Alsop, B., Bell, E., Goley, M., Jenkinson, J., & Liu, B. (2017). Bringing new plant varieties to market: Plant breeding and selection practices advance beneficial characteristics while minimizing unintended changes. *Crop Science, 57*, 2906–2921. https://doi.org/10.2135/cropsci2017.03.0199

Grassman, W., & Dyckerhoff, H. (1928). Subcellular localization and levels of aminopeptidases and dipeptidase in *Saccharomyces cerevisiae. Journal of Physiological Chemistry, 179*, 41–66. https://doi.org/10.1016/0005-2744(78)90253-X

Grudkowska, M., & Zagdanska, B. (2004). Multifunctional role of plant cysteine proteinases. *Acta Biochimica Polonica, 51*, 609–624.

Guimaraes-Ferreira, C. A., Rodrigues, E. G., Mortara, R. A., Cabral, H., Serrano, F. A., & Ribeiro-dos-Santos, R. (2007). Antitumor effects in vitro and in vivo and mechanisms of protection against melanoma B16F10-Nex2 cells by fastuosain, a cysteine proteinase from *Bromelia fastuosa. Neoplasia, 9*, 723–733. https://doi.org/10.1593/neo.07427

Gupta, R., Beg, Q. K., & Lorenz, P. (2002). Bacterial alkaline proteases: Molecular approaches and industrial applications. *Applied Microbiology and Biotechnology, 59*, 15–32. https://doi.org/10.1007/s00253-002-0975-y

Heap, I. (2019). *The international survey of herbicide resistant weeds*. www.weedscience.org.

Johnson, K. A., & Goody, R. S. (2011). The original Michaelis constant: Translation of the 1913 Michaelis–Menten paper. *Biochemistry, 50*, 8264–8269. https://doi.org/10.1021/bi201284u. A modern translation, commentary and re-analysis of the original 1913 paper.

Kuo, L. C., & Shafer, J. A. (1994). Retroviral proteases. *Methods in Enzymology, 241*, 3–178. ISBN: 978-0-12-182142-5, ISSN: 0076-6879.

Leary, D., Vierros, M., Hamon, G., Arico, S., & Monagle, C. (2009). Marine genetic resources: A review of scientific and commercial interest. *Marine Policy, 33*, 183–194. https://doi.org/10.1016/j.marpol.2008.05.010

Lecaille, F., Kaleta, J., & Brömme, D. (2002). Human and parasitic papain-like cysteine proteases: Their role in physiology and pathology and recent developments in inhibitor design. *Chemical Reviews, 102*, 4459–4488. https://doi.org/10.1021/cr0101656

Lindberg, R. A., Eirich, L. D., Price, J. S., Wolfinbarger, L., Jr., & Drucker, H. (1981). Alkaline protease from *Neurospora crassa. The Journal of Biological Chemistry, 256*, 811–814.

Linderstrom-Lang, K. (1949). Structure and enzymatic breakdown of proteins. *Cold Spring Harbor Symposia on Quantitative Biology, 14*, 117–126. https://doi.org/10.1101/SQB.1950.014.01.016

Maddela, N. R., Garcia, L. C., & Chakraborty, S. (Eds.). (2021). *Advances in the domain of environmental biotechnology*. Springer Nature Singapore Pte Ltd. https://doi.org/10.1007/978-981-15-8999-7; ISBN: 978-981-15-8999-7 (pp XVIII, 717).

Majerle, A., & Jerala, R. (2003). Protein inhibitors form complexes with procathepsin L and augment cleavage of the propeptide. *Archives of Biochemistry and Biophysics, 417*, 53–58. https://doi.org/10.1016/S0003-9861(03)00319-9

Mangalassery, S., Kalaivanan, D., & Philip, P. S. (2019). Effect of inorganic fertilisers and organic amendments on soil aggregation and biochemical characteristics in a weathered tropical soil. *Soil and Tillage Research, 187*, 144–151. https://doi.org/10.1016/j.still.2018.12.008

Mc Kerrow, J. H., Caffrey, C., Kelly, B., Loke, P., & Sajid, M. (2006). Proteases in parasitic diseases. *Annual Review of Pathology, 1*, 497–536. https://doi.org/10.1146/annurev.pathol.1.110304.100151

Morihara, K. (1974). Comparative specificity of microbial proteinases. *Advances in Enzymology, 41*, 179–243. https://doi.org/10.1016/0003-9861(74)90142-8

Murachi, T. (1983). In W. Y. Cheung (Ed.), *Intracellular protease and its inhibitor protein: Calpain and calpastatin in calcium and cell function* (Vol. 4, pp. 377–410). Academic Press. https://doi.org/10.1016/0065-2571(81)90026-1

North, M. J. (1982). Comparative biochemistry of the proteinases of eukaryotic microorganisms. *Microbiological Reviews, 46*, 308–340.

Novoigt, E. (2008). Progress in metabolic engineering of *Sacharomyces cerevisiae. Microbiology and Molecular Biology Reviews, 72*(3), 379–412. https://doi.org/10.1128/MMBR00025-07

Otsuki, N., Dang, N. H., Kumagai, E., Kondo, A., Iwata, S., & Morimoto, C. (2010). Aqueous extract of *Carica papaya* leaves exhibits anti-tumor activity and immunomodulatory effects. *Journal of Ethnopharmacology, 127*, 760–767. https://doi.org/10.1016/j.jep.2009.11.024

Pande, M., Dubey, V. K., Yadav, S. C., & Jagannadham, M. V. (2006). A novel serine protease cryptolepain from Cryptolepis buchanani: Purification and biochemical characterization. *Journal of Agricultural and Food Chemistry, 54*, 10141–10150. https://doi.org/10.1021/jf062206a

Perpetuo, E. A., Juliano, L., & Lebrun, I. (2003). Biochemical and pharmacological aspects of two bradykinin-potentiating peptidases obtained from tryptic hydrolysis of casein. *Journal of Protein Chemistry, 22*, 601–606. https://doi.org/10.1023/B:JOPC.0000008724.98339.ff

Rajesh, R., Shivaprasad, H. V., Raghavendragowda, C. D., Nataraju, A., Dhananjaya, B. L., & Vishwanath, B. S. (2007). Comparative study on plant latex proteases and their involvement in hemostasis: A special emphasis on clot inducing and dissolving properties. *Planta Medica, 73*, 1061–1067. https://doi.org/10.1055/s-2007-981575

Richter, G., Hans, P. S., Friedrich, D., & Peter, L. (2002). Activation and inactivation of human factor X by proteases derived from *Ficuscarica. British Journal of Haematology, 119*, 1042–1051. https://doi.org/10.1046/j.1365-2141.2002.03954.x

Sielecki, A. R., Fujinaga, M., Read, R. J., & James, M. N. G. (1991). Refined structure of porcine pepsinogen at 1.8A resolution. *Journal of Molecular Biology, 219*, 671–692. https://doi.org/10.1016/0022-2836(91)90664-R

Siffert, O., Emod, I., & Keil, B. (1976). Interaction of clostripain with natural trypsin inhibitors and its affinity labeling by Nα-p-nitrobenzyloxycarbonyl arginine chlormethyl ketone. *FEBS Letters, 66*, 114–119. https://doi.org/10.1016/0014-5793(76)80598-4

Silva, S. V., & Malcata, F. X. (2005a). Partial identification of water-soluble peptides released at early stages of proteolysis in sterilized ovine cheese-like systems: Influence of type of coagulant and starter. *Journal of Dairy Science, 88*, 1947–1954. https://doi.org/10.3168/jds.S0022-0302(05)72870-8

Silva, S. V., & Malcata, F. X. (2005b). Studies pertaining to coagulant and proteolytic activities of plant proteases from *Cynara cardunculu. Food Chemistry, 89*, 19–26. https://doi.org/10.1016/j.foodchem.2004.01.074

Stoll, E., Weder, H. G., & Zuber, H. (1978). Aminopeptidase II from *Bacillus sterothermophilus. Biochimica et Biophysica Acta, 438*, 212–220. https://doi.org/10.1016/0005-2744(76)90237-0

Sumantha, A., Larroche, C., & Pandey, A. (2006). Microbiology and industrial biotechnology of food grade proteases: A perspective. *Food Technology and Biotechniques, 44*, 211–220.

Tai, H. H., Tai, C. L., & Hollander, C. S. (1976). Biosynthesis of prostaglandins in rabbit kidney medulla, properties of prostaglandin synthase. *Biochemistry Journal, 154*, 257–264. https://doi.org/10.1042/bj1540257

Taylor, A. I., Pinheiro, V. B., Smola, M. J., Morgunov, A. S., Peak-Chew, S., Cozens, C., Weeks, K. M., Herdewijn, P., & Holliger, P. (2015). Catalysts from synthetic genetic polymers. *Nature, 518*, 427–430.

Upadhyay, R. K. (2011). Plant latex: A natural source of pharmaceuticals and pesticides. *International Journal of Green Pharmacy, 5*, 169–180. https://doi.org/10.22377/ijgp.v5i3.199

Vander Hoorn, R. A. L., & Jones, J. D. G. (2004). The plant proteolytic machinery and its role in defence. *Current Opinion in Plant Biology, 7*, 400–407. https://doi.org/10.1016/j.pbi.2004.04.003

Vashist, S., Zheng, D., Al-Rubeaan, K., Luong, J. H. T., & Sheu, F. S. (2011). Technology behind commercial devices for blood glucose monitoring in diabetes management: A review. *Analytica Chimica Acta, 703*, 124–136. https://doi.org/10.1016/j.aca.2011.07.024

Waldschmidt-Leitz, E. (1931). The mode of action and differentiation of proteolytic enzymes. *Physiological Reviews, 11*, 358. https://doi.org/10.1152/physrev.1931.11.3.358

Walsh, G. (2002). *Proteins, biochemistry and biotechnology* (p. 420). John Wiley and Sons Ltd. https://doi.org/10.1002/cbf.987

Weaver, L. H., Kester, W. R., & Matthews, B. W. A. (1977). Crystallographic study of the complex of phosphoramidon with thermolysin. A model for the presumed catalytic transition state and for the binding of structures. *Journal of Molecular Biology, 114*, 119–132. https://doi.org/10.1016/0022-2836(77)90286-8

Yonaha, K., Elliot, S. D., & Liu, T. Y. (1982). Primary structure of zymogen of streptococcal proteinase. *Journal of Protein Chemistry, 1*, 317–334. https://doi.org/10.1007/BF01039555

Biotechnological Applications of Essential Oils: Post-harvest and Food Preservation

Virginia Monserrate López-Zambrano, Alex Alberto Dueñas-Rivadeneira, José Daniel Zambrano-Veliz, Carlos Alfredo Cedeño-Palacios, and María Hipatia Delgado-Demera

1 Introduction

Foods are frequently prone to cross-contamination by pathogens, which results in significant losses in quality, quantity, and nutritional composition. Registered cases of loss of fruits, vegetables, grains, nuts, meats, and processed foods in poor condition are reported in figures equivalent to metric tons of food each year (FAO, 2015).

Food security represents a major problem, especially in low-income countries, affecting hundreds of millions of people as a result of the increased incidence of foodborne illness (Reyes et al., 2016). In order to extend the shelf life of food, a large number of food products require protection against spoilage during the stages of: preparation, storage, and distribution (OMS, 2007). Another important problem is

V. M. López-Zambrano
Maestría en Agroindustria, Instituto de Posgrado, Universidad Técnica de Manabí, Portoviejo, Ecuador

A. A. Dueñas-Rivadeneira (✉)
Departamento de Procesos Agroindustriales, Facultad de Ciencias Zootécnicas, Universidad Técnica de Manabí, Chone, Manabí, Ecuador
e-mail: alex.duenas@utm.edu.ec

J. D. Zambrano-Veliz
Carrera de Industrias Agropecuarias, Facultad de Ciencias Zootécnicas, Universidad Técnica de Manabí, Chone, Ecuador

C. A. Cedeño-Palacios
Departamento de Procesos Químicos, Facultad de Ciencias Matemáticas, Físicas y Químicas, Universidad Técnica de Manabí, Portoviejo, Ecuador

M. H. Delgado-Demera
Departamento de Ciencias Veterinarias, Facultad de Ciencias Veterinarias, Universidad Técnica de Manabí, Portoviejo, Ecuador

N. R. Maddela, L. C. García (eds.), *Innovations in Biotechnology for a Sustainable Future*, https://doi.org/10.1007/978-3-030-80108-3_4

due to the persistence of pathogenic microorganisms to survive post-harvest, causing quality deterioration (Martínez, 2018).

Essential oils are hydrophobic liquids made up of aromatic compounds that are present in various parts of vegetables, a number of them are used as flavorings or aroma enhancers in cosmetics, perfumes, and soaps (Pandey et al., 2017). Some studies present results that indicate that essential oils have isolated components and antibacterial, antiviral, antifungal, and insecticidal activity (Manion & Widder, 2017).

The development of these postharvest preservation methods for fruits, vegetables, and processed foods arise from the need to extend the shelf life and improve storage conditions, and for this reason the objective is to analyze the usefulness of essential oils in post-harvest and conservation to have a safe food for human consumption.

2 Food Preservation

Foods are substances that nature provides, which have elemental components for the maintenance of vital functions, and therefore it must be ensured that they are kept in good condition and therefore natural and chemical preservatives are used.

The preservation of food is a topic of great interest to the food industry since its consumption must ensure safety; In its origins, man was a hunter and gatherer of fruits, his food should be consumed immediately to prevent it from decomposing, then preservation means were applied such as smoking, cooking by game and drying in the sun that allowed them to keep them free of pathogens for a longer time (Barreiro & Sandoval, 2006).

The growing demand by consumers and the great openness to new products generates the need for these to be safe before which chemical synthesis preservatives are used and can affect the safety of the product, before this the use of natural preservatives is an alternative that has led to exhaustive research on conservation techniques with a tendency to improve the quality of food without the presence of microorganisms, while maintaining its nutritional and organoleptic properties (Da Silva & Mendonça, 2012). In particular, the consumption of natural preservatives has increased, such as essential oils, since in addition to protecting the product from the presence of microorganisms, the nutrients remain stable so that the consumer can take full advantage of them.

3 Essential Oils as a Food Preservative

The interest of current consumers is that a healthy product is offered without chemical additives that last longer, which makes them look for alternatives such as essential oils which are a group of substances obtained from plant material such as stems, roots, leaves, flowers or seeds that contain aromatic compounds (Oliva et al.,

2010). Essential oils have a somewhat oily liquid structure with strong odors; some are pleasant and others contain sulfur components which make their aroma unpleasant (Pichersky et al., 2006). Essential oils have components such as sesquiterpenes, terpenes, and aliphatic components in their structure, which make them have great antibacterial and antifungal power, which protects the food from harmful agents (López et al., 2020).

Food safety represents an important problem, which directly and indirectly affects hundreds of people as a consequence of the increase in the incidence of foodborne diseases, which have caused the death of several people (OMS, 2007). Another problem is due to the great variety of microorganisms and the great capacity they have to reproduce and adapt in all environments, which cause deterioration and thus affect food, the preservation of processed or semi-processed products arises from the need to extend the shelf life and improve storage conditions, for this, alternative methods have been extended including non-thermal inactivation technology, high pressure, radiation, modified atmosphere, biopreservation, and plant-derived compounds (Witkowska et al., 2013).

4 Food Biotechnology

The development of food science and technology dates back a few decades and its influence on processing and conservation constitute an accelerating factor in food production, from this context biotechnology is defined as "the application of organisms, systems and processes biological to the production of goods and services for the benefit of man in order to increase productivity, performance, and solve problems of food supply to the general population" (Villanueva & Mejía, 2016; Maddela et al., 2021).

Biotechnology is a tool available to the researcher, the advances in this specialty of science and its applications have been useful for the food production sector to currently achieve technological development comparable to any other industrial branch, therefore food biotechnology is very important to the processes involved and their impact, which is related to food production and nutrition (Ortega et al., 2013).

For thousands of years, man applies genetics to improve raw materials and finished food products, therefore biotechnology is an association of scientific and technological knowledge that uses part of it to produce goods and services for humanity and the environment, with the creation of products that are made at the laboratory level using resources that are not traditional to produce superfoods (Vargas et al., 2018).

Food biotechnology stands out fundamentally in the areas of molecular biology of plants and industrial microbiology, being the contribution in animal biotechnology and in the production of natural food additives a new alternative (Ramón, 2014).

This biotechnological science offers new biological preservation systems through essential oils that have the ability to provide food preservation without using synthetic additives (Ramírez, 2020).

New foods are transformed thanks to biotechnology, which offers resources to create new quality nutrients for man with a positive image for health from cheap raw materials, such as plant material from which bioactive substances are extracted by various methods.

5 Essential Oil Extraction Methods

For the extraction of essential oils, different methods (conventional and emerging) are used to separate them from the plant material. Their composition may vary according to the extraction method used. Conventional extraction methods have been used throughout history, however it has been shown that they have higher energy consumption, which increases production costs. Emerging methods seek to increase industrial profitability, be sustainable, improve the characteristics of essential oils, and be respectful with the environment.

5.1 Conventional Methods

5.1.1 Steam Drag

It is the most common method to carry out the extraction, where the generally fresh or dry vegetable sample is cut into small pieces, is enclosed in an inert chamber and subjected to a stream of superheated water vapor; the essence thus entrained is later condensed, collected, and separated from the aqueous fraction. This process is widely used for fluid essences, implemented for perfumery. It is used at an industrial level due to its high performance, the purity of the oil obtained and it does not require sophisticated technology. During the development of this extraction, collateral processes such as polymerization and chemical changes of terpenes can be generated; as well as hydrolysis of esters and thermal destruction of some components (Díaz, 2018) (Fig. 1).

5.1.2 Cold Pressing Extraction

In this method, the essential oil, when extracted by cold pressing, maintains its natural characteristics, is of high quality, healthy, and has valuable nutritional properties for the body and therefore it is recommended to eat it raw (Garcés, 2018). This type of extraction is carried out exclusively by a mechanical process and without the addition of heat, it can also be washed, sedimented, filtered, and

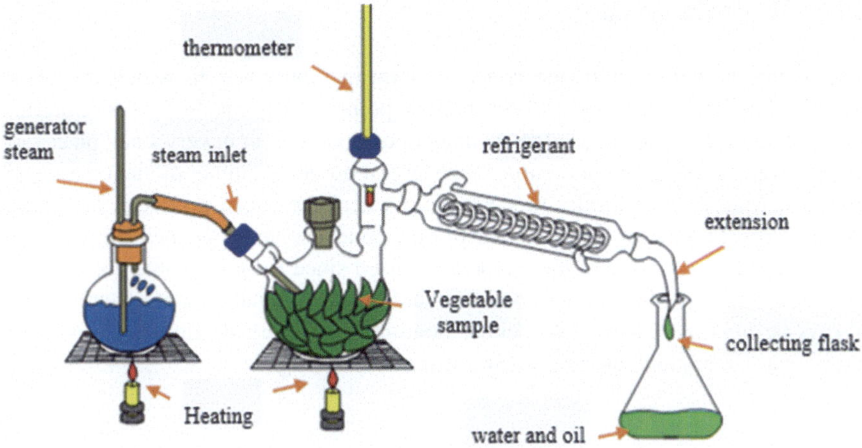

Fig. 1 Essential oil extraction process by the steam entrainment method

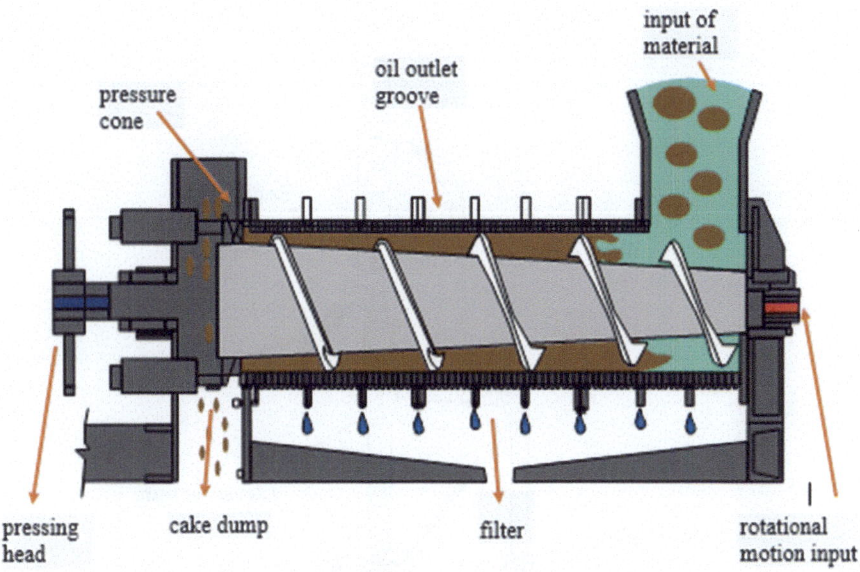

Fig. 2 Essential oil extraction method by cold pressing

centrifuged, but without altering the characteristics of the oil (Codex Alimentarius, 2015).

The process consists of applying an external force that causes changes at the microscopic (cellular) and macroscopic level, which compresses. This causes the cell walls to be destroyed, causing the oil to escape to the outside due to the pressure generated and the reduction of the physical space available (Martínez, 2015) (Fig. 2).

5.1.3 Hydrodistillation

The hydrodistillation technique consists of raising the water in which the plant material has been submerged to the boiling point, which can be leaves, flowers, roots, and stems which are used fresh, dehydrated, powder, or in very small pieces so that the water vapor exerts its action on the largest possible number of plant particles; this technique is very similar to steam drag where the steam produced drags the essential oils to another container where they condense and separate. This process has some drawbacks due to the temperature used since it causes some compounds present in plants to degrade and be lost, therefore the aromatic plant material must always be in contact with water, either floating or submerged, through in order to avoid overheating and charring it (Stashenko, 2009) (Fig. 3).

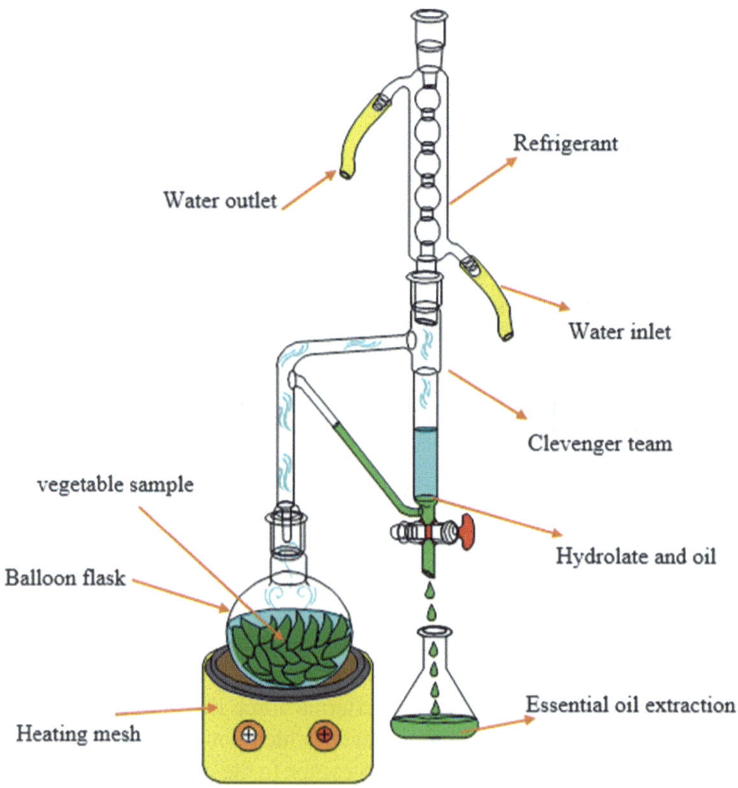

Fig. 3 Method of extraction of essential oil by hydrodistillation

5.1.4 Solvent Extraction Methods

Solvent extraction has been implemented for fragile or delicate flowers, which are not tolerant to the heat of the distillation process. Different substances have been used such as acetone, hexane, petroleum ether, methanol, or ethanol (Mora, 2014). In this process, the solvent is mixed with the raw material and then heated to extract the essential oil, later a filtration process is carried out. Later, the filtrate is concentrated by evaporation of the solvent; the concentrate is a resin or mixture of waxes, fragrances, and essential oil. The concentrate is mixed with pure alcohol to extract the essential oil and distilled at a low temperature. The alcohol absorbs the volatile compounds and when the alcohol evaporates, the essential oil is obtained. However, this method is time consuming and the commercial value of solvents is high, making essential oils more expensive than those obtained by other methods (Eslava & Fajardo, 2020) (Fig. 4).

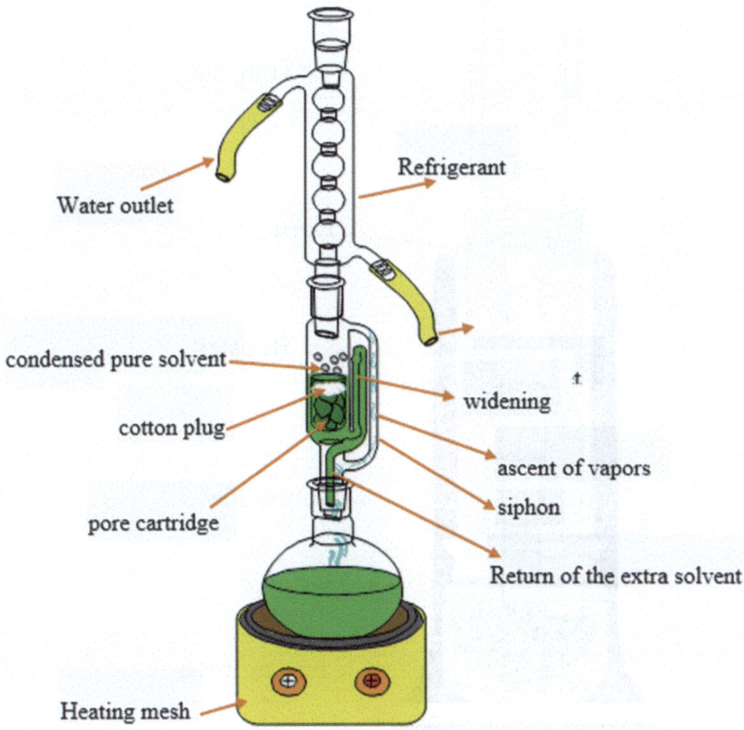

Fig. 4 Solvent extraction method of essential oil

5.2 *Emerging Methods*

5.2.1 Use of Ultrasound

The improvement in the efficiency of the extraction of organic compounds by ultrasound is attributed to the phenomenon of cavitation, produced in the solvent by the passage of an ultrasonic wave. Cavitation bubbles are produced and compressed during the application of ultrasound, this is applied as an extraction alternative or to assist in extraction processes of volatile plant components, including essential oils. The proportion in the composition of the extracts and their performance depends on the temperature at which the process is carried out and the solvent used, the application of this technique increases the efficiency of the extraction, decreases the time and reduces the risk of thermal degradation (López et al., 2009) (Fig. 5).

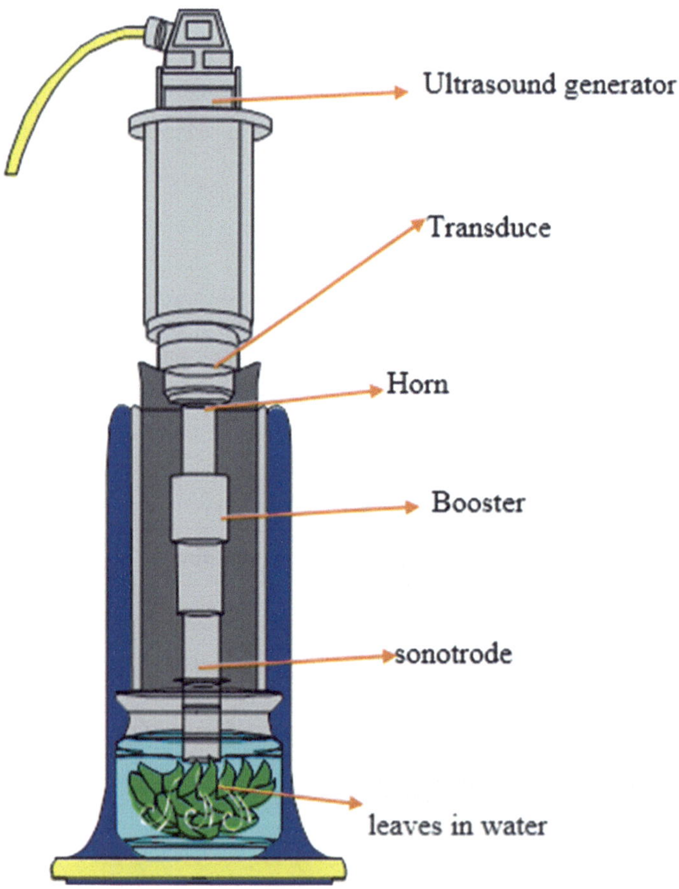

Fig. 5 Essential oil extraction method using ultrasound

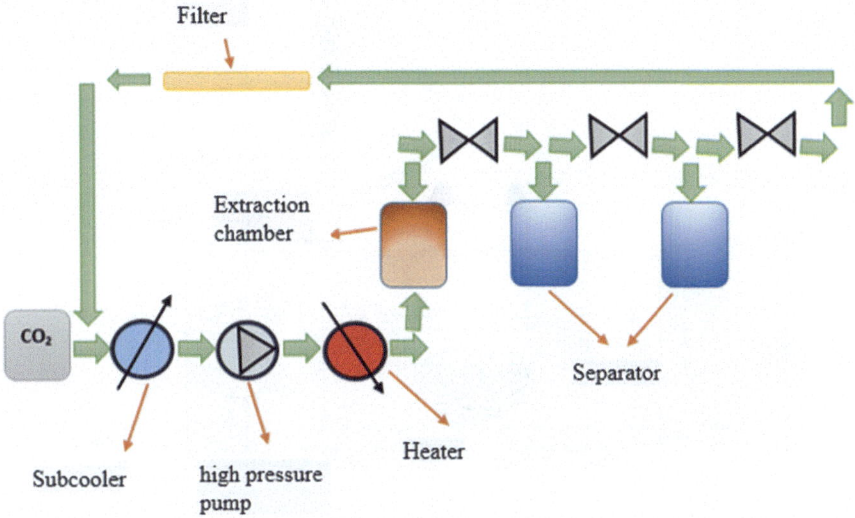

Fig. 6 Extraction method of essential oil with supercritical CO_2

5.2.2 Supercritical CO_2 Extraction

Supercritical fluids have been considered as an alternative means of essential oil extraction. Carbon dioxide is the most widely used supercritical fluid due to its critical conditions (critical temperature = 31.1 °C and critical pressure = 72.8 atm). At high pressure, CO_2 becomes liquid, so it can be used as a safe and inert medium to extract aromatic molecules from the plant to be studied. There are no solvent residues in the final product obtained because the liquid CO_2 simply reverts to gas and evaporates at atmospheric pressure and temperature conditions. Despite the high solubility of essential oil components in CO_2, the extraction speed is relatively slow (Fornari et al., 2012) (Fig. 6).

5.2.3 Microwave-Assisted Hydrodistillation Extraction

It is an alternative for the extraction of essential oil, an emerging method that can be adapted to a conventional microwave, making a hole in the upper part that connects a flat bottom flask which is adapted for extraction, condensers are attached with the microwave system. The performance depends on the temperature in the water circulation, in addition, the vegetal sample must be dry and in small pieces in contact with the water. Microwave extraction offers a benefit with a considerable reduction in time and energy consumption (Ventura, 2017) (Fig. 7).

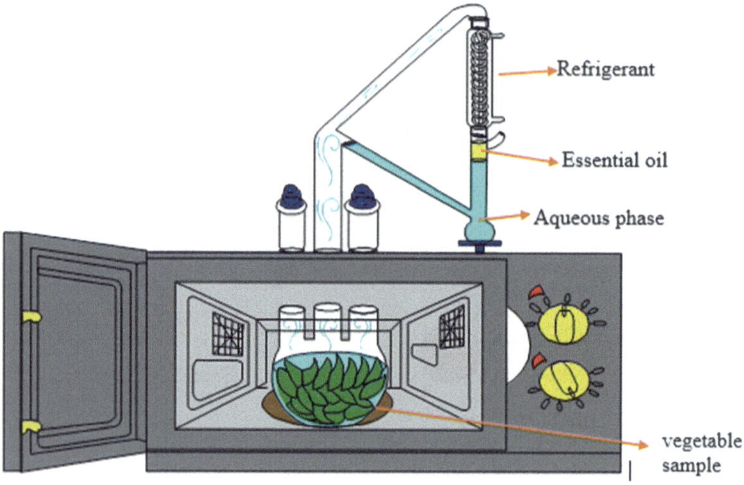

Fig. 7 Essential oil extraction method by microwave-assisted hydrodistillation

6 Essential Oils and Their Biotechnological Applications in Post-harvest

Food is frequently attacked by microbial agents, due to cross-contamination, storage, transportation, before or after its post-harvest, which raises the concern of giving more meaning to the harvest of the products since having a good product will have great results (FAO, 2015). The World Health Organization states that every one in three people a year worldwide becomes ill from contaminated food or water (Prakash, 2015), leading to 2.2 million deaths, especially in children, which generates great concern and therefore a natural alternative for conservation is sought, such as essential oils (Chen et al., 2016).

Post-harvest is a set of principles, standards, activities, and technical recommendations that are aimed at obtaining a safe product ensuring the protection of employees, consumers and the environment (Balaguera & Palacios, 2018). Safe products are those that are free from physical, chemical, and pathogenic microorganisms contamination, in this way the consumer will be able to access the product and enjoy the benefits of its nutritional properties (Bouzayen et al., 2010) (Table 1).

Postharvest losses of fruits and vegetables caused by microorganisms are in the order of 5–25% in developed countries and 40% in developing countries, these being the most common cause of postharvest diseases in fruits and vegetables, damaging their product and thus contaminating the health of those who consume it as a result of a microorganism that alters the flora intestinal of people (Gatto et al., 2011).

It is estimated that postharvest losses of fruit and vegetable products that are produced in the world exceed 20%, this is due to the presence of microbiological and also physiological aspects, these damages occur as a consequence of technological factors such as inadequate collection process, inappropriate packaging and

Table 1 Application of essential oil in post-harvest

Essential oil	Fruit or vegetable	Microorganisms	Applications	Author
Thymus vulgaris, Origanum vulgare, *Mentha piperita, Melaleuca alternifolia*	Apple	*Penicillium* spp.	Fumigation	Silva et al. (2015)
Oregano (Lippia graveolens *Kunth*) and thyme (*Thymus vulgaris* L.)	Avocado	Combat the antioxidant system	Covering	Rivera (2019)
Thyme (*Thymus vulgaris* L.)	Tomato	*Pectobacterium carotovorum*	Nano coatings	Correa et al. (2019)
Thyme	Citrics	Fungicide	Food coating	Ortiz (2020)
Cinnamon	Uvilla (*Physalis peruviana*)	Mushroom *Botritys* sp.	Immersion	González et al. (2020)
Cinnamon and sodium alginate	Blueberry	*Staphylococcus aureus and Escherichia coli* O157:H7	Nanoemulsion	Segura (2019)
Melaleuca alternifolia	Peach	*Monilinia fructicola and Rhizopus* spp.	Fumigation	Mitidieri (2020)
Cymbopogon nardus L.	Potato (Solanum tuberosum L.)	*Rhizoctonia solani*	Laboratory	Vaillant et al. (2009)
Orange (Citrus sinensis L.)	*Carica papaya* L.	*Colletotrichum gloeosporioides, Penicillium indicum,* Fusarium solani, *Rhizopus stolonifer and Aspergillus flavus*	Covering	Guédez et al. (2014)
Chitosan 1% + bee wax 0.5% and chitosan + beeswax + essential oils of cinnamon 0.025% or + lemon	Fig	*Aspergillus* and *Penicillium*	Covering	Baldoni et al. (2016)
Muña (*Minthostachys spicata*)	Tree tomato (*Cyphomandra betacea*)	Molds and yeasts	Rosio	Merma (2014)
Cloves, oregano, ginger and rosemary	Mangos (Mangifera indica L.)	*Colletotrichum gloeosporioides*	Biofilm	Rico (2013)
Oregano (*Lippia berlandieri Schauer*)	Tomato (Lycopersicon esculentum *Mill.*)	*Fusarium oxysporum*	Covering	Cueto (2010)
Ocotea quixos and Piper aduncum	Agricultural crops		Fumigation	

(continued)

Table 1 (continued)

Essential oil	Fruit or vegetable	Microorganisms	Applications	Author
		Aspergillus oryzae, Cladosporium cladosporioides, Fusarium solani, Rhyzopus stolonifer, Moniliophthora roreri and *Phytophthora* sp.		Scalvenzi et al. (2016)
Base of aloe vera (*aloe vera*) and essential oil of cinnamon (*C*innamomum verum)	Tomato (*Lycopersicum esculentum Mill*)	*Rhizoctonia solani*	Covering	Molocho and Orbegoso (2017)
Syzygium aromaticum	Fruits and vegetables	*Salmonella typhi, Salmonella paratyphi A* and Bacillus cereus	Laboratory	Gamboa and Vásquez (2015)
Citrus (orange, lemon and grapefruit)	Strawberry	Antifungal effect	Biofilm	Alarcón et al. (2015)
Turmeric (Curcuma longa L.)	Squash	*AAAPenicillium* sp. and *Cladosporium* sp.	Covering	Mahecha and Andrade (2017)
Oregano	Mango (*Mangifera indica* L.)	*Salmonella* sp., *Staphylococcus aureus, Escherichia coli* and Saccharomyces cerevisia*e*	Covering	Pardo (2017)
Cinnamon and nutmeg	Blackberry from castile (*Rubus glaucus Benth*)	*Rhizopus* and *Colletotrichum*	Covering	Herrera (2011)
Tangerine	Citrics	Mushrooms *Penicillium digitatum* and P. italicum	Fumigation	Velásquez et al. (2015)
Oregano	Papaya '*Maradol*'	*Salmonella* spp.	Covering	Pontigo et al. (2015)
Cinnamomum zeylanicum, Syzygium aromaticum	Papaya (*Carica papaya*).	*Fusarium* sp.	Laboratory	Necha and Barrera (2008)
	Lettuce (*Lactuca sativa*)		Laboratory	

Lemon (*Citrus latifolia*) and orange (*Citrus sinensis* L.)		White mold *Sclerotinia sclerotiorum*		Delgado (2019)
Oregano (*Origanum vulgare*)	Papaya *Maradol*	Hongo *Fusarium* spp.	Covering	Castillo et al. (2018)
Cinnamon, Orange and lemon and chitosan based coatings	Strawberries, cucumbers, jujube, bell pepper and mango	*Nauphoeta cinerea, Penicillium citrinum, oomycete Phytophthora drechsleri*	Covering	Anaya et al. (2020)
Thyme, cinnamon, or cloves	Papaya	*Rhizopus stolonifer, Colletotrichum gloeosporioides,* Alternaria alternata, *Fusarium oxysporum* and *Penicillium digitatum*	Biofilm	Hernández et al. (2018)
Oregano	Cucumber (*Cucumis sativus* L.)	*Salmonella typhimurium* (inoculated) and *Escherichia coli* (inoculated)	Covering	Pacheco (2015)
Cinnamon	Sweet pepper	*E. coli, salmonella and Staphylococcus*	Micro encapsulation	López et al. (2018)
Cinnamon	Pear	*Molds and yeasts*	Covering	García et al. (2017)

insufficient routes for transportation, among others, which translates into a short storage period (Almeida et al., 2012).

Farmers fight every day with these inconveniences that cause them great economic losses, so what is desired is to avoid or minimize the adverse effects of the aforementioned factors and jointly prolong the postharvest life of horticultural products for which different technologies have been implemented. These include storage at low temperatures, application of gamma and ultraviolet radiation, biological control, conservation by controlled atmosphere, the use of plastic packaging, the use of films and the application of edible coatings, among others (Núñez et al., 2012).

Essential oils have recently become useful in post-harvest as bactericides and fungicides, which is gaining great importance, demonstrating that it has antifungal properties against microorganisms that affect the growth and yield of fruits, fungi affect their composition causing irreversible damage and a different flavor is its maturity, molds affect crops repeatedly causing the death of fruit trees (Ochoa et al., 2019).

7 Physicochemical Parameters of Essential Oils

The referential physicochemical parameters that are carried out on essential oils are refractive index, solubility, and density, these analyzes are used to determine their quality and may require other more complex analyzes.

7.1 Solubility

This analysis is carried out to verify the affinity that essential oils have with a solvent to be soluble, this analysis is expressed in percentage of solute or in units such as moles per liter or gram per liter. It is vitally important to note that not all substances dissolve in the same solvents since the higher the concentration of the solvent, the faster the solubility will be; the factors that affect this analysis are temperatures and pressure (Araujo et al., 2020).

7.2 Density

Density analysis is an internal property of essential oil; it does not depend on the amount of the substance to be analyzed, but on the temperature. The density of a substance is defined as the quotient of its mass for each unit of volume. Therefore, if we know the mass and volume of a substance (liquid, solid, and gaseous) (Argote et al., 2017).

7.3 *Refractive Index*

The index of refraction is called the relationship between the speed of light in a vacuum with respect to another medium or material, this analysis is symbolized by the letter "*n*". This analysis is carried out to know how much the speed of light is reduced when passing through it in the essential oil analyzed as the material is denser, the value of the refractive index increases (Castro et al., 2019) (Table 2).

Essential oils are extracted from various parts of the vegetable, obtaining positive results from their physical-chemical properties; the density is between 0.8 and 0.9 although there are flower oils that have a density lower than 0.5 and oils from citrus peels that have a higher density between 0.9 and 1; the refurbishment index is between 1.4 and 1.5. The solubility of essential oils is determined by ethanol, petroleum ether, hexane, methanol, chloroform. There are oils that are soluble while there are others that do not present solubility.

8 Essential Oils as Secondary Metabolites

When plants are in the soil, they are attacked by various microorganisms that cause them damage, which have developed various defense strategies against conditions of biotic and abiotic beings. Plants synthesize enzymes that degrade the cell wall of microorganisms or that have the ability to inactivate toxins of microbial origin to defend themselves from the damage they cause (Toro et al., 2017).

The composition and structure of the walls of the plants form rigid and thick walls less digestible for insects; this response in turn is united with the components of the predator's development such as spines, spikes, trichomes, and hairs. Glandular that are on the outside of the plant covering the trunk and protecting it from damage to growth. Likewise and as part of chemical protection, plants use secondary metabolites with antimicrobial activity, against herbivores or with antimicrobial activity (Croteau et al., 2000).

Secondary metabolites are low molecular weight compounds that not only have great ecological importance because they participate in the adaptation processes of plants to a suitable environment but also help as a component of food; they also attract many pollinating insects and seed dispersers; these components are activated against attack by microorganisms (viruses, bacteria, and fungi), attack by the consumption of herbivorous animals (arthropods and vertebrates), and it also helps to compete for nutrients in the soil (Robles et al., 2016).

In plants, we find primary and secondary metabolites; primary metabolites are those chemical components that plants have to survive in growth, development, and survival while secondary metabolites are compounds derived from the primary metabolism of plants, these compounds play an ecological role important as it serves as a defense (Pérez & Jiménez, 2011).

The vast majority of plant species present secondary metabolites such as terpenes, phenolic compounds, glycosides, and alkaloids, of which about 800 polyphenols,

Table 2 Physical-chemical analysis of different essential oils

Root essential oils	Physical parameters			
	Density	Refractive index	Solubility	Authors
Salvia trifilis Epling	0.86 g/mL	1.4750	Soluble in 96% ethanol and is volatile	Díaz (2018)
Zingiber officinale	0.88 g/mL	1.52	Soluble in 96% ethanol	Cuellar et al. (2017)
Cúrcuma longa L.	0.8983 g/m^3	1.4995	Soluble in 96% alcohol, *N*-hexane and ethyl ether	Aguirre and Gutierrez (2016)
Renealmia thyrsoidea	0.873 mg/mL	–	Does not present solubility	Rivera et al. (2017)
Essential oil from the leaf				
Schinus molle L.	0.872 ± 0.931	1.476 ± 1.484	Soluble in 95% alcohol	Plaza and Ricalde (2015)
Lantana cámara	0.845 ± 0.001	1.423 ± 0.029	–	Valdéz et al. (2018)
Ocimum basilicum L.	0.8744 kg/m^3	1.4156	Soluble in ethanol at 60%	Araujo (2018)
Eucalyptus globulus Labill	0.901 g/mL	1.4751	Soluble in ethanol at 70%	Alarcón et al. (2019)
Oreganum vulgare	0.890	1.4375	–	Silupu et al. (2019)
Essential oil from flowers				
Senecio nutans Sch.	0.8616 g/mL	1.4432	Soluble 96% ethanol and methanol	Flores and Mily (2018)
Tagetes minuta	0.8373	1.4960	Soluble in hexane, chloroform, ethers and 96% alcohol	Mendoza and Espinosa (2019)
Matricaria chamomilla L.	0.508	1.365	70% ethanol solubility	AlarcÃn et al. (2017)
Trapaeolum majus L.	0.946	–	Solubility in 70% ethanol and alcohols	Juscamaita et al. (2017)
Clinopodium weberbaueri	0.886 g/mL	1.537	Solubility in ethanol at 80% and 90%; ethyl alcohol 70%	Oscco (2018)
Essential oil from the rind of the fruit				
Citrus sinensis	0.8423 g/mL	1.4710	Soluble in ethanol (70% and 90%)	Leónet al. (2015)
Citrus paradisi	1.030 g/mL	1.425	Miscible in aqueous medium	Cabrera (2019)
Annona squamosa L.	1.015	1.337	–	Cala et al. (2018)
Citrus reticulata L.	0.88086	1.4700	Solubility in ethanol at 96 g/L	Chávez and Gomez (2016)

(continued)

Table 2 (continued)

Root essential oils	Physical parameters Density	Refractive index	Solubility	Authors
			Petroleum ether and methanol	
Citrus aurantium Engl	0.89	1.478	Does not present solubility	Murillo et al. (2018)
Essential oil from seeds				
Azadirachta indica	0.8 g/mL	1.463	–	Martínez et al. (2016)
Moringa oleífera	–	1.4678	–	Fernández et al. (2018)
Pimpinella anisum L.	0.98 g/mL	1.5534	Solubility in hexane, dichloromethane, ethanol	Espinoza (2018)
Coriandrum sativum	0.9400 g/mL	1.513	–	Condori (2019)
Curcuvita	0.96 g/mL	1.46	–	Valenzuela et al. (2018)

270 non-protein amino acids, 32 cyanogens, 10,000 alkaloids, various saponins and steroids have been reported (Ávalos & Pérez, 2009) These metabolites are an active source for the elaboration of medicines and valuable chemical products whose function is as analgesic, antibacterial, antioxidant, antiviral, antitumor, fungicide, among others (Agustin et al., 2011).

9 Antimicrobial Activity of Essential Oils

Plants in their structure have a phyllosphere area and especially in the leaves where a large number of cells and microorganisms are found in them, a large part of the microorganisms are part of the plant ecology. Consequently, naturally occurring plants produce more than 100,000 low molecular weight natural products, also known as secondary metabolites (Calcina, 2020) that unlike the primary ones that do not protect the plants; this diversity is very rich in the evolutionary process in order to acquire a defense against the attack of microorganisms, insects, and other animals.

These substances can be divided into two large groups: phytoanticipins that are constitutively present in plants and phytoalexins whose presence increases the form considerably in response to microbial invasion, the differences between them are that phytoanticipins are present in the plant while that phytoalexins are produced as a response to microbial invasion. However, this definition is not entirely true, since some compounds may be phytoanticipins in ones and phytoalexins (Perveen et al., 2012).

Table 3 Most important chemical group obtained from plants as conservation

Chemical group	Compounds	Plants	Activity
Phenols	Anthocyanins. Thymol, Carvacol, Terpenoids, Flacones, flavonones, flavanols, and flavanonols. Flavanols, condensed tannins, and lignans	Cattail, ginseng, ginkgo, eleutherococcus, anamu, grape, eucalyptus, periwinkle, devil's claw, tangerine, grapefruit, lemon, orange, rosemary, agrimony, calendula, and oats	*Staphylococcus aureus*, *Escherichia coli*, *Salmonella enterica Choleraesuis*, and *Listeria monocytogenes*
Quinones	Hypericin, *o*-benzoquinone, *p*-benzoquinone, naphthoquinone, naphthoquinones, anthraquinones	*Hypericum perforatum*	HIV *Escherichia coli*, *Pseudomonas aeruginosa*, *Staphylococcus aureus*, Klebsiella
Tannins	Phenolic acid esters, gallotannins, ellagitannins, complex tannins and condensed tannins	These are polyphenolic compounds found in many dicotyledonous plants, especially forage legumes of temperate and tropical regions	Bacteria, Fungi, viruses
Flavonoids	Catechin, Isoflavone, Keratin	It is found in almost all vegetables and flowers with which essential oils are extracted	It is used in plants to reduce the bacterial and nicotic load that affects its post-harvest, it is used for allergies, it inhibits bacteria *Staphylococcus aureus*, *Escherichia coli* y *Bacillus typhosus*
Alkaloids	Coca, Piperine, Mesealine	Found in medicinal plants	Big positive coconuts, Fungi, lactobacilli
Coumarins	Umbeliferona, Esculetina, Herniarin, Psoralenos, Imperatorina	Chamomile Turmeric Ginseng	Virus

The plant kingdom offers great possibilities to discover new compounds that could serve as a good candidate for the development of standardized antimicrobial agents and/or phytomedicines that will be very useful in food, medicines, crops and more in order to provide protection. Plant extracts are able to circumvent resistance mechanisms due to their limited use compared to traditional antibacterials (López & Domingo, 2003).

An important point to note is that a large proportion of the products synthesized with antimicrobial character which show in vitro susceptibility tests a minimum inhibitory concentration high (100–1000 mg/L) compared to those obtained by conventional antimicrobials. These are related to the existence of antipathic compounds that act as a true expulsion pump for a wide spectrum of substances, including those antimicrobial actions. The use of chemical inhibitors causes them to cause irreversible damage to the health of consumers and products from their postharvest period, while natural inhibitors cause great advantages.

Around 12,000 compounds have been isolated from plant organisms and it is estimated that they constitute only 10% of secondary metabolites, a significant percentage having some activity against microorganisms. Plants have an unlimited capacity to synthesize compounds, most of them related to phenol and its derivatives (Table 3).

The production of metabolites by plants is considered an adaptive capacity to cope with stressful limitations during a challenging and changing growing environment that may involve the production of complex chemical types and interactions in structural and functional stabilization through of processes (Yang et al., 2018) that ensures proper growth and total protection of plants.

References

Aguirre, Y., & Gutierrez, C. (2016). *Características fisicoquímicas y determinación del porcentaje relativo de sus componentes hidrocarbonados y oxigenados de la raíz de cúrcuma longa l. (Cúrcuma) procedente de la selva peruana–Madre de Dios*. Universidad Nacional de Trujillo.

AlarcÃn, T., Conde, G., Pajaro, P., & Tovar, T. (2017). EvaluaciÃn de la actividad antioxidante del aceite esencial de *Matricaria chamomilla L. Revista Cubana de Farmacia, 51*(1), 12–21.

Alarcón, Y., Trejo, A., Corrales, A., & Bustamante, S. P. (2015). Desarrollo de un envase activo adicionado con aceites esenciales de cítricos para el control de la podredumbre en fresa. *Revista Iberoamericana de Tecnología Postcosecha, 16*(2), 222–227.

Alarcón, T., Conde, G., & Méndez, L. (2019). Extracción, caracterización y actividad antioxidante del aceite esencial de *Eucalyptus globulus Labill. Revista Cubana de Farmacia, 52*(1), 10–25.

Almeida, A., Reis, D., Santos, D., Vieira, T., & da Mariana, C. (2012). Estudio de la conservación de la papaya (*Carica papaya L.*) asociado a la aplicación de películas comestibles. *Venezolana de Ciencia y Tecnología de Alimentos, 2*(1), 12.

Anaya, L., Pérez, A., Ruvalcaba, M., Sánchez, A., Romero, R., & Montalvo, E. (2020). Funcionalización de los recubrimientos a base de quitosano para la conservación postcosecha de frutas y hortalizas. *TIP. Revista especializada en ciencias químico-biológicas, 23*, e20200241.

Araujo, C. (2018). Parámetros de extracción de aceite esencial de albahaca (*Ocimum basilicum L.*) por arrastre de vapor. Universidad Nacional del Altiplano. http://repositorioslatinoamericanos.uchile.cl/handle/2250/3277328

Araujo, C., Altamirano, C., Barreto, V., & Sarapura, J. (2020). Evaluación del rendimiento y características físico-químicas del aceite esencial de *Satureja serícea. Revista Ciencia Nor@ndina, 3*(2), 117–122.

Argote, E., Montenegro, S., Delgado, T., Alvarez, P., Hurtado, A., & Ospina, J. (2017). Evaluación de la capacidad inhibitoria de aceites esenciales en *Staphylococcus aureus* y *Escherichia coli. Biotecnología en el Sector Agropecuario y Agroindustrial: BSAA, 15*(2), 52–60.

Augustin, J., Kuzina, V., Andersen, B., & Bak, S. (2011). Molecular activities, biosynthesis and evolution of triterpenoid saponins. *Phytochemistry, 72*(6), 435–457.

Ávalos, G., & Pérez, C. (2009). Metabolismo secundario de plantas. *Reduca (Biología), 2*(3), 119–145.

Balaguera, E., & Palacios, A. (2018). Comportamiento poscosecha de frutos de mandarina (Citrus reticulata Blanco) var. Arrayana: efecto de diferentes tratamientos térmicos. *Revista Colombiana de Ciencias Hortícolas, 12*(2), 369–378.

Baldoni, D., Ventura, I., Hernández, M., Corona, L., Barrera, L., Correa, Z., & Bautista, S. (2016). Calidad postcosecha de higos 'black mission' tratados con cubiertas naturales. *Revista Iberoamericana de Tecnología Postcosecha, 17*(2), 267–275.

Barreiro, A., & Sandoval, A. (2006). *Operaciones de conservación de alimentos por bajas temperaturas*. Equinoccio.

Bouzayen, M., Latché, A., Nath, P., & Pech, J. (2010). Mechanism of fruit ripening. In E. C. Pua & M. R. Davey (Eds.), *Plant developmental biology—Biotechnological perspectives* (Vol. 1, pp. 319–339). Springer-Verlag.

Cabrera, E. (2019). Actividad antimicrobiana de un sistema a base de un extracto vegetal y tres aceites esenciales. *Ciencia e Investigación, 22*(1), 21–25.

Cala, L., Fernández, G., Sánchez, E., & Vadell, C. (2018). Estudio farmacognóstico preliminar de la especie *Annona squamosa L. Revista Cubana de Plantas Medicinales, 23*(2). http://www.revplantasmedicinales.sld.cu/index.php/pla/article/view/637

Calcina, R. R. (2020). Efecto antimicótico in vitro de decocciones e infusiones de Minthostachys setosa y Xanthium catharticum en diferentes concentraciones sobre Candida albicans 2017. http://repositorio.unap.edu.pe/handle/UNAP/13792

Castillo, A., Salazar, K., Mosquera, A., & Rengifo, E. (2018). Efecto de recubrimientos de almidón modificado de yuca, proteína aislada de soya y aceite esencial de orégano aplicados a la papaya. *Revista U.D.C.A Actualidad & Divulgación Científica, 21*(1), 71–80.

Castro, M., Chávez, G., Auquiñivín, A., Fernández, B., Cruz, L., Rodríguez, N., & Sepúlveda, D. (2019). Aceites esenciales de plantas nativas del Perú: Efecto del lugar de cultivo en las características fisicoquímicas y actividad antioxidante. *Scientia Agropecuaria, 10*(4), 479–487.

Chávez, N., & Gomez, M. (2016). *Características fisicoquímicas del aceite esencial del pericarpio de citrus reticulata l. y determinación del porcentaje relativo de sus componentes hidrocarbonados y oxigenados*. Universidad Nacional de Trujillo.

Chen, P., Lee, J., & Chang, I. (2016). Essential oils from Taiwan: Chemical composition and antibacterial activity against Escherichia coli. *Journal of Food and Drug Analysis, 24*(3), 464–470.

Codex Alimentarius. (2015). Norma para aceites vegetales especificados Codex Stan 210–1999.

Condori, D. (2019). Rendimiento y caracterización fisicoquímica del aceite esencial de (*Coriandrum sativum*) cilantro extraído por arrastre de vapor en un equipo modular.

Correa, N., García, D., Bautista, S., & Corona, L. (2019). Efecto de nanorecubrimientos de quitosano-aceite esencial de tomillo sobre la calidad postcosecha en frutos de jitomate. *Revista Mexicana de Fitopatología, Mexican Journal of Phytopathology, 37*(1), 29–36.

Croteau, R., Kutchan, M., & Lewis, G. (2000). Natural products (secondary metabolites). In B. Buchanan, W. Gruissem, & R. L. Jones (Eds.), *Biochemistry and molecular biology of plants* (Vol. 24, pp. 1250–1318). American Society of Plant Physiologists. 1367 p.

Cuellar, T., Wilches, S., Rivera, R., Rivera, R., Perdomo, L., Romero, B., & Murillo, D. (2017). Evaluación de la vida útil de quesos semimaduros con recubrimientos comestibles utilizando aceite esencial de jengibre (*Zingiber officinale*) como agente antimicrobiano. *Revista Colombiana de Investigaciones Agroindustriales, 4*(1), 78–87.

Cueto, C. (2010). *Determinación del efecto inhibitorio del aceite esencial y diferentes extractos de orégano (Lippia berlandieri Schauer) sobre el crecimiento de Fusarium oxysporum tanto in vitro como en plántula de tomate* (Doctoral dissertation, Universidad Autónoma de Nuevo León).

Da Silva, J., & Mendonça, N. (2012). Association between antimicrobial resistance and virulence in Escherichia coli. *Virulence, 3*(1), 18–28.

Delgado, L. (2019). *Evaluación in vitro de los aceites esenciales de naranja (citrus sinensis l.) y limón (citrus latifolia), frente a sclerotinia sp., agente causal del moho blanco en lechuga* (Doctoral dissertation).

Díaz, M. (2018). Determinación del rendimiento a diferentes tiempos de extracción de aceite esencial de la raíz *Salvia trifilis Epling* (mejorana) por el método de arrastre de vapor. *Agroindustrial Science, 7*(2), 73–77.

Eslava, A., & Fajardo, D. (2020). *Obtención de un aceite esencial a partir de la semilla del mango utilizando el método de extracción con solventes* (Bachelor's thesis, Fundación Universidad de América).

Espinoza, J. (2018). *Extracción y caracterización del aceite esencial de anis (pimpinella anisum l.) considerando distintas procedencias y diferentes tiempos de almacenamiento* (Doctoral dissertation).

Fernández, J., Pascual, G., Silva, I., Salvá, B., Guevara, A., & Encina, C. (2018). Efecto del tratamiento enzimático de la semilla de moringa (Moringa oleifera) sobre las características físico-químicas del aceite obtenido por extracción con prensa expeller. *Scientia Agropecuaria, 9* (3), 371–380.

Flores, B., & Mily, N. (2018). *Evaluación de la capacidad antimicrobiana, antioxidante y propiedades físicas, del aceite esencial de chachacoma (Senecio nutans Sch) en queso fresco tipo paria*. Universidad Nacional del Altiplano.

Fornari, T., Ruiz, A., Vicente, G., Vázquez, E., Rodríguez, M., & Reglero, G. (2012). Kinetic study of the supercritical CO2 extraction of different plants from Lamiaceae family. *Journal of Supercritical Fluids, 64*, 1–8.

Gamboa, J., & Vásquez, M. (2015). Efecto del aceite esencial de *Syzygium aromaticum* sobre la supervivencia de Salmonella typhi, Salmonella paratyphi A y Bacillus cereus. *Revista Rebiolest, 3*(1), 42–51.

Garcés, L. (2018). Aceites de Primera Presión en Frío: Ventajas y Obtención. https://www.biomanantial.com/aceites-de-primera-presion-en-frio-ventajas-y-obtencion/

García, C., Posligua, E., Mantuano, L., Basurto, M., Montes, G., & Delgado, L. (2017). Recubrimiento comestible de quitosano, almidón de yuca y aceite esencial de canela para conservar pera (Pyrus communis *L.* cv. "Bosc"). *La Técnica: Revista de las Agrociencias*, Edición Especial, 42–53. https://revistas.utm.edu.ec/index.php/latecnica/article/view/970/910

Gatto, A., Ippolito, A., Linsalata, V., Cascarano, A., Nigro, F., Vanadia, S., & Di Venere, D. (2011). Activity of extracts from wild edible herbs against postharvest fungal diseases of fruit and vegetables. *Postharvest Biology and Technology, 61*, 72–82.

González, V., Cabrera, M., Herrera, T., & Armando, P. (2020). Determinación de la capacidad conservante del aceite esencial de canela sobre uvilla (*Physalis peruviana*) como tratamiento postcosecha. *ConcienciaDigital, 3.2*(1), 210–230.

Guédez, C., Cañizalez, L., Avendaño, L., Scorza, J., Castillo, C., Olivar, R., & Sánchez, L. (2014). Actividad antifúngica del aceite esencial de naranja (*Citrus sinensis L.*) sobre hongos postcosecha en frutos de lechosa (*Carica papaya L.*). *Revista de la Sociedad Venezolana de Microbiología, 34*(2), 81–87.

Hernández, M., Guillén, J., Bautista, S., & Guillén, D. (2018). Evaluación de películas biodegradables en el control de hongos postcosecha de la papaya. *Cultivos Tropicales, 39*(1), 52–60.

Herrera, N. M. (2011). Evaluación de aceites esenciales de canela y de nuez moscada en un recubrimiento comestible para la conservación de frutos de mora de castilla (Rubus glaucus Benth). https://ciencia.lasalle.edu.co/ing_alimentos/86

Juscamaita, L., Pérez, T., Espinoza, C., Quispe, M., Hinostroza, G., Flores, O., & Manyari, G. (2017). Evaluación de la estabilidad de carotenoides y actividad antioxidante de la flor de mastuerzo (Tropaeolum majus L.) en la microencapsulación por Spray-Drying. *Revista de la Sociedad Química del Perú, 83*(3), 282–293.

León, G., Osorio, & Martínez, S. (2015). Comparación de dos métodos de extracción del aceite esencial de *Citrus sinensis L. Revista Cubana de Farmacia, 49*(4), 742–750.

López, M., & Domingo, D. (2003). Plantas con acción antimicrobiana. *Revista Española de Quimioterapia, 16*(4), 385–393.

López, A., Peredo, H., & Paloy, E. (2009). *Aceites esenciales: métodos de extracción* (p. 27). Universidad de las Américas Puebla.

López, C., López, L., & González, A. (2018). *Elaboración y caracterización de microcápsulas con aceite esencial de canela para su aplicación en la fabricación de películas de almidón* (Doctoral dissertation).
López, V., Dueñas, A., Cuenca, J., & Rodríguez, J. (2020). Caracterización fitoquímica, actividad antioxidante y antibacteriana del aceite esencial y extractos de *Tagetes patula* sobre *Staphylococcus aureus*. *Revista de la Facultad de Agronomía, 37*, 347–367.
Maddela, N. R., Garcia, L. C., & Chakraborty, S. (Eds.). (2021). *Advances in the domain of environmental biotechnology*. Springer Nature Singapore Pte Ltd. https://doi.org/10.1007/978-981-15-8999-7; ISBN: 978-981-15-8999-7, (pp XVIII, 717).
Mahecha, V., & Andrade, M. (2017). Aceite esencial de cúrcuma (Curcuma longa L.) como agente antifúngico en recubrimientos comestibles aplicados a zapallo (Cucurbita maxima) mínimamente procesado. *Revista de Ciências Agrárias, 40*(3), 641–654.
Manion, C., & Widder, R. (2017). Essential of essential oils. *American Journal of Health-System Pharmacy, 74*(9), e153–e162.
Martínez, M. (2015). *Aceites vegetales no tradicionales: Guía para la producción y evaluación de calidad*. Encuentro Grupo Editor.
Martínez, E. (2018). Evaluación antimicrobiana de aceites esenciales de plantas contra microorganismos patógenos: Estudio in vitro del aceite de orégano combinado con conservadores alimenticios convencionales. *Acta Universitaria, 28*(4), 10–18.
Martínez, M., Parra, J., Vera, A., & Vera, A. (2016). Parámetros de calidad del Aceite de las Semillas de Azadirachta indica (neem). *Revista CENIC Ciencias Químicas, 47*, 70–74.
Mendoza, N., & Espinosa, R. (2019). *Características fisicoquímicas del aceite esencial y determinación del porcentaje relativo de sus componentes hidrocarbonados y oxigenados de la hoja de Tagetes minuta L. (HUACATAY)*. Universidad Nacional de Trujillo.
Merma, B. (2014). *Conservación del kétchup de tomate de árbol (cyphomandra betacea) mediante la utilización del aceite esencial de muña (minthostachys spicata)*. Universidad Nacional Del Altiplano.
Mitidieri, M. (2020). *Evaluación de alternativas a los fungicidas de síntesis químicas para el control de enfermedades de poscosecha en durazno*. INTA Ediciones.
Molocho, V., & Orbegoso, L. (2017). *Evaluación del efecto de un recubrimiento a base de sábila (Aloe vera) y aceite esencial de canela (cinnamomum verum) en el tiempo de vida útil del tomate (lycopersicum esculentum mill). Lambayeque–2016*. Universidad Señor de Sipán.
Mora, A. (2014). *Diseño de una planta para la extracción del aceite esencial del palo santo mediante destilizacion por arrastre de vapor*. Escuela Politécnica Nacional.
Murillo, E., Correa, J., Cerquera, C., & Méndez, J. (2018). Potencial antimicrobiano y citotóxico del aceite esencial de Citrus aurantium Engl (naranja agria) y Swinglea glutinosa Merr (limón de cerco). *Revista Cubana de Plantas Medicinales, 23*(3). http://www.revplantasmedicinales.sld.cu/index.php/pla/article/view/795
Necha, B., & Barrera, G. (2008). Actividad antifúngica de aceites esenciales y sus compuestos sobre el crecimiento de Fusarium sp. aislado de papaya (Carica papaya). *Revista científica UDO agrícola, 8*(1), 33–41.
Núñez, K., Castellano, G., Ramírez, R., Sindoni, M., & Marin, C. (2012). Efecto del cloruro de calcio y una cubierta plástica sobre la conservación de las propiedades organolépticas de la fresa (Fragaria x ananassa Duch). *Revista Iberoamericana de Tecnología Postcosecha, 13*(1), 21–30.
Ochoa, M., Hernández, A., Delgado, C., Hernández, O., Cerna, E., Aguirre, L., & Tapia, M. (2019). Control orgánico in vitro de Phytophthora cinnamomi con aceites esenciales de orégano y clavo. *Revista mexicana de ciencias agrícolas, 10*(4), 961–968.
Oliva, M., Beltramino, E., Gallucci, N., Casero, C., Zygadlo, J., & Demo, M. (2010). Actividad antimicrobiana de aceites esenciales de *Aloysia triphylla* (L'Her.) Británica de diferentes regiones de Argentina. *Boletín Latinoamericano y del Caribe de Plantas Medicinales y Aromáticas, 9*(1), 29–37.
Organización De Las Naciones Unidas Para La Agricultura Y La Alimentación (FAO). (2015). *Pérdidas y desperdicio de alimentos el mundo*. FAO. 42 p.

Organización Mundial de la Salud (OMS). (2007). Seguridad alimentaria y enfermedades transmitidas por alimentos. Organización Mundial de la Salud, Ginebra, Suiza. *Hoja informativa, 237*, 1–4.

Ortega, M., Rubio, E., & López, G. (2013). *Biotecnología y alimentación*. Editorial UNED.

Ortiz, M. (2020). *Nuevos materiales con propiedades antifúngicas para el tratamiento postcosecha de cítricos*. Universitat Politècnica de València.

Oscco, A. (2018). *Evaluación de las propiedades fisicoquímicas y sensoriales durante el proceso de extracción del aceite esencial del runtuwayra (Clinopodium weberbaueri)*. Universidad Nacional José María Arguedas.

Pacheco, G. (2015). Propiedades de barrera a la humedad y al crecimiento microbiano de recubrimientos de quitosano-cera de carnauba-aceite esencial de orégano en frutos de pepino.

Pandey, K., Kumar, P., Singh, P., Tripathi, N., & Bajpai, K. (2017). Essential oils: Sources of antimicrobials and food preservatives. *Frontiers in Microbiology, 7*, 2161.

Pardo, A. (2017). *Aplicación de recubrimientos comestibles para la conservación de frutas y efectos de su utilización para preservar la calidad del mango (Mangifera indica L.)*. Universidad Miguel Hernández.

Pérez, A., & Jiménez, E. (2011). Producción de metabolitos secundarios de plantas mediante el cultivo in vitro. *Biotecnología Vegetal, 11*(4), 195–211.

Perveen, K., Bokhari, N., & Soliman, D. (2012). Antibacterial activity of Phoenix dactylifera L. leaf and pit extracts against selected gram negative and gram positive pathogenic bacteria. *Journal of Medicinal Plant Research, 6*(2), 296–300.

Pichersky, E., Noel, P., & Dudareva, N. (2006). Biosíntesis de volátiles vegetales: diversidad e ingenio de la naturaleza. *Science, 311*(5762), 808–811. https://doi.org/10.1126/science.1118510

Plaza, M., & Ricalde, M. (2015). Establecer parámetros de control de calidad físico-químicos del aceite esencial del Schinus molle l. obtenido por arrastre de vapor. *Revista Ciencia, Tecnología e Innovación, 11*(12), 693–696.

Pontigo, G., Trejo, A., & Lira, A. (2015). Desarrollo de un recubrimiento con efecto antifúngico y antibacterial a base de aceite esencial de orégano para conservación de papaya 'Maradol'. *Revista Iberoamericana de Tecnología Postcosecha, 16*(1), 58–63.

Prakash, B. (2015). Plant essential oils as food preservatives to control moulds, mycotoxin contamination and oxidative deterioration of agri-food commodities—Potentials and challenges. *Food Control, 47*, 381–389.

Ramírez, L. (2020). Biotecnología: un bien para la agricultura y el mundo. *Revista Neuronum, 6*(3), 180–183.

Ramón, D. (2014). Avances en biotecnología de alimentos. *Arbor, 190*(768), 151.

Reyes, Y., Vergara, I., Torres, O., Lagos, D., & Jimenez, E. (2016). Contaminación por metales pesados: Implicaciones en salud, ambiente y seguridad alimentaria. *Ingeniería Investigación y Desarrollo, 16*(2), 66–77.

Rico, F. (2013). *estudio de la aplicación de recubrimientos comestibles de quitosano y su combinación con aceites esenciales sobre la vida útil del mango (Mangifera indica L.) mínimamente procesado*. Facultad de Ciencias.

Rivera, R. R. (2019). Efecto de la aplicación pos cosecha de un recubrimiento natural y aceites esenciales en el sistema antioxidante y metabolismo secundario de aguacate (persea americana mill., cv.'hass') orgánico. http://ri-ng.uaq.mx/handle/123456789/1346

Rivera, N., Paredes, A., Gómez, D., Lueckhoff, A., Almeida, A., & Suarez, E. (2017). Composición química y actividad antimicrobiana del aceite esencial de los rizomas de Renealmia thyrsoidea (Ruiz & Pav) Poepp. & Eddl (shiwanku muyu). *Revista Cubana de Plantas Medicinales, 22*(2). http://www.revplantasmedicinales.sld.cu/index.php/pla/article/view/505/276

Robles, A., Aguilar, J., Gutiérrez, M., Rodríguez, F., Morales, A., Guerrero, P., & Del Toro, L. (2016). Identificación cualitativa de metabolitos secundarios y determinación de la citotoxicidad de extractos de tempisque (*sideroxylum capiri pittier*)/qualitative identification of secondary metabolites and cytotoxicity determination of tempisque extracts (Sideroxylum capiri PITTIER). *Biotecnia, 18*(3), 3–8.

Scalvenzi, L., Yaguache, B., Cabrera, P., & Guerrini, A. (2016). Actividad antifúngica in vitro de aceites esenciales de Ocotea quixos (Lam.) Kosterm. y Piper aduncum L. *Bioagro, 28*(1), 039–046.

Segura, M. (2019). Aplicación de un recubrimiento comestible a base de nanoemulsión de aceite esencial de canela (*Cinnamomun verum*) y alginato de sodio en la calidad postcosecha del arándono azul (*Vacinium corymbosum*). http://hdl.handle.net/20.500.12840/3162

Silupu, E., Muñoz, S., Silvera, R., & Salcedo, P. (2019). Composición química, características fisicoquímicas y capacidad antioxidante de aceites esenciales de cinco hierbas aromáticas. *Repositorio de revistas de la universidad privada de pucallpa, 4*(2), 12–12.

Silva, E., Freitas, R., Balbi, I., Terumi, A., Clemente, E., & Stangarlin, R. (2015). Control del moho azul en poscosecha de manzana con productos naturales. *Idesia (Arica), 33*(2), 57–63.

Stashenko, E. (2009). *Aceites esenciales*. Universidad Industrial de Santander, Centro Nacional de Investigaciones para la Agroindustrialización de Especies Vegetales Aromáticas y Medicinales Tropicales Cenivam y Departamento Administrativo de Ciencias, Tecnología e Innovación Colciencias.

Toro, D., Martínez, Y., Rodríguez, R., Pupo, G., Rosabal, O., & Olmo, C. (2017). Análisis preliminar de los metabolitos secundarios de polvos mixtos de hojas de plantas medicinales. *Revista Cubana de Plantas Medicinales, 22*(1), 1–9.

Vaillant, D., Romeu, C., Ramos, E., González, M., Ramírez, R., & González, J. (2009). Efecto inhibitorio in vitro de cinco monoterpenos de aceites esenciales sobre un aislado de Rhizoctonia solani en papa (Solanum tuberosum L.). *Fitosanidad, 13*(3), 197–200.

Valdéz, A., Delgado, E., & Ramírez, J. (2018). Actividad adulticida y composición química del aceite esencial de hojas de Lantana camara sobre Drosophila melanogaster. *Maskana, 9*(1), 21–30.

Valenzuela, M., Gimenez, C., & Soro, S. (2018). Caracterización química y cuantificación de fenoles totales en aceite de semillas de *Cucurbita spp. Dominguezia, 34*(1), 33–38.

Vargas, D., Basso, A., Rodrigues, V., Silva, B., Gatzke, M., & Frizzo, N. (2018). Biotecnologia e alimentos geneticamente modificados: uma revisão. *Revista Contexto & Saúde, 18*(35), 19–26.

Velásquez, A., Álvarez, M., Tamayo, J., & Carvalho, P. (2015). Evaluación in vitro de la actividad fungistática del aceite esencial de mandarina sobre el crecimiento de Penicillium sp. *Ciencia y Tecnología Agropecuaria, 15*(1), 7–14.

Ventura, E. (2017). *Comparación de tres métodos en la extracción de aceite esencial de orégano silvestre (Lippia ssp.)*. UNTRM.

Villanueva, D., & Mejía, F. (2016). Aplicación de la Biotecnología Moderna a la Agricultura en Colombia. PATROCINADOR OFICIAL, 238. http://www.socolen.org.co/images/stories/pdf/44_congreso.pdf#page=270

Witkowska, A., Hickey, D., Alonso-Gomez, M., & Wilkinson, M. (2013). Evaluación de las actividades antimicrobianas de extractos comerciales de hierbas y especias contra bacterias seleccionadas transmitidas por los alimentos. *Revista de investigación alimentaria, 2*(4), 37–54. https://doi.org/10.5539/jfr.v2n4p37

Yang, L., Wen, K-S., Ruan, X., Zhao, Y-X., Wei, F., Wang, Q. (2018). Response of Plant Secondary Metabolites to Environmental Factors. *Molecules, 23*(4), 762. https://doi.org/10.3390/molecules23040762

Use of Waste from the Citrus Industry for the Production of Unicellular Biomass

Andrea Guadalupe Flores-Valdes, José L. Martínez-Hernández, Anna Ilyina, Cristóbal N. Aguilar, and Mónica L. Chávez-González

1 Introduction

Citrus is one of the main fruits produced and highly consumed around the world. In Mexico, citrus is one of the most popular fruits and is one of the main produced and export products. At the global level, Mexico is strongly positioned as a producer of citrus, with the 5th place as a producer of orange, 2nd in lemon, 4th in grapefruit and 13th in tangerine (SAGARPA, 2018). In the foreign market, it is positioned within the five countries in terms of citrus exports, with 729,650 tons of lemon, and 75,644 tons of orange per year (SIAP, 2019). The orange is one of the citrus commonly consumed throughout the country due to its extensive hectares sown and harvested (58.9% ha) (60.5% ha), its high vitamin C content, and the taste for its flavor (SIAP, 2019; Bastías & Cepero, 2016).

Within the citrus industry, the orange is used mainly for the extraction of its juice contained in the pulp, which generates a large amount of waste composed of the orange peel and fiber tissue. These residues have a high content of carbohydrates known as "fiber," which is mainly composed of celluloses, hemicelluloses, and

A. G. Flores-Valdes · M. L. Chávez-González (✉)
Bioprocesses and Bioproducts Research Group, Food Research Department, School of Chemistry, Autonomous University of Coahuila, Saltillo, Coahuila, Mexico

Nanobioscience Group, School of Chemistry, Autonomous University of Coahuila, Saltillo, Coahuila, Mexico
e-mail: monicachavez@uadec.edu.mx

J. L. Martínez-Hernández · A. Ilyina
Nanobioscience Group, School of Chemistry, Autonomous University of Coahuila, Saltillo, Coahuila, Mexico

C. N. Aguilar
Bioprocesses and Bioproducts Research Group, Food Research Department, School of Chemistry, Autonomous University of Coahuila, Saltillo, Coahuila, Mexico

N. R. Maddela, L. C. García (eds.), *Innovations in Biotechnology for a Sustainable Future*, https://doi.org/10.1007/978-3-030-80108-3_5

lignins, or also called lignocellulosic material. Most of the time, lignocellulosic material is insoluble and is not degraded in the small intestine of mammals by digestive enzymes (Micó Ballester, 2014), so it is not suitable for human consumption, and it is automatically discarded, attributing to the increased pollution from waste disposal. Organic waste generation accounts for almost 50% of the waste that is generated in the country. These, of no be well managed, ends in the dumps and landfills decomposing, releasing toxic gases and greenhouse, leached, between other impacts that affect health and environment (SEMARNAT, 2019). Proper treatment of these waste materials is necessary to protect our environment from pollution (Cheng, 2017).

To reduce the environmental impact and take advantage of the lignocellulosic composition of this pulp waste, it is necessary to develop alternatives for their management and treatment so that they can be used in bioprocesses for their transformation and final obtaining of high-value products and by-products. Consequently, it must go through different methodologies for its acquisition, one of the most viable is fermentation, a process by which raw material is transformed into different products through the biological action of microorganisms. The use of agro-industrial waste as substrates in fermentation provides an alternative for their use or revaluation (Subramaniyam & Vimala, 2012). The result of this transformation is biomass, an accumulated matter composed of biological material and different components with great added value.

Since before the development of modern technology and methods, some species of bacteria, fungi and other microorganisms have been used to produce biomass and other compounds of interest through fermentation (Selim et al., 2018; Maddela et al., 2021). One of the microorganisms most used in fermentation is yeast. Which have proven to be microorganisms without high nutritional requirements, provide high growth yields, and their cultivation does not depend on geographical or environmental conditions (Santos et al., 2013). One of the yeasts recognized as GRAS (Generally recognized as safe) by the FDA (Food and Drug Administration) and which has multiple applications in the food industry is *Candida utilis*. This yeast is another typical eukaryotic probiotic strain that can produce edible protein from various wastes (Liu et al., 2019).

Although bioconversion of lignocellulose by yeast fermentation has been reported as efficient and economical, the lignocellulosic fermentation process is still a challenge due to multiple process parameters involved for bioprocess design and optimization (Unrean, 2016). As well, for the use of lignocellulosic material for its bioconversion through fermentation, it needs to be processed through physical, chemical, or enzymatic treatments, to obtain monomeric sugars so it is easily assimilated by yeast.

This study seeks to take advantage of the orange peel generated and discarded as a final residue of the citrus industry, for its use in a bioprocess as a substrate rich in carbohydrates and thus be able to obtain biomass high in protein and other useful compounds. Caring and looking for the ideal medium, for the efficient growth of yeast at the time of fermentation.

2 Materials and Methods

An experimental work divided into two stages was carried out; in the first, the characterization of the composition of the orange residues (peel) was carried out in order to know if it could be used in the fermentation as a substrate. For this, the total sugar content, nutritional, mineral and moisture content of the orange peel residue were evaluated. In the second stage, the best conditions for the biomass production process of *C. utilis* were defined. To do this, the strain *C. utilis* was first subjected to adaptation tests on the substrate. Subsequently, the fermentations were carried out in a bioreactor with 50 mL of stock mineral medium and the inoculated strain, evaluating different ranges of pH, temperature, and substrate concentration. After fermentation, the fermented material was taken, the material was stirred and then it was filtered for the quantification of biomass, determination of total sugars, and pH.

2.1 *Treatment of Lignocellulosic Material*

The orange peels were cut into 4 parts and dried for 72 h in an oven at 50 °C. Then material was milled to a fine powder and stored in hermetic plastic bags at room temperature (25 ± 2 °C). The moisture of the powder was measured to verify that it was dry.

2.2 *Chemical Composition of Citrus Peel Powder*

The orange peel powder was subjected to different methodologies to estimate its content of minerals, total and reducing sugars, total protein, fiber, and lipid content.

2.2.1 Determination of Total Sugars

To estimate the total number of sugars present in the orange peel powder, the phenol-sulfuric method was used, where sucrose was used as standard (DuBois et al., 1956). Samples were measured at 480 nm. The sample was prepared with 1 g of orange peel powder in 100 mL of distilled water.

2.2.2 Determination of Lipid Content

The determination of the lipid (fat) content was carried out by Soxhlet extraction, 3 g of moisture-free sample was weighed on a filter paper to be analyzed. C_6H_{14} (hexane) was used as a solvent for the extraction. The extraction was maintained for 3 h. The flask was dried at 100 °C for 24 h. The content of lipids was calculated by weight difference. The official method of the AOAC (2016) was taken as reference.

2.2.3 Determination of Total Protein

For the measurement of nitrogen content and total protein, the fat-free sample was treated using the Kjeldahl method via multiplication of total nitrogen by 6.25 (AOAC, 2016), taking into account the modifications made by Moreno Dávila (2018).

2.2.4 Determination of Crude Fiber

The crude fiber content was determined by treating 2 g of fat-free sample with an acid solution (0.225N sulfuric acid (0.25%)) and another boiling alkaline solution (0.313N sodium hydroxide). The method was carried out as described by Ochoa Reyes (2018).

2.2.5 Determination of Mineral Content

In order to know the different solid inorganic compounds (minerals) present in the orange peel, a sample of dried and crushed orange peel was taken, it was introduced in the X-ray spectrophotometry equipment "Epsilon 1" from Malvern Panalytical®.

2.3 Candida utilis *Strain and Culture Medium*

Candida utilis strain from the collection of the National University of Tucumán (Faculty of Exact Sciences and Technology) (Argentina) was reactivated on PDA (BD Bioxon®), and it was preserved in tubes with this semi-slanted agar or flute beak, in an incubator at 30 °C.

The mineral medium that was used for the growth of the strain before fermentation was described by Nishio and Nagai (1981), composed of four salts, and it was modified by adding anhydrous dextrose as a 5% carbon source (Table 1). The medium was prepared in triplicate in 500 mL Erlenmeyer flasks with 100 mL of culture medium.

The fermentation kinetics began with a previous stage of adaptation of the *Candida utilis* in stock medium incubating at 30 °C and 150 rpm for 24–48 h. Subsequently, a cell count of 1×10^6 cells was performed in a Neubauer chamber and inoculated into the mineral medium with orange peel as substrate and carbon

Table 1 Components of the mineral culture medium (stock) and glucose, used for the growth of *C. utilis*

Reagent	Molecular formula	g/L
Ammonium sulfate	$(NH_4)_2SO_4$	1.5
Monopotassium phosphate	KH_2PO_4	0.75
Dipotassium phosphate	K_2HPO_4	0.75
Magnesium sulfate	$MgSO_4 \cdot 7H_2O$	0.05
Anhydrous dextrose	$C_6H_{12}O_6$	5

source, sterile and hydrolyzed with steam explosion. It was incubated at 30 °C at 150 rpm in triplicate, a sample was taken every 24 h for 96 h.

2.3.1 Pretreatment of Lignocellulosic Material

Two pretreatments of lignocellulosic material (orange peel powder) were carried out to select the best pretreatment to obtain the highest biomass production. One of these pretreatments was a heat treatment by steam explosion, this treatment allowed for the simultaneous sterilization and hydrolysis of substrate. The culture medium was autoclaved at 121 °C (15 lb pressure) with an exposure time of 15 min, then it was cooled to room temperature. The second pretreatment was irradiation of the dried orange peel, orange peel powder was irradiated with ultraviolet light of 10 W for 10 min, inside a closed chamber. Once irradiated, it was added to 50 mL of sterile mineral medium. In both cases, culture medium was inoculated with a concentration of 1×10^6 cells of the *Candida utilis* strain and an adjustment pH 6 was made, the inoculated medium was incubated at 30 °C at 150 rpm in triplicate, taking a sample every 24 h for 4 days.

2.4 *Fermentation: Variables in the Growth Kinetics of* C. utilis

For the bioprocess, three variables were measured to obtain the highest yield in biomass production, so different fermentations were carried out with different variations in: (1) substrate concentration, (2) variance in temperature, and (3) variance in the pH. In the effect of substrate concentration on biomass production, five concentrations of the carbon source were evaluated; 1%, 2%, 3%, 4%, and 5% in mass/volume ratio of orange peel. The effect of temperature was evaluated in 4 different incubation temperatures; 20, 25, 30, and 35 °C. The effect of pH was also determined (pH 4, 5, and 6), adding 0.1N KOH or 0.01N HCl to the fermentation medium as the case may be. Fermentation was carried out after inoculation of culture medium as described in Sect. 2.3. Flasks were incubated at 30 °C at 150 rpm for 96 h, sampling every 24 h.

2.5 Determination of Biomass by Dry Weight

Each sample recollected during fermentation was placed in 50 mL Falcon™ conical centrifuge tubes and centrifuged at 5000 rpm for 10 min at 10 °C. Biomass was separated and dried at 60 °C for 24 h. Then biomass production was determined by dry weight. A sample of the supernatant was recollected and it was kept frozen. Likewise, the pH of the supernatant was measured.

2.6 Determination of Total Sugar Content

From the frozen supernatant from the fermentation samples, the sugar content was measured by the methodology described in Sect. 2.2 was taken as a reference.

2.7 Total Protein Measurement

Dry biomass was treated with the previously described lipid determination method (Sect 2.2) to eliminate the fat. Once the sample was fat-free, a 15–40 mg sample was taken from it for the measurement of nitrogen content and total protein (percentage).

3 Results and Discussion

3.1 Chemical Composition of Citrus Peel Powder

Knowledge of the chemical composition of substrates is an important step to determine if lignocellulosic material can be useful in bioprocesses as a carbon source or as a raw material to obtain products of interest. Table 2 summarizes the physico-chemical composition of the dried orange peel used in the fermentation medium to obtain unicellular protein by *C. utilis*.

The percentage of lipid content present in the dried orange peel used was 3.65% this result agrees with the report by De la Rosa Delgado (2005) who establishes that, in general, it has a low content of lipids (3–4% on the dry matter). The lipids in the

Table 2 Chemical composition of citrus peel powder

Determinations	Content %
Humidity	5.90 ± 0
Lipid (fat)	3.65 ± 0.83
Raw fiber	13.97 ± 0.53
Protein	7.90 ± 0.13
Total sugars	68.58 ± 0.9

orange peel can be an antimicrobial factor, it is known that essential oils extracted from plants of the *Rutaceae* family (orange, lemon, tangerine) have shown antimicrobial effectiveness (Shankar Raut & Mohan Karuppayil, 2014).

The moisture content of citrus peel powder was 5.9%. This result is almost nil, but in the same way, it serves as an indicator that the dehydration pretreatment carried out on the orange peel is good and helps to eliminate the oily substances from the fresh peel. The orange contains essential oils and 90% of the total of these essential oils is D-limonene, a compound with antimicrobial activity, which means that it causes growth inhibition. This essential oil affects the yeast cell wall and alters the transfer of H^+ and K^+ ions in glycolysis (Boluda-Aguilar et al., 2010). Despite its high content in essential oils, including D-limonene, they are very sensitive to oxidation degradation reactions. The stability of limonene decrease with a temperature of 55 °C (Torres Alvarez, 2018), and the temperature used to dry the orange peel was 60 °C for 3 days, this means that the dehydration pretreatment allowed to eliminate the oily substances and the moisture contained.

The total protein content from the citrus powder by the Kjeldahl method was 7.9%. According to De la Rosa Delgado (2005), it has a low crude protein content of around 7–9% on dry matter. The average of crude fiber was 13.97%; the percentage obtained is slightly high compared to other authors who record having obtained 10–11% (Virreira Flores Bach & Góngora Pereira, 2014).

In the mineral analysis (Table 3), the orange peel presented a great variety of them, the main one in percentage was calcium with 55.40%, offering itself as the main component of the metallic elements of citrus.

The chemical composition of the dry orange peel, allowed to know that it is possible to use it as a substrate rich in nutrients to be used in fermentation process. This because, the principal component of the dry orange peel are carbohydrates, including the raw fiber and the total sugars (Table 2), which represent more than 80% of its chemical composition. The total sugars are known to be "simple sugars" because they group the monosaccharides and disaccharides, which are the biomolecules with the potential to use in fermentation process. Added to this, the method for drying the orange peel, allowed to obtain a material with a lipid content suitable for its use, preventing oily substances from acting as an antimicrobial factor during fermentation. Likewise, the low moisture content makes the orange peel easy to preserve, since having almost zero humidity, prevents it from being contaminated by unwanted microorganisms.

3.2 *Selection of Pretreatment of the Orange Peel for the Fermentation Medium*

It was compared two pretreatments for the orange peel before the fermentation; irradiation and exposure to steam explosion. These two treatments were compared to choose which of the two produce a higher concentration of biomass and the best

Table 3 Minerals present in the dried orange peel

Compound	Ca	K	S	P	Cl	Sr	Fe	Er	Sc	Zn
Concentration (%)	55.398	38.915	1.826	1.322	0.684	0.486	0.313	0.205	0.142	0.135
Compound	**Cu**	**Ba**	**Mn**	**Te**	**Sn**	**Rb**	**Ti**	**Sb**	**Mo**	**Co**
Concentration (%)	0.096	0.096	0.093	0.055	0.049	0.030	0.026	0.022	0.012	0.011
Compound	**Na, Mg, Al, Si, V, Cr, Ni, Ga, Ge, As, Se, Br, Y, Zr, Nb, Tc, Ru, Rh, Pd, Cd, In, I, Cs, Hf, Ta, W, Re, Os, Ir, Pt, Au, Hg, Tl, Pb, Bi**									
Concentration (%)	0.1									

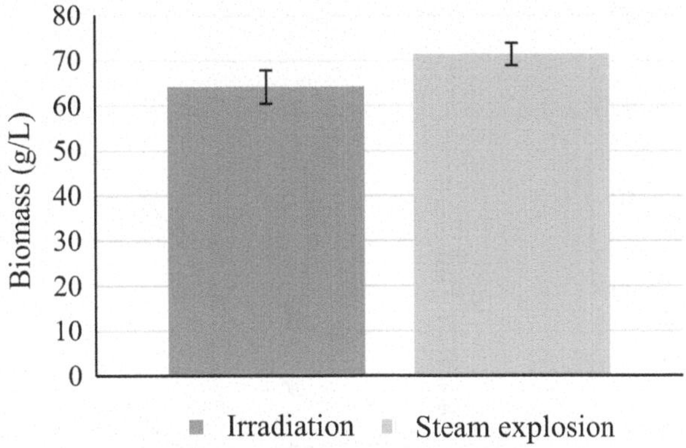

Fig. 1 Comparison of the biomass production of *C. utilis* using two methods of sterilization of the material: irradiation vs. steam explosion using 10% substrate

sterilization treatment of the culture medium. The biomass production by *C. utilis* using the two pretreatments methods for the orange peel as substrate are shown graphically in Fig. 1.

Both methods allow the growth of *C. utilis* but the higher production of biomass is obtained by steam explosion treatment (71.25 g/L ± 2.47). As indicated by Boluda-Aguilar et al. (2010), the steam explosion is used as a pretreatment of lignocellulosic raw material to release the different polymers contained in lignin, as well as its solubilization, also for the partial depolymerization of cellulose and the solubilization of hemicellulose. The heat treatment helps to break down the lignin for the release of fermentable carbohydrates. In the irradiation method, a biomass concentration of 64.07 g/L ± 3.70 was obtained, in this case, radiation acts on biological material that contains DNA and eliminates undesirable microorganisms on the material without hydrolysis of lignocellulose present in the substrate. It was decided to continue applying the steam explosion as an orange peel pretreatment in the following fermentation.

3.2.1 Kinetics of Obtaining Biomass from *Candida utilis*

The growth kinetics with the measurement of different parameters such as pH, temperature, and substrate, allowed to evaluate of the capacity of *Candida utilis* to degrade the orange peel as a specific substrate and unique carbon source.

3.3 Obtaining an Optimal Level of Substrate Concentration

The kinetics of biomass production by *C. utilis* with different concentrations of orange peel as substrate are shown graphically in Fig. 2. The biomass production on culture medium with 1% of citrus peel powder, it can be observed an increase in biomass production during the first 24 h, followed by a slight decrease in biomass

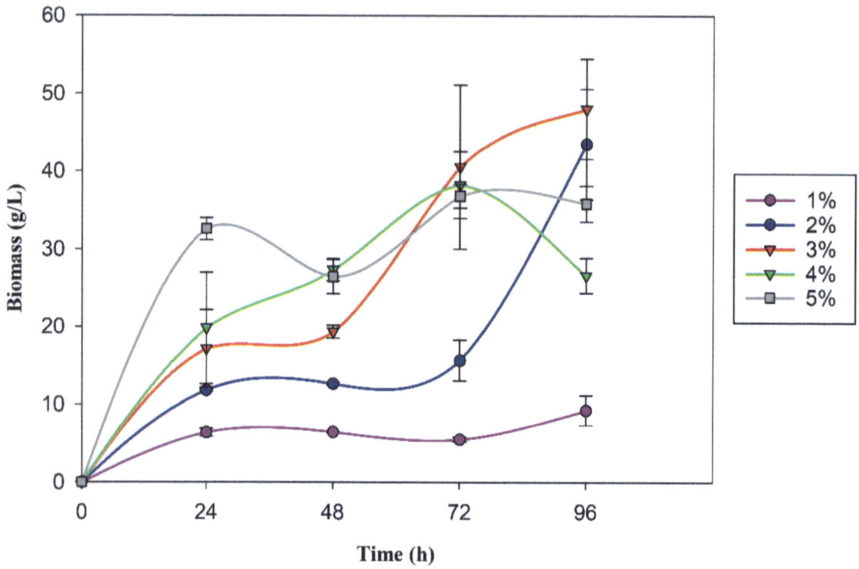

Fig. 2 Kinetic of biomass production by *C. utilis*, with 1, 2, 3, 4, and 5% of orange peel

production during the next hours, this is because *C. utilis* is assimilating the culture medium and enters to a stationary phase, since, as only the orange peel is used as the unique carbon source, it is likely that it contains complex sugars that are difficult to break down in order to assimilate and use them for growth. Then, at 96 h there was an increase, this indicates that the yeast takes time to assimilate the different carbohydrates in the culture medium.

Analyzing the behavior with 2% of substrate, a phenomenon of two phases of growth and biomass production by *C. utilis* can be observed, the first given at 24 h and the second at 72 h. And between these two times, at 24 and 48 h, a stationary phase, where the limiting substrate is finished. This phenomenon is known as diauxic-type growth, which is associated with a biphasic microbial growth that takes place when two different substrates are present in the culture medium (Clark et al., 2009). One of them is easier to metabolize by *C. utilis* and, therefore, it is consumed first, which leads to rapid growth (at 24 h), followed by a stationary phase (between 24 and 48 h). In this stationary phase, the yeast prepares to assimilate and metabolize the second sugar, which can be seen represented by another phase of growth or biomass production at 72 h. This indicated that perhaps the sugars contained in the orange peel are difficult to assimilate for *C. utilis*, so it requires time to consume it. A subsequent time optimization study was not carried out, which could serve to determine whether the sugars present in the orange peel are complex and difficult for yeast to assimilate.

For the concentration of 3%, 4%, and 5% of orange peel substrate, a phenomenon similar to the previous one can be seen and it was possible to appreciate a diauxic growth again. With a concentration of 3%, growth can be seen during the first 24 h followed by a slight stationary phase. To later, have a noticeable increase between 48 and 96 h, obtaining the highest production of biomass.

During the fermentation with 4% of the substrate, it is seen a growth from 24 to 72 h, where it reaches the maximum biomass production. Then, it can be seen a decrease at 96 h, this can be attributed to two reasons: (1) the limiting substrate term that indicates a phase of death or (2) the beginning of assimilation from some other carbon source where *C. utilis* enters an adaptive phase.

On the other hand with a concentration of 5% of orange peel, there are 2 exponential phases; a very small one during 24 and 48 h, and another between 48 and 72 h. This means that between 48 h there are 2 phases; a cell death phase and an exponential or adaptation phase. The first phase of death (24–48 h) followed by an exponential one indicates an adaptive behavior on the part of the yeast, which tries to assimilate the different carbon sources that the orange peel contains as the only substrate.

These results indicate that by increasing the substrate concentration, there is greater assimilation and production of biomass by *C. utilis*.

Kurcz et al. (2018), used two sources of carbon in fermentation to obtain unicellular protein and observed a phenomenon of diauxic growth caused by the two carbon sources (potato wastewater and glycerol). In the kinetic (Fig. 2), it was observed that during the first hours (24 h) it was a growth phase, and then it enters to a new phase of adaptation, to later begin again a growth phase. *C. utilis* used reducing sugars from the culture medium, to achieve the completion of the adaptation phase and entering again to a growth phase. Modeling a biochemical process knowing the kinetic parameters allows to obtain some advantages in the simulation of the process and reduction of costs in industrial experimentation for the optimization of the process (Volesky Von & Votruba, 1992). Constructing the kinetics of the biomass production by *C. utilis*, allowed to know the behavior of yeast on the orange peel as the only carbon source, and opens the possibility of remodeling the process for its optimization.

The biomass production kinetics by *C. utilis* with different concentrations of orange peel as substrate, allowed us to analyze and evaluate the best carbon source concentration to achieve the higher development of *C. utilis*. Thus, with the results obtained, it was decided to select the concentration of 5% orange peel as a carbon source in future fermentations.

3.3.1 Measurement of pH in the Kinetics of Variance in Substrate Concentration

The changes in pH during the different fermentations were favorable, as it was shown to be related to the production of biomass and cells of *C. utilis*. Table 4 describes the growing pH values of *C. utilis* in a mineral medium control medium plus the different concentrations of orange peel. It was possible to appreciate that the concentration of 1–3% there is an increase in the pH and when the concentration was increased from 4% of substrate, the initial pH became slightly more acid.

Table 4 Initial pH of culture medium with different concentrations of orange peel

Concentration OP (%)	1	2	3	4	5
pH	4.85 ± 0.02	5.58 ± 0.23	6.1 ± 0.14	4.73 ± 0.05	4.63 ± 0.02

OP orange peel

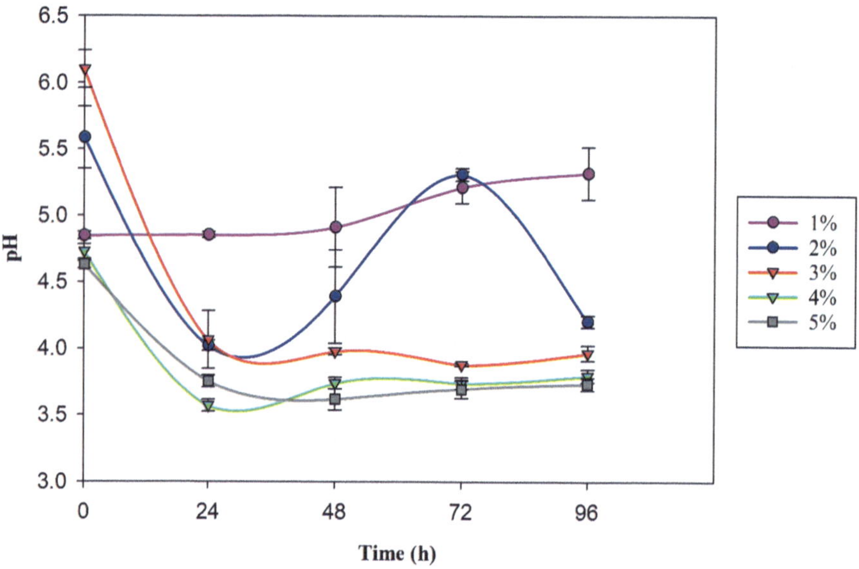

Fig. 3 Kinetics of pH changes of *C. utilis* growing in different substrate concentrations

One of the characteristics of *C. utilis* is its ability to grow in media with a pH between 4 and 6. One of the advantages that yeasts have in fermentation processes is that they can grow at a slightly acidic pH, which can promote to prevent other unwanted microorganisms from growing in the medium, thereby inhibiting the growth of most Gram-positive and Gram-negative bacteria (Contreras, 2014).

The kinetics of pH change by *C. utilis* with different concentrations of orange peel as substrate are shown graphically in Fig. 3. In the kinetic with 1% orange peel, the pH of the medium became slightly more acid with the passage of 96 h, showing a pH of 4.8 at 24 h and rising to 5.5 at 96 h. This means that *C. utilis* can grow with this pH even if it was slightly acid. Regarding the pH in the kinetic with 2% orange peel, slightly abrupt changes are observed, starting at 24 h with a pH of 4 on average, so that between 48 and 72 it increases and ends the kinetics with a slightly acidic pH. This could be related to diauxic growth, as observed in Fig. 2, where it can be seen that at 96 h there is an increase of biomass production and, a decrease in pH to 4.2 (Fig. 3).

In the kinetics with 3, 4, and 5% of orange peel, a decrease in pH begins to be observed in the first 24 h and remains constant between the values of 3.5–3.9. Kurcz et al. (2018) begin their experimental medium with a pH of 5.0, which is ideal for the growth of *C. utilis* in the kinetics of obtaining unicellular protein using potato wastewater as a substrate.

In general, the pH showed a stable behavior. The decrease of the pH between 3.5 and 3.9, occurred due to the adaptation of the yeast with the substrate. The decrease in pH indicates that *C. utilis* is adapting to the medium of orange peel. This is favorable for the process, as it indicates that the yeast adapts the medium for its growth and keeps it constant. In addition, acidic conditions can inhibit the growth of unwanted microorganisms, avoiding the contamination of the fermentation medium.

In this way, the decision to continue working with the concentration of 5% orange peel as a substrate was reinforced, because it showed stability in terms of the change in pH, although it started with an initial pH that was more acid than the other initial pH of the different concentrations of orange peel (Table 4).

3.3.2 Measurement of Total Sugars in the Variance Kinetics in Substrate Concentration

The kinetics of sugar consumption by *C. utilis* with different concentrations of orange peel as substrate is shown graphically in Fig. 4. The initial concentration of sugars is similar even if we use different concentrations of substrate. Obtaining values between 93 and 95 g/L for concentrations of 1 and 3% orange peel, and values between 116 and 123 g/L for 2, 4 and 5% orange peel. This means that despite the variance of the initial substrate concentration, the amount of sugar capable of being fermented remains the same into the orange peel.

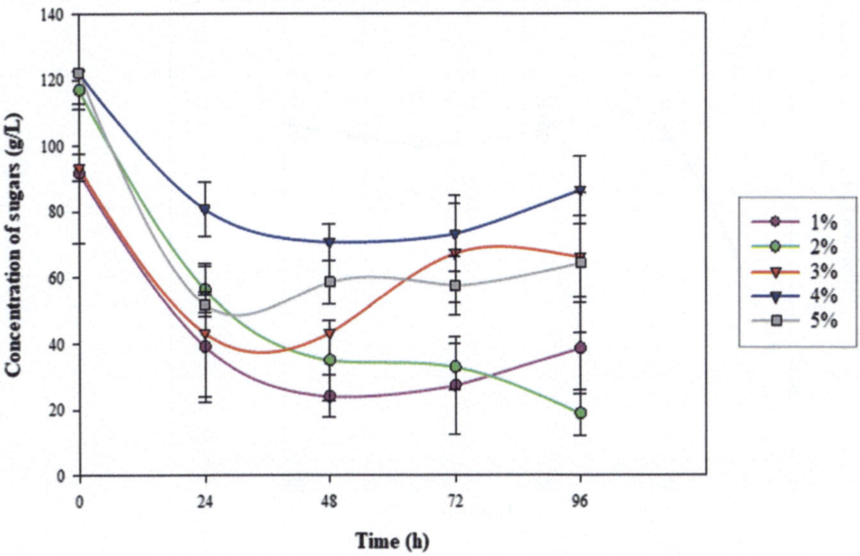

Fig. 4 Sugar consumption in the different kinetics with variance in the substrate

The concentration of sugars in the culture medium decreases as fermentation progresses. Observing that the substrate is diminished from 24 h, agreeing with the biomass production at different concentrations (Fig. 2). When there is more biomass production, there is a greater reduction in the concentration of sugars in the growing medium. Biomass production is proportional to sugar consumption. With the concentration of 5% orange peel, started with a sugar concentration of 122.20 ± 0 g/L to end with 64.37 ± 2.4 g/L, this indicates that nearly half of the initial sugar content was used and metabolized by *C. utilis* during the 96 h of fermentation.

3.4 Obtaining the Optimum Level of Temperature

Candida utilis has a high growth speed, in a substrate rich in sugars; needing factors such as temperature; that must be adequately controlled (Giraldo & López, 2008). In this stage, the results obtained in the aerobic fermentation of *C. utilis* are observed.

3.4.1 Effect of Variance on Temperature

Temperature was evaluated since it is one of the most important factors in the growth of the microorganism, mainly increases its productivity of biomass rich in unicellular protein (Rao et al., 2010). The kinetic of biomass production by *C. utilis* with different temperature measurements are shown graphically in Fig. 5. There is no

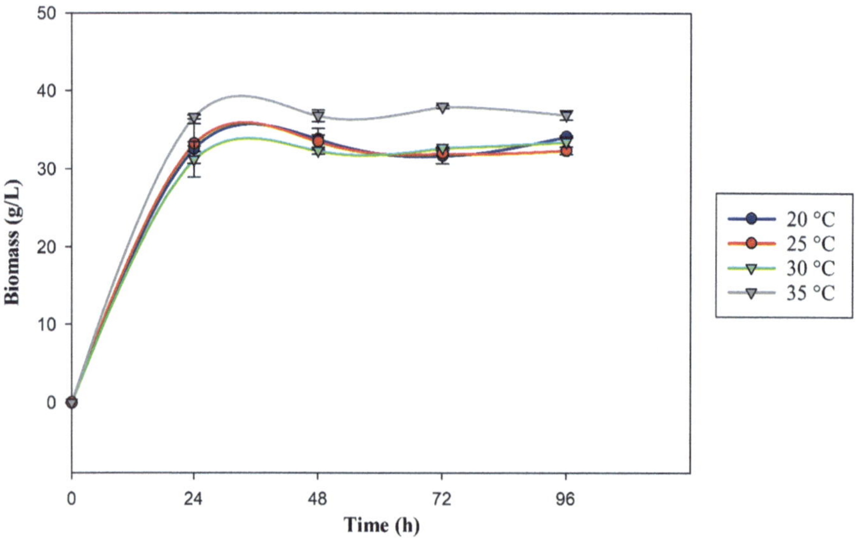

Fig. 5 Kinetic of biomass production of *C. utilis* in 50 mL of mineral medium with 5% orange peel at different temperature measurements

Table 5 *C. utilis* biomass production obtained during 96 h growing on mineral medium and 5% of orange peel at temperatures of 25, 27.5, and 30 °C

Temperature (°C)	Average of biomass (g/L)
25	26.20 ± 0.83
27.5	26.34 ± 1.49
30	25.92 ± 0.69

significant difference between the temperatures evaluated. Also did not show an increase in the yield of biomass produced.

Authors report that between 25 and 30 °C there is an optimal growth of *C. utilis* and a greater production of biomass rich in unicellular protein is observed. Kurcz et al. (2018) used a temperature of 28 °C, that is, neutral among the aforementioned, to obtain unicellular protein-rich biomass with a mineral medium and residues of potato water and glycerol as a carbon source and nitrogen. Observing in Fig. 5, the two optimum temperatures most used by different authors (25 and 30 °C), there is no significant difference between them, since the biomass content produced is similar in terms of its average.

An average temperature (27.5 °C) between these two was also taken into account (Table 5), and this temperature was taken from the substrate concentration kinetics (Fig. 2).

The majority of yeasts are mesophilic, with a maximum growth temperature between 24 and 48 °C. This is confusing because different authors report that the optimal growth temperature of *C. utilis* is 30 °C for the production of biomass rich in unicellular protein. However, Zhao et al. (2010) report that for some microorganisms such as *C. utilis*, a temperature between 33 and 35 °C is optimal for it, in the same way, they reported having a percentage of 48.2% of crude protein by inoculating *C. utilis* at 30 °C. Rajoka et al. (2006) reported that the biomass production yield of *Candida utilis* increased at 35 °C and after this, it began to decrease.

The fact that there is more biomass production at a temperature of 35 °C could be attributed to two reasons: (1) it is the optimal growth temperature for *C. utilis* and (2) as the temperature increases, the orange peel solubilizes part of the sugars present in itself. According to Bekhta and Marutzky (2007) the stability of cellulose and lignin is dependent of the temperature applied, it decreases with an increase in temperature.

3.4.2 Measurement of pH During the Kinetics of Variance in Temperature

For the growth kinetics and biomass production of *C. utilis* at different temperatures, the pH was measured in each sampling.

There was a pH change in the bioprocess by *Candida utilis* (Fig. 6). It is also observed that for each kinetic the initial pH was the same, which is why its variance is better appreciated over time. The change in pH in a bioprocess indicates that the

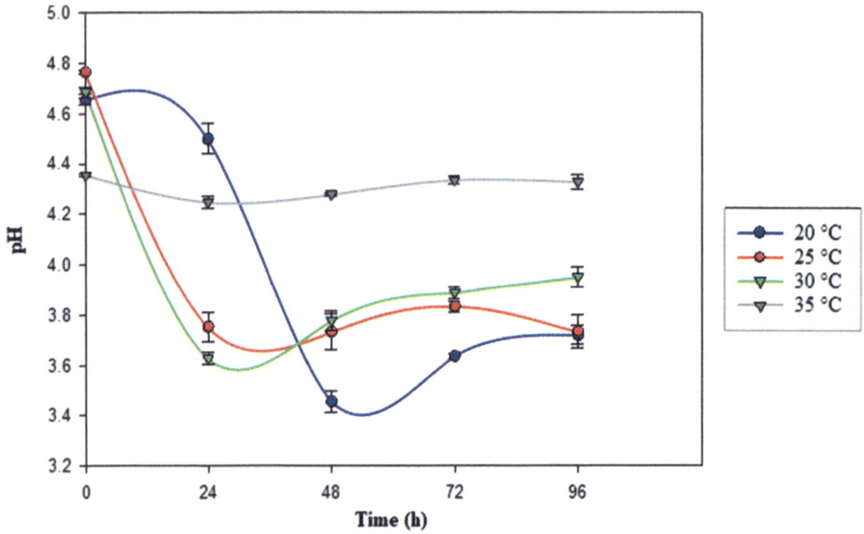

Fig. 6 Kinetics of pH change of *C. utilis growing* at different temperatures with 5% orange peel

microorganism used is performing various metabolic actions to take advantage of the medium in an optimal way, in this case, it is possible to appreciate how *C. utilis* with a lower temperature changes the pH drastically to be able to grow optimally, as we know that the growth range of this yeast is 25–38 °C, which means that it may have undergone an adaptation stage.

3.5 Effect of Variance on Initial pH

One of the characteristics that *C. utilis* presents is its ability to grow in media with a pH between 4 and 6, for this reason, it was possible to measure with which of the pH on this scale the best biomass production is achieved by this yeast.

It was observed that for each medium at the end of the steam explosion treatment, a decrease in pH was seen to approximately 4.6, so the hydroxide was used in a greater proportion, using only the acid if the concentration of the base was very high. In Fig. 7, it can be seen that there was a slightly higher yield when adjusting the initial pH of the medium to 6.0.

Munawar et al. (2010), optimized submerged fermentation for biomass production and reported when evaluating the initial pH of a process that the highest percentage of biomass obtained was found by leveling the pH to 6.0 at the beginning of the fermentation system.

The results obtained, show that there is no great difference in the biomass production yield. Fatemeh et al. (2018), concluded that for the studies carried out with different microorganisms, among them *Candida utilis*, the initial pH value was

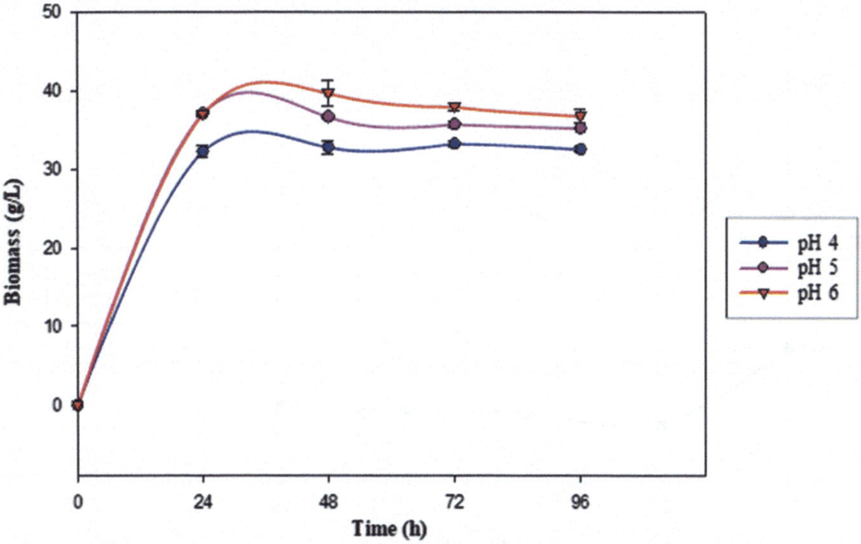

Fig. 7 Growth kinetics of *C. utilis* growing at different initial pH with 5% orange peel

not among the independent variables to evaluate the optimal condition for the high production of biomass rich in unicellular protein. This is beneficial for future processes since the fact that there is no significant relevance in the pH for the biomass production is favorable since it indicates the lower use of additives to level the pH and therefore there is no increase in the economy to optimize the process.

3.5.1 pH Measurement During Initial pH Variance Kinetics

During the kinetic study to measure the initial pH as a variable in the optimization of the unicellular biomass production process, the pH was recorded during each sample collection. This variable is not necessary to optimize the process (Fig. 8), although when adjusting the pH to 6.0, there is a decrease in the pH of the system, which means that *Candida utilis* metabolizes and changed the composition of the medium to assimilate the substrate more effectively.

The change in pH in an initially adjusted system indicates that the microorganism used is performing different metabolic actions to be able to assimilate or take advantage of the nutrients in the environment.

3.6 Total Protein Measurement

The total protein content of the biomass with a higher yield was evaluated. The sample was taken from the biomass obtained during fermentation with 5% orange

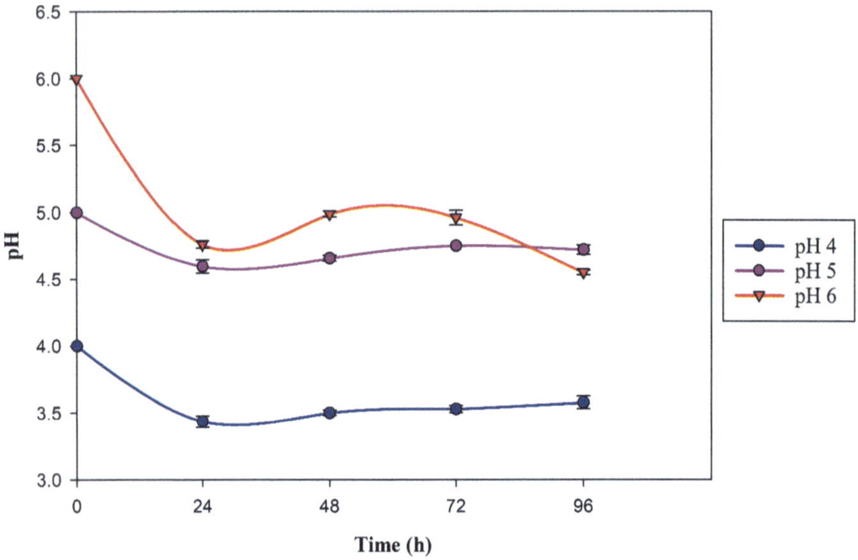

Fig. 8 Kinetics of pH change of *C. utilis* growing at different initial pH with 5% orange peel

peel. First, it was evaluated the biomass generated at 24 h of fermentation, of which 5% of total protein was obtained. On the other hand, the sample taken at 48 h had 7.4% of total protein. The result obtained is low if we compare it with other studies that obtained the unicellular protein in a similar way. Cajo et al. (2011), obtained a total protein concentration in yeast of 52.5%.

The fact that a low yield was obtained in biomass rich in cellular protein is because a nitrogen source was not optimized in the medium. If we observe in the stock medium (Table 1), the only source of this essential nutrient is ammonium sulfate. The orange peel waste do not contain mineral elements which could help to increase the growth and adaptation of the yeast. Molk et al. (2002) named the unicellular protein or bioprotein that which, produced by some microorganism, assure an efficient growth rate, through the use of cheap substrates composed primarily of carbon, nitrogen and phosphorus. Various authors such as Munawar et al. (2010), report that within their optimization for obtaining biomass rich in unicellular protein, the effect of the nitrogen source with the fermentation medium is evaluated. The source of nitrogen is fundamental for the upturn of efficiency and economics of microbial fermentation (Nancib et al., 2001).

Zhao et al. (2010) evaluated different nitrogen sources in the fermentation medium for the production of biomass rich in unicellular protein by *Candida utilis* 1769 at 30 °C, and argued that, among the added inorganic nitrogen sources, the addition of fermented corn liquor, yeast extract and peptone significantly increased biomass production. In contrast to these, the addition of ammonium sulfate and ammonium chloride has an adverse effect on them. Knowing this, an evaluation of the nitrogen content in the fermentation medium is recommended for further studies.

4 Conclusions

In the present work, the viability of using orange peel residues to be used as a substrate for the production of unicellular biomass of *C. utilis* was demonstrated, the nutritional content of the residue showed that orange peels are a rich material and that it can be used as a fermentation substrate. Of the different variables evaluated, it was highlighted that the substrate concentration was important in the process, having the highest biomass production titers with 5% orange peel in the fermentation medium, reaching a production of 35.84 g/L $\pm$ 2.28 at 96 h of fermentation.

From the evaluation of the effect of temperature on biomass production, there was no difference between the temperatures evaluated, having a similar biomass production, *C. utilis* could grow well in a range of 20–35 °C. Regarding the evaluation of the growth of *C. utilis* over different pH, it was found that growth was positively influenced in a pH range of 5–6 with maximum biomass of 39.72 g/L $\pm$ 1.61 at 48 h of fermentation and with a substrate concentration of 5%.

Although it was possible to produce a good concentration of biomass of *C. utilis*, there are some other variables that could be studied to optimize the single-celled biomass production process, such as the content and type of nitrogen in the medium, the range of aeration, the age of the inoculum. These are important factors that could have an effect on the production of single-cell protein.

References

AOAC. (2016). *The official methods of analysis of AOAC International* (20th ed.). George W. Latimer Jr. 3172p. http://www.eoma.aoac.org

Bastías, J. M., & Cepero, Y. (2016). The vitamin C as an effective micronutrient in the fortification of foods. *Revista Chilena de Nutrición, 43*, 81–86. https://doi.org/10.4067/S0717-75182016000100012

Bekhta, P., & Marutzky, R. (2007). Bending strength and modulus of elasticity of particleboards at various temperatures. *Holz als Roh- und Werkstoff, 65*(2), 163–165.

Boluda-Aguilar, M., García-Vidal, L., González-Castañeda, F. P., & López-Gómez, A. (2010). Mandarin peel wastes pretreatment with steam explosion for bioethanol production. *Bioresource Technology, 101*, 3506–3513.

Cajo, L., Nizama, L., & Carreño, C. (2011). Effect of the inoculum concentration and the molasses as a supplement of the vinaza from the destilery for the biomass production by *Candida utilis* native. *Scientia Agropecuaria, 2*(2011), 65–72.

Cheng, J. J. (2017). *Biomass to renewable energy processes* (2nd ed.). CRC Press. https://doi.org/10.1201/9781315152868

Clark, D. P., Martinko, J. M., Madigan, M. T., & Dunlap, P. V. (2009). *Brock. Biology of the microorganism* (12th ed.). Editorial Pearson.

Contreras, R. (2014). Growth mediums for yeast. *The guide of biology*. https://goo.gl/nR93NC

De la Rosa Delgado, B. (2005). *Silage in wet areas and its quality indicators. IV Animal feeding day*. Laboratory of Mouriscade Lollín (Puntepiedra).

Dubois, M., Gilles, K. A., Hamilton, J. K., Robers, P. A., & Smith, F. (1956). Colorimetric method for the determination of sugars and related substances. *Analytical Biochemistry, 28*, 350–356.

Fatemeh, S., Reihani, S., & Khosravi-Darani, K. (2018). Influencing factors on single cell protein production by submerged fermentation: A review. *Electronic Journal of Biotechnology, 37*, 34–40. https://doi.org/10.1016/j.ejbt.2018.11.005

Giraldo, V., & López, P. (2008). *Production of single cell protein from agroindustrial wastes*. Revista VirtualPro. No. 82. ISSN 1900-6241.

Kurcz, A., Stanisław, B., Kot, A. M., Bzducha-Wróbel, A., & Kieliszek, M. (2018). Application of industrial wastes for the production of microbial single-cell protein by fodder yeast Candida utilis. *CrossMark, Waste Biomass Valor, 9*, 57–64.

Liu, J., Shi, P., Ahmad, S., Yin, C., Liu, X., Liu, Y., Zhang, H., Xu, Q., Yan, H., & Qingxiao, L. (2019). Co-culture of Bacillus coagulans and *Candida utilis* effciently treats Lactobacillus fermentation wastewater. *AMB Express, 9*, 15. https://doi.org/10.1186/s13568-019-0743-3

Maddela, N. R., Garcia, L. C., & Chakraborty, S. (Eds.). (2021). *Advances in the domain of environmental biotechnology*. Springer Nature Singapore Pte Ltd. https://doi.org/10.1007/978-981-15-8999-7; ISBN: 978–981–15-8999-7, (pp XVIII, 717).

Micó Ballester, M. J. (2014). Methods of fiber analysis and physicochemical determinations in citric for the module of food control of the training cycle of dietetics. *Ciencias y letras, 3Ciencias, 69*(28), pp. 28.

Molk, J. N., Schuyler, S. C., Liu, J. Y., Evans, J. G., Salmon, E. D., Pellman, D., & Bloom, K. (2002). The differential roles of budding yeast Tem1p, Cdc15p, and Bub2p protein dynamics in mitotic exit. *Molecular Biology of the Cell, 15*, 1519–1532.

Moreno Dávila, I. M. (2018). *Determination of proteins*. Dpto. de investigación en Alimentos, Facultad de Ciencias Químicas, Universidad Autónoma de Coahuila.

Munawar, R. A., Irfan, M., Nadeem, M., Syed, Q. A., & Siddique, Z. H. (2010). Biosynthesis of single cell biomass of *Candida utilis* by submerged fermentation. *Pakistan Journal of Science, 62*(1), 1–2.

Nancib, N., Nancib, A., Boudjelal, A., Benslimane, C., Blanchard, F., & Boudrant, J. (2001). The effect of supplementation by different nitrogen sources on the production of lactic acid from date juice by Lactobacillus casei subsp. Rhamnosus. *Bioresource Technology, 78*, 149–153.

Nishio, N., & Nagai, S. (1981). Single cell protein production from Mandarin Orange Peel. *European Journal of Applied Microbiology and Biotechnology, 11*, 156–160.

Ochoa Reyes, E. (2018). *Determination of fiber*. Dpto. de investigación en Alimentos, Facultad de Ciencias Químicas, Universidad Autónoma de Coahuila.

Rajoka, M. I., Khan, S., Jabbar, M. A., Awan, M. S., & Hashmi, A. S. (2006). Kinetics of batch single cell protein production from rice-polishings with *Candida utilis* in continuously aerated tank reactors. *Bioresource Technology, 97*, 1934–1941.

Rao, M., Varma, A. J., & Deshmukh, S. S. (2010). Production of single cell protein, essential amino acids, and xylanase by *Penicillium janthinellum*. *BioResources, 5*, 2470–2477.

Santos, C. A., Caldeira, M. L., Lopes da Silva, T., Novais, J. M., & Reis, A. (2013). Enhanced lipidic algae biomass production using gas transfer from a fermentative Rhodosporidium toruloides culture to an autotrophic Chlorella protothecoides culture. *Bio-resource Technology, 138*, 48–54.

Secretary of Agriculture y Rural Development (SAGARPA). (2018). *Agrifood Atlas 2012–2018*. https://cutt.ly/rxsZffM

Secretary of the Environment and Natural Resources (SEMARNAT). (2019). *Vision towards a sustainable management: zero wastes*. https://cutt.ly/IxsZ4kd

Selim, K. A., El-Ghwas, D. E., Easa, S. M., Hassan, A., & Mohamed, I. (2018). Bioethanol a microbial biofuel metabolite; new insights of yeasts metabolic engineering. *Fermentation, 4*(1), 16. https://doi.org/10.3390/fermentation4010016

Shankar Raut, J., & Mohan Karuppayil, S. (2014). A status review on the medicinal properties of essential oil. Revision. *Industrial Crops and Products, 62*(2014), 250–264.

SIAP. (2019). *Agricultural Production México Sagarpa*. https://www.gob.mx/siap/acciones-y-programas/produccion-agricola-33119

Subramaniyam, R., & Vimala, R. (2012). Solid state and submerged fermentation for the production of bioactive substances: A comparative study. *International Journal of Science and Nature, 3*, 480–486.

Torres Alvarez, C. (2018). *Microencapsulation of orange essential oil and its concentrates through emulsions and molecular inclusion for the stabilization and preservation of their activity antimicrobial and antioxidant* (PhD thesis). University of New Lion, Faculty of biological sciences, Monterrey, México.

Unrean, P. (2016). Bioprocess modelling for the design and optimization of lignocellulosic biomass fermentation. *Bioresources and Bioprocessing, 3*, 1. https://doi.org/10.1186/s40643-015-0079-z

Virreira Flores Bach, J. J., & Góngora Pereira, O. A. G. (2014). *Physicochemical characterization of orange peels (Citrus Sinensis L.) and grapefruit (Citrus grandis) to obtain bioethanol* (Undergraduate thesis). National University of the Amazonía Peruana, Faculty of chemistry engineer.

Volesky Von, B., & Votruba, J. (1992). *Modeling and optimization of fermentation processes* (Vol. 66, p. 547). Elsevier.

Zhao, G., Zhang, W., & Zhang, G. (2010). Production of single cell protein using waste capsicum powder produced during capsanthin extraction. *Letters in Applied Microbiology, 50*(2), 187–191. https://doi.org/10.1111/j.1472-765X.2009.02773

Organic Waste: A Cheaper Source for Probiotics Production

G. Vidya Sagar Reddy, Ch. Vijaya, Bellamkonda Ramesh, Srinivasan Kameswaran, Somavarapu Silpa, M. Subhosh Chandra, Ch. Venkatrayulu, and M. Srinivasulu

1 Organic Wastes and Probiotics

Probiotics are microorganisms when given in sufficient quantity provide health advantage to the host (Kechagia et al., 2013). They are bioactive microorganisms with beneficiary effects. They have been found to be preventive to certain diseases and are considered as essential component in human nutrition. Probiotics provide necessary nutrients, enhance the growth and immunity of the host by inhibiting pathogenic microorganisms, they also contribute the enzymes that help in digestion (Ashraf & Shah, 2014). In recent times, there is an increase in demand for probiotics due to its health promoting abilities. Probiotics were also reported to have ability toward prevention of cancer, treatment of bacterial vaginosis, management of atopic eczema, compensation for lactase insufficiency, prevention of traveler's diarrhea, etc., in humans (Isolauri et al., 2000; Kumar et al., 2010; Falagas et al., 2007; Vrese et al., 2001; McFarland, 2007). The application of probiotics is not only confined to

G. V. S. Reddy
Department of Biotechnology, Vikrama Simhapuri University, Nellore, Andhra Pradesh, India

C. Vijaya
Department of Marine Biology, Vikrama Simhapuri University, Nellore, Andhra Pradesh, India

B. Ramesh (✉) · S. Silpa · C. Venkatrayulu
Department of Food Technology, Vikrama Simhapuri University, Nellore, Andhra Pradesh, India

S. Kameswaran
Department of Botany, Vikrama Simhapuri University PG Centre, Kavali, Andhra Pradesh, India

M. S. Chandra
Department of Microbiology, Yogi Vemana University, Kadapa, Andhra Pradesh, India

M. Srinivasulu
Department of Biotechnology, Yogi Vemana University, Kadapa, Andhra Pradesh, India

N. R. Maddela, L. C. García (eds.), *Innovations in Biotechnology for a Sustainable Future*, https://doi.org/10.1007/978-3-030-80108-3_6

humans but also used in animal husbandry. Probiotics are given to animals in their feed for the production of healthier animals in animal husbandry (Awad et al., 2009). It was reported that, *Saccharomyces cerevisiae* RC016, a probiotic strain isolated from pig intestine has ability to absorb several mycotoxins such as aflatoxin, zearalenone, fumonisin, and ochratoxin (Zoghi et al., 2014). Probiotics are considered to be generally regarded as safe (GRAS) microorganisms and approved to use for consumption (Hempel et al., 2012). Application of probiotics as a feed additive intended for production animals is very promising and entails minimal risk (Collins et al., 2009).

The production of probiotics in bulk scale is money consuming process where raw materials alone call for more than 50% of the production cost. Considering the huge demand, scientists are in search for new supportive components which should be of low cost, growth promoting and stability enhancer of probiotics. Organic wastes were found to be cost effective production medium for probiotics and functions as alternate to classical defined media. Organic wastes contain essential nutrients required for the production of the probiotic organisms (Sagar et al., 2018). A group of materials including corn extract, sugar, plant extracts, cassava starch, sugarcane bagasse, date waste are tested to improve growth and stability of probiotics (Ezejiofor et al., 2014; Vandenberghe et al., 2021).

Organic waste is any material that is biodegradable and comes from either a plant or an animal. It can be compostable, broken into carbon dioxide, methane or simple organic molecules. Examples of organic waste include green waste from plants, paper, cardboard, food waste, timber—wood waste, dairy wastes, meat and fish processing wastes, etc. (Salah & Hala, 2017). Based on their source, the possible organic wastes that can be used for the production of probiotics can be classified as Agricultural waste, Industrial waste, Municipal and house hold waste (Cheng & Hu, 2010).

2 Agricultural Wastes as Substrate for Probiotics

Agricultural waste generally contains livestock manure and diversified biodegradable materials, such as grain stalks, husk, shells, etc. In general farmers use agricultural waste as manure but due to huge production quantities, time shortage between the two crops, much of the agri-waste is burnt and raising environmental concerns (Bhuvaneshwari et al., 2019). Moreover decrease in land availability, production, and transporting the materials for the composting process makes the process slow and costlier. Agro-industrial residues could be effectively utilized as low cost substrates for the production of microorganisms (Naggar et al., 2014), this plays a major role in lowering the production costs of enzymes and also probiotics. Sugarcane bagasse is one of the widely available agro-industrial wastes which serves as an ideal habitat for microbial growth and yields many value-added products. Sugarcane bagasse constitutes of 50% cellulose, remaining part includes hemicellulose, lignin, and others that has ability to support growth of several microorganisms (Sidana & Farooq, 2014). Further, its pre-treatment by a variety of physical and chemical agents makes it more favorable for microbial growth and increased productivity of different metabolites. In a research, Sugarcane bagasse was treated with *Acinetobacter calcoaceticus* and the formed fermented hydrolysate which was further used as growth inducer to develop probiotic organisms such as *Lactobacillus* sp. and *Bifidobacterium* sp. (Soren et al., 2020). In a study, reported by Hu et al. (2020), when cattle were fed with rice straw co-fermented with various probiotics and enzymes, it altered the rumen bacterial community and increased the relative abundance of variety of microbial species like *Ruminococcus*, *Lachnoclostridium*, *Fibrobacteres*, *Bacteroidetes*, *Saccharofermentans*, *Treponema*, *Pseudobutyrivibrio* and *Ruminobacter* species. It also elevated the propionate production and changed the rumen fermentation pattern (Hu et al., 2020).

3 Industrial Wastes as Substrate for Probiotics

Organic wastes generated by industries are the byproducts of processes in a variety of industries; they can be classified as: vegetable processing wastes which include wash water, skins, rinds, pulp, and other organic waste from fruit and vegetable cleaning, processing, cooking, and canning. Meat processing wastes which include, grease, fat, oils, wash water, cooking waste, dripping, hair, and feathers from slaughtering, butchering, cooking, and packaging of fish, chicken, beef, and all other meat products. Dairy and egg processing wastes which include wash water and process waste from egg and milk processing, drying, bottling, and packaging.

Miscellaneous food processing organic wastes include wastewater from soda or fruit or juice bottling, bakeries, breweries, distilleries, sugar and grain processing, and animal food production (Table 1).

Table 1 Types of waste and their origin in the industry (Source: Yiu Fai Tsang et al., 2019)

Type of waste	Origin of waste
Waste from the preparation, processing, and rendering of meat, fish, and other food stuffs originating from animals	Slaughter house, butcher shops, fish processing plants, egg processing plants, tallow processing plants
Waste from the preparation and processing of fruit, vegetables, grain, edible oil, cocoa, coffee, and tobacco, production of canned foods	Fruit and vegetable processing plant, starch manufacturers, malt houses, grist and husting mill, oil mills, manufacturers of coffee, tea, cocoa, and canned foods, tobacco processing plants
Waste from sugar production	Sugar manufacturers
Waste from milk processing	Dairies
Waste from production of baked foods and sweets	Bakeries, confectioners, candy producers
Waste from the production of both non-alcoholic and alcoholic beverages	Breweries, wineries, liqueur producers, distilleries, nonalcoholic beverage, and fruit juice producers

3.1 Organic Wastes from Fruits and Vegetable Processing

There are numerous industries which are based on fruits as well as vegetables e.g., juice industries, pickle industries, oil industries, etc. These industries process the products to increase their shelf life by using canning, drying, freezing, and preparation of juices, jams, jellies, etc. The main solid waste constitutes organic materials, including discarded fruits vegetables, peel/skin, seeds, etc. whereas the effluents contain liquid waste of juice and wash waters (Sadh et al., 2018). Fruit and vegetable losses and waste do not represent only the wasting of food commodities, but also indirectly include wasting of critical resources such as land, water, fertilizers, chemicals, energy, and labor. These immense quantities of lost and wasted food commodities also contribute to immense environmental problems as they decompose in landfills and emit harmful greenhouse gases (Yahaya & Mardiyya, 2019). Followed by household garbage, fruit, and vegetable processing units commonly produce the highest wastes into the environment (Negi & Anand, 2016).

India is the largest producer of bananas. It produces more bananas in a year than the rest of the world produces for export. Most of the bananas grown in India are for the domestic market (Gumisiriza et al., 2017). Banana peel constitutes about 30% of fresh banana by weight. Ripe banana peels contain up to 8% Crude Protein (CP), 13.8% soluble sugars, and 4.8% total phenolics. Banana peels are rich in trace elements; the ripened peel has approximately 30% free sugars. Green plantain peels contain 40% starch. A commercial medium such as de Man, Rogosa and Sharpe (MRS), or Corn Steep Liquor is usually too costly for commercial production of probiotics. Therefore, exploring locally available sources as culture media for probiotics from various agro-industrial wastes could be a better alternative for reducing the cost of production. Hence, the applicability of Banana Peel as a nutritional source was explored for cultivation of *Lactobacilli* strains, which are potential probiotics (Farees et al., 2017).

Pineapple is the second most consumed and produced fruit after bananas, contributing to more than 20% of the world production of tropical fruits. More than 70% of pineapple is consumed as fresh fruit in producing countries. Pineapple waste is a byproduct of the pineapple processing industry and it consists of residual pulp, peels, and skin (Campos et al., 2020). About 30% of the pineapples are turned into waste during the canning operation. These wastes can cause environmental pollution problems if not utilized because it still contains high content of carbohydrates as well as high fiber and low protein contents. Based on the physicochemical properties of the pineapple waste, it can be potentially used as carbon sources for production of lactic acid by microbial systems. Using waste pineapple material for probiotic production would be optimal both economically as well as environmentally (Hassan Pyar et al., 2014).The addition of apple, banana, and passion fruit peel powder in probiotic yogurt improved the rheological properties and enhanced the growth of *Lactobacillus casei, Bifidobacterium animalis* subsp. *lactis*, *Lactobacillus acidophilus,* and *Lactobacillus paracasei* (Sah et al., 2016). The effect of milk supplementation with mango peels on the kefir microorganism's growth rates and antioxidant properties were also estimated in fermented products (Vicenssuto & De Castro, 2020). *Lactobacillus* is used in food industry for fermentation, preservation and as food additives, in pharmaceutical industry as probiotic and antimicrobial substances (Gowe, 2015). For production of *Lactobacillus* using fruit peel wastes, Shweta Hardia and Sanjida Iqbal (2014) tested banana peels, orange peels, and papaya fruit peels for maximum growth of *Lactobacillus*. Ten percent concentration of crushed fruit peels were tested for production media and observed that papaya peels extract supported the highest growth of *Lactobacillus* (Hardia & Iqbal, 2014). Therefore, application of agro-industrial waste and co-products in bioprocesses provides an alternative way to replace the refined and costly raw materials in the process of fermentation. In addition, the bulk use of agro-industrial waste residues will help to solve environmental issues. Microbial biomass production by using distillers dried grains with solubles (DDGs) is an alternative of great interest for reuse of this industrial byproduct. It can serve as a protein concentrate as well as energy source for microbial growth. Such waste substrates can be used as medium component which can induce biomass production. The large amount of waste DDGs, produced during bioethanol production can be used to generate *S. cerevisiae* biomass and it can be applied in feed additive formulations. The use of agro-industrial waste that pollutes the environment can contribute to sustainable development of the process and reduce the costs of large-scale production (Fochesato et al., 2018).

3.2 *Dairy Wastewaters*

Dairy waste water is a mixture of industrial wastewaters from tank truck, pack, and equipment washing. Dairy processing effluents mostly include milk or milk products lost in the technological cycles (spilled milk, spoiled milk, skimmed milk, and curd pieces); starter cultures used in manufacturing; byproducts of processing operations (whey, milk, and whey permeates); contaminants from the washing of milk trucks, tanks, cans, equipment, bottles, and floors (Table 2); reagents applied in CIP procedures, cooling of milk and milk products, for sanitary needs, in equipment damage or operational problems; and various additives introduced in manufacturing (Cristian, 2010; Karadag et al., 2015). Milk loss in wastewater is around 0.5–2.5% of processed milk, but it can increase to 3–4% (Slavov, 2017).

Lactic acid bacteria (LAB) are currently of great importance given their increasing use in the improvement of human and animal health and nutrition (Vieco-Saiz et al., 2019). They exhibit complex nutritional requirements, which is the reason why their production costs are high. Research efforts are being made aimed at evaluating different substrates for their production as well as the production of valuable metabolites from them (Ahlberg et al., 2015). Oscar et al. proposed the use of milk, industrial cheese whey, cane molasses, hydrolyzed starches, lingo-cellulosic materials, organic food waste and bovine blood plasma, among others, have been

Table 2 Composition of milk processing effluents (Source: Slavov, 2017)

Milk processing effluent	Active reaction (pH)	γ/(g/L)							
		BOD_5	COD	FOG	TS	TSS	TN	TP	Alkalinity as $CaCO_3$
Mixed dairy	4–11	0.24–5.9	0.5–10.4	0.02–1.92	0.71–7	0.06–5.80	0.01–0.66	0–0.6	0.32–1.2
Milk reception	7.18	0.8	2.54	1.06	–	0.65	–	–	–
Dairy/sewage = 7:3	9.1 ± 6.7	1.08–2.81	2.04–4.73	0.24–0.29	–	0.53–1.13	–	0.02–0.03	–
Fluid milk	5–9.5	0.5–1.3	0.95–2.4	–	–	0.09–0.45	–	–	–
Yoghurt	4.53	–	6.5	–	–	–	–	–	–
Butter	12.08	0.22–2.65	8.93	2.88	–	0.7–5.07	–	–	–
Ice cream	5.1–6.96	2.45	5.2	–	3.9	3.1	–	0.014	0.22
Cheese	3.38–9.5	0.59–5	1–63.3	0.33–2.6	1.92–53.2	0.19–2.5	0.018–0.83	0.005–0.28	–
Cottage cheese	7.83	2.6	17.65	0.95	–	3.38	–	–	–
Cheese whey	3.92–6.5	27–60	50–102.1	0.9–14	55–70.9	1.27–22.15	0.2–1.76	0.12–0.53	–
Hard cheese whey	5.8	9.48	73.45	0.99	–	7.15	–	–	–
Soft cheese whey	5.35	26.77	58.55	0.49	–	8.31	–	–	–
Cottage cheese whey	4.5	–	79	–	68	–	2	–	–
Cheese whey wastewater	4.6	35	–	0.8	–	–	–	0.64	–
Whey processing effluent	5–9	0.59–1.21	1.07–2.18	–	–	0.08–0.44	–	–	–
Milk permeate	5.55–6.52	–	52.94–57.46	–	11.61–15.39	1.94–3.4	0.3–0.4	0.35–0.45	2.5
Condensate	8.3	–	–	–	–	–	0.0006	0.0001	–
Washing wastewater	10.37	3.47	14.64	3.11	–	3.82	–	–	–

BOD_5 biological oxygen demand for 5 days, *COD* chemical oxygen demand, *FOG* fat, oil, and grease, *TS* total solids, *TSS* total suspended solids, *TN* total nitrogen, *TP* total phosphorus, Gamma Per Milliliter

proposed for *Lactobacillus* cultivation with the purpose of reducing costs and increasing performance in their production (Oscar et al., 2019).

The availability of carbohydrate reservoir of lactose in whey and presence of other essential nutrients for the growth of microorganisms makes the whey one of the potential substrates for the production of different bioproducts through biotechnological means. Nadeem et al. (2016) investigated the applicability of whey water-waste as a nutritional source for cultivation of lactobacilli strains, which are potential probiotics. It was reported that, the growth of *Lactobacillus sporogenes* and *Lactobacillus acodophillus* in whey water medium was comparable to that of MRS medium. Therefore, whey water from cheese production industries can be used as a cost effective and cheap substitute for the growth of probiotic *Lactobacilli* (Nadeem et al. 2016).

3.3 Fish Processing Wastes

Both marine and freshwater fish processing generate wastes that include scales, skins, visceral mass (viscera, air bladder, gonads, and other organs), head, and fins. Unlike the marine fish processing sector, freshwater fish processing is not well organized and hence presents a different level of waste generation and disposal. Fish waste is generally not used for any alternative purpose and therefore dumped, which in turn leads to serious environmental problems (Arvanitoyannis & Kassaveti, 2008). Among different types of waste generated, fish viscera alone contributes 15–25% of the total body weight. Fish industry waste stands for a good source of recoverable biomolecules. Various researchers are focusing on methods to recover the biomolecules from fish waste in order to reduce organic loading on the environment as well as decrease the pollution related problems (Ghaly et al., 2019). The potential of fish waste effluent as a fermentation medium for production of antibacterial compounds, by lactic acid bacteria was evaluated by Tahajod and Rand (1996). Costas Malvido et al. (2018), produced probiotic biomass and synthesis of nisin by *L. lactis* by using culture media prepared from whey wastes (liquid remaining after the first cheese pressing) and mussel processing waste which were obtained from local dairy and mussel processing coastal plant (Costas Malvido et al., 2018).

3.4 Fermented Silages

Fish silage is a liquid product made from whole fish or fish waste plus acid. Natural methods of fermented fish silage production depend on the biological production of lactic acid by bacteria to lower the pH. In general, lactic acid bacteria such as *Lactobacillus plantarum* ferment sugars to organic acids (primarily lactic), thus lowering the pH of the mixture. Fish contain only small quantities of fermentable carbohydrates and it is usually necessary to add suitable carbohydrates for the bacteria to convert to acid (Palkar et al., 2017). Addition of mixtures of malt and cereal meal, molasses and cereal meal, malt and tapioca meal, and molasses and tapioca meal have all proved successful.

The fermentation process for conversion of carbohydrate to lactic acid involves the break down complex carbohydrates to maltose to glucose by alpha and beta amylase. Further, maltose is broken down to glucose by maltase. Finally, glucose is converted to lactic acid by bacteria. Small amounts of other substances such as acetic acid and alcohol are also formed during the process (Rehana et al., 2018).

Generally lactic acid bacteria can be divided into two types: (a) homofermentative, which convert one molecule of glucose to two of lactic acid, and (b) heterofermentative, which convert one molecule of glucose to one molecule of lactic acid plus ethyl alcohol and water. It is, therefore, better to use a homofermentative bacterium if possible. Since fish do not contain many lactic acid bacteria themselves, it is essential to add a starter culture, usually of lactobacilli, for successful fermentation. In addition, it is also necessary to add a source of amylase since the first step in the fermentation relies on the hydrolysis of carbohydrate (Ozyurt et al., 2017). In most processes, the amylase is provided by the addition of malt to the mixture since malt is a rich source of amylases. The fermentation should be carried out in full airtight containers so that conditions are anaerobic and successful fermentation is indicated by a rapid drop in pH, as the lactic acid is formed, and the production of gas (Akhtar et al., 2016).

3.5 Wastes from Meat Processing

The majority of the waste, in the meat industry is produced during slaughtering. Slaughter house waste consists of the portion of a slaughtered animal that cannot be sold as meat or used in meat products. Such waste includes bones, tendons, skin, contents of the gastrointestinal tract, blood, and internal organs. Such wastes from the meat processing industry contain a large number of organic compounds, which makes their disposal an environmental issue. Wastewater from meat processing plants contains highly concentrated fats, nitrogen, phosphorus, and potassium. Therefore the meat extracts and waste water could be effectively utilized for fermentation medium to grow aerobic and anaerobic microbes (Jayathilakan et al., 2012).

Fermented sausages (such as dry sausages) are favorable meat products with probiotic bacteria and such meat products has been shown to be one of the good means for the supply of probiotic bacteria to the host. The most commonly used species of probiotic microorganisms in fermented meat products are: *Lactobaccillus casei, L. paracasei, L. plantarum, L. rhamnosus, L. sakei, Pediococcus acidilactici,* and *P. pentosaceus* (Aguero et al., 2020).

4 Municipal Wastes as Substrate

Municipal waste is the trash that is collected by every city and town across each nation. Organic components make up over half of municipal waste, and include a diverse biosolids ranging from lawn clippings and trimmed branches, to spoiled food and a wide assortment of paper and cardboard products (Alam & Kulkarni, 2016). Organic waste is being generated daily and is so abundant that dealing it is challenging. Much of the municipal waste is from the house hold waste such as kitchen waste (Ferronato & Torretta, 2019). Kitchen waste is a rich-nutrition resource. It contains sugar, lipids, proteins, cellulose, and other compounds (Wang, 2012). Yin et al. (2013) used five strains of microorganisms including one strain of *Lactobacillus*, two strains of Bacillus, and three strains of yeast, respectively and were mixed at the same ratio and cultured using the kitchen waste as culture medium at pH of 7.2 and temperature of 37 °C. After 24 h, the total count of the viable cells reached 2.24×10^{10} CFU/g, which was higher than that obtained in any single probiotic strain pure culture. It was found that the presence of yeasts and *Bacillus* species enhanced the growth of *Lactobacillus* strain. Bench scale experiments were also done in a self-designed rotating drum type bioreactor. The experimental results indicate that there was a good possibility of utilizing kitchen waste for the economic production of probiotics (Yin et al., 2013). Surplus bread is considered a major waste problem for bakeries and food retailers. While some unsold bread is donated to charities, most are resold as low-value animal feed. A team of food scientists from the National University of Singapore (NUS) found a solution to reduce bread waste by using a novel fermentation process to 'upcycle' surplus bread into a beverage fortified with gut-friendly microorganisms.

Biotechnology Resource Centre (BRC), Mumbai, novel three-tier state-of-the-art biotechnology treatment, the municipal biosolid waste is now converted into useful soil conditioners and highly nutritive biofertilizer. Specialized microbial brigades boost the composting process and successfully convert the biosolid waste in to probiotic soil conditioner. Biomass is inoculated with the specific microbial biocultures including degraders, deodorizers, enrichers, fixers, and solubilizers, with inoculation of nitrogen fixers, phosphate solubilizers, and potassium enrichers to attain minimum probiotic count or C.F.U. of 2×10^{14}. The probiotic soil conditioners developed using specialty treatment developed by BRC, have been successfully tested for their increase in the productivity along with biosafety on Sweet Sorghum crop at field level (Rastogi et al., 2020).

References

Aguero, d. L. N., Frizzo, L. S., Ouwehand, A. C., Aleu, G., & Rosmini, M. R. (2020). Technological characterisation of probiotic lactic acid bacteria as starter cultures for dry fermented sausages. *Food, 9*(5), 596.

Ahlberg, S. H., Joutsjoki, V., & Korhonen, H. J. (2015). Potential of lactic acid bacteria in aflatoxin risk mitigation. *International Journal of Food Microbiology, 207*, 87–102.

Akhtar, A., Zamal, H., Naser, M. N., Islam, M. S., Bhuyan, M. S., & Fakruddin, M. (2016). Production of microbial silages from animal wastes as fishmeal replacer in the aquaculture diets. In *Proceeding of the 3rd International Conference on Fisheries and Aquaculture* (Vol. 3, pp. 8–22).

Alam, T., & Kulkarni, K. (2016). Municipal solid waste management and its energy potential in Roorkee City, Uttarakhand, India. *Journal of the Institution of Engineers India Series, 97*, 9–17.

Arvanitoyannis, I. S., & Kassaveti, A. (2008). Fish industry waste: Treatments, environmental impacts, current and potential uses. *International Journal of Food Science & Technology, 43*, 726–745.

Ashraf, R., & Shah, N. P. (2014). Immune system stimulation by probiotic microorganisms. *Critical Reviews in Food Science and Nutrition, 54*, 938–956.

Awad, W., Ghareeb, K., Abdel-Raheem, S., & Bohm, J. (2009). Effects of dietary inclusion of probiotic and synbiotic on growth performance, organ weights, and intestinal histomorphology of broiler chickens. *Poultry Science, 88*, 49–56.

Bhuvaneshwari, S., Hettiarachchi, H., & Meegoda, J. N. (2019). Crop residue burning in India: Policy challenges and potential solutions. *International Journal of Environmental Research and Public Health, 16*(5), 832.

Campos, D. A., Gomez-Garcia, R., Vilas-Boas, A. A., Madureira, A. R., & Pintado, M. M. (2020). Management of fruit industrial by-products—A case study on circular economy approach. *Molecules, 25*(2), 320.

Cheng, H., & Hu, Y. (2010). Municipal solid waste (MSW) as a renewable source of energy: Current and future practices in China. *Bioresource Technology, 101*, 3816–3824.

Collins, J. W., La Ragione, R. M., Woodward, M. J., & Searle, L. E. J. (2009). Application of prebiotics and probiotics in livestock. In D. Charalampopoulos & R. A. Rastall (Eds.), *Prebiotics and probiotics science and technology*. Springer. https://doi.org/10.1007/978-0-387-79058-9_30

Costas Malvido, M., Alonso Gonzalez, E., Outeirino, D., & Perez Guerra, N. (2018). Production of a highly concentrated probiotic culture of *Lactococcus lactis* CECT 539 containing high amounts of nisin. *3 Biotech, 8*(7), 292.

Cristian, O. (2010). Characteristics of the untreated wastewater produced by food industry. *Analele Universității din Oradea, Fascicula: Protecția Mediului, 15*, 709–714.

Ezejiofor, T. N., Enebaku, U. E., & Ogueke, C. (2014). Waste to wealth-value recovery from agrofood processing wastes using biotechnology: A review. *British Biotechnology Journal, 4* (4), 418–481.

Falagas, M. E., Betsi, G. I., & Athanasiou, S. (2007). Probiotics for the treatment of women with bacterial vaginosis. *Clinical Microbiology and Infection, 13*(7), 657–664.

Farees, N., Abateneh, D. D., Geneto, M., & Naidu, N. V. (2017). Evaluation of banana peel waste as growth medium for probiotic *lactobacillus species*. *International Journal of Applied Biology and Pharmaceutical Technology, 8*(4), 19–23.

Ferronato, N., & Torretta, V. (2019). Waste mismanagement in developing countries: A review of global issues. *International Journal of Environmental Research and Public Health, 16*(6), 1060.

Fochesato, A. S., Galvagno, M. A., Cerrutti, P. C., & Gonzalez Pereyra, M. L. (2018). Optimization and production of probiotic and antimycotoxin yeast biomass using bioethanol industry waste *via* response surface methodology. *Advances in Biotechnology & Microbiology, 8*(1), 555727.

Ghaly, A. E., Ramakrishnan, V., Brooks, M. S., Budge, S. M., & Dave, D. (2019). Fish processing wastes as a potential source of proteins, amino acids and oils: A critical review. *Journal of Microbial and Biochemical Technology, 5*, 107–129.

Gowe, C. (2015). Review on potential use of fruit and vegetables by-products as a valuable source of natural food additives. *IISTE-Food Science and Quality Managemen, 45*, 47–61.

Gumisiriza, R., Hawumba, J. F., & Okure, M. (2017). Biomass waste-to-energy valorization technologies: A review case for banana processing in Uganda. *Biotechnology for Biofuels, 10*, 11.
Hardia, S., & Iqbal, S. (2014). Production of the best natural health supplements using fruit waste materials. *International Journal of Innovative Research & Development, 3*(5), 131–133.
Hempel, S., Newberry, S. J., & Maher, A. R. (2012). Probiotics for the prevention and treatment of antibiotic-associated diarrhea a systematic review and meta-analysis. *The Journal of the American Medical Association, 307*(18), 1959–1969.
Hu, Y., He, Y., & Gao, S. (2020). The effect of a diet based on rice straw co-fermented with probiotics and enzymes versus a fresh corn Stover-based diet on the rumen bacterial community and metabolites of beef cattle. *Scientific Reports, 10*, 10721.
Isolauri, E., Arvola, T., Sutas, Y., Moilanen, E., & Salminen, S. (2000). Probiotics in the management of atopic eczema. *Clinical and Experimental Allergy, 30*(11), 1604–1610.
Jayathilakan, K., Sultana, K., Radhakrishna, K., & Bawa, A. S. (2012). Utilization of byproducts and waste materials from meat, poultry and fish processing industries: A review. *Journal of Food Science and Technology, 49*, 278–293.
Karadag, D., Koroglu, O. E., Ozkaya, B., & Cakmakci, M. (2015). A review on anaerobic biofilm reactors for the treatment of dairy industry wastewater. *Process Biochemistry, 50*, 262–271.
Kechagia, M., Basoulis, D., Konstantopoulou, S., Dimitriadi, D., Gyftopoulou, K., Skarmoutsou, N., & Fakiri, E. M. (2013). Health benefits of probiotics: A review. *ISRN Nutrition, 2013*, 481651. https://doi.org/10.5402/2013/481651
Kumar, M., Kumar, A., & Nagpal, R. (2010). Cancer-preventing attributes of probiotics: An update. *International Journal of Food Sciences and Nutrition, 61*(5), 473–496.
McFarland, L. V. (2007). Meta-analysis of probiotics for the prevention of traveler's diarrhea. *Travel Medicine and Infectious Disease, 5*(2), 97–105.
Naggar, N.-A., Abdelwahed, N. A., Saber, W. I., & Mohamed, A. A. (2014). Bioprocessing of some agro industrial residues for endoglucanase production by the new subsp.; Streptomyces albogriseolus subsp. cellulolyticus strain NEAE-J. *Brazilian Journal of Microbiology, 45*(2), 743–756.
Nadeem, F. A., Chetana, A. E., Prasad, M., Birajdar, R., & Naidu, N. V. (2016). Evaluation of Whey water as growth medium for Lactobacillus species. *International Journal of Applied Biology and Pharmaceutical Technology, 8*(1), 38–42.
Negi, S., & Anand, N. (2016). Factors leading to losses and wastage in the supply chain of fruits and vegetables sector in India. *Conference Proceedings of International Conference on Management of Infrastructure (ICMI) 2016 Energy, Infrastructure and Transportation: Challenges and Way Forward*, 89–105.
Oscar, J. Sanchez., Pedro, J. Barragan., & Liliana, Serna. (2019). Review of Lactobacillus in the food industry and their culture media. *Revista Colombiana de Biotecnologia*, 21, 63–76.
Ozyurt, G., Ozkutuk, A., Boga, M., Durmus, M., & Boga, K. E. (2017). Biotransformation of seafood processing wastes fermented with natural lactic acid bacteria; the quality of fermented products and their use in animal feeding. *Turkish Journal of Fisheries and Aquatic Sciences, 17*, 543–555.
Palkar, N., Koli, J., Patange, S. B., Sharangdhar, S. T., Sadavarte, R. K., & Sonavane, A. E. (2017). Comparative study of fish silage prepared from fish market waste by using different techniques. *International Journal of Current Microbiology and Applied Sciences, 17*(6), 3844–3858.
Pyar, H., Liong, M.-T., & Peh, K. K. (2014). Potentials of pineapple waste as growth medium for lactobacillus species. *International Journal of Pharmacy and Pharmaceutical Sciences, 6*(1), 142–145.
Rastogi, M., Nandal, M., & Khosla, B. (2020). Microbes as vital additives for solid waste composting. *Heliyon, 6*(2), e03343.
Rehana, R., Raju, C. V., Lakshmisha, I. P., & Jagpal. (2018). Nutritional and biochemical properties of fish silage prepared as an ingredient in poultry feed. *International Journal of Current Microbiology and Applied Sciences, 7*(5), 423–428.

Sadh, P. K., Kumar, S., Chawla, P., & Duhan, J. S. (2018). Fermentation: A boon for production of bioactive compounds by processing of food industries wastes (by-products). *Molecules, 23*(10), 2560.

Sagar, N. A., Pareek, S., Sharma, S., Yahia, E. M., & Lobo, M. G. (2018). Fruit and vegetable waste: Bioactive compounds, their extraction, and possible utilization. *Comprehensive Reviews in Food Science and Food Safety, 17*, 512–531.

Sah, B. N., Vasiljevic, T., McKechnie, S., & Donkor, O. N. (2016). Effect of pineapple waste powder on probiotic growth, antioxidant and antimutagenic activities of yogurt. *Journal of Food Science and Technology, 53*(3), 1698–1708.

Salah, E., & Hala, O. (2017). Sustainable and cost-effective use of organic waste. *Current Trends in Biomedical Engineering & Biosciences, 7*(4), 74–79.

Sidana, A., & Farooq, U. (2014). Sugarcane bagasse: A potential medium for fungal cultures. *Chinese Journal of Biology, 2014*, 840505. https://doi.org/10.1155/2014/840505. 5 pages.

Slavov, A. K. (2017). General Characteristics and Treatment Possibilities of Dairy Wastewater – A Review. *Food Technology and Biotechnology, 55*(1), 14–28.

Soren, J. P., Paul, T., & Banerjee, A. (2020). Exploitation of agricultural waste as sole substrate for production of bacterial L-glutaminase under submerged fermentation and the proficient application of fermented hydrolysate as growth promoting agent for probiotic organisms. *Waste and Biomass Valorization, 11*, 4245–4257.

Tahajod, A. S., & Rand, A. G. (1996). Seafood waste potential to support antimicrobial compound production by lactic acid bacteria. In *IFT Annual Meeting: Book of Abstracts* (pp. 31–32). https://www.newfoodmagazine.com/news/108638/food-scientists–upcycle–bread–waste–into–nondairy probiotic-drink/

Tsang, Y. F., Kumar, V., PallabiSamadar, Y. Y., Lee, J., Ok, Y. S., Song, H., Kim, K.-H., Kwon, E. E., & Jeon, Y. J. (2019). Production of bioplastic through food waste valorization. *Environment International, 127*, 625–644.

Vandenberghe, L. P. S., Pandey, A., & Carvalho, J. C. (2021). Solid-state fermentation technology and innovation for the production of agricultural and animal feed bioproducts. *Systems Microbiology and Biomanufacturing, 1*, 142–165. https://doi.org/10.1007/s43393-020-00015-7

Vicenssuto, G. M., & De Castro, R. J. S. (2020). Development of a novel probiotic milk product with enhanced antioxidant properties using mango peel as a fermentation substrate. *Biocatalysis and Agricultural Biotechnology, 24*, 101564.

Vieco-Saiz, N., Belguesmia, Y., & Raspoet, R. (2019). Benefits and inputs from lactic acid bacteria and their bacteriocins as alternatives to antibiotic growth promoters during food-animal production. *Frontiers in Microbiology, 10*, 57.

Vrese, M., Stegelmann, A., Richter, B., Fenselau, S., Laue, C., & Schrezenmeir, J. (2001). Probiotics compensation for lactase insufficiency. *American Journal of Clinical Nutrition, 73* (2), 421S–429S.

Wang, Z. G. (2012). *Study on utilizing probiotics to treat kitchen wastes*. MS thesis, University of Science and Technology Beijing, Beijing, China.

Yahaya, S. M., & Mardiyya, A. Y. (2019). Review of post-harvest losses of fruits and vegetables. *Biomedical Journal of Scientific & Technical Research, 13*(4), 002448.

Yin, C., Dong, X., & Lv, L. (2013). Economic production of probiotics from kitchen waste. *Food Science and Biotechnology, 22*, 59–63.

Zoghi, A., Khosravi-Darani, K., & Sohrabvandi, S. (2014). Surface binding of toxins and heavy metals by probiotics. *Mini Reviews in Medicinal Chemistry, 14*, 84–98.

Agro-Industrial Waste as an Option for the Sustainable Development of Bioplastic

María Antonieta Riera and Silvina Maldonado

1 Introduction

Plastic is a material whose characteristics favor its application in different sectors, as reflected in the increasing levels of sales reported annually by this industry. Most plastics are of synthetic origin, that is, they are synthesized from raw materials from gas or oil. However, in recent years, the environmental commitment has motivated the development of new materials from raw materials that are sustainable. One option is waste of agro-industrial origin, which, being organic in nature, has a wide variety of useful compounds for obtaining biomaterials. In this regard, there are some investigations where the use of some agro-industrial waste is recorded to obtain bioplastics. This, in addition to being a solution to the environmental problems caused by these wastes, is an alternative for their revaluation in the framework of the circular economy and the bioeconomy.

M. A. Riera (✉)
Facultad de Ciencias Matemáticas, Físicas y Químicas, Universidad Técnica de Manabí, Portoviejo, Ecuador

Doctorado en Ingeniería Industrial, Universidad Nacional de Cuyo, Mendoza, Argentina
e-mail: maria.riera@utm.edu.ec

S. Maldonado
Facultad de Ingeniería Industrial, Universidad Nacional de Jujuy, San Salvador de Jujuy, Argentina
e-mail: silvinamaldonado@fi.unju.edu.ar

N. R. Maddela, L. C. García (eds.), *Innovations in Biotechnology for a Sustainable Future*, https://doi.org/10.1007/978-3-030-80108-3_7

2 Consumer Society Versus Sustainable Production

Since the industrial revolution and up to the present, scientific and technological advances have made possible the development of new products to meet the needs of consumers. Today there is a great variety of articles on the market, in different brands, models, and prices, accessible to most of the population. But this wide variety of available products, together with the concept of planned obsolescence adopted as a commercial strategy in almost all value chains, has fostered the establishment of consumer societies.

The creation of a fast-fashion society, controlled by the different types of obsolescence, leads to improve the characteristics of new products concerning for to their predecessors (function obsolescence), to design products that wear out shortly after the end of a minimal warranty (planned obsolescence) and systematically educating consumers to appreciate the newest as the best (obsolescence of desirability), leads to disposable patterns of behavior, accelerates resource depletion and contributes to pollution environmental (Hellmann & Luedicke, 2018). The rise of these consumer societies has made the tendency to reuse objects that have already served their usefulness disappears, which in some way is a way to value them and reduce the rate of waste generation (Kedzierski et al., 2020).

This consumer behavior beyond generating sales revenue, makes manufacturers become environmental aggressors, by progressively requiring more resources to meet the demands of their production processes. Man has been based on the consumption of this material, first experimenting with natural polymers, horn, waxes, natural rubber, and resins, until the nineteenth century, when the development of modern thermoplastics began (Andrady & Neal, 2009). Worldwide, the per capita consumption of plastic was 11 kg in 1980, 30 kg for 2005, and it was estimated that it was 45 kg for 2015; with greater participation for the countries of the NAFTA zone and Western Europe with a per capita consumption of 139 and 136 kg, respectively (PlasticsEurope, 2008).

Since its inception, the plastics industry has been in constant growth, registering for the year 1950 a production of 1.5 million metric tons (MMt) and for the year 2018, a total of 359 MMt produced (Statista, 2020). In recent years, this increase is influenced by single-use plastics, invented for modern society with the purpose of use and disposal, whose main application is food packaging, shopping bags, or disposable tableware (Chen et al., 2021). Both the production and consumption of this material contribute greatly to environmental deterioration, not only due to the number of fossil resources required for the manufacturing process but also due to the various pollutants that are released into ecosystems during disposal end of wastes of this type and their prolonged permanence in them, given their slow degradation process.

In the face of existing environmental pressures and the face of imminent climate change, international commitments have been made in favor of the planet. An example of this is the Sustainable Development Goals (SDGs) of the 2030 Agenda, where at least seven of them directly or indirectly address environmental issues.

Fig. 1 Types of bioplastics

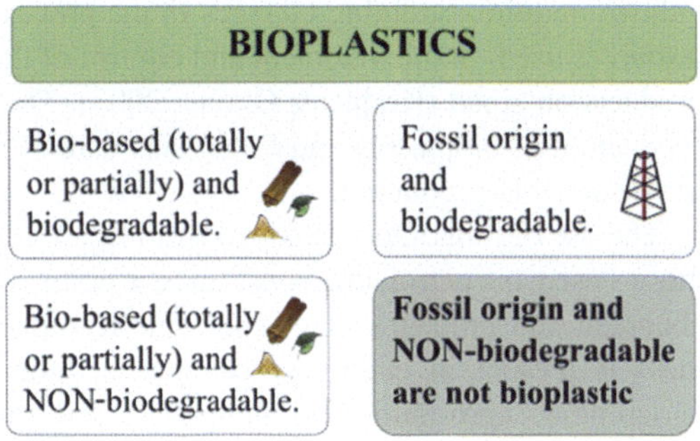

Specifically, objective 12 of responsible production and consumption aims to decouple economic growth from environmental degradation, while increasing resource efficiency and promoting sustainable lifestyles (ONU, 2018).

A strategy related to the above is the adoption of new economic models focused on circularity and the use of raw materials of biological origin, to obtain new products. Within this approach are bioplastics, which are defined as a plastic that is bio-based, biodegradable, or that meets both criteria (European Bioplastics, 2018), and given their characteristics, they offer the possibility of introducing an alternative to the problems caused by conventional plastics.

A bio-based plastic is obtained totally or partially from biomass (Fig. 1). It includes starch, cellulose, proteins, lignin, chitosan, polylactic acid (PLA), and polyhydroxyalkanoates (PHAs)/polyhydroxybutyrates (PHBs). A wide variety of biomass of plant origin (complete plants or their residues, wood, dry grass) or animal (for example, bird feathers) is used for its manufacture, which is a resource with great potential to be used, rich in carbon, capable of being processed by microbial methods, for the production of bio-based polymers, a mixture of biopolymers and various chemicals (Brodin et al., 2017a; Maraveas, 2020). The main bio-based plastics that are currently marketed are thermoplastic starch (TPS), polylactic acid (PLA), polyhydroxyalkanoates (PHAs), polyethylene (bio-PE), propylene (bio-PP), and polyethylene terephthalate (bio-PET) bio-based, containing at least some renewable carbon (Lackner, 2015).

For a plastic to be biodegradable, it must decompose by the action of microorganisms or suffer a decrease in its molecular weight due to biological activity, producing CO_2, H_2O, CH_4 (depending on the environment in which it is carried out), mineral salts, in addition to biomass (Reddy et al., 2013; Vert et al., 2012). It should be noted that the biodegradability of the material does not depend on the source of origin but the structure of the polymer. That is why there are biodegradable plastics of natural origin, synthesized from renewable resources and petroleum-based, but there are also bio-based nonbiodegradable plastics (Jiang & Zhang, 2017; Reddy et al., 2013).

The development of new bio-based materials under this bioeconomic model represents an important factor to achieve the sustainable growth of the bio-based plastics industry, which are also biodegradable. Sustainable development in any industry requires adopting changes in the processes, in the type and quantity of the resources used, in the treatment and control of the waste generated, as well as in the products obtained (Krajnc & Glavic, 2003). The bio-based industry and with it the production of bioplastics must also take care of its resources, processes, and waste, for sustainable production.

The use of monomers obtained from lignocellulose biomass as a replacement for those based on petroleum constitutes a point in favor of sustainability as it is an abundant and biodegradable renewable resource. Specifically, lignocellulosic biomass is present in energy crops, marine biomass, forestry, as well as forestry, agricultural, agro-industrial, industrial, and municipal solid waste (Al-Battashi et al., 2019). In this particular, the use of lignocellulosic waste is of special interest for its recovery through the obtaining of various products, one of them being bioplastics.

Regarding transformation processes, the production of bioplastics is carried out mainly by fermentation routes through biotechnological procedures, generally expensive and with a low performance from an industrial point of view. Efficient processes are required for the fractionation and purification of biomass, cost-effective routes for conversion to monomers and platform molecules (Brodin et al., 2017b), in addition to low-cost substrates, which could include some waste generated in the agro-industry.

Some pretreatments of lignocellulose biomass incorporate the use of steam or dilute acids and subsequent enzymatic treatments to break it up into simpler sugars. Although it is a practical approach, it represents economic limitations in large-scale processes. An important advance in the bioplastics industry is the development of a new bioprocessing system, which uses thermophilic microorganisms for the one-step conversion of lignocellulose into polyhydroxyalkanoate (PHA), excluding the chemical and enzymatic pretreatment steps (Govil et al., 2020).

Another consideration in the sustainable production of bioplastics from lignocellulosic waste is the energy requirements of the process. To reduce the consumption of public services in this area and satisfy the total heating requirements, an integrated process is proposed that takes advantage of the calorific value of the total biomass waste, the biogas generated by the wastewater treatment, in addition to a network of heat exchangers between hot and cold process streams (Kim et al., 2020).

Regarding the generation of waste, it is expected that bio-based plastics have a reduced carbon footprint about those produced from oil since they are in complete harmony with the rates and the time scale of the biological carbon cycle (Narayan, 2011). However, it should be considered that the use of food crops such as corn, sugar cane, rice, etc., in addition to competing with the population's food needs, represents a threat to the total substitution of plastic containers of fossil origin by bioplastics, since according to the evaluation of the impact of the life cycle of bioplastics in terms of greenhouse gas emissions and land and water environmental footprint, it would represent a considerable increase in the use of land and water (Brizga et al., 2020).

Waste treatment must address the degradation routes available for the bioplastic generated. Although there is the possibility of using mechanical recycling, or using

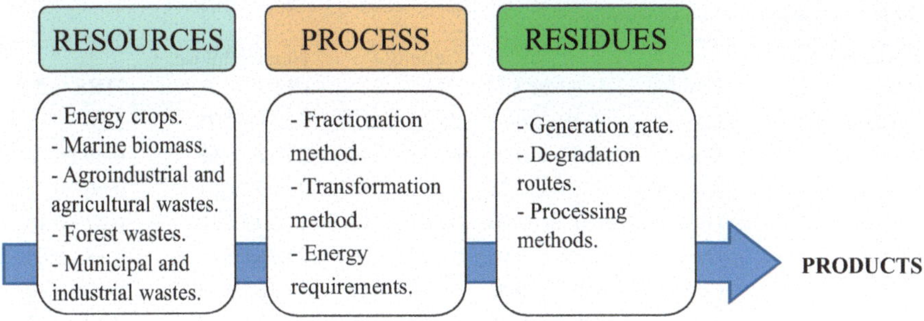

Fig. 2 Aspects in the sustainable production of bioplastics

chemical treatments, including hydrolysis, pyrolysis, or alcoholysis, to depolymerize bioplastics such as polylactic acid (PLA) and thus generate value-added products, biodegradation is one of the most discussed aspects for this type of material. In this topic, it is important to define the environmental conditions necessary to ensure the decomposition of the biopolymer. There are specific conditions such as temperature, humidity, presence of microorganisms, that favor or counteract this biological reaction.

To achieve sustainable and economically attractive production, biomass should not be wasted under any circumstances, but rather should be used in a closed loop so that all waste streams are reintroduced into the value chain with a new function (Márquez Luzardo & Venselaar, 2012). Figure 2 shows the aspects to consider for the sustainable production of bioplastics.

3 Valorization of Agro-Industrial Waste

Agro-industrial waste such as husks, seeds, whey, waste liquids, molasses, bagasse, among others, are generated in the processing of agricultural products (Panesar et al., 2015). The harvest is also part of the food production (Pfaltzgraff et al., 2013), therefore, the waste generated at this stage can be included within the agribusiness supply chain.

The Food and Agriculture Organization of the United Nations (FAO) in its record of burning crop residues, reported that in 2018 alone, about 460 MMt of dry biomass from four crops (rice husks) were burned, sugar cane, corn, and wheat), worldwide (Food and Agricultural Organization of the United Nations, 2020). All this available biomass, instead of being burned, could be used under a cascade economy model, to obtain a wide variety of products. The efficient use of biomass from both an ecological and economic point of view assumes that it must be used mainly in high-cost and low-volume applications, and then use the residues at a next level in applications, until reaching those of lower value and large volumes (Márquez Luzardo & Venselaar, 2012).

Table 1 Characterization of residues

Agricultural product	Residue	Amylose (%)	Cellulose (%)	Hemicellulose (%)	[a]TRS (g/L)
Plantain	Peel	8.59 ± 0.82	58.89 ± 0.76	3.47 ± 0.76	ND
Rice	Dust	0.87 ± 0.02	60.28±5.48	25.57±0.63	0.69±0.24
	Arrocillo	8.23 ± 0.11	40.19±3.16	23.75±1.27	0.66±0.22
	Husk	8.51 ± 0.003	45.67±3.16	28.50±2.19	0.87±0.10
Corn	Dust	22.35 ± 0.05	51.15±3.16	28.86±0.63	0.18±0.04
	Cob	6.93 ± 0.05	56.63±3.16	19.36±0.63	0.52±0.32
	Silks	6.92 ± 0.05	63.93±3.16	21.56±1.67	0.89±0.12
Cassava	Peel	20.2 ± 0.4	ND	ND	1.05 + 0.18

[a]For alkaline hydrolysis at 3% (w/v). *ND* not determined

Agro-industrial waste with a high production rate throughout the world, in addition to being biodegradable, has great potential as a primary or secondary raw material for the production of biopolymers, as they are rich in useful substances (such as fermentable sugars, carbohydrates, lipids, polysaccharides, pigments, and aromatic compounds) (Heredia-Guerrero et al., 2017; Panesar et al., 2015; Ranganathan et al., 2020). However, they are rarely recovered and, on the contrary, are disposed of without any type of control, generating damage to the environment and economic losses (Beltrán-Ramírez et al., 2019; Bilo et al., 2018).

From the lignocellulose material present in some food waste, cellulose, and hemicellulose fractions can be extracted (De et al., 2020). Also, agro-industrial waste rich in starch has the potential to be used in obtaining thermoplastic starch, polyhydroxyalkanoates, and PLA (Chan et al., 2021; Grewal et al., 2020; Tsang et al., 2019). Residues from banana, rice, corn, and cassava have shown the presence of compounds of this type, useful in the production of bioplastics (Table 1).

As in synthetic polymers made up of a chain of monomers, a starch polymer (composed of amylose and amylopectin) is made up of chains of sugar monomers connected by glucosidic bonds. Thus, bioplastic is a polymer made up of simple sugars and can be synthesized from bio-based materials. Starch-based bioplastics are a promising substitute due to the abundance, renewability, sustainability, and biodegradability of this compound (Samer et al., 2019; Shafqat et al., 2020). PLA, PHAs, and polybutylene succinate (PBS) are promising bioplastics with bio-based raw materials and biodegradability properties that are produced by bacterial fermentation of sugars from carbohydrate sources (Changwichan et al., 2018). On the other hand, lignocellulosic fibers have also been studied as reinforcing material in bioplastics, exhibiting properties that are compatible with the polymeric matrix and thus the possibility of replacing synthetic fibers in bioplastics (Yang et al., 2019).

The trend of agricultural waste as a source of bioplastics production is increasing due to the amount generated per year, its low cost, and availability. The limitations of use are related to the lack of standardized definitions for the management of food waste, the scarce information regarding the quantities generated and the low production performance compared to the food raw material. The outlook for the use of agricultural waste as a raw material in bioplastics is expected to improve, with the advancement of biotechnology, product life cycle analysis, prioritization of value chains, investments in a future circular economy, and the intervention of the government with the establishment of legislation that favors its use (Chan et al., 2021; Teigiserova et al., 2019).

4 Biorefineries and Transformation Processes

A biorefinery is an industrial facility (or network of facilities), which use a combined set of technologies and conversion routes, to use the available biomass comprehensively and sustainable, to simultaneously produce biofuels, energy, materials, and other chemicals with added value (Morais & Bogel-Lukasik, 2013). It consists of an industrial complex that emulates the processing carried out in a traditional refinery, but unlike this one, instead of using oil as raw material, it uses biomass from different sources.

Currently, most of the biorefineries operating and under construction are located in North America and European countries (IEA Bioenergy, 2021). Although it is a relatively new production model that is still under investigation, interest in its implementation increases every day given the need to replace fossil resources with others of renewable origin.

The main objective of biorefineries is to use biomass to produce small quantities of a greater quantity of bio-based products and downstream, to use secondary waste, for the production of energy destined for internal or external use (Bell et al., 2014). Depending on the raw material or processing technology used, biorefineries can be classified into lignocellulosic, whole culture, green, marine, platform, conventional, chemical, and thermochemical (Cherubini et al., 2009). Another classification indicates that biorefineries can be a first and second generation or integrated (Trigo et al., 2012). In Table 2, information corresponding to each of them is shown.

For the efficient transformation of biomass, in addition to pretreatment activities, two conversion methods are used: biochemical and thermochemical. Biochemical processes are carried out by the action of microorganisms (fungi, bacteria, and yeasts), through biochemical reaction mechanisms; consider fermentation and anaerobic digestion, which produces biofuels and other chemicals, as well as biogas and biofertilizer. Thermochemical transformations are carried out at high temperatures and include combustion, gasification, and pyrolysis processes to produce thermal energy, synthesis gas, and bioproducts (bio-oil and bio-carbon). The conversion rate in these processes depends on the operating conditions: temperature, pressure, feed rate, heating time, biomass particle size, catalytic activity (Ferreira, 2017; Vaz, 2019).

Bioplastic production would be a value-added co-product within biorefining, as happens in oil refineries with the production of plastics and chemicals. The challenge, in this case, is to find biomass compatible with the biomaterial to be obtained in a biorefinery scenario where the markets justify its production (Snell & Peoples, 2009). The implementation of a biorefinery platform that uses food waste as raw

Table 2 Classification of biorefineries

Type of biorefinery	Raw material used	Processing methodology	Products obtained	Advantages/disadvantages	Referencess
Lignocellulose (LC)	Lignocellulosic biomass in its natural state (straw, cane, grass, wood) or its residues.	Fractionation of biomass and extracting its compounds (cellulose, hemicellulose, and lignin).	Biomaterials, chemical compounds, biofuels, and energy.	Lignocellulosic biomass is an abundant material with great potency. Conversion products have a good position concerning to those of traditional petrochemicals.	Ferreira (2017), Kamm and Kamm (2004), Trigo et al. (2012)
Whole crop	Whole grain crops (rye, wheat, triticale, corn).	Mechanical separation of grain and straw, mainly by dry or wet milling.	Fuels, bioplastics, adhesives, binders, cement, modified starch, and other high value-added products.	The straw obtained can be processed in an LC type biorefinery. It is also a starting material for the production of synthesis gas using pyrolysis technologies. The whole crop can compete with the feeding needs of the population.	Kamm and Kamm (2004), Salazar and Cárdenas (2013)
Green	Wet biomass such as grass from the cultivation of permanent pastures, closed fields, canned or green crops (alfalfa, clover), immature cereals from extensive crops.	Wet fractionation to isolate components, generating as waste products a cake rich in fiber and a green juice rich in nutrients.	Dyes and pigments, proteins, enzymes, lactic acids, fibers, and energy.	The exploitation of grasslands can have secondary effects related to the conservation of landscapes.	Kamm and Kamm (2004), Kromus et al. (2004)
Marine	Marine biomass (plants, algae).	Microalgae cultures in open ponds or photobioreactors. For dehydration of the biomass, flocculation,	Bioproducts (biofuels, proteins, polysaccharides, pigments, biopolymers, fertilizers) and energy.	Marine biomass does not compete with arable land and has high productivity. They do not compete with food production by using	Ferreira (2017), Harun et al. (2010)

		centrifugation, and filtration are used.		marginal areas unsuitable for agriculture. The cultivation and processing of this type of biomass are still under investigation.	
Platform	Sugars and syngas platforms.	The biomass is subjected to gasification (extreme temperature process in the presence of oxygen or air).	After syngas is cleaned, it can be used to produce energy, bio-alcohols, or fuels.	The current fermentation processes have been developed from edible crops rich in sugar, with a considerable loss of carbon in the fermentation process is substantial. Improvements in conversion processes are required.	Jung et al. (2020), Salazar and Cárdenas (2013), Trigo et al. (2012)
Conventional	The ones used in existing industries: beet, cane (sugar industry), wheat, cassava, potato (starch industry), soybeans, rapeseed, sunflower (vegetable oil industry), forest resources (pulp and paper industry).	Physical and mechanical operations and chemical transformations are required in the industry depending on the process.	Sugar, molasses, starch, vegetable oils and fats, pulp, and paper.	Efforts are required to produce a wide variety of value-added products (biochemicals or biofuels). It does not fit the new biorefinery concept.	Diep et al. (2012), Cherubini et al. (2009)
Chemical and thermochemical	All types of biomass.	Based on a mixture of various technologies (torrefaction, pyrolysis, gasification, HTU, product separation, catalytic synthesis).	Solvents, a variety of chemicals, biofuels, energy.	The production of various types of products makes it difficult to calculate energy efficiency, assess the sustainability and profitability of the biorefinery. Its design requires the	Diep et al. (2012), Haro et al. (2014)

(continued)

Table 2 (continued)

Type of biorefinery	Raw material used	Processing methodology	Products obtained	Advantages/disadvantages	Referencess
				implementation of tools and perspectives different from biorefineries aimed at obtaining a single product.	-
First generation	Sugars and oils from agricultural resources through the use of the entire plant.	Processes with biotransformations (enzymology, microorganisms) and fermentation.	Biofuels (ethanol and biodiesel) and some chemical products.	These biorefineries are not designed for efficient use of biomass, minimize energy consumption, and recycle waste. The biomass used can compete with food production.	Octave and Thomas (2009), Salazar and Cárdenas (2013), Trigo et al. (2012)
Second generation	Lignocellulose biomass.	Processes with biotransformations (enzymology, microorganisms) and fermentation.	Biofuels (ethanol and biodiesel) and some bioproducts.	It seeks to increase sustainability, by maximizing the use of biomass and energy efficiency. Biomass works in symbiosis, so as not to compete with the food sector.	Octave and Thomas (2009), Salazar and Cárdenas (2013), Trigo et al. (2012)
Integrated	Different from biomass (energy crops, agricultural, forestry, industrial, and municipal waste).	Various thermochemical (combustion, gasification, pyrolysis) or biochemical (anaerobic digestion, fermentation) conversion technologies.	Energy, fuels, chemical products, polymers, and chemical platforms.	The wide range of manufactured products optimizes the use of raw materials and improves the economy of the process. On the other hand, the development of specific technologies and the adaptation of raw materials for their proper use are required.	Ferreira (2017), Maity (2015), Salazar and Cárdenas (2013), Trigo et al. (2012)

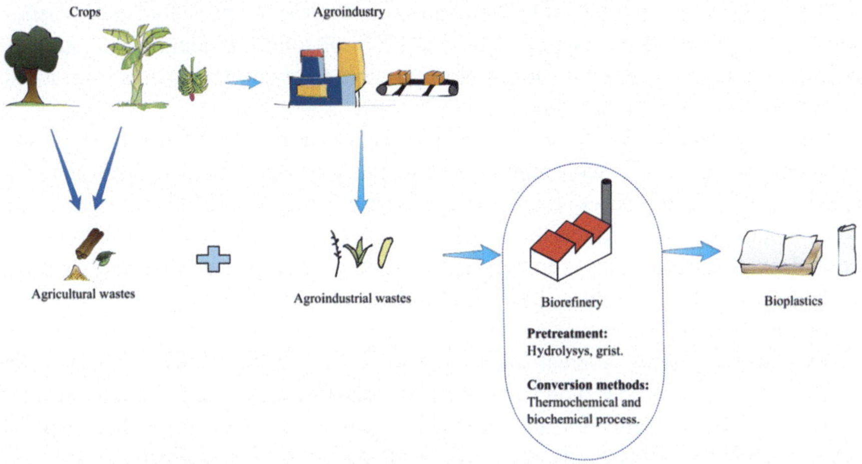

Fig. 3 Biorefining for the production of bioplastics

material is an interesting option (Tsang et al., 2019). This concept is possible for the production of thermoplastic starches, as well as bacterial polymers. The selection of biomass, the processing methodologies, and the correct integration in products and co-products to be obtained, would contribute to a viable biorefinery from an economic and environmental point of view.

It is aimed at an integrated biorefinery, where all the flows generated in the process are used. Concerning agribusiness, waste from the starch industry (reject raw materials, shells, seeds) could be used to extract the remaining starch and use it in the manufacture of thermoplastic starch. A similar case would happen in the manufacture of juices and beverages, where it is possible to use the waste generated (seeds, peels, skins) to obtain the sugars present and use them as platform molecules or lowcost substrates in subsequent fermentation processes. A representation of what has been described is shown in Fig. 3.

Biorefining is a promising concept that seeks to close loops to value biomass in a circular economy framework and comprehensively address the economic, environmental, and social aspects of the industrial sector (Lindorfer et al., 2019). Although it represents a great challenge, adequate integration of technologies and raw materials will allow the establishment of future sustainable production chains for the production of different bioproducts, which are also competitive in the market and which lead to the progressive substitution of the products obtained by the industry oil company (Cherubini, 2010; Ubando et al., 2020).

5 The Future of Bioplastics

The use of environmentally friendly materials is a topic that has gained interest in recent years, both at an environmental and commercial level. The plastics industry immersed in this reality has shown some experiences in which petroleum derivatives are being replaced by others of renewable origin, to produce partially based plastics such as biopolyethylene terephthalate (bio-PET) or totally biodegradable such as polylactic acid. In other cases, it has been possible to obtain bioplastics from fossil resources, such as polycaprolactone.

Currently, bioplastics have a market share close to 1% of the total plastic produced in the world, with annual growth rates between 20 and 30% and this behavior is expected to continue for the next 5 years. Dadas las altas tasas de crecimiento en el uso de ácido poliláctico, polipropileno de base biológica y polihidroxialcanoatos, se estima que la producción de este material pase de 2,11 MMt en el año 2020 a 2,87 MMt para el año 2025 (European Bioplastics, 2020; Lackner, 2015).

But this represents a great challenge not only for the industry but for the community in general. Beyond the biomass transformation technologies and the operational requirements for the production processes to be profitable, aspects related to the environmental management of the waste generated must be taken into account, the establishment of clear legislation by the governments, as well as consumer trends by users.

On the environmental side, it must be taken into account that not all bioplastics are biodegradable. About half of the current bioplastics market is nonbiodegradable and their end-of-life disposal will be problematic if not properly addressed. The main nonbiodegradable bioplastics are biopolyethylene, biopolypropylene, biopolyethylene terephthalate, and biopolyamide. For biopolyethylene, biopolypropylene, given their chemical structure, they recommend using them as raw materials for catalytic pyrolysis and from them producing liquid hydrocarbons. Instead, for biopolyethylene and PLA, they suggest it as potential raw materials for the gasification process (Rahman & Boi, 2021). It should be remembered that bio-based materials or biodegradable materials such as PLA have great potential to be compostable (Sidek et al., 2019).

Knowledge about the biodegradability of bioplastics is a starting point for legislators to assess environmental impacts and create legislation to limit these impacts as much as possible. Within this context, worldwide guidelines should be developed for the production and use of bioplastics, as well as waste management (Polman et al., 2021; Sidek et al., 2019).

As far as users are concerned, communication is needed between societies and markets on how to put bioplastics into service in the future. It is necessary to promote sustainable consumer behavior and determine the factors that influence their behavior in terms of purchasing the purchases they make so that these revolve around sustainability (Thakur et al., 2018; Zwicker et al., 2020).

Finally, bioplastics represent an opportunity to tackle the problem caused by resource depletion and plastic pollution. Current obstacles and challenges must be overcome to reach the production goals foreseen for the coming years.

6 Conclusions

The use of agro-industrial waste for the production of bioplastic is an option that has gained interest in recent years, as a strategy for its recovery and to obtain an emerging material that is respectful with the environment. Although the current market for bioplastics is low when compared to the production of synthetic plastic, there are some experiences of biorefineries with favorable results for the production of bioplastics. Future trends indicate that the production of bioplastics will increase in the coming years, to progressively replace traditional plastics. For this reason and given the composition of agribusiness residues, they represent an attractive option for them to become their raw material, within a framework of circular economy and bioeconomy.

References

Al-Battashi, H. S., Annamalai, N., Sivakumar, N., Al-Bahry, S., Tripathi, B. N., Nguyen, Q. D., & Gupta, V. K. (2019). Lignocellulosic biomass (LCB): A potential alternative biorefinery feedstock for polyhydroxyalkanoates production. *Reviews in Environmental Science and Biotechnology, 18*(1), 183–205.

Andrady, A. L., & Neal, M. A. (2009). Applications and societal benefits of plastics. *Philosophical Transactions of the Royal Society B: Biological Sciences, 364*(1526), 1977–1984. https://doi.org/10.1098/rstb.2008.0304

Bell, G., Schuck, M., Jungmeier, S., Wellisch, G., Felby, C., Jørgensen, H., Stichnothe, M., Clancy, H., De Bari, S., Kimura, I., van Ree, R., & Jong, D. (2014). *IEA bioenergy task42 biorefining*. IEA Bioenergy.

Beltrán-Ramírez, F., Orona-Tamayo, D., Cornejo-Corona, I., González-Cervantes, J. L. N., de Jesús Esparza-Claudio, J., & Quintana-Rodríguez, E. (2019). Agro-industrial waste revalorization: The growing biorefinery. In *Biomass for bioenergy—Recent trends and future challenges*. IntechOpen.

Bilo, F., Pandini, S., Sartore, L., Depero, L. E., Gargiulo, G., Bonassi, A., Federici, S., & Bontempi, E. (2018). A sustainable bioplastic obtained from rice straw. *Journal of Cleaner Production, 200*, 357–368. https://doi.org/10.1016/j.jclepro.2018.07.252

Brizga, J., Hubacek, K., & Feng, K. (2020). The unintended side effects of bioplastics: Carbon, land, and water footprints. *One Earth, 3*(1), 45–53. https://doi.org/10.1016/j.oneear.2020.09.004

Brodin, M., Opedal, M. T., Opedal, M. T., & Chinga-carrasco, G. (2017a). Lignocellulosics as sustainable resources for production of bioplastics. *Journal of Cleaner Production, 162*, 646–664. https://doi.org/10.1016/j.jclepro.2017.05.209

Brodin, M., Vallejos, M., Opedal, M. T., Area, M. C., & Chinga-Carrasco, G. (2017b). Lignocellulosics as sustainable resources for production of bioplastics—A review. *Journal of Cleaner Production, 162*, 646–664. https://doi.org/10.1016/j.jclepro.2017.05.209

Chan, J. X., Wong, J. F., Hassan, A., & Zakaria, Z. (2021). 8—Bioplastics from agricultural waste. In N. Saba, M. Jawaid, & M. Thariq (Eds.), *Biopolymers and biocomposites from agro-waste for packaging applications* (pp. 141–169). Woodhead Publishing.

Changwichan, K., Silalertruksa, T., & Gheewala, S. H. (2018). Eco-efficiency assessment of bioplastics production systems and end-of-life options. *Sustainability (Switzerland), 10*(4), 952. https://doi.org/10.3390/su10040952

Chen, Y., Awasthi, A. K., Wei, F., Tan, Q., & Li, J. (2021). Single-use plastics: Production, usage, disposal, and adverse impacts. *Science of the Total Environment, 752*, 141772.

Cherubini, F. (2010). The biorefinery concept: Using biomass instead of oil for producing energy and chemicals. *Energy Conversion and Management, 51*(7), 1412–1421. https://doi.org/10.1016/j.enconman.2010.01.015

De, D., Sai, M. S. N., Aniya, V., & Satyavathi, B. (2020). Strategic biorefinery platform for green valorization of agro-industrial residues: A sustainable approach towards biodegradable plastics. *Journal of Cleaner Production, 290*, 125184. https://doi.org/10.1016/j.jclepro.2020.125184

Diep, N. Q., Sakanishi, K., Nakagoshi, N., Fujimoto, S., Minowa, T., & Xuan, D. T. (2012). Biorefinery: Concepts, current status, and development trends. *International Journal of Biomass & Renewables, 1*(2), 1–8.

European Bioplastics. (2018). What are bioplastics? *Bioplastics*. https://www.european-bioplastics.org/bioplastics/

European Bioplastics. (2020). Bioplastics market development update 2020. Berlin.

Ferreira, A. F. (2017). Biorefinery concept. In *Biorefineries* (Lecture notes in energy). Springer International Publishing.

Food and Agricultural Organization of the United Nations. (2020). FAOSTAT. *Burning—Crop residues*. http://www.fao.org/faostat/en/#data/GB

Cherubini, F., Wellisch, M., Willke, T., Skiadas, I, Ree, R. V., & de Jong, E. (2009) Toward a common classification approach for B.... *Biofuels, Bioproducts and Biorefining, 3*, 534–546.

Govil, T., Wang, J., Samanta, D., David, A., Tripathi, A., Rauniyar, S., Salem, D. R., & Sani, R. K. (2020). Lignocellulosic feedstock: A review of a sustainable platform for cleaner production of nature's plastics. *Journal of Cleaner Production, 270*, 122521.

Grewal, J., Sadaf, A., Yadav, N., & Khare, S. K. (2020). Agroindustrial waste based biorefineries for sustainable production of lactic acid. In *Waste biorefinery* (pp. 125–153). Elsevier.

Haro, P., Perales, Á. L. V., Arjona, R., & Ollero, P. (2014). Thermochemical biorefineries with multiproduction using a platform chemical. *Biofuels, Bioproducts and Biorefining, 8*(2), 155–170. https://doi.org/10.1002/bbb.1465

Harun, R., Singh, M., Forde, G. M., & Danquah, M. K. (2010). Bioprocess engineering of microalgae to produce a variety of consumer products. *Renewable and Sustainable Energy Reviews, 14*, 1037–1047. https://doi.org/10.1016/j.rser.2009.11.004

Hellmann, K. U., & Luedicke, M. K. (2018). The throwaway society: A look in the back mirror. *Journal of Consumer Policy, 41*(1), 83–87. https://doi.org/10.1007/s10603-018-9371-6

Heredia-Guerrero, J. A., Heredia, A., Domínguez, E., Cingolani, R., Bayer, I. S., Athanassiou, A., & Benítez, J. J. (2017). Cutin from agro-waste as a raw material for the production of bioplastics. *Journal of Experimental Botany, 19*(9), 5401–5410. https://doi.org/10.1093/jxb/erx272

IEA Bioenergy. (2021). Facilities. *Global database of biomass conversion facilities, including advanced biofuels, combustion, gasification and pyrolysis plants*. https://www.ieabioenergy.com/installations/

Jiang, L., & Zhang, J. (2017). Biodegradable and biobased polymers. In *Applied plastics engineering handbook: Processing, materials, and applications* (2nd ed., pp. 127–143). William Andrew.

Jung, S., Kim, H., Tsang, Y. F., Lin, K. Y. A., Park, Y. K., & Kwon, E. E. (2020). A new biorefinery platform for producing (C2-5) bioalcohols through the biological/chemical hybridization process. *Bioresource Technology, 311*, 123568.

Kamm, B., & Kamm, M. (2004). Biorefinery—Systems. *Chemical and Biochemical Engineering Quarterly, 18*(1), 1–6. https://doi.org/10.1016/b978-0-444-59561-4.00014-0

Kedzierski, M., Frère, D., Le Maguer, G., & Bruzaud, S. (2020). Why is there plastic packaging in the natural environment? Understanding the roots of our individual plastic waste management behaviours. *Science of the Total Environment, 740*, 139985.

Kim, H., Lee, S., Ahn, Y., Lee, J., & Won, W. (2020). Sustainable production of bioplastics from lignocellulosic biomass: Technoeconomic analysis and life-cycle assessment. *ACS Sustainable*

Chemistry and Engineering, 8(33), 12419–12429. https://doi.org/10.1021/acssuschemeng.0c02872

Krajnc, D., & Glavic, P. (2003). Indicators of sustainable production. *Clean Technologies and Environmental Policy, 5*, 279–288. https://doi.org/10.1007/s10098-003-0221-z

Kromus, S., Wachter, B., Koschuh, W., Mandl, M., Krotscheck, C., & Narodoslawsky, M. (2004). The Green Biorefinery Austria—Development of an integrated system for green biomass utilization. *Chemical and Biochemical Engineering Quarterly, 18*(1), 7–12.

Lackner, M. (2015). Bioplastics. In *Kirk-Othmer encyclopedia of chemical technology* (pp. 1–41). John Wiley & Sons.

Lindorfer, J., Lettner, M., Hesser, F., Fazeni, K., Rosenfield, D., Annevelink, B., & Mandl, M. (2019). *Technical, economic and environmental assessment of biorefinery concepts: Developing a practical approach for characterisation.* IEA Bioenergy.

Maity, S. K. (2015). Opportunities, recent trends and challenges of integrated biorefinery: Part I. *Renewable and Sustainable Energy Reviews, 43*, 1427–1445. https://doi.org/10.1016/j.rser.2014.11.092

Maraveas, C. (2020). Production of sustainable and biodegradable polymers from agricultural waste. *Polymers, 12*(5), 1127.

Márquez Luzardo, N. M., & Venselaar, J. (2012). Bio-based targeted chemical engineering education; role and impact of bio-based energy and resource development projects. *Procedia Engineering, 42*, 214–225.

Morais, A. R. C., & Bogel-Lukasik, R. (2013). Green chemistry and the biorefinery concept. *Sustainable Chemical Processes, 1*, 1–3. https://doi.org/10.1186/2043-7129-1-18

Narayan, R. (2011). Carbon footprint of bioplastics using biocarbon content analysis and life-cycle assessment. *MRS Bulletin, 36*, 716–721. https://doi.org/10.1557/mrs.2011.210

Octave, S., & Thomas, D. (2009). Biorefinery: Toward an industrial metabolism. *Biochimie, 91*(6), 659–664. https://doi.org/10.1016/j.biochi.2009.03.015

ONU. (2018). *Objetivo 12: Garantizar Modalidades de Consumo y Producción Sostenibles* (pp. 26–27).

Panesar, R., Kaur, S., & Panesar, P. S. (2015). Production of microbial pigments utilizing agro-industrial waste: A review. *Current Opinion in Food Science, 1*, 70–76.

Pfaltzgraff, L. A., De Bruyn, M., Cooper, E. C., Budarin, V., & Clark, J. H. (2013). Food waste biomass: a resource for high-value chemicals. *Green Chem, 15*(2), 307–314. https://doi.org/10.1039/C2GC36978H

PlasticsEurope. (2008). Los Plásticos En 2007.

Polman, E. M. N., Gruter, G. J. M., Parsons, J. R., & Tietema, A. (2021). Comparison of the aerobic biodegradation of biopolymers and the corresponding bioplastics: A review. *Science of the Total Environment, 753*, 141953. https://doi.org/10.1016/j.scitotenv.2020.141953

Rahman, M. H., & Boi, P. R. (2021). An overview of non-biodegradable bioplastics. *Journal of Cleaner Production, 294*, 126218. https://doi.org/10.1016/j.jclepro.2021.126218

Ranganathan, S., Sayantani, D., Moses, J. A., & Anandharamakrishnan, C. (2020). Utilization of food waste streams for the production of biopolymers. *Heliyon, 6*(9), e04891. https://doi.org/10.1016/j.heliyon.2020.e04891

Reddy, M. M., Vivekanandhan, S., Misra, M., Bhatia, S. K., & Mohanty, A. K. (2013). Biobased plastics and bionanocomposites: Current status and future opportunities. *Progress in Polymer Science, 38*(10–11), 1653–1689. https://doi.org/10.1016/j.progpolymsci.2013.05.006

Salazar, R. A., & Cárdenas, G. J. (2013). La Bioeconomía y Las Biorrefi Nerías. *Avance Agroindustrial, 34*(3), 31–34.

Samer, M., Khalefa, Z., Abdelall, T., Moawya, W., Farouk, A., Abdelaziz, S., Soliman, N., Salah, A., Gomaa, M., & Mohamed, M. (2019). Bioplastics production from agricultural crop residues. *Agricultural Engineering International: CIGR Journal, 21*(3), 190–194.

Shafqat, A., Tahir, A., Mahmood, A., Tabinda, A. B., Yasar, A., & Pugazhendhi, A. (2020). A review on environmental significance carbon foot prints of starch based bio-plastic: A substitute of conventional plastics. In *Biocatalysis and agricultural biotechnology* (p. 101540). Elsevier.

Sidek, I. S., Draman, S. F. S., Abdullah, S. R. S., & Anuar, N. (2019). Current development on bioplastics and its future prospects: An introductory review. *INWASCON Technology Magazine, 1*, 3–8. https://doi.org/10.26480/itechmag.01.2019.03.08

Snell, K. D., & Peoples, O. P. (2009). PHA bioplastic: A value-added coproduct for biomass biorefineries. *Biofuels, Bioproducts and Biorefining, 3*(4), 456–467. https://doi.org/10.1002/bbb.161

Statista (2020). 70 años de "boom" del plástico. https://es.statista.com/grafico/20441/produccion-de-plastico-a-nivel-mundial/

Teigiserova, D. A., Hamelin, L., & Thomsen, M. (2019). Review of high-value food waste and food residues biorefineries with focus on unavoidable wastes from processing. *Resources, Conservation and Recycling, 149*, 413–426. https://doi.org/10.1016/j.resconrec.2019.05.003

Thakur, S., Chaudhary, J., Sharma, B., Verma, A., Tamulevicius, S., & Thakur, V. K. (2018). Sustainability of bioplastics: Opportunities and challenges. *Current Opinion in Green and Sustainable Chemistry, 13*, 68–75.

Trigo, E., Regúnaga, M., Acquaroni, M., Jimenez, F., & Peña-Farinaccia, J. (2012). *Biorrefinerías En La República Argentina: Análisis Del Mercado Potencial Para Las Principales Cadenas de Valor*. MINCyT.

Tsang, Y. F., Kumar, V., Samadar, P., Yang, Y., Lee, J., Ok, Y. S., Song, H., Kim, K. H., Kwon, E. E., & Jeon, Y. J. (2019). Production of bioplastic through food waste valorization. *Environment International, 127*, 625–644. https://doi.org/10.1016/j.envint.2019.03.076

Ubando, A. T., Felix, C. B., & Chen, W. H. (2020). Biorefineries in circular bioeconomy: A comprehensive review. *Bioresource Technology, 299*, 122585. https://doi.org/10.1016/j.biortech.2019.122585

Vaz, S. (2019). Sugarcane-biorefinery. *Advances in Biochemical Engineering/Biotechnology, 166*, 125–136. https://doi.org/10.1007/10_2016_70

Vert, M., Doi, Y., Hellwich, K. H., Hess, M., Hodge, P., Kubisa, P., Rinaudo, M., & Schué, F. (2012). Terminology for biorelated polymers and applications (IUPAC recommendations 2012). *Pure and Applied Chemistry, 84*(2), 377–410. https://doi.org/10.1351/pac-rec-10-12-04

Yang, J., Ching, Y. C., & Chuah, C. H. (2019). Applications of lignocellulosic fibers and lignin in bioplastics: A review. *Polymers, 11*(5), 751. https://doi.org/10.3390/polym11050751

Zwicker, M. V., Nohlen, H. U., Dalege, J., Gruter, G. J. M., & van Harreveld, F. (2020). Applying an attitude network approach to consumer behaviour towards plastic. *Journal of Environmental Psychology, 69*, 101433. https://doi.org/10.1016/j.jenvp.2020.101433

Part III
Agricultural Biotechnology

Flow and Distribution of Phosphorus in Soils from a Geochemical and Agronomic Approach

Gregorio Vásconez Montúfar, Dante Pinochet Tejos, Ronald Oswaldo Villamar-Torres, Carlos Alberto Molina Hidrovo, Verónica Segovia Motesdeoca, and Seyed Mehdi Jazayeri

G. Vásconez Montúfar (✉)
Universidad Técnica Estatal de Quevedo, Quevedo, Ecuador

Graduate School, Faculty of Agricultural Sciences, University Austral of Chile, Valdivia, Chile
e-mail: gvasconez@uteq.edu.ec

D. Pinochet Tejos
Institute of Agricultural Engineering and Soils, Faculty of Agricultural Sciences, University Austral of Chile, Valdivia, Chile

Centro de investigación de Suelos Volcánicos (CISVO), Universidad Austral de Chile, Valdivia, Chile
e-mail: dpinoche@uach.cl

R. O. Villamar-Torres
Universidad Técnica Estatal de Quevedo, Quevedo, Ecuador

Instituto Superior Tecnológico "Ciudad de Valencia" – Tecnología en Producción Agrícola y Tecnología en Procesamiento de Alimentos, Quevedo, Ecuador
e-mail: rvillamart@uteq.edu.ec

C. A. Molina Hidrovo
Instituto Nacional de Investigaciones Agropecuarias, Guayaquil, Ecuador
e-mail: carlos.molina@iniap.gob.ec

V. Segovia Motesdeoca
Universidad Técnica Estatal de Quevedo, Quevedo, Ecuador

S. M. Jazayeri
Faculty of Biology, University-College of Science, University of Tehran, Tehran, Iran

Departamento de Biología, Facultad de Ciencias, Universidad Nacional de Bogotá, Bogotá, Colombia
e-mail: smjazayeri@ut.ac.ir; smjazayeri@unal.edu.co

N. R. Maddela, L. C. García (eds.), *Innovations in Biotechnology for a Sustainable Future*, https://doi.org/10.1007/978-3-030-80108-3_8

Abbreviations

ACPwithoutP	Absorption of the crop in the plot without P
ACPwithP	Absorption of the crop in the plot with P
CPA	Amount of P added to the plot
DAP	Diammonium phosphate
MAP	Monoammonium phosphate
P	Phosphorus
Pi	Inorganic phosphorus
Po	Organic phosphorus
PR	Phosphate rock
Pt	Total phosphorus
RP	Recovery of P
TSP	Triple superphosphate

1 Introduction

Crop productivity is determined by potentially defining factors (e.g., solar radiation, temperature, CO_2 concentration, plant characteristics) that usually limit it in ecosystems (e.g., water, soil fertility) and/or the factors that reduce it (e.g., diseases, pests, weeds), which is manifested in the growth rate of a crop (van Ittersum et al., 2003), where soil fertility plays a determinant role in the different agroecosystems (Henao & Baanante, 1999; Drechsel et al., 2001; Mueller et al., 2012). From this point the importance is taken to consideration, the definition of an adequate level of fertility contemplates calculating a concentration or range of this that does not produce deficiencies in the growth of crops, but at the same time that does not imply a risk of contamination of the water or air of the ecosystems (Pinochet, 1995; Johnston & Dawson, 2005).

The world population is increasing geometrically, which is estimated to reach nine billion inhabitants in the year 2050 (FAO, 2009). This implies an increase in the demand for food, fibers, biofuels (FAO, 2009; Roberts, 2009; Godfray et al., 2010) and, consequently, the need to develop strategies to increase crop productivity will become evident, where the management of soil fertility, especially N, P, and K, will be the key to obtain high crop yields, with the least risk of contamination of water or air in ecosystems (van der Wiel et al., 2019), having in perspective the level of soil fertility is the consequence of a state of equilibrium originating from the balance between the addition, removal and transformations of nutrients in the soil system (Hartemink, 2006; Phong et al., 2011; Zhan et al., 2015; Dayton et al., 2020).

Under natural conditions, nutrient addition occurs through solid and liquid atmospheric deposition, biological fixation of gaseous nutrients, and the entry of nutrients from sediments and effluents. Removal involves leaching from the soil, gaseous losses (denitrification and others), surface or subsurface runoff (nutrient dissolved in water), and erosion (loss of soil plus nutrient). The transformation represents the income from the weathering of the minerals (transfer of reservoirs). As is the case of the release of cations from the primary minerals and the mineralization of organic matter (release of N, P and S), which constitute the native income from the soil and other forms of transformation are given by the chemical and physical reactions that determine the retention of minerals in colloids through adsorption, the formation of precipitated products and the transmission of mineral sources from inorganic to organic forms by the microbial action, such as the immobilization of N.

Depending on agronomic practices and the physical-chemical relationship of nutrients with soil colloids, the relationship between the processes of addition and removal of nutrients suggests that agricultural systems can accumulate (input of nutrients > export of nutrients in crops) or de-accumulate nutrients (export of nutrients in crops > nutrient inputs). A predominance of de-accumulation can lead to a progressive reduction in soil fertility and a predominance of accumulation can result in over-fertility of the soil and even increase the potential risk of contamination (Pinochet, 1995; Johnston & Dawson, 2005), especially of water bodies by eutrophication processes of aquatic ecosystems (Bennett et al., 2001; Sharpley, 2016).

In this context, soil fertility is a matter of concern for the agricultural policies of all countries, especially when referring to phosphorous fertility, since the availability of P in many of the agricultural systems constitutes a strong limitation of the production of crops (Sharpley & Tunney, 2000; Tóth et al., 2014). The physical-chemical relationship between phosphate ions and colloids in the soil, and the field evidence shows the accumulation capacity of P in soils and the potential to maintain or build their fertility (Whalen & Chang, 2001; McLaughlin et al., 2011; Zhang et al., 2019). However, the unbalanced P balances that are registered in many localities of the world, which can manifest themselves in low fertility or in a potential risk of contamination, have created the urgency to develop strategies aimed at increasing the efficient use of P (Buerkert et al., 2001; Simpson et al., 2011; MacDonald et al., 2011; Van Dijk et al., 2016), such as adjusting the fertility level according to the supply required to satisfy the demand of the crop, replenishing what is taken out of the system, selecting the correct source and place of addition, and identifying cultivars based on their strategies for the use of P (Sharpley et al., 2005; Norton, 2014; Shi et al., 2015). Faced with these challenges, it is essential to review the existing scientific information regarding the dynamics of P in soils.

2 Residual Effect of P on Soils

At the beginning of the twentieth century, interest grew in the effect of the nutrients that were applied to the soil via fertilizers. Thus, the effect of fertilization has been evaluated by quantifying the relative yield of the crops during the years following the initial application of the fertilizer, which have been compared with the yields obtained in unfertilized plots (Syers et al., 2008). At that time, it was considered that P was lost or irreversibly fixed in the soil and therefore with little or no capacity to increase soil fertility. This conclusion was reached because researchers at the time possibly tried to measure the residual effect of one or more small doses of P, in soils with a low level of available P and high retention capacity. However, some studies have shown that the P retained in the soil can be recovered if the concentration of available P is reduced (Barrow, 1983a), a situation that was later corroborated through the analysis of a series of field experiments that showed that It is possible to recover over 80% of the P added to the soil by removing it in the harvest of successive crops (Syers et al., 2008).

Concern about precipitation implied that it was not considered that much of P could be subject to adsorption reactions on soil colloids (Syers et al., 2008). According to the works with an agronomic approach that have described the adsorption of ions on characteristic surfaces of dominant colloids in soils (Posner & Barrow, 1982) today it is recognized that the P retained in the soil can be released, reflecting in a high accumulated recovery originating from successive crops. Subsequently, this process was simulated in a reaction model between P and soil (Barrow, 1983b), which was based on three assumptions: (1) adsorption reactions occur between phosphate ions and a reaction surface; (2) the properties of the reaction surface are normally distributed and; (3) the initial adsorption induces a diffusion gradient toward the interior of the reaction surface. The model adequately described adsorption/desorption when exposing the soil to different concentrations of P, pH, temperature and contact time between P and the soil.

The crops, during their vegetative and productive cycle, do not use all the P added to the soil for the season, leaving a fraction of absorbable P for the subsequent crop. Thus, the P available for crop production in seasons after the addition of P has been called the residual effect of P (Howard, 2006; Syers et al., 2008; Li et al., 2011). Therefore, it can be inferred that the amount of residual P present in the soil is basically determined by two processes: (1) the absorption of the crop and subsequent export of P at harvest and (2) the reactions that determine its retention in the soil, which according to Javid and Rowell (2002) can be classified as fast and slow reactions. The fast reactions would be being dependent on the colloids and their proportion in the soil, and the slow ones would respond to the laws that govern the diffusion of ions.

The evaluation of the residual effect of P in short-term experiments led to the conclusion that P is used inefficiently in agriculture, with recovery percentages of added P that remains between 10 and 20%. Currently, the recognition of the accumulation of P in the soil and its reversibility has reoriented the analyses done by the researchers. Thus, Johnston and Syers (2008) argue that the traditional method by difference to calculate efficiency is not suitable for P (Eq. 1) since it was initially developed for nitrogen (N). The N contained in fertilizers rarely remains in the soil as inorganic, nitrate that was not absorbed by the crop or immobilized by soil organisms, is potentially lost through leaching or denitrification (Riley et al., 2001; Zhang et al., 2015). However, on rare occasions a very small amount of added P is lost from the soil, but mostly tends to accumulate as a reserve (Zhang et al., 2019). When the efficiency of P is measured by the balance method (Eq. 2), the P removed by successive crops, during a suitable period of time, can exceed 80% of the P added to the soil via fertilizer (Ibrikci et al., 2005; Sá et al., 2017).

$$\mathrm{RP} = \frac{\mathrm{ACP\ with\ P} - \mathrm{ACP\ without\ P}}{\mathrm{APAP}} \tag{1}$$

$$\mathrm{RP} = \frac{\sum \mathrm{ACP\ with\ P}_t - \sum \mathrm{ACP\ without\ P}_t}{\mathrm{APAP}} \tag{2}$$

In Eq. (1): ACPwithP = absorption of the crop in the plot with P, ACPwithoutP = absorption of the crop in the plot without P, APAP = amount of P added to the plot. Equation (2) considers the recovery of P by successive sowings during a time (t) of recovery of P from successive crops. In both Eqs. (1) and (2), RP refers to the recovery of P by the crop.

The reactions that occur over time when a phosphate fertilizer comes into contact with the soil largely determine the residual effect of P. Results have suggested that low solubility phosphate products originate when the soil solution reaches high concentrations of P, a favored condition at the beginning of each fertilization event (Lindsay et al., 1962; Ghosh et al., 1996). By adding P to the soil, the fertilizer granule is hydrated and subsequently the diffusion of phosphate ions begins, delimiting a zone of high P concentration in the closest part to the fertilizer and another zone of low concentration in the distal limit of influence of the same limit and, according to McLaughlin et al. (2011), in the zone of higher P concentration the precipitation reactions would be dominant and in the zone of low concentration the adsorption reactions would dominate (Fig. 1).

The initial reactions that are triggered when P is added to the soil depend largely on its particular properties of the fertilizer, where the reactivity of the soil determines the fate of P (Fig. 2). The current sources of P used in agricultural systems are manufactured basically from phosphate rock (PR), whose P content generally fluctuates between 9 and 17% of P. By treating PR with sulfuric acid, simple superphosphate can be produced (7–10% P) or with phosphoric acid to produce triple superphosphate (17–24% P). From phosphoric acid triple superphosphate (TSP: 0-46-0), diammonium phosphate (DAP: 18-46-0), or monoammonium (MAP: 15-52-0) is produced, depending on whether it is combined with RF or with ammonia. These sources of P seem to end up all forming dicalcium phosphate in soil, regardless of their manufacturing process, and the main difference lies in the pH and the concentration of P in the perimeter of influence of the fertilizer granule. The latter was evidenced in an acid oxisol (pH ~ 4) and a calcareous inceptisol (pH ~ 8), collected in Rondonopolis—Brazil and Idaho—USA, respectively. In the first soil, the pH increases by adding MAP or DAP, while in the second soil, the pH is reduced. In the first soil the total P added was distributed in the first 28 mm of diameter, while in the second soil the total P added was distributed in the first 50 mm of diameter (Pierzynski et al., 2014).

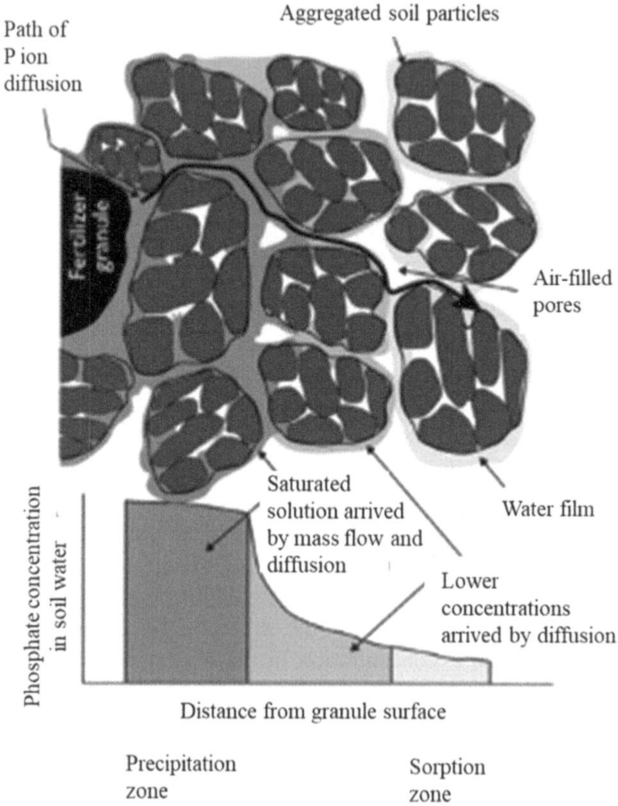

Fig. 1 Representative diagram of the movement of P by diffusion from a soluble fertilizer granule through the pores of a well-aggregated soil (McLaughlin et al., 2011)

The development of soil analysis methods opens the opportunity to measure the residual effect of nutrients considering their availability to potentially be absorbed by crops, under specific site conditions (soil, climate) and management (crop rotation, crop management of residues, fertilization techniques). In the case of P, the analysis methods do not strictly determine the concentration of available P, but provide an index of the amount of P that can be absorbed by the culture (significant correlation between the availability index and the absorption of the culture) (Kumar et al., 1991; Tiessen & Moir, 1993). These methods have been developed considering specific characteristics of the soils, which considerably limits their generalization (Kumar et al., 1991; Watson & Mullen, 2007; Jordan-Meille et al., 2012; Ziadi et al., 2013).

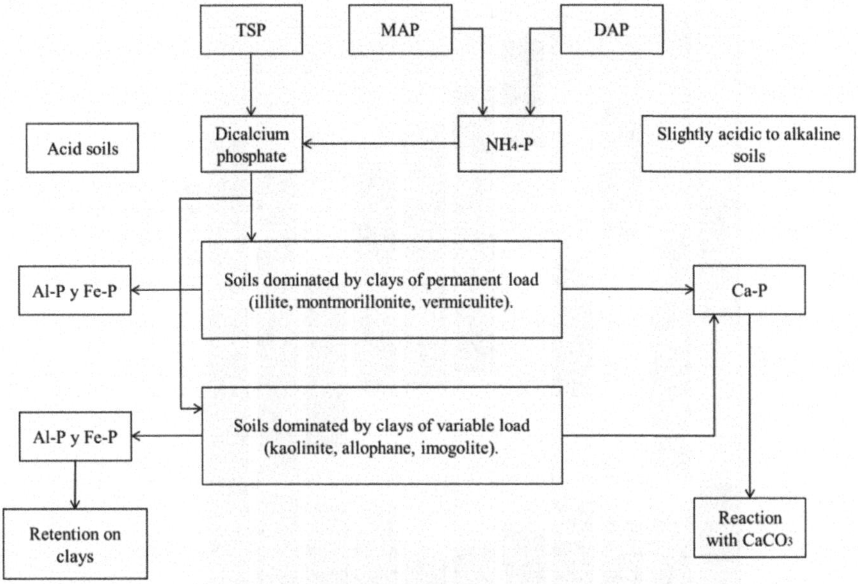

Fig. 2 Dynamics of phosphorous fertilizer in the soil. (Adapted from Fixen, 1992)

3 Phosphorus Shapes in Soil

The amount of P contained in soils exceeds widely the amount absorbed by crops, regardless of the degree of soil evolution, or of its pedogenesis (Yang & Post, 2011), which is contained in organic (Po) and inorganic (Pi) reservoirs (Johnston & Syers, 2008). However, from an agronomic perspective, the availability of P in the soil is limiting for crop production in many regions of the world (Lynch, 2011), which predicts an increase in the demand for phosphate fertilizers and the constant development of strategies that allow increasing the efficient use and recycling of P (Sharpley et al., 2005; Norton, 2014).

It has been found that the level of total P (Pt) in oxisols and ultisols can be quite low, not exceeding 18 mg P kg^{-1}, while in soils derived from volcanic ash it can be much higher with 3300 mg P kg^{-1} (Fassbender, 1993). Po represents between 20 and 75% of Pt (Brady & Weil, 1999), consisting essentially of polyphosphates (ATP, nucleic acids), phosphonates (phosphonic acid), diester phosphate (phospholipids), monoester phosphate (carbohydrate phosphates, inositol phosphate) (Fuentes et al., 2006). In a meta-analysis by Yang and Post (2011), it became clear that the proportion of each P form is related to parent material and time. The so-called labile Pi constitutes a small fraction of Pt, in all soil orders, through all stages of development, even in oxisols this form did not represent more than 5% of Pt. Secondary Pi, defined as Pi adsorbed on the surface of secondary minerals of Al and Fe, represents a small fraction that increases with the degree of weathering, but in no case it exceeded 20%. The apatite content decreases with weathering, thus, entisols exhibit

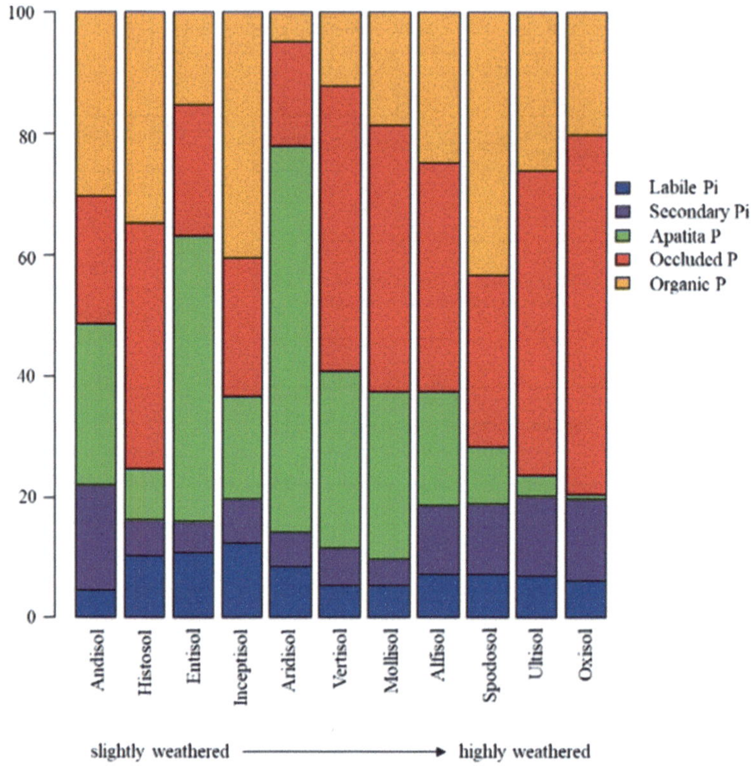

Fig. 3 Forms of P expressed as a percentage of total soil P (Yang & Post, 2011)

apatite content greater than 40% and oxisols do not exceed 3%. Aridisols, with an intermediate stage of development, exhibit apatite content greater than 60% attributable to dry climate conditions where these soils predominate (Fig. 3).

The reduced fertility of P that soils generally exhibit in various regions of the world (Lynch, 2011), basically responds to the naturally low content of P in the soil in reference to other elements, low or no solubility of most of the compounds of P, and progressive adsorption of the soluble P added via fertilizer, and even formation of poorly soluble secondary minerals. In addition, soil fertility problems are exacerbated by P outputs from the system annually: 5–50 kg P ha^{-1} are removed by plants in the harvested biomass, 0.1–10 kg P ha^{-1} by wind erosion of mineral and organic particles, 0.01–3 kg P ha^{-1} is lost by surface runoff, and due to the low solubility of P, losses by leaching are not very quantifiable (Brady & Weil, 1999).

3.1 Soil Inorganic Phosphorus

Soil Pi represents an important fraction of Pt, which under alkaline conditions is forming compounds linked to Ca, under acidic conditions compounds linked to Al,

Fe, and Mn, and under intermediate conditions phosphate ions tend to replace hydroxyl groups located in the edges of the clays (Brady & Weil, 1999; Penn & Camberato, 2019). The soil solution also constitutes a small reservoir of Pi, since it can reach an average concentration of 0.2 mg P L^{-1} (Bolland et al., 2003), a situation that would represent 60 g P ha^{-1} at 20 cm depth, with an apparent density of 0.7 g cm^{-3}, and with 21% humidity.

Plants take up P from the soil solution. However, the amount of P in solution (60 g P ha^{-1}) is far from satisfying the demand for crops such as wheat or peas, which can achieve yields of 10,000 in southern Chile (Sandaña et al., 2009; Valle et al., 2009) and 7000 kg ha^{-1} (Valle et al., 2009), respectively. If we consider a wheat yield of 10,000 kg ha^{-1} and an average P concentration in the vegetable dry matter of 2 mg g^{-1} (Schachtman et al., 1998; Sharma et al., 2013) it can be argued that wheat crop is able to remove 20 kg P at harvest, which means that the crop totally exhausts the solution (60 g P ha^{-1}) from the soil more than 300 times. It has been suggested that precipitation/dissolution, adsorption/desorption, and immobilization/mineralization reactions control recovering P concentration in the soil solution (Frossard et al., 2000; Penn & Camberato, 2019).

P sorption isotherms are derived from describing the amount of P adsorbed per unit mass of soil as a function of the concentration of P in solution. From the analysis of P sorption isotherms it has been possible to establish the maximum retention capacity of the soil and the existence of a balance between P adsorption and desorption (Anghinoni et al., 1996; Jiao et al., 2008; Wang et al., 2013), determining that a certain level of P in solution is linked to a specific amount of P adsorbed in the soil. That is, if the concentration of P in solution is increased (addition of fertilizers), the amount of P adsorbed in the soil increases and, if the concentration of P in solution is depleted (export of P in the harvest of crops) it decreases the amount of P adsorbed in the soil. This approach has reasonably explained how small amounts of P in solution can cover the demand for P by crops.

Adsorption can occur at the surface and intraparticle level, in the first case the P in solution is retained on the surface of the soil colloids and in the second case the initially retained P diffuses into the microaggregates, as a consequence of a gradient of unbalanced concentration (Barrow, 1983a, b; Bolland et al., 2003). Desorption refers to the passage of P adsorbed to solution, mainly caused by a depletion of P in solution (Bolland et al., 2003). The adsorbed P can be trapped between the clay minerals of the soil and the precipitated oxides of Al and Fe, this condition is called occluded P (McLaren & Cameron, 1996). In Fig. 4 the adsorption and occlusion of Pi is shown.

In the soil, the phosphate ion is subject to reactions that produce low solubility compounds depending on the pH, and therefore the P contained in these compounds is less susceptible to being absorbed by plants (Bohn, 1993; Busman et al., 2002; Hopkins & Ellsworth, 2005). Generally, the reactions by which the phosphate ions are removed from the soil solution give rise to: Al, Fe, and Mn phosphates, compounds that are formed when P reacts with metal ions in solution (Fig. 5a); reversible external sphere adsorption that favors anion exchange, where the phosphate ion in solution is subject to exchange with OH^- and SO_4^{2-} ions (Fig. 5b);

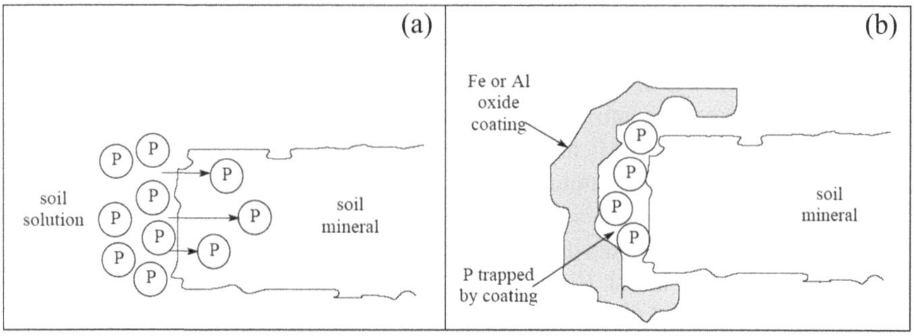

Fig. 4 Surface and intraparticle adsorption (**a**) and occlusion (**b**) of P in the mineral fraction of the soil. (Adapted from McLaren & Cameron, 1996)

retention of the external sphere on the surface of hydrated oxides where the availability of P tends to be much lower, since the reaction can be repeated, binding the phosphate ion to two adjacent hydrated oxides (Fig. 5c).

In alkaline soils, the availability of P is linked to the solubility of the compounds that tend to form when the phosphate ion interacts with divalent bases (Ca, Mg). In alkaline soils, the phosphate ion reacts rapidly with Ca, giving rise to the formation of a systematic sequence of compounds that decrease in solubility (Naeem et al., 2013). The highly soluble monocalcium phosphate [Ca $(H_2PO_4)_2 \cdot 2H_2O$] reacts with the calcium carbonate present in the soil ($CaCO_3$). Monocalcium phosphate plus water and calcium carbonate originate dicalcium phosphate dihydrate, subsequently under the same domain conditions of calcium carbonate, tricalcium phosphate originates. In each transition there is release of carbon dioxide and the insolubility decreases in each transition with respect to monocalcium phosphate (Table 1).

3.2 Soil Organic Phosphorus

The soil Po has a high agronomic and ecological significance since it can represent more than 50% of Pt (Haygarth et al., 2018), which is corroborated by observing a significant contribution of Po to the P availability that is estimated through routine laboratory methods (Cade-Menun et al., 2018). Soil Po has been difficult to study due to the fragility of the molecules that contain it, a situation that limits its extraction with acids or bases. Chromatographic partitioning is a technique that has made it possible to objectively study the Po of the soil. However, much of the scientific knowledge of Po is due to nuclear magnetic resonance (NMR) as it is a better known technique than chromatographic partition.

Four organic molecules that contain P in their structure are recognized in the soil: polyphosphates (ATP), phosphonates (phosphonic acid), phosphate monoester (inositol phosphate) and phosphate diester (phospholipid) (Cade-Menun, 2005; Cheesman et al., 2014). The nature of P associated with humic materials in the

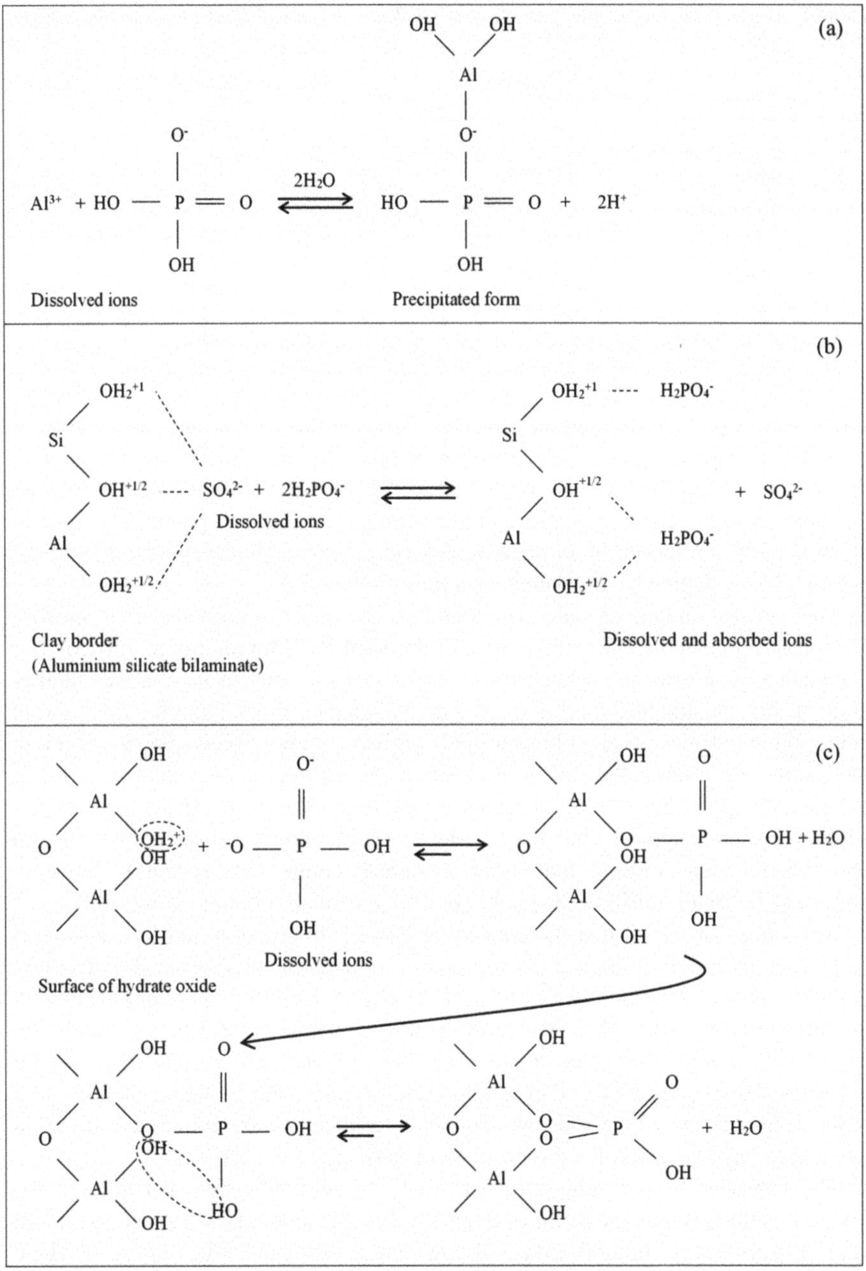

Fig. 5 Precipitation reaction (**a**), anion exchange (**b**) and adsorption on the surface of oxides (**c**). (Adapted from Brady & Weil, 1999)

Table 1 Minerals formed by precipitation of P in calcareous soils (Adapted from Brady & Weil, 1999)

Mineral	Formula	Transition (weeks)	Insolubility[a]
Monocalcic phosphate	$Ca(H_2PO_4)_2 \cdot 2H_2O$		
Dicalcic dehydrate phosphate	$CaHPO_4 \cdot 2H_2O$	2–3	60
Dicalcic phosphate	$CaHPO_4$		
Octocalcic phosphate	$Ca_4H(PO_4)_3 \cdot 2.5H_2O$		
Tricalcic phosphate	$Ca_3(PO_4)_2$	8–10	900
Apatite hydroxide	$Ca_5(PO_4)_3OH$	52–104	
Fluorapatite	$Ca_5(PO_4)_3F$		

[a]Referential insolubility with respect to the solubility of monocalcium phosphate

soil is little known, a situation that does not diminish their importance as a source of P for plants. These organic molecules can be grouped into two organic fractions of the soil, a fraction that actively participates in plant nutrition and another fraction that appears to be relatively stable and therefore not available to plants. The amount of Po is more influenced by climate, biological activity, relief, vegetation and time and to a lesser degree by the parent material of the soil.

The most abundant organic compound in the soil that contains P is inositol phosphate representing more than 50% of the total Po (McLaren et al., 2015). The abundance of inositol phosphate in the soil is possibly due to its relative stability under acidic and alkaline conditions, and active interaction with humic acids present in the soil. Nucleic acids and phospholipids probably do not exceed 2% of total Po in most soils; these reach the soil in relatively high amounts as remains (residues) or secondary metabolites of microorganisms, plants and animals. However, the low amount present suggests that they undergo rapid mineralization unlike inositol phosphate, which suggests that these phosphate compounds appear to be more important for plant nutrition than suggested by the small amounts in the soil.

It has been observed that the concentration of P in solution and in leachates is high when there are biological depositions of animals in soils, where P is forming part of organic compounds (Azevedo et al., 2018). Po forms are more mobile than Pi in soils, possibly due to their low reaction with minerals present in the soil. In the deep horizons of soils that receive biological depositions from animals, it is common to see that Po exceeds 50% of P in solution. This explains the greater availability of P in the deep horizons of the soils when they are fertilized by Po sources, compared to when they have been fertilized with Pi sources (Rigo et al., 2019).

The forms of Po are subject to pass to Pi by mineralization and through the immobilization process the forms of Pi pass to Po. The amount of Pi added to the soil via fertilizers that is immobilized is unknown but is estimated to be important. There is evidence that some of the forms of Po product of immobilization are too stable to be used by plants. It is possible to reduce the immobilization rate by modifying the pH of the soil, going from acid to slightly acidic, but the accumulation of di and tricalcium phosphate increases. Thus, it has been observed that calcium amendment applications can reduce the need for phosphorous fertilizers in some cases by increasing the mineralization/immobilization ratio.

4 Fractionation of P Contained in Soils

As a result of the development of techniques to analyze the contained P in soils, the sequential extraction of P has been possible. This way of analyzing P has been shown to reasonably support the hypothesis of P reservoirs in equilibrium (Johnston & Syers, 2008; Syers et al., 2008). This hypothesis assumes that P is retained by soil components with different degrees of energy (depending on the type of bond that forms between P and soil colloids) and consequently they would determine several P reservoirs with different availability grade. One of the first methods of sequential extraction was proposed by Chang and Jackson (1957), which quickly stopped being used due to its low representation in agronomic terms. Subsequently, other procedures have been proposed with the same objective (Table 2). Thus, in order to have a practical explanation, conceptual diagrams of soil P reservoirs categorized according to biological availability have been suggested.

The method of Hedley et al. (1982) has shown to be reasonably reproducible and sufficiently sensitive to changes in conditions in agroecosystems (Tiessen & Moir, 1993; Cross & Schlesinger, 1995). The strength of this method is that it allows the extraction of inorganic and organic forms of P with labile to nonlabile characteristics. Several modifications of this sequential extraction method have been proposed, among the best known are that of Condron and Goh (1989) and that of Tiessen and Moir (1993). In the first case, the microbial P was excluded, the 0.1 M NaOH extraction plus ultrasound was replaced by extraction with 0.5 M NaOH, and the

Table 2 Sequential extraction methods for P

Method	Extraction	Designation
Chang and Jackson (1957)	1. 1.0 M NH_4Cl	P labile
	2. 0.5 M NH_4F	P-Al
	3. 0.1 M NaOH	P-Fe
	4. 0.25 M H_2SO_4	P-Ca
	5. Citrate-ditionite	P-Fe reduced
	6. 0.1 M NaOH	P-Al and P-Fe occluded
Bowman and Cole (1978)	1. 0.5 M $NaHCO_3$	P labile
	2. 1.0 M H_2SO_4	P moderately labile
	3. 0.5 M NaOH	P resistant
Hedley et al. (1982)	1. Resin	P very labile
	2. 0.5 M $NaHCO_3$	P labile
	3. Fumigation, 0.5 M $NaHCO_3$	P microbial
	4. 0.1 M NaOH	P-Al and P-Fe
	5. 0.1 M NaOH + ultrasound	P intra added
	6. 0.1 M HCl	P-Ca
	7. H_2SO_4 and H_2O_2	P residual
Ivanoff et al. (1998)	1. 0.5 M $NaHCO_3$	P labile
	2. Fumigation, 0.5 M $NaHCO_3$	P microbial
	3. 1.0 M HCl	P moderately labile
	4. 0.5 M NaOH	P no labile
	5. Ignition, 1.0 M H_2SO_4	P residual

Table 3 Geochemical and ecological significance of the P fractions resulting from the sequential extraction of Hedley et al. (1982) modified by Tiessen and Moir (1993)

Fraction of P	Geochemical Significance	Ecological
Resin-Pi	Absorbed in the surface of crystalline compounds. Not occluded.	Immediately available para las plantas, directly interchangeable con la solution of soil. Quick rotation.
$NaHCO_3$-Pi	Absorbed in the surface of crystalline compounds and in soil colloids. Not occluded.	Easily available for plants. Quick rotation.
$NaHCO_3$-Po	Weakly adsorbed to humic and fulvic acids. Not occluded.	Easily mineralizable. Quick rotation.
NaOH-Pi	Chemically adsorbed to amorphous and crystalline Al and Fe compounds. Not occluded.	Low availability for plants. Slow rotation.
NaOH-Po	Strongly associated with humic and fulvic acids. Not occluded.	Low availability for plants. Slow rotation.
HCl-Pi	Associated with calcium compounds.	Low availability for plants.
[HCl]-Pi	Minerals rich in phosphorus with low solubility such as apatite. Occluded.	Low availability protected phosphorus for plants. Very slow rotation.
[HCl]-Po	Associated with organic matter not extractable in alkali. Occluded.	Low availability protected phosphorus for plants. Very slow rotation.
[H_2SO_4]-P	Highly resistant inorganic and organic forms. Occluded.	Low availability protected phosphorus for plants. Very slow rotation

[] refers to a highly concentrated acid

0.5 M NaOH extraction was introduced after the 1 M HCl extraction. In the second case, the main modification was that the 0.5 M NaOH extraction was replaced by extraction with concentrated HCl.

After the works of Hedley et al. (1982), Tiessen and Moir (1993), Tiessen et al. (1984), Cross and Schlesinger (1995), it is possible to give a geochemical and ecological meaning to the P extracted sequentially (Table 3) according to the extraction scheme presented in Fig. 6. On this basis, conceptual models of the soil P system have been proposed (Fig. 7).

5 Considerations for Studies of Phosphorous Fertility of Soils

The behavior of P from different fertilizer sources is subject to reactions that determine its accumulation in specific reservoirs, depending on the characteristics of the soils, which have originated under particular conditions of topography, vegetation, and temperature and humidity regimes. In such circumstances, it is necessary to evaluate the residual effect of the fertilizer P, identify the destinations (reservoir fractions) of the added P and their relationship with its availability. Unlike

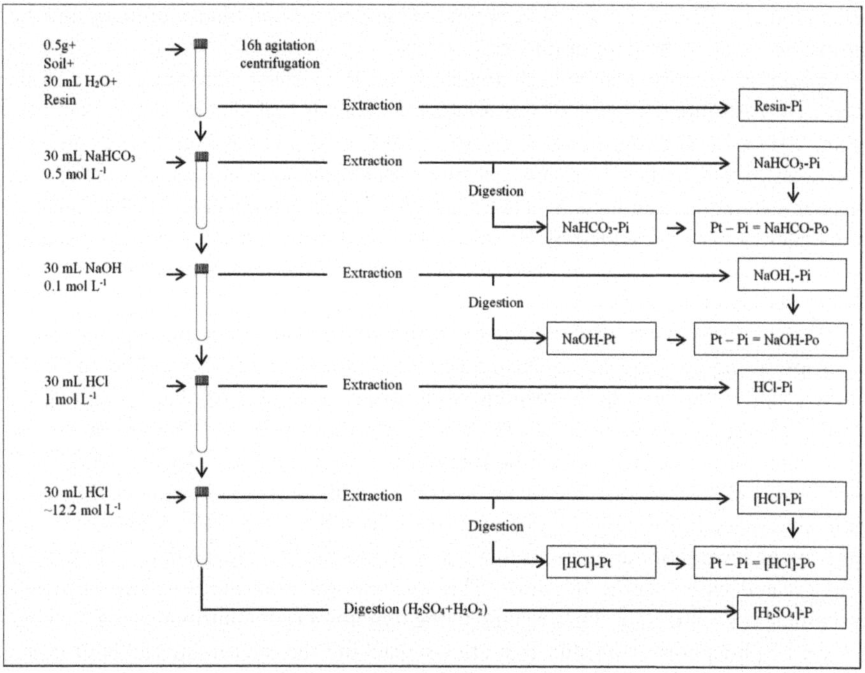

Fig. 6 Sequential extraction of P with the method of Hedley et al. (1982) and modified by Tiessen and Moir (1993). [] refers to a highly concentrated acid

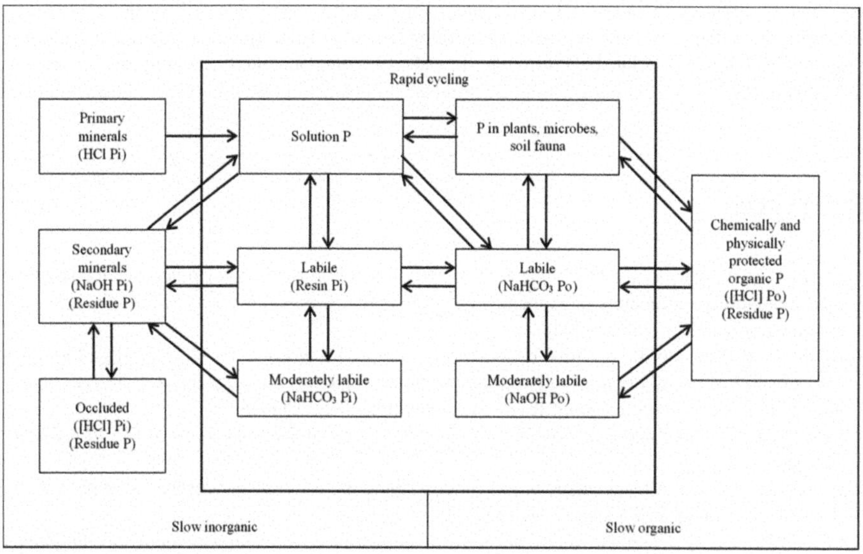

Fig. 7 Conceptual flow diagram between soil P fractions and their respective associated extractants. (Adapted from Tiessen et al., 1984; Tiessen & Moir, 1993). [] refers to a highly concentrated acid

other elements, P does not cycle between the biosphere (soil, plants, animals and the atmosphere), as is the case of nitrogen, so that very often soils exhibit low levels of phosphorous fertility and a high demand for P in most agroecosystems. This threatens the future depletion of P from the reservoirs (Cordell et al., 2009; Schröder et al., 2011). In addition, in agroecosystems, P can be lost through runoff and pass into lakes, rivers, and even oceans, causing contamination of aquatic ecosystems. In this context, the International Soil Information and Reference Center (ISRIC), in one of its latest reports presented by Batjes (2011), declares the need to initiate research programs aimed at improving the understanding of P behavior regarding its availability and dynamics in soils.

The laboratory methods developed to estimate the P available for the absorption by crops during growing season have made it possible to reasonably predict the yield of the crops or the productive response to a certain dose of P fertilizer (Ziadi et al., 2013). Therefore, these methods for estimating the available P should show the biological importance of each of the P fractions, when correlated by the available P with the different fractions that make up the P of soils. It would be expected that the estimated availability with the different methods (Olsen, Bray-P1, Mehlich-3) is a function of several of the fractions that make up the soil P and even that each method is associated with specific fractions. This information contributes to have a higher and efficient P utilization. Data related to the dynamics of P contributes to achieving a level of phosphorous fertility that allows reaching the maximum yields of crops and pastures, and not increasing (pollution), it beyond what is necessary or reducing it (degradation) once, it has reached.

Acknowledgments Corresponding author acknowledges the Ministry of Higher Education, Science and Technology (SENESCYT) of Ecuador for the scholarship received for doctoral study. Similarly, the authors express special thanks Editor Dr. Naga Raju Maddela (Main Professor the Universidad Técnica de Manabí, Portoviejo, Ecuador) for guidance and accepting our request to write this chapter.

References

Anghinoni, I., Baligar, V. C., & Wright, R. J. (1996). Phosphorus sorption isotherm characteristics and availability parameters of Appalachian acidic soils. *Communications in Soil Science and Plant Analysis, 27*(9), 2033–2048.

Azevedo, R. P., Salcedo, I. H., Lima, P. A., da Silva Fraga, V., & Lana, R. M. Q. (2018). Mobility of phosphorus from organic and inorganic source materials in a sandy soil. *International Journal of Recycling of Organic Waste in Agriculture, 7*(2), 153–163.

Barrow, N. J. (1983a). On the reversibility of phosphate sorption by soils. *Journal of Soil Science, 34*, 751–758.

Barrow, N. J. (1983b). A mechanistic model for describing the sorption and desorption of phosphate by soil. *Journal of Soil Science, 34*, 733–750.

Batjes, N. H. (2011). *Overview of soil phosphorus data from a large international soil database. Report 2011/01*. Plant Research International (PRI)/ISRIC - World Soil Information. 56 p.

Bennett, E., Carpenter, S., & Caraco, N. (2001). Human impact on erodable phosphorus and eutrophication: A global perspective. *Bioscience, 51*(3), 227–234.

Bohn. (1993). *Química del suelo. . Suelos ácidos*. Limusa Noriega Editores. 370 p.
Bolland, M. D. A., Allen, D. G., & Barrow, N. J. (2003). *Sorption of phosphorus by soils. How it is measured in Western Australia*. Government of Western Australia, Department of Agriculture, Bulletin 4591.
Bowman, R. A., & Cole, C. V. (1978). Transformations of organic P substrates in soils as evaluated by Na-HCO_3 extraction. *Soil Science, 125*, 49–54.
Brady, N. C., & Weil, R. R. (1999). *The nature and properties of soils* (12th ed.). Prentice Hall. 881 p.
Buerkert, A., Bationo, A., & Piepho, H. (2001). Efficient phosphorus application strategies for increased crop production in Sub-Saharan West Africa. *Field Crops Research, 72*, 1–15.
Busman, L., Lamb, J., Randall, R., Rehm, G., & Schmitt, M. (2002). *The nature of phosphorus in soils*. University of Minnesota.
Cade-Menun, B. (2005). Characterizing phosphorus in environmental and agricultural samples by ^{31}P nuclear magnetic resonance spectroscopy. *Talanta, 66*, 359–371.
Cade-Menun, B. J., Elkin, K. R., Liu, C. W., Bryant, R. B., Kleinman, P. J. A., & Moore, P. A., Jr. (2018). Characterizing the phosphorus forms extracted from soil by the Mehlich III soil test. *Geochemical Transactions, 19*, 1–17.
Chang, S., & Jackson, M. (1957). Fractionation of soil phosphorus. *Soil Science Society of America Journal, 84*, 133–144.
Cheesman, A. W., Turner, B. L., & Reddy, K. R. (2014). Forms of organic phosphorus in wetland soils. *Biogeosciences, 11*(23), 6697–6710.
Condron, L. M., & Goh, K. M. (1989). Effects of long-term phosphatic fertilizer applications on amounts and forms of phosphorus in soils under irrigated pasture in New Zealand. *Journal of Soil Science, 40*(2), 383–395.
Cordell, D., Drangert, J. O., & White, S. (2009). The story of phosphorus: Global food security and food for thought. *Global Environmental Change, 19*, 292–305.
Cross, A. F., & Schlesinger, W. H. (1995). A literature review and evaluation of the Hedley fractionation: Application to the biogeochemical cycle of soil phosphorus in natural ecosystems. *Geoderma, 64*, 197–214.
Dayton, E. A., Shrestha, R. K., Fulford, A. M., Love, K. R., Culman, W., & S. and Lindsey, L. E. (2020). Soil test phosphorus and phosphorus balance trends: A county-level analysis in Ohio. *Agronomy Journal, 112*(3), 1617–1624.
Drechsel, P., Gyiele, L., Kunze, D., & Cofie, O. (2001). Population density, soil nutrient depletion, and economic growth in sub-Saharan Africa. *Ecological Economics, 38*, 251–258.
FAO (Food and Agriculture Organization of the United Nations). (2009). *Global agriculture towards 2050. High level expert forum—How to feed the world in 2050*, Rome, IT.
Fassbender, H. W. (1993). *Modelos edafológicos de sistemas agroforestales* (2nd ed.). Centro de Agronomía Tropical y Enseñanza. 491 p.
Fixen, P. (1992). *Dinámica suelo-cultivo del fósforo y manejo de los fertilizantes fosforados (Parte II)*. Informaciones Agronómicas No. 17.
Frossard, E., Condron, L. M., Oberson, A., Sinaj, S., & Fardeau, J. C. (2000). Processes governing phosphorus availability in temperate soils. *Journal of Environmental Quality, 29*, 15–23.
Fuentes, B., Bolan, N., Naidu, R., & Mora, M. d. l. L. (2006). Phosphorus in organic waste–soil systems. *Journal of Soil Science and Plant Nutrition, 6*(2), 64–83.
Ghosh, G. K., Mohan, K. S., & Sarkar, A. K. (1996). Characterization of soil-fertilizer P reaction products and their evaluation as sources of P for gram (*Cicer arietinum* L.). *Nutrient Cycling in Agroecosystems, 46*, 71–79.
Godfray, H. C., Beddington, J. R., Crute, I. R., Haddad, L., Lawrence, D., Muir, F., Pretty, J., Robinson, S., Thomas, S. M., & Toulmin, C. (2010). Food security: The challenge of feeding 9 billion people. *Science, 327*, 812–818.
Hartemink, A. (2006). Assessing soil fertility decline in the tropics using soil chemical data. *Advances in Agronomy, 89*, 179–225.

Haygarth, P. M., Harrison, A. F., & Turner, B. L. (2018). On the history and future of soil organic phosphorus research: A critique across three generations. *European Journal of Soil Science, 69* (1), 86–94.

Hedley, M. J., Stewart, J., & Chauhan, B. (1982). Changes in inorganic and organic phosphorus fractions induced by cultivation practices and by laboratory incubations. *Soil Science Society of America Journal, 46*, 970–976.

Henao, J., & Baanante, C. A. (1999). *Nutrient depletion in the agricultural soils of Africa*. International Food Policy Research Institute.

Hopkins, B., & Ellsworth, J. (2005). Phosphorus availability with alkaline/calcareous soil. In *Western Nutrient Management Conference* (pp. 88–93).

Howard, A. E. (2006). Agronomic thresholds for soil phosphorus in Alberta: A review. In *Alberta Soil Phosphorus Limits Project*, Alberta, Canada, 42 p.

Ibrikci, H., Ryan, J., Ulger, A. C., Buyuk, G., Cakir, B., Korkmaz, K., Karnez, E., Ozgenturk, G., & Konuskan, O. (2005). Maintenance of phosphorus fertilizer and residual phosphorus effect on corn production. *Nutrient Cycling in Agroecosystems, 72*(3), 279–286.

Ivanoff, D. B., Reddy, K. R., & Robinson, S. (1998). Chemical fractionation of organic phosphorus in selected histosols. *Soil Science, 163*, 36–45.

Javid, S., & Rowell, D. L. (2002). A laboratory study of the effect of time and temperature on the decline in Olsen P following phosphate addition to calcareous soils. *Soil Use and Management, 18*, 127–134.

Jiao, Y., Hendershot, W. H., & Whalen, J. K. (2008). Modeling phosphate adsorption by agricultural and natural soils. *Soil Science Society of America Journal, 72*(4), 1078–1084.

Johnston, A. E., & Dawson, C. J. (2005). *Phosphorus in agriculture and in relation to water quality*. Agricultural Industries Confederation. 71 p.

Johnston, A. E., & Syers, J. K. (2008). A new approach to assessing phosphorus use efficiency in agriculture. *Better Crops, 93*(3), 14–16.

Jordan-Meille, L., Rubaek, G. H., Ehlert, P. A. I., Genot, V., Hofman, G., Goulding, K., Recknagel, J., Provolo, G., & Barraclough, P. (2012). An overview of fertilizer-P recommendations in Europe: Soil testing, calibration and fertilizer recommendations. *Soil Use and Management, 28* (4), 419–435.

Kumar, V., Gilkes, R. J., & Bolland, M. D. A. (1991). Residual phosphate fertilizer compounds in soils. II. Their influence on soil tests for available phosphate. *Fertilizer Research, 30*, 31–38.

Li, H., Huang, G., Meng, Q., Ma, L., Yuan, L., Wang, F., Zhang, W., Cui, Z., Shen, J., Chen, X., Jiang, R., & Zhang, F. (2011). Integrated soil and plant phosphorus management for crop and environment in China. A review. *Plant and Soil, 349*, 157–167.

Lindsay, W. L., Frazier, A. W., & Stephenson, H. F. (1962). Identification of reaction products from phosphate fertilizers in soils. *Soil Science Society of America Journal, 26*(5), 446–452.

Lynch, J. (2011). Root phenes for enhanced soil exploration and phosphorus acquisition: Tools for future crops. *Plant Physiology, 156*, 1041–1049.

MacDonald, G. K., Bennett, E. M., Potter, P. A., & Ramankutty, N. (2011). Agronomic phosphorus imbalances across the world's croplands. *Proceedings of the National Academy of Sciences, 108* (7), 3086–3091.

McLaren, R. G., & Cameron, K. C. (1996). *Soil science, sustainable production and environmental protection* (2nd ed.). Oxford University Press. 304 p.

McLaren, T. I., Smernik, R. J., McLaughlin, M. J., McBeath, T. M., Kirby, J. K., Simpson, R. J., Guppy, C. N., Doolette, A. L., & Richardson, A. E. (2015). Complex forms of soil organic phosphorus—A major component of soil phosphorus. *Environmental Science & Technology, 49* (22), 13238–13245.

McLaughlin, M., McBeath, T., Smernik, R., Stacey, S., Ajiboye, B., & Guppy, C. (2011). The chemical nature of P accumulation in agricultural soils-implications for fertilizer management and design: An Australian perspective. *Plant and Soil, 349*, 69–87.

Mueller, N. D., Gerber, J. S., Johnston, M., Ray, D. K., Ramankutty, N., & Foley, J. A. (2012). Closing yield gaps through nutrient and water management. *Nature, 490*(7419), 254–257.

Naeem, A., Akhtar, M., & Ahmad, W. (2013). Optimizing available phosphorus in calcareous soils fertilized with diammonium phosphate and phosphoric acid using Freundlich adsorption isotherm. *Scientific World Journal, 2013*, 1–5.

Norton, R. (2014). Combating climate change through improved agronomic practices and input-use efficiency. *Journal of Crop Improvement, 28*, 575–618.

Penn, C. J., & Camberato, J. J. (2019). A critical review on soil chemical processes that control how soil pH affects phosphorus availability to plants. *Agriculture, 9*(6), 120.

Phong, L. T., Stoorvogel, J. J., Van Mensvoort, M. E. F., & Udo, H. M. J. (2011). Modeling the soil nutrient balance of integrated agriculture aquaculture systems in the Mekong Delta, Vietnam. *Nutrient Cycling in Agroecosystems, 90*, 33–49.

Pierzynski, J., Hettiarachchi, G., & Khatiwada, R. (2014). Can soil chemical changes influence plant growth? *The Fluid Journal, 22*(1), 9–12.

Pinochet, D. (1995). *The residual effect of applications of phosphate fertilizer measured by the Olsen method* (Thesis of Doctor of Philosophy, The University of Reading).

Posner, A. M., & Barrow, N. J. (1982). Simplification of a model for ion adsorption on oxide surfaces. *Journal of Soil Science, 33*, 211–217.

Rigo, A. Z., Corrêa, J. C., Mafra, Á. L., Hentz, P., Grohskopf, M. A., Gatiboni, L. C., & Bedendo, G. (2019). Phosphorus fractions in soil with organic and mineral fertilization in integrated crop-livestock system. *Revista Brasileira de Ciência do Solo, 43*, e0180130.

Riley, W. J., Ortiz-Monasterio, I., & Matson, P. A. (2001). Nitrogen leaching and soil nitrate, nitrite, and ammonium levels under irrigated wheat in Northern Mexico. *Nutrient Cycling in Agroecosystems, 61*, 223–236.

Roberts, T. L. (2009). The role of fertilizer in growing the world's food. *Better Crops, 93*(2), 12–15.

Sá, J. M., Jantalia, C. P., Teixeira, P. C., Polidoro, J. C., Benites, V. d. M., & Araújo, A. P. (2017). Agronomic and P recovery efficiency of organomineral phosphate fertilizer from poultry litter in sandy and clayey soils. *Pesquisa Agropecuária Brasileira, 52*(9), 786–793.

Sandaña, P. A., Harcha, C. I., & Calderini, D. F. (2009). Sensitivity of yield and grain nitrogen concentration of wheat, lupin and pea to source reduction during grain filling. A comparative survey under high yielding conditions. *Field Crop Research, 114*, 233–243.

Schachtman, D. P., Reid, R. J., & Ayling, S. M. (1998). Phosphorus uptake by plants: From soil to cell. *Plant Physiology, 116*, 447–453.

Schröder, J. J., Smit, A. L., Cordell, D., & Rosemarin, A. (2011). Improved phosphorus use efficiency in agriculture: A key requirement for its sustainable use. *Chemosphere, 84*, 822–831.

Sharma, S. B., Sayyed, R. Z., Trivedi, M. H., & Gobi, T. A. (2013). Phosphate solubilizing microbes: Sustainable approach for managing phosphorus deficiency in agricultural soils. *Springerplus, 2*, 1–14.

Sharpley, A. (2016). Managing agricultural phosphorus to minimize water quality impacts. *Science in Agriculture, 73*(1), 1–8.

Sharpley, A., & Tunney, H. (2000). Phosphorus research strategies to meet agricultural and environmental challenges of the 21st century. *Journal of Environmental Quality, 29*, 176–181.

Sharpley, A. N., Withers, P. J. A., Abdalla, C. W., & Dodd, A. R. (2005). Strategies for the sustainable management of phosphorus. In J. T. Sims & A. N. Sharpley (Eds.), *Agriculture and the environment* (pp. 1069–1101). American Society of Agronomy; Crop Science Society of America; Soil Science Society of America.

Shi, L., Shen, M., Lu, C., Wang, H., Zhou, X., Jin, M., & Wu, T. (2015). Soil phosphorus dynamic, balance and critical P values in long-term fertilization experiment in Taihu Lake region, China. *Journal of Integrative Agriculture, 14*(12), 2446–2455.

Simpson, R., Oberson, A., Culvenor, R., Ryan, M., Veneklaas, E., Lambers, H., Lynch, L., Ryan, P., Delhaize, E., Smith, F., Smith, S., Harvey, P., & Richardson, A. (2011). Strategies and agronomic interventions to improve the phosphorus-use efficiency of farming systems. *Plant and Soil, 349*, 89–120.

Syers, J. K., Johnston, A. E., & Curtin, D. (2008). *Efficiency of soil and fertilizer phosphorus use: Reconciling changing concepts of soil phosphorus behaviour with agronomic information*. FAO Fertilizer and Plant Nutrition. Bulletin 18.

Tiessen, H., & Moir, J. O. (1993). Characterization of available phosphorus by sequential extraction. In M. R. Carter (Ed.), *Soil sampling and methods of analysis* (pp. 75–86). Canadian Society of Soil Science/CRC Press.

Tiessen, H., Stweart, J., & Cole, C. V. (1984). Pathways of phosphorus transformations in soils of differing pedogenesis. *Soil Science Society of America Journal, 48*, 853–858.

Tóth, G., Guicharnaud, R. A., Tóth, B., & Hermann, T. (2014). Phosphorus levels in croplands of the European Union with implications for P fertilizer use. *European Journal of Agronomy, 55*, 42–52.

Valle, S., Carrasco, J., Pinochet, D., & Calderini, D. F. (2009). Grain yield, above-ground and root biomass of Al-tolerant and Al-sensitive wheat cultivars under different soil aluminum concentrations at field conditions. *Plant and Soil, 318*, 299–310.

van der Wiel, B. Z., Weijma, J., van Middelaar, C. E., Kleinke, M., Buisman, C. J. N., & Wichern, F. (2019). Restoring nutrient circularity: A review of nutrient stock and flow analyses of local agro-food-waste systems. In *Resources, Conservation and Recycling: X* (p. 100014).

Van Dijk, K. C., Lesschen, J. P., & Oenema, O. (2016). Phosphorus flows and balances of the European Union Member States. *Science of the Total Environment, 542*, 1078–1093.

van Ittersum, M., Leffelaar, P. A., Van, K. H., Kropff, M. J., Bastiaans, L., & Goudriaan, J. (2003). On approaches and applications of the Wageningen crop models. *European Journal of Agronomy, 18*(3), 201–234.

Wang, X., Liu, F., Tan, W., Li, W., Feng, X., & Sparks, D. L. (2013). Characteristics of phosphate adsorption-desorption on to ferrihydrite: Comparison with well-crystalline Fe (Hydr) oxides. *Soil Science, 178*, 1–11.

Watson, M., & Mullen, R. (2007). *Understanding soil tests for plant available phosphorus*. Ohio State University Extension.

Whalen, J. K., & Chang, C. (2001). Phosphorus accumulation in cultivated soils from long-term annual applications of cattle feedlot manure. *Journal of Environmental Quality, 30*, 229–237.

Yang, X., & Post, W. M. (2011). Phosphorus transformations as a function of pedogenesis: A synthesis of soil phosphorus data using Hedley fractionation method. *Biogeosciences, 8*, 2907–2916.

Zhan, X., Zhang, L., Zhou, B., Zhu, P., Zhang, S., & Xu, M. (2015). Changes in Olsen phosphorus concentration and its response to phosphorus balance in black soils under different long-term fertilization patterns. *PLoS One, 10*(7), e0131713.

Zhang, W., Zhan, X., Zhang, S., Ibrahima, K. H. M., & Xu, M. (2019). Response of soil Olsen-P to P budget under different long-term fertilization treatments in a fluvo-aquic soil. *Journal of Integrative Agriculture, 18*(3), 667–676.

Zhang, X., Wang, Q., Xu, J., Gilliam, F. S., Tremblay, N., & Li, C. (2015). In situ nitrogen mineralization, nitrification, and ammonia volatilization in maize field fertilized with urea in Huanghuaihai Region of Northern China. *PLoS One, 10*(1), e0115649.

Ziadi, N., Whalen, J. K., Messiga, A. J., & Morel, C. (2013). Assessment and modeling of soil available phosphorus in sustainable cropping systems. *Advances in Agronomy, 122*, 85–126.

Environmental Factors Enhance Production of Plant Secondary Metabolites Toward More Tolerance and Human Health: Cocoa and Coffee Two Model Species

Seyed Mehdi Jazayeri, Byron Oviedo-Bayas, Raquel Guerrero-Chuez, Yenny Torres-Navarrete, and Ronald Oswaldo Villamar-Torres

Abbreviations

ABA	Abscisic acid
CHO	Carbohydrates
CNS	Central nervous system
FC	Field capacity
GWAS	Genome wide association study
GxE	Genetics by environment
IPP	Isoprene or isopentenyl diphosphate
NA	Nucleic acid
PGR	Plant growth regulator
PM	Primary metabolites
QTL	Quantitative trait loci

S. M. Jazayeri
Faculty of Biology, University-College of Science, University of Tehran, Tehran, Iran

Departamento de Biología, Facultad de Ciencias, Universidad Nacional de Bogotá, Bogotá, Colombia
e-mail: smjazayeri@ut.ac.ir; smjazayeri@unal.edu.co

B. Oviedo-Bayas · R. Guerrero-Chuez · Y. Torres-Navarrete
Universidad Técnica Estatal de Quevedo, Quevedo, Ecuador
e-mail: boviedo@uteq.edu.ec; rguerrero@uteq.edu.ec; ytorres@uteq.edu.ec

R. O. Villamar-Torres (✉)
Universidad Técnica Estatal de Quevedo, Quevedo, Ecuador

Instituto Superior Tecnológico "Ciudad de Valencia" – Tecnología en Producción Agrícola y Tecnología en Procesamiento de Alimentos, Quevedo, Ecuador
e-mail: rvillamart@uteq.edu.ec

N. R. Maddela, L. C. García (eds.), *Innovations in Biotechnology for a Sustainable Future*, https://doi.org/10.1007/978-3-030-80108-3_9

ROS	Reactive oxygen species
SM	Secondary metabolites
TF	Transcription factor

1 Introduction

Plants species distributed around the earth, i.e., vascular plants, are over 390,000, of which about 369,000 species (or 94%) are flowering plants, as revealed by the Kew Garden. The World Health Organization has listed 21,000 medicinal plants (Chakraborty, 2018) while the Kew garden estimates over at least 28,000 plant species are currently recorded as being of medicinal use of which 16% (4478) of the species are used in plant-based medicines as cited in medicinal regulatory publications (Royal Botanic Gardens Kew, 2017). The use of medicinal plants based on the first written Papyrus Ebers document dates back to 1800 BC (Häkkinen et al., 2012). About 60% of plants have found medicinal use in the post-Neolithic human history (Hao & Xiao, 2015). What make plants medicinal are secondary metabolites (SM) in an acceptable concentration that are estimated hundreds of thousands whose structure, function, and utility are yet to be explored (Wang et al., 2019). All plants produce secondary metabolites in their ontogeny and lifespan. This leads us to consider that all plants are medicinal but their pharmaceutical features and medicinal attributes are unknown and remain to reveal. Plants like wheat, barley, maize, potato, tomato are used in cosmetic and pharmaceutical industries although they are not known as medicinal plants. But when they are used in given quantities and/or based on their SMs they are considered not only food but also plants with medicinal purposes like lycopene of tomato that is used against hypertension or hordenine of barley when used as adrenergic agent and weight loss. Therefore, we hypothesize that as all plants contain pharmaceuticals and medicinal molecules that remain to be discovered to recognize all are medicinal/ pharmaceutical.

There is a high need to reveal plant SM compositions and curing characteristics as well as their relevant functions in plant life as plant SMs have not been fully studied to disclose such attributes (Ashraf et al., 2018). Although several studies are available and have been done but because of highly complex properties of SMs altogether the performed research up to now is the tip of iceberg. The ancient medicines such as Avicenna, Galen, Razi, Jorjani, etc. categorized plants in three use levels: food, pharmaceutical, and toxin. Therefore, the consuming quantity of primary metabolites (PM) and the concentration of SMs determine if the plant should be considered as food, medicine, or poison. The higher concentration of SMs can make a plant toxic or medicinal. Thus, the consuming dosage is a very important factor to nominate it in each of three levels. However, all plants have this capability to produce trace quantity of SMs to supply their need for living, health, and survival. The amount required for a SM to make it medicinal is not clear, though effective suggested dosages and useful consuming quantity for some SMs are available.

On the other hand, the functions of SMs in plants might be different from their medicinal property and pharmaceutical application. Plants produce SMs to handle physiological processes and biochemical reactions. Secondary metabolites are produced by plants as an intrinsic system when required in response to the condition they are exposed like undesired abiotic stress, insect attack, pathogen infection. This usage difference remains to be revealed and needs serious efforts to concord functionalities of SMs in plants and other living organisms that use them as their energetic or non-energetic sources. However, as an interesting point, biological similarities between human, animals, and plants make it modellable and predictable to predict properties of SMs. An antioxidant in plants that can scavenge ROS molecules, function as an agent against cancer to reduced ROS produced during cancer progress. Therefore, it is a reciprocal field that disclosed medicinal attributes of SMs can suggest/indicate their function in plants and functionality of a SM can be extended to have similar effect in human and animals. If the effects of a SM have been revealed in human and/or animals and its function is not clear yet in plants, the function can be developed biologically functionally to plants as putative.

Various SMs of many plant species have been studied because of their medicinal properties but gradually scientists and researchers have found that they possess multifunctionality involving in growth and development as well as plant response to surrounding ambient (Wang et al., 2019). First SMs were considered as waste and undesirable (by)products (Hartmann, 2007) but by revealing their physiological and biological effects in plants they are considered now as necessary and fundamental metabolites in plant life and evolution. By progressing in agriculture and plant sciences, scientists found that SMs have evolved during time to defend plants against insects, pests, pathogens while facilitating pollinator attraction (Wink, 2018). Therefore, many functions have come out by advances in biology, medicine, and agriculture emphasizing that SMs are complicated molecules that possess several benefits for plants, animals, and human and subsequently are primordial. These make plant SMs foremost.

Modeling a real environment for SM accumulation is a way that can make plant produce in a better manner their SMs. Although plant photosynthesis and production are affected when exposed to stresses toward less yield in terms of biomass (Jazayeri et al., 2015) but SMs are altered in terms of quantity and quality to enhance their potential response to stresses (Jazayeri, 2000; Kleinwächter & Selmar, 2015). One of the best ways to make plants produce more SMs even cultivated ones is to simulate natural conditions in their habitat. However, this type of simulation needs more research and deep studies to disclose optimum condition for SM production in plants taking into account species specific requirements, plant habitat and SMs as well as simultaneous occurrence of stresses.

Production, type, and concentration(s) of SMs in plants are determined by species, genotype, physiology, developmental stage, and environmental factors. In addition, plant groups and families that possess resistance quality to stresses share similar SMs indicating that there is a relation between plant genotypes, SMs, and resistance. This can help us understand how plants tolerate stresses and how stresses enhance SM production. The environmental factors including local geo-climatic and seasonal changes, external conditions of temperature, light, humidity and developmental processes, among others, impact biomass production and biosynthesis of SMs (Isah, 2019). Each environmental cue can affect SM yield based on plant genetics suggesting GxE effect should be considered a particular case in plant SM quality and quantity.

Growing conditions affect plant vegetative and reproductive phases and the concentrations of various secondary plant products are strongly depending on the growing conditions and it is obvious that especially stress situations have a strong impact on the metabolic pathways responsible for the accumulation of the related natural products (Selmar, 2008). In one study on *Thymus daenensis*, the authors found that concentration of essential oils, thymol and carvacrol were changed in different ecological populations growing in varied zones (Rustaiee et al., 2011) indicating that natural habitat facing various stresses is of importance to quality and quantity of SMs. Natural growing conditions give fruitful lessons on plant ability to manufacture SMs while artificial selection of plants with higher potential of producing SMs can be an option to domesticate wild plants for medicinal purposes. This can help plant-based drug discovery and development as well as the sustainable utilization of vegetal pharmaceutical resources (Fakhr Tabatabei, 1999). However, several studies have been performed toward disclosing optimum conditions for plant SM production but it remains to further studies to have a better understanding of plant SM knowledge. Environmental cues are grouped into biotic and abiotic. Each of them has its own impacts on SM production. As in nature they occur simultaneously modeling multi-stress studies are essential to understand how plant respond to adverse stresses in term of SM production.

Genomics can reveal the genes and their copy number in plants and transcriptomics can disclose if these genes expressed in the acceptable amount that is worthy to use for medicinal purpose and plant breeding as well. On the other hand, the concentration of metabolites is very important to make a plant medicinal. Metabolomics studies the metabolites of living organisms to see the relation between genes and environment. This can be done by a new branch of omics as pharmaceomics by which the pharmaceome that is the whole SMs of a plant are studied to reveal and predict its pharmaceutical/medicinal properties. Pharmaceomics is the biological field by which plants are studied to reveal their pharmaceutical value. Like genomics, proteomics, and metabolomics, pharmaceomics will open new pavements for scrutinizing SMs and PMs. This method starts with an extraction of metabolites as a whole and then results in their effects on plant life, adaptation and contribution considering their roles in plants, concentration, and composition. It will be of help for interatomic studies.

In the following sections we present plant response to stress via SMs, SM functions in plant, their explanations as pharmaceuticals and in plant breeding, with cocoa and coffee as two examples because of their importance and climate change impacts on them. Finally, we conclude ecological factors and environmental stresses can be of assistance for obtaining more SM concentrations while having more tolerant plants and stress induction can enhance SM production in plants.

2 Stress and Plant Response

Plants respond to stress by different mechanisms and approaches that are cellular and molecular. These responses result in different levels toward the whole plant. At the time of perceiving a stress by receptors, at molecular level some reactions occur like change of metabolite production and composition, gene expression pattern, and at cellular level other relevant actions are subsequently done like altering cell shape, internal turgor, stomata closure/opening, or cell wall composition. All these reactions, thus, are found out in the whole plant as turgidity or wilt appearance, hard or soft leaves and shoots, big or small fruits, more or less growth/reproduction, etc. Therefore, each plant response in whatever level seems to be reflected by the whole plant. Studying plant responses in different levels is done by different biological sciences. Ecology is the science of the whole plant responses to environment and ecophysiology determines molecular and cellular impacts while many biological science branches are involved in each study like genomics, transcriptomics, proteomics, biochemistry, physiology, histology, morphology, etc. This makes such kind of study and research very interesting and at the same time complicated (Fakhr Tabatabei, 1999).

As shown in Fig. 1, in metabolic level and genomic level plants can employ different metabolites known as PM, SM, phytohormones to tolerate a stress. Primary metabolites are affected in a way that can control the situation in favor of plant survival but secondary metabolites are produced to enhance plant tolerance to undesirable conditions. Accumulation of SMs under drought stress is seen but this positive effect might be appeased or compensated by other impacts and effects as other environmental factors exist simultaneously (Kleinwächter & Selmar, 2015). For example, proteins are degraded to accumulate branched-chain amino acids in osmotic stress (Huang & Jander, 2017) and nucleic acids are damaged during stress

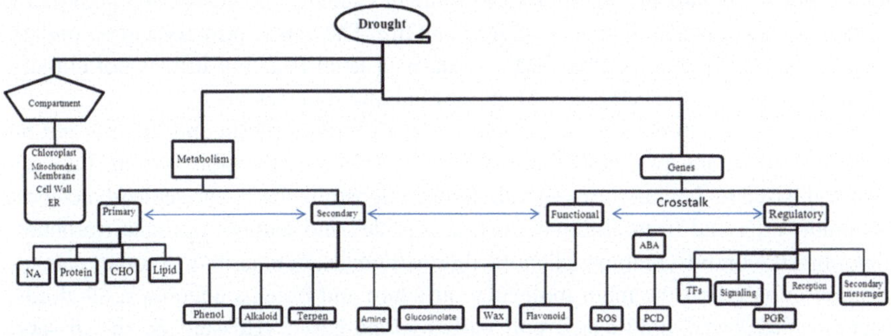

Fig. 1 Plants adopt different approaches to stresses by cellular or molecular responses. These approaches are demonstrated in the whole plant as various effects of stress like changes in composition of hydrocarbons, lipids and fatty acids, damaged DNA and proteins for PMs. The figure is taken from the first author doctoral thesis (Jazayeri, 2015). *NA* nucleic acid, *CHO* carbohydrates, *ROS* reactive oxygen species, *ABA* abscisic acid, *TF* transcription factor, *PGR* plant growth regulator

as signaling mechanism (Nisa et al., 2019). Terpenes are changed to modify plant ability and water potential to tolerate an undesirable condition like drought or salinity. Therefore, plants choose how to reply to environmental factors by different approaches via modifying PMs and SMs toward more tolerance to an adverse environmental situation that can include different undesirable conditions as simultaneously occurring stresses. While phytohormones are influenced by stress but as a feedback they can lead plants to more tolerance to stresses. Thus, phytohormones, because of their multifunctionalities, can impact the plant responses to stresses by controlling various approaches happening in plant life as modifying metabolites shown in Fig. 1.

Plant secondary metabolism as shown in Fig. 1 is altered by stresses (in the Fig. 1 Drought is an example of stresses and environmental cues. It can be replaced by any kind of stress). Each molecule or compartment that is altered by stress affects its downstream parts. However, finding upstream receptors and involving keys that perceive stress and initiate plant response is an issue that remains to further discoveries.

3 Secondary Metabolites in Plants

Plants produce primary and secondary metabolites. Primary metabolites include proteins, carbohydrates, nucleic acids, and lipids and fatty acids. They are generated as the essential molecules without which plants are not created and cannot survive; i.e., they are basic life elements. Secondary metabolites were known as not necessarily vital molecules in continuation of life in plants (Jamwal et al., 2018) at the first glance but without them plant life does not continue. Therefore, SMs are considered as essential compounds originated from PMs to complete their functionalities and their existence is indispensable for plant ontogeny and life processes although many authors do not consider them as vital for plant life. The names primary and secondary metabolites imply this fact that SMs are derived from PMs but it does not classify them as two separated categories as primordial and unnecessary.

Secondary metabolites are classified under different criteria and there is not an exact categorization or classification for them. In a classification system, they are grouped based on having nitrogen (alkaloids, glucosinolates, cyanogenic glycosides, alkamides) derived from amino acids and peptides and without nitrogen (terpenes, saponins, flavonoids, tannins, phenols, lignins/lignans, coumarins). Another categorization is based on the main molecular structure and base as phenolics, alkaloids, saponins, terpenes, lipid-based and carbohydrate-based (Hussein & El-Anssary, 2019). As another one, plant SMs are categorized into terpenes, phenolics, N containing compounds and S containing compounds (Jamwal et al., 2018). Phenolic and polyphenolic compounds, terpenoids, alkaloids and sulfur-containing compounds are the four major classifications of secondary metabolites (Hounsome et al., 2008).

Thus, it is better to employ different systems and consider different plant aspects for classifying SMs to facilitate understanding their function and production in plants as normally plants are classified sometimes based on their SM types like aromatic plants (Lamiaceae, Apiaceae, Pinacease, and Verbenaceae), alkaloid plants (Ephedraceae, Papaveraceae, Solanaceae, Ranunculaceae, Amaryllidaceae), sulfur-containing plants (Brassicaceae, Liliaceae), etc. By such multi-dimension and poly-level classification, they are not grouped only based on their chemical structure but on their plant origin, pharmaceutical effects, and any other attribute that help differentiate them.

Terpenoids and terpenes (well-known as aromatic and volatile compounds) are one of the largest and most diverse class of plant metabolites as they derive to different molecules due to their simple unit; isoprene or isopentenyl diphosphate (IPP) (Tholl, 2015) that can be converted to complex metabolites. Previously, they were called essential oils or essences due to their flammability and insolubility in water while soluble in alcohol, ether and fixed oils (Dhifi et al., 2016) and mainly some families like Lamiacease (Labiatae), Pinacease, Verbenaceae, Apiaceae are known as strong odor plants with well-known oily extracts that are used in many products like perfumes, toothpastes, and analgesic creams. Alkaloids and phenolic acids and their derivatives are of diverse classes and most overrepresented in plant kingdom that possess phenol unit and nitrogen-based structure respectively (Hussain et al., 2018; Mandal et al., 2010). Phenolic containing plants are known due to their bitter taste (specially in fruits at the early stages) and analgesic character like Cameliaceae, Rutaceae, Rubiaceae. Alkaloid plants that are represented by narcotics are famous because of their latex like Berberidaceae, Papaveraceae, Ranunculaceae, and Euphorbiaceae families. Flavonoids contain plant pigments with different color (Samanta et al., 2011) and possess capacity to modulate key cellular enzyme function (Panche et al., 2016). Flower and fruit pigments and colors come from flavonoids. They are phenolic derivatives but are preferably classified as a distinct group because of their antioxidant (ROS scavenging) and colorful characters. Saponins are a glycoside derivates (a skeleton derived of the 30-carbon precursor oxidosqualene to which glycosyl residues are attached) (Kregiel et al., 2017) with a soap-making property that makes them a base for vaccines and pharmaceuticals as anticancer and anticholesterol (Guclu-Ustundag & Mazza, 2007). There are 11 main classes of saponins named based on their plant origin (Vincken et al., 2007) that mainly are present in the plant families including Agavaceae, Alliaceae, Amaryllidaceae, Asparagaceae, Bromeliaceae, Dioscoreaceae, Liliaceae, Palmae, and Scrophulariaceae.

Table 1 shows the plant metabolite groups and their chemical units with their pharmaceutical attributes and functions in plants. In addition, the name form for different plant metabolite groups is explained in to make reader more familiar with them. However, this nomenclature is used to more easily recognize them by their origin. In addition, IUPAC names considering their chemical structures are used in literature.

Secondary metabolites can be classified based on their effects on plants and on human and animals. These categories are somehow similar depending on the SM

Table 1 The different SM categories in plant based on their chemical and pharmaceutical properties

Metabolite	Pharmaceutical/ medicinal effects	Chemical unit, name character	Example molecules	Plant function
Alkaloids	Analgesic, sedative, dietary supplement, antidiabetes, anticancer, antibacterial, antiarrhythmic, antifungal, antiviral, protein precipitation, astringent, digestive agent	Nitrogen-containing organic molecules originating from amino acids normally. The name of alkaloids generally terminates with –ine as suffix while the name base comes from plant name	Actinidine, atropine, berberine, caffeine, cocaine, cytisine, ephedrine, fumaricine, galantamine, glaucine, hyoscyamine, isoquinoline, morphine, nicotine, papaverine, piperine, psilocin, quinidine, quinine, reserpine, scopolamine, vinblastine, vincamine, vincristine	Nitrogen reservoirs, protective agents against herbivores, growth regulators, balancing osmotic pressure, protective in pathogen attacks, prevention of insect and fungi penetration, water balance, alleviating oxidative stress
Flavonoids	Antioxidant, Anti-cholinesterase steroid-genesis antioxidative, anti-inflammatory, antimutagenic, anticarcinogenic, preventing PMs (DNA, protein) damage	Phenolic derivatives, the names terminate in—in or—one generally	Anthocyanin, arbutin, betacyanin, betaxantin, chalcone, cyanidin, delphnidin, flavonol, luteolin, malvidin, orientin, quercetin, pelargonidin, phloretin, vicenin	Stress protectors, ROS scavengers, UV-filter, signal molecules, antimicrobial defensive agents, insect pollinator colorful attractants, prevent PMs (DNA, protein) damage
Saponins	Nutrient absorption, digestion, drug surfactant base, anticancer and anticholesterol	Aglycone (glycoside-free). The name termination is –ane	Cucurbitane, cycloartane, dammarane, hopane, lanostane, lupane, oleanane, steroid, taraxasterane, tricullane, ursane	Antifeedant, antimicrobe, stress tolerance
Terpenes and terpenoids (essential oils)	Antiseptic, antioxidant, antimicrobial, antifungal, anti-inflammatory, anticancer, skin care, antileprotic, antifilarial, antibacterial, antiviral	Isoprene unit (IPP), C_5H_8. Their names terminate to –en, –ene, and –ol as suffix with a plant name base at the first part	Azadirachtin, artemisinin, carotene, carotenoid, carvacol, cineole, cymene, eucalyptol, eugenol, limonene, linalool, linolen, menthol, terpineol, tetrahydrocannabinol	Antioxidant, growth and development, physiological and biochemical reactions, deterring herbivores, predator attractant, abiotic and biotic protectors, adaptogenic, antifeedant, ameliorating water balance increased

(continued)

Table 1 (continued)

Metabolite	Pharmaceutical/ medicinal effects	Chemical unit, name character	Example molecules	Plant function
				in pest and pathogen attack to defend plant cells
Phenol derivatives	Antioxidant, anti-cancer, Antibleeding, anticoagulant, antimicrobial, antifungal	Phenol chain. The name normally is based on a plant name derivative and acid. Or sometimes the name suffix is –in	4-hydroxybenzaldehyde, 4-methoxycinnamic acid, apigenin, caffeic acid, cinnamic acid, circimaritin, coumaric acid (coumarin), ferulic acid, gallic acid, isothymusin, lignin, phenolic acid, rosmarinic acid, salicylic acid, syringic acid, tannic acid, vanillic acid, vanillin	Chemical defense against pests, IAA production, phytoanticipiant, chemo attractant, insect repellent, symbiosis agents with microbes, rhizobium growth and nodulation

structure. However, the effects on animal and human cells are different from the effects on plants as physiologically they are different organisms. Despite of their differences, the effects might be scrutinized to find which effects they have on animals and human based on plant physiology and biological mechanisms and processes. They are antibiotic, antibacterial, antiviral, antifungal, anti-inflammatory, antitumor, antianaphylactic, antimutagenic, choleretic, and bronchodilatory actions to protect plants against biotic (pests and pathogens) and abiotic (drought, salinity, UV, light excess, flooding, hypoxia, and so on) stresses. These effects may be similar when they are used as pharmaceutical agents.

Interestingly, some SMs can be stress signals causing plants react to them as abnormal condition like methyl jasmonate, jasmonic acid, salicylic acid, polyamines, nitric oxide, melatonin, serotonin, brassino steroids, abscisic acid (Ramakrishna & Ravishankar, 2011). Such study of SMs generates more details but it is more complicated. That is why studying SMs in plants responding to stress is a tricky research.

4 Plant SM Production Under Environmental Cues and Stresses

Plants require a minimum of inputs as water, light, air, micro/macronutrients, under an optimum climate that permit them to grow, develop, and reproduce adequately. The factors involved in plant production in the nature are not under control and therefore not being optimum always. On the other hand, plants are sessile and cannot move in order to escape from undesirable conditions occurring naturally out of control. Therefore, during evolution they have adopted various adaptation strategies in order to live and survive via producing SMs as protective agents like flavonoids (antioxidants) for preventing damages of cellular structure and DNA and as storage reservoirs like alkaloids and terpenoids, and as regulators of biological reactions involved in growth, development, and reproduction like (poly)amines and alkaloids. The difference between plants and other moving living organisms is these plant-specific SMs that plants produce and employ while other non-plant organisms are not able to generate them.

Generally, plant cultivation triggers high biomass and yield (depending on crop product: foliage, bark, trunk, flower, fruit, seed) while for medicinal plants biomass increase should be taken into account with—predictable—SM extract profiles. The value of medicinal plants is based on SMs and phytochemicals as pharmaceuticals that make them more valuable economically and medicinally. Hence, medicinal plant production needs a combination of developing strategies toward more SM accumulation meanwhile not experiencing biomass decrease that affect SM productivity. Subsequently, stress influence on SM rendering becomes more important than biomass yield in plants with pharmaceutical and medicinal interests.

Exposed to environmental constraints (biotic and abiotic) plants produce SMs toward more adaptation. Stress conditions are not optimum for plant growth and development but they provoke expression or repression of stress-gene networks that result in the production of SMs in the form of regulators, detoxifiers, transporters, chaperons, proteases (Isah, 2019). Stress induction and stimulation on plants results generally positively in SM changes contrasting with likely negative effects on biomass and vegetable production. Accordingly, it is hypothesized that stresses can positively alter SM production in terms of quantity and quality either with decrease in biomass or without change in plant biomass and growth and development. Plants under optimum growth condition produce their biomass in a desirable manner. They have intrinsic abilities to produce SM that is altered by conditions in which plants grow and reproduce.

Plants produce SMs in varied concentration depending on genetics and environment. However, it is a key factor in plant SM production to obtain as maximum SMs as possible from wild and/or cultivated plants. Theoretically farms provide as optimum as possible the conditions and requirements for plant growth, development, and production to reach more yield but the optimum condition does not mean that plants maximize SM productivity. On the other hand, plant breeding is aimed to create plants that can produce more while being more resistant to undesirable conditions. It is believed that wild plants produce more SMs than cultivated plants (Selmar, 2008). This idea comes from this fact that in the nature plants, at the same time, are faced with several environmental factors with whom they have to adapt like light competing, water, etc. Thus, the nature and wild plants are the best sources for evaluating optimum life conditions and genetic pools enabling further improvements in plant production and breeding. Studying and modeling natural conditions for each plant to evaluate its SM productivity potential is the best way to find an optimized cultivation method.

Plants produce more metabolites in response to abiotic stresses as revealed in many stress studies like drought (Jazayeri, 2000), salinity (Aghaei & Komatsu, 2013), light (Thoma et al., 2020), herbivores (Villamar-Torres et al., 2018). In one study antioxidant activity and content of phenols, flavonoids proline and malondialdehyde (MDA) of four *Achillea* species under drought as four field capacity (FC) levels (100%, 75%, 50%, and 25% as control, low, moderate and sever stress, respectively) fluctuated by different water level and among the species. This study indicated that the optimum condition to obtain phenol and flavonoid content is when moderate drought stress treatment applied and the highest antioxidant activity is obtained in severe stress condition (25% FC). Drought stress promoted the production of phenolic compounds and increased 40-fold uliginosin B in *Hypericum polyanthemum*. In the study, drought helped the accumulation of SMs and in some stress treatments caused an increase of *H. polyanhemum* biomass as well (de Matos Nunes et al., 2014).

Terpenes are emitted in response to various stresses specially herbivore or pathogen attack. They are known as phytoalexins for dissuading malign attackers. In one study, their compositions and quantity were shown that were changed when the cotton plants were subjected to insect attacks (Villamar-Torres et al., 2018). In another study by Zhang et al. their involvement and production as chemical defense agent in *Santalum album* was confirmed. The authors scrutinized three terpene synthase genes and showed that enzyme activity and terpene production resulted in accumulation of terpenes upon stress (Zhang et al., 2019).

Table 2 shows plant SMs responding to different stress conditions by enhancing their production. This leads us to think about this fact that stress induction per se can be effectively useful for plants to fabricate more SMs toward more tolerating point during their life from seed to seed. Taking into account that plant responses are not same always but somehow similar, modeling from similar studies can reveal which environmental cue(s) are of help or effective to produce more SMs. On the other hand, plant requirements in each life cycle from seed and plantlet to adult stage and next generation seed vary and this point is essential to understand how to apply enhancing condition for SM production while considering plant need in each life stage. This conclusion can be determined better by combining genomic, transcriptomic, metabolomic, and pharmaceomic studies. The references presented in Table 2 show that plant SM production enhanced by stresses has been of interest and under consideration for a long time, maybe as long as agricultural history.

Advances in technology permits to perform more sophisticated methods in order to enhance SM production and improve SM quality taking into account that due to stress influence often plant biomass might be altered negatively but sometimes positively. For plant breeding it is suggested to study natural genetic pool in wild plants and compare their genome and transcriptome—and pharmaceome—with cultivated/domesticated/bred plants to retrieve the best (if possible) genotype. In such effort, GWAS in population, QTL data, gene expression profiles, metabolome composition, and pharmaceome details trigger the variants by which potential SM capacity comes out. One method is combination of different stresses at the same time to evaluate plant reaction by evaluating gene behavior by RNA-Seq, GWAS studies,

Table 2 Stress response and SMs in plants. In this table, the secondary metabolite production increased upon imposed stress. For most cases, SM production enhances in response to stresses but there are also SMs that decrease upon stress. However, it requires modeling or experimental confirmation to determine increase/decrease SM production as stress response depends on plant genetics and environmental cues. Another point, some SMs increase in a stressed plant and others decrease, thus SM composition vary. It means that plant SM quality and quantity should be taken into account as a whole

Secondary metabolites	Plant species	Stress response	References
Alkaloids	*Phalaris aquatica*	Drought	Ball and Hoveland (1978)
Allicin	*Allium sativum*	UV	Jeong et al. (2013)
Andrographolide	*Andrographis paniculata*	Drought (population effects)	Saravanan et al. (2009)
Anthocyanins	*Pisum sativum*	Drought	Nogués et al. (1998)
Anthocyanins	*Grevillea* sp.	Salinity	Kennedy and De Filippis (1999)
Anthocyanins	*Grevillea* sp.	Salinity	Parida and Das (2005)
Artemisinin	*Artemisia annua*	Salinity, drought, water excess	Vashisth et al. (2018)
Asiaticoside and madecassoside	*Centella asiatica*	Low temperature and water dehydration	Plengmuankhae and Tantitadapitak (2015)
Betulinic acid	*Hypericum brasiliense*	Drought	De Abreu and Mazzafera (2005)
Camptothecin	*Camptotheca acuminata*	Nitrogen, drought and anti-transpiration agents	Feng et al. (2002), Sun and Tan (2008)
Capsaicin	*Capsicum* sp.	Salinity	Gammoudi et al. (2019)
Chinolizidin alkaloids	*Lupinus angustifolius*	Drought	Christiansen et al. (1997)
Chlorogenic acid	*Helianthus annuus*	Drought	del Moral (1972)
Choline	*Trifolium alexandrinum*	Salinity	Varshney and Gangwar (1988)
Codeine	*Papaver somniferum*	Drought	Szabó et al. (2003)
Cyanogenic glucosides	*Manihot esculenta*	Drought	De Bruijn (1973)
Cyanogenic glucosides	*Manihot esculenta*	Drought	Okogbenin et al. (2003)
Cyanogenic glucosides	*Triglochin maritima*	Drought	Clawson and Moran (1937)
Cyanogenic glucosides	*Eucalyptus cladocalyx*	Drought	Woodrow et al. (2002)
Di- and polyamines	*Oryza sativa*	Salinity	Krishnamurthy and Bhagwat (1989)
Epicatechins	*Camellia sinensis*	Drought	Hernández et al. (2006)

(continued)

Table 2 (continued)

Secondary metabolites	Plant species	Stress response	References
Essential oils, eucalyptol, menthol, linoleol, carvacol	*Eucalyptus camaldulensis*	Drought, salinity	Jazayeri (2000)
Flavonoids	*Pisum sativum*	Drought	Nogués et al. (1998)
Flavonoids	*Hordeum vulgare*	Salinity	Ali and Abbas (2003)
Flavonoids	*Hordeum vulgare*	Salinity	Ali and Abbas (2003)
Flavonoids	*Prisms sativum*		Larson (1988)
GABA	*Sesamum indicum*	Salinity, drought, heat, heavy metal	Bor et al. (2009)
Glucosinolates	*Brassica napus*	Drought	Jensen et al. (1996)
Glucosinolates	*Brassica napus*	Drought	Bouchereau et al. (1995)
Glycine betaine	*Triticum aestivum*	Salinity	Krishnamurthy and Bhagwat (1989)
Glycine betaine	*Trifolium repens*	Salinity	Varshney and Gangwar (1988)
Glycosides	*Scrophularia ningpoensis*		Wang et al. (2010)
Glycyrrhyzin	*Glycyrrhiza glabra*	Drought	Nasrollahi et al. (2014)
Hyoscyamine and scopolamine	*Atropa belladonna*	Chromium	Vakili et al. (2012)
Indole alkaloids (vindoline, vinblastine and catharanthine)	*Catharanthus roseus*	Drought	Liu et al. (2017)
Jasmonic acid	*Lycopersicon esculentum*	Salinity	Pedranzani et al. (2003)
Morphine alkaloids	*Papaver somniferum*	Drought	Szabó et al. (2003)
Phenolic compounds	*Thymus capitatus*	Drought	Delitala et al. (1986)
Polyamines	*Oryza sativa*	Salinity	Krishnamurthy and Bhagwat (1989)
Polyphenol	*Cakile maritima*	Salinity	Ksouri et al. (2007)
Proline	*Camptotheca acuminata*	Low temperature	Jiancan et al. (2002)
Proline	*Atropa belladonna*	Chromium	Vakili et al. (2012)
Proline	*Theobroma cacao*	Drought	M'bo Kacou et al. (2016)
Pyrrolizidine alkaloids	*Senecio longilobus*	Drought	Briske and Camp (1982)
Salvianolic acid B	*Salvia miltiorrhiza*	Drought	Liu et al. (2011)

(continued)

Table 2 (continued)

Secondary metabolites	Plant species	Stress response	References
Rutin	*Dimorphandra mollis*	Drought, fooding, and salinity	Lucci and Mazzafera (2009)
Rutin	*Hypericum brasiliense*	Drought	De Abreu and Mazzafera (2005)
Saikosaponins	*Bupleurum chinense*	Drought, watering and re-watering	Yang et al. (2019), Zhu et al. (2009)
Sorbitol	*Lycopersicon esculentum*	Salinity	Tari et al. (2010)
Stevioside	*Stevia rebaudiana*	Hydroponic culture, salt stress	Aghighi Shahverdi et al. (2019), Srivastava and Srivastava (2014), Zeng et al. (2013)
Tannin, alkaloids, terpenoids	*Mentha piperita*	Drought, heat	Alhaithloul et al. (2019)
Tannin, alkaloids, terpenoids	*Catharanthus roseus*	Drought, heat	Alhaithloul et al. (2019)
Total phenols	*Echinacea purpurea*	Drought	Gray et al. (2003)
Total phenols	*Prunus persica*	Drought	Kubota et al. (1988)
Total phenols, dihydroxy-xanthone, betulinic acid and rutine	*Hypericum brasiliense*	Drought	De Abreu and Mazzafera (2005)
Trigonelline	*Glycine max*	Salinity	Cho et al. (1999)
Trigonelline strong	*Glycine max*	Drought	Cho et al. (2003)
Tropane alkaloids	*Datura innoxia*	Salinity	Brachet and Cosson (1986)
Valepotriates	*Valeria* sp.	Drought, salinity	Ebrahimzadeh et al. (2008)
Zealexins and kauralexins	*Zea mays*	Drought	Vaughan et al. (2015)

and QTL methods as well as metabolomic approaches. Stress combination lets us scrutinize more precisely plants' comportment as a whole. This can improve our understanding of relationship between plant and stress. Hence, multi-stress simulation to study plant response to environmental cues is a way to enhance plant SM production. This, then, can be applied in real cultivations by changing seed sowing period, irrigation time, pest and disease management, agronomic practices to move forward an ameliorated cultivation. By using encapsulated bioactive reagents, the content of phenolics and flavonoids in *Lactuca sativa* was increased without any significant impact on the morphology and moisture content (Jurić et al., 2020). This method fortifies plants' defense system against pests and increases the tolerance to damaging environmental cues. Such modern method can be applied for plants under stress to obtain more SMs in response to the stress. Combination of such technical approaches and multi-stress strategy is recommended for further plant selection and breeding research and agricultural procedures.

5 Multifunctionality of SMs: Plants Resist Stresses by Secondary Metabolites, Thus Why Human Not?

Generally speaking, the properties of SMs can be modelized between plants and other organisms considering PMs as common metabolites among all living organisms. In all living organisms, nucleic acids (DNA or RNA) contain and transfer inheritable genetic matter from a generation to another, proteins regulate biological reactions and construct structures, carbohydrates, fatty acids, and lipids are energetic compounds. Thus, all function in a similar way, however, they vary in composition, quantity, and quality depending on each living organism.

Plants respond to stress by repertoire of molecular, cellular cross-talk, and signaling approaches when perceiving specific or combined biotic or abiotic stress that may result in the induction of SMs (Wink, 1988). Functions of SMs might be different for plants as necessary molecules and for human as pharmaceuticals with medicinal purposes. However, there is very few studies on functionality of SMs in plants and animals and human all together. It is known that SMs are not of primary metabolism molecules but they are required for survival and responding to environmental factors as functional, signaling and interacting molecules (Pagare et al., 2015). For example, while plants producing alkaloids like morphine use them as molecules for conducting their response to environmental factors, augmenting tolerance to stress and nitrogen assimilation, human uses it as a relaxant and appeasing drug but not as nitrogen sink. Such differences between plants and human and plants and animals make multifunctionality of SMs an extremely interesting but enough complicated to evaluate similar and varied effects of SMs in different organisms; plants as producers and animals and human as consumers. The prospective for this kind of research; i.e., studying effects of SMs in plants as models for cure similar problems (stresses and diseases) in human. An example can clarify this type of studies. If a molecule like Trehalose in plants function to decrease the effects of drought stress, adjust osmosis and help plants increase cellular pressure and therefore balance water need, it can be used where maintenance of cellular pressure is required as reported in ophthalmology that Trehalose can prevent damage to mammalian eyes caused by desiccation and oxidative insult (Luyckx & Baudouin, 2011).

Plants and animals and human share conserved genes (as sequences) and proteins with the same or very similar function. Although plants do photosynthesis particularly that differentiates them from animals and human but there are many conserved noncoding elements (CNEs) common between plants and vertebrates (Burgess & Freeling, 2014). The COP9 signalosome shares a common evolutionary ancestor in structural composition with closely relation with the lid subcomplex of the 26S proteasome. A multifaceted role of the COP9 signalosome is associated kinase activity as well as the involvement of its subunits in regulating multiple cell-signaling pathways and cell-cycle progression in cell-signaling processes. These functions are conserved between plants and human (Wei & Deng, 1999). Using mass spectrometry of the recovered complexes, around 120 proteins were identified as potential in vivo 14-3-3 interacting proteins in *Arabidopsis*. Comparison of these *Arabidopsis* proteins with the 14-3-3 interacting proteins in human embryonic kidney cell cultures disclosed 8 interacting proteins that represent fundamental 14-3-3 interaction complexes being highly conserved across all eukaryotes. Based on biochemical function, many additional similarities in the human and *Arabidopsis* were revealed that possess conserved functional interactions, while also leaving many proteins uniquely identified in either *Arabidopsis* or human cells (Paul et al., 2009).

Plant antioxidants act as first line of defense against oxidative damage induced by different environmental cues (salinity, drought, temperature, heavy metal, UV, and hypoxia stress). Oxidative damage result from usually increasing reactive oxygen species (ROS) caused by environmental stresses. In human and animals, diseases like cancer and adverse stresses impose ROS production that is mitigated and detoxified by antioxidants (Liou & Storz, 2010). Plants containing antioxidants like cyanidins, genistein, fisetin, gingerol, lycopene, rosmarinic acid are used as remedy or enhancer of medicines against diseases like cancer, blood pressure, as anticancer, anti-inflammation, antitumor, and anti-proliferation as they attenuate impacts of ROS and subsequently lessen disease and stress impacts on human body and cells (Wang et al., 2012).

The antioxidants include low molecular weight compounds such as GSH, ascorbate (ASC), α-tocopherol, carotenoids, polyphenols, and enzymes including CAT, SOD, and the thiol peroxidases of the PRX and glutathione peroxidase (GPX) type (Laxa et al., 2019). The goal of plants from producing such antioxidants is very close to the application they have as pharmaceuticals. Plants produce antioxidants in order to prevent ROS molecules like superoxide, hydroxyl and nitric oxide radicals that can damage DNA during stresses. Considering these compounds as pharmaceuticals, their biological effects, including but not all as anti-inflammatory, anti-aging, anti-atherosclerosis and anticancer have similar effects on human that help cells to prevent likely damage for DNA (Xu et al., 2017). Thus, antioxidants in plants, animals, and human function similarly.

Melatonin was explored in plants in 1995 after it had been found in human in 1958 (Lerner et al., 1958). It is an antioxidant that can protect plants in stress conditions and also play a role as osmoregulatory metabolite. It is active in redox network that alleviate stress effects on membrane integrity that is hampered by elevated cellular levels of ROS (Arnao & Hernández-Ruiz, 2019). In one study, Najafi et al. showed that various concentrations of melatonin decreases intracellular ROS levels and stabilizes membrane integrity in human spermatozoa (Najafi et al., 2018). Thus, the studies on human could help understand how melatonin acts in plants.

Terpenes constitute the main SMs of *Centella asiatica*; compounds like asiatic acid, madecassic acid, asiaticoside, madecassoside, brahmoside brahmic acid, brahminoside, thankiniside, isothankunisode, centelloside, madasiatic acid, centic acid, and centellicac. They are known as medicinal molecules having medicinal properties as antileprotic, antifilarial, antibacterial, adaptogenic, antifeedant, and antiviral (Rao et al., 2015). As terpenes are involved in defense response in plants, they can ameliorate condition for plants under stress due their antioxidant behavior for example upon exposure to UVB (Hashim et al., 2011). Thus, according to effects of *C. asiatica* on human and the background of these terpenoid components, it is possible to predict their functionality in plants themselves although there are no many studies on functionality of such terpenes in *C. asiatica*.

Flavonoids or coloring agents in plants are considered as antioxidant, anti-inflammatory, antimutagen, and anticancer. They are also cardiac medicines with low cardiovascular mortality and CHD prevention attribute (Panche et al., 2016). They function in plants, animals, and human under similar roles as their ROS scavenging property can prevent cancer and cardiac issues in human and stress protection in plants.

Searching the literature, unfortunately such studies revealing similarities between plants and human gene; protein and metabolite functions as well as biological pathways and molecular complexes are not currently pursued despite of their importance in biology, medicine, and agriculture. Plant metabolites as subsequent products of genes and proteins may have close function in plants, animals and human. The above examples lead us to various points: plants can be good models for studying SM medicinal and pharmaceutical effects and research results about PMs and SMs can be modelized among different living organisms. These accelerate studies in medicine and pharmacology on one side and on other side improve our knowledge about undiscovered properties of different SMs not only in plant but also in human and animal. This type of modeling can help us predict how one metabolite function along with all living organisms. Plants grow more easily than animals, they are sessile and controlling them is more achievable. This plant special character let us predict and hypothesize how SMs can affect human as pharmaceuticals. In Table 1, comparing SM function in plants and their similar effects in human as pharmaceuticals are presented.

If we would have found a new metabolite in plants, which we know its role and function in plants, subsequently we could predict how it can affect human and animals even before experimental tests. This hypothesis promotes studies on pharmaceutical and medicinal properties of plants and also assists to perform reverse modeling from human to plants. This remains to be detailed by further research practically while theoretically it is understandable and seems achievable.

6 Cocoa and Coffee Two Species as Models for Potential SM Production and Adaptation

Coffee (*Coffea* sp.) and cocoa (*Theobroma cacao* L.) are two of the most important species cultivated in tropical areas and in Latin American countries in particular. The cultivation of cocoa and coffee has a highly influence on global economy as millions of people depend on them from production to consumption as farmers, workers, agroindustrial, and hygienic manufacturers, etc. They seem to be in the core of food and cosmetic industry with global turnover of ~305 billion US dollars (Pipitone, 2019). Their agronomic development takes place in heterogeneous ecosystems, subjected to different environmental conditions. Cocoa and coffee suffer from drought and higher temperatures resulting from climate change and the studies showed that their tolerance and stress response need further studies to achieve and prove genetically their adaptation to adverse environmental cues (Medina & Laliberte, 2017). Caffeine, caffeic acid, and theobromine and their derivatives are well-known metabolites that admit high value to cocoa and coffee as a quality index. They are involved in plant stress adaptation and their concentration is changed by stresses.

In the case of cocoa, the flavor and aroma of its beans were the main reasons that allowed its domestication in pre-Columbian times in Mesoamerica (Motamayor et al., 2002). Many chemical compounds in almonds are responsible for sensory quality, but polyphenols and alkaloids stand out, compounds that directly affect the flavor and palatability of almonds and indirectly on aroma precursors. The composition of the alkaloids is related to bitterness, and their concentration is related to the variety and changes with processing. On the other hand, polyphenols are responsible, along with other molecules for astringency (undesirable in chocolates), but also for antioxidant properties, the latter being highly desired among consumers from different markets. The composition of the cocoa bean is related to the SM production, and is the result of the interaction of genetic, environmental, and management factors.

Phenolic compounds in cocoa from different tissues including tegument, cotyledons, whole seed and seed without mucilage have been reported as chlorogenic acid, ferulic acid, caffeic acid, *p*-cumaric acid, *p*-hydroxycinnamic acid, *p*-hydroxybenzoic acid, syringic acid, vanilic acid, protocathetic acid, catechin, epicatechin while caffeic acid, *p*-coumaric acid and protocathetic acid have been found in all studies tissues. The phytosterols characterization of seed parts showed a qualitative homogenous distribution with a quantitative predominance of β-sitosterol in the tissues of tegument, cotyledon and embryo. The concentrations of the single compounds were significantly higher in tegument in comparison to the other seed parts, except for β-sitosterol and Δ5-avenasterol that were detected in significantly larger amount in the embryo-axis (Cerri et al., 2019).

Research studies have shown that cocoa is sensitive to water deficit and waterlogging (De Almeida et al., 2016) and specially its bean yield but it can mitigate drought effects to tolerate drought and higher temperatures by osmotic adjustment (Moser et al., 2010). Previously some studies have shown that stressed plants enhanced higher production SMs that include terpenes, complex phenols, and alkaloids through the induction of ionic or osmotic stress (Isah, 2019). Polyamines (ornithine, spermidine, spermine) are induced upon drought stress and *Phytophthora megakarya* infection as cocoa responses to mitigate adverse effects imposed by stress (Bae et al., 2008). Polyamines prevent human aging and overall, cardiovascular and cancer-related mortality (Madeo et al., 2018) and positively impact cellular functions in plants in increasing longevity, increasing pro-health carotenoids such as lycopene, recalling physiological memory, enhancing carbon and nitrogen resource allocation/signaling, as well as in plant development and responses to extreme environment in similar way to those of human taking into account that their high concentration act negatively (Handa et al., 2018). These can justify why cocoa is one of the plants used for promoting heart health, preventing aging, and empowering memory as its polyphenolic and polyamine compounds affect cells in a similar way between plants and human (Handa et al., 2018).

Terpenoids and phenolic compounds (arjunolic acid, 3,4 dihydroxyacetophenone and 4-hydroxyaceto-phenone) play roles in cocoa resistance mechanisms to pathogens like *Verticillium dahliae* (Resende et al., 1996). I-limonene, *p*-ethylguaiacol and 2,3-dihidrobenzofuran are produced by cocoa upon infection caused *Ceratobasidium theobromae* acting as potential phytoalexins (Iman Santoso et al., 2017). Various SMs, purine alkaloids, polyphenols, flavonoids, serotonin, are accumulated in cocoa seeds upon abiotic and biotic stresses by switching first to secondary metabolism to give resistant attribute to cocoa seeds (Wang et al., 2016). These SMs are of those metabolites endowing taste and benefits to cocoa seeds.

Thus, the cocoa SM reservoirs supply the means to tolerate and protect from abiotic and biotic stresses while such SMs are medicinally and nutritionally useful compounds for human. Cocoa by its highly genetic variation can adapt to adverse stresses although its bean production might be affected (Lahive et al., 2019). This suggests that how cocoa trees can tolerate climate change adverse conditions by improving their SMs content especially those with human interest, i.e., theobromine, caffeine, and their derivatives although their yield might have been decreased. Of note, cocoa diversity and genetic variation are of importance in its tolerance along with domestication as high number of cocoa genotypes are available (Cornejo et al., 2018) those have not been used yet in domestication and breeding programs and this invaluable genetic pool accentuates it as an effective source for studies of plant response to stress, SM storage, and medicinal properties. However, this remains to reveal these potential capacities of cocoa hidden in its genetic variation, SM content and responses to climate change.

As in the entire plant kingdom, coffee plants allocate a significant amount of assimilated carbon and energy to the synthesis of a wide variety of organic molecules that do not seem to have a direct function in photosynthetic and respiratory processes, assimilation of nutrients, solute transport or synthesis of proteins, carbohydrates or lipids. As the main well-known metabolites of Arabica Coffee (*Coffea arabica*) alkaloids and polyphenols (caffeine, caffeic and palmitic acids, tannins among others) have their famous influence in medicine and pharmacology. All emissions of SMs are mostly related to environmental effects, which allow an increase or decrease in the quantities of produced SMs. These metabolites have their own function in plants as different biotic and abiotic stress protective agents. The role of caffeine and caffeic acid as insecticide of coffee plants has also been shown via their accumulation or secretion in cells (Phankaen et al., 2017). Caffeine paralyzes and kills many insects as a central nervous system (CNS) stimulant, having the effect of warding off drowsiness and restoring alertness. Coffee has repellent, insecticidal, antifeedant, and growth regulatory properties against various insect pests like *Leptinotarsa decemlineata*, *Streptomyces scabies*, and *Clavibacter michiganensis*, *Aedes aegypti*, and *Ochlerotatus notoscriptus* larvae (Bedmutha et al., 2011; Derraik & Slaney, 2005; Laranja et al., 2003). Positively by inotropic and chronotropic effects with their locomotor activity stimulation and anxiogenic-like effects, coffee and caffeine-containing products affect human cardiovascular system and CNS (Cappelletti et al., 2014). Application of coffee extract on plant leaves and caffeine transgenic plants have been shown that it can act a vaccination for plant to repel insects and pathogens and increase tolerance to stress (Kim et al., 2010) mitigating stress effect by antioxidant activity and being effective in oxidative stress. In one study on human endometriosis, caffeic acid reduced oxidative stress by alleviating complications associated with endometriosis (Jamali et al., 2019). Thus, these findings and applications suggest that coffee SMs specially caffeine, caffeic acid, their intermediates, derivates and other purine alkaloids having medicinal effects are involved in plant adaptation and their increase in coffee plants can be advantageous for plant tolerating stress and richness of pharmaceutical products derived from them (Farah & Donangelo, 2006).

These SMs, therefore, are essential for plant life in response to stress and also play similar roles in human disease and stressful conditions. On the other hand, coffee plants generate them to ameliorate adverse condition upon exposure to environmental cues. Altogether, stress can enhance SM production in plants in favor of promoting tolerance while at the same time it can be useful for more SM yield in plants that benefits human health.

7 Conclusions

All plants produce SMs as their essential compounds for living and survival, therefore, all plants are medicinal/pharmaceutical. Pharmaceomics studies pharmaceome that is the whole SMs. Plant SMs benefit human health and wealth as basic ingredients for medicines and remedy. However, how we apply and use plants and their products (containing SMs) determines whether they are medicinally utilized as pharmaceutical in traditional medicines and pharmacology. Usage of SMs is a human directed model as remedy while SMs have their own attributes and benefits for plants whose roles may be different from those used by human as pharmaceuticals while sharing similar effects and pathways. Production of SM in plants is a response to basically environmental cues. It is important to model new methods from nature to use them on plant SM production and cultivation to obtain better quality and quantity of pharmaceutics. Functionalities of SMs are likely conserved between plants and human considering their organismic evolutionary differences and they play close similar roles. This leads us to model plants for finding solutions via SMs for human diseases and issues coming from similar factors.

Plant SMs play a significant role in plant adaptation to various environmental cues that cause alteration in plant growth and the biosynthesis of SMs. Secondary metabolism resulting in SM production regulate various plant life processes like growth, development, reproduction, senescence, and apoptosis. It acts as a reservoir of key phytochemicals protecting plants against multiple environmental constraints. For human, plant SMs are of importance for nutritional and pharmaceutical purposes. Despite of their importance, plant SM production and attributes are unknown and remain to be disclosed. On the other hand, more research and efforts are required to understand proteins and genes involved in SM biosynthesis toward plant tolerance and plant pharmaceomic selection. In this chapter have been detailed for two model species as cocoa and coffee latest highlights regarding SMs production and adaptation mean that plants have acquired. It has been explained in overall, that plants under stress as well as wild plants can be of help to find adequate cultivation condition to produce more SMs, under adverse cues imposed by climate change. Strategies considering integrated omics including genomics, transcriptomics, proteomics, metabolomics, interactomics, and pharmaceomics should be boosted to enhance plant SM production by ecological factors and environmental constraints to overcome climate change issues. Stress studies using such tools direct us to better programmed plant improvement and breeding. Stresses and climate change are two opportunities to take advantages of in order to accelerate plant SM production.

References

Aghaei, K., & Komatsu, S. (2013). Crop and medicinal plants proteomics in response to salt stress. *Frontiers in Plant Science, 4*, 8. https://doi.org/10.3389/fpls.2013.00008

Aghighi Shahverdi, M., Omidi, H., & Tabatabaei, S. J. (2019). Stevia (Stevia rebaudiana Bertoni) responses to NaCl stress: Growth, photosynthetic pigments, diterpene glycosides and ion content in root and shoot. *Journal of the Saudi Society of Agricultural Sciences, 18*(4), 355–360. https://doi.org/10.1016/j.jssas.2017.12.001

Alhaithloul, H. A., Soliman, M. H., Ameta, K. L., El-Esawi, M. A., & Elkelish, A. (2019). Changes in ecophysiology, osmolytes, and secondary metabolites of the medicinal plants of Mentha piperita and Catharanthus roseus subjected to drought and heat stress. *Biomolecules, 10*(1), 43. https://doi.org/10.3390/biom10010043

Ali, R. M., & Abbas, H. M. (2003). Response of salt stressed barley seedlings to phenylurea. *Plant Soil Environment, 49*(4), 158–162.

Arnao, M. B., & Hernández-Ruiz, J. (2019). Melatonin and reactive oxygen and nitrogen species: A model for the plant redox network. *Melatonin Research, 2*(3), 152–168. https://doi.org/10.32794/11250036

Ashraf, M. A., Iqbal, M., Rasheed, R., Hussain, I., Riaz, M., & Arif, M. S. (2018). In P. Ahmad, M. A. Ahanger, V. P. Singh, D. K. Tripathi, P. Alam, & M. N. Alyemeni (Eds.), *Plant metabolites and regulation under environmental stress*. Academic Press.

Bae, H., Kim, S.-H., Kim, M. S., Sicher, R. C., Lary, D., Strem, M. D., Natarajan, S., & Bailey, B. A. (2008). The drought response of Theobroma cacao (cocoa) and the regulation of genes involved in polyamine biosynthesis by drought and other stresses. *Plant Physiology and Biochemistry, 46*(2), 174–188. https://doi.org/10.1016/j.plaphy.2007.10.014

Ball, D. M., & Hoveland, C. S. (1978). Alkaloid levels in Phalaris aquatica L. as affected by environment. *Agronomy Journal, 70*(6), 977–981. https://doi.org/10.2134/agronj1978.00021962007000060021x

Bedmutha, R., Booker, C. J., Ferrante, L., Briens, C., Berruti, F., Yeung, K. K. C., Scott, I., & Conn, K. (2011). Insecticidal and bactericidal characteristics of the bio-oil from the fast pyrolysis of coffee grounds. *Journal of Analytical and Applied Pyrolysis, 90*(2), 224–231. https://doi.org/10.1016/j.jaap.2010.12.011

Bor, M., Seckin, B., Ozgur, R., Yılmaz, O., Ozdemir, F., & Turkan, I. (2009). Comparative effects of drought, salt, heavy metal and heat stresses on gamma-aminobutryric acid levels of sesame (Sesamum indicum L.). *Acta Physiologiae Plantarum, 31*(3), 655–659. https://doi.org/10.1007/s11738-008-0255-2

Bouchereau, A., Clossais-Besnard, N., Bensaoud, A., Leport, L., & Renard, M. (1995). Water stress effects on rapeseed quality. *European Journal of Agronomy, 5*, 19–30.

Brachet, J., & Cosson, L. (1986). Changes in the total alkaloid content of Datura innoxia Mill. subjected to salt stress. *Journal of Experimental Botany, 37*, 650–656. https://doi.org/10.2307/23691492

Briske, D. D., & Camp, B. J. (1982). Water stress increases alkaloid concentrations in threadleaf groundsel (Senecio longilobus). *Weed Science, 30*(1), 106–108. https://doi.org/10.1017/s0043174500026278

Burgess, D., & Freeling, M. (2014). The most deeply conserved noncoding sequences in plants serve similar functions to those in vertebrates despite large differences in evolutionary rates. *Plant Cell, 26*(3), 946–961. https://doi.org/10.1105/tpc.113.121905

Cappelletti, S., Daria, P., Sani, G., & Aromatario, M. (2014). Caffeine: Cognitive and physical performance enhancer or psychoactive drug? *Current Neuropharmacology, 13*(1), 71–88. https://doi.org/10.2174/1570159x13666141210215655

Cerri, M., Reale, L., & Zadra, C. (2019). Metabolite storage in Theobroma cocoa L. seed: Cyto-histological and phytochemical analyses. *Frontiers in Plant Science, 10*, 1599. https://doi.org/10.3389/fpls.2019.01599

Chakraborty, P. (2018). Herbal genomics as tools for dissecting new metabolic pathways of unexplored medicinal plants and drug discovery. *Biochimie Open, 6*, 9–16. https://doi.org/10.1016/J.BIOPEN.2017.12.003

Cho, Y., Lightfoot, D. A., & Wood, A. J. (1999). Trigonelline concentrations in salt stressed leaves of cultivated glycine max. *Phytochemistry, 52*(7), 1235–1238. https://doi.org/10.1016/S0031-9422(99)00410-0

Cho, Y., Njiti, V. N., Chen, X., Lightfoot, D. A., & Wood, A. J. (2003). Trigonelline concentration in field-grown soybean in response to irrigation. *Biologia Plantarum, 46*(3), 405–410. https://doi.org/10.1023/A:1024390522259

Christiansen, J., Jørnsgard, B., Buskov, S., & Olsen, C. (1997). Effect of drought stress on content and composition of seed alkaloids in narrow-leafed lupin, Lupinus angustifolius L. *European Journal of Agronomy, 7*, 307–314.

Clawson, A. B., & Moran, E. A. (1937). Toxicity of arrowgrass for sheep and the reme-144 dial treatment. *Technical Bulletin/USDA, 580*, 1–16.

Cornejo, O. E., Yee, M. C., Dominguez, V., Andrews, M., Sockell, A., Strandberg, E., Livingstone, D., Stack, C., Romero, A., Umaharan, P., Royaert, S., Tawari, N. R., Ng, P., Gutierrez, O., Phillips, W., Mockaitis, K., Bustamante, C. D., & Motamayor, J. C. (2018). Population genomic analyses of the chocolate tree, Theobroma cocoa L., provide insights into its domestication process. *Communications Biology, 1*(1), 1–12. https://doi.org/10.1038/s42003-018-0168-6

De Abreu, I. N., & Mazzafera, P. (2005). Effect of water and temperature stress on the content of active constituents of Hypericum brasiliense Choisy. *Plant Physiology and Biochemistry, 43*(3), 241–248. https://doi.org/10.1016/j.plaphy.2005.01.020

De Almeida, J., Tezara, W., & Herrera, A. (2016). Physiological responses to drought and experimental water deficit and waterlogging of four clones of cocoa (Theobroma cacao L.) selected for cultivation in Venezuela. *Agricultural Water Management, 171*, 80–88. https://doi.org/10.1016/j.agwat.2016.03.012

De Bruijn, G. H. (1973). The cyanogenic character of cassava (Manihot esculenta). In B. Nestel & R. MacIntyre (Eds.), *Chronic Cassava Toxicity: Proceedings of an Interdisciplinary Workshop, London, England, 29-30 January 1973* (pp. 43–48). Internat Development Research.
de Matos Nunes, J., Bertodo, L. O. O., da Rosa, L. M. G., Von Poser, G. L., & Rech, S. B. (2014). Stress induction of valuable secondary metabolites in Hypericum polyanthemum acclimatized plants. *South African Journal of Botany, 94*, 182–189. https://doi.org/10.1016/j.sajb.2014.06.014
del Moral, R. (1972). On the variability of chlorogenic acid concentration. *Oecologia, 9*(3), 289–300. https://doi.org/10.1007/BF00345238
Delitala, L.-F., Gessam, C., & Solinas, V. (1986). Water stress and flexibility of phenolic metabolism in thymus capitatus. *Fitoterapia, 57*(6), 401–408.
Derraik, J. G. B., & Slaney, D. (2005). The toxicity of used coffee grounds to the larvae of Ochlerotatus (Finlaya) notoscriptus (Skuse) (Diptera: Culicidae). *The Annals of Medical Entomology, 14*, 14–24.
Dhifi, W., Bellili, S., Jazi, S., Bahloul, N., & Mnif, W. (2016). Essential oils' chemical characterization and investigation of some biological activities: A critical review. *Medicine, 3*(4), 25. https://doi.org/10.3390/medicines3040025
Ebrahimzadeh, H., Radjabian, T., Ekhteraei Tousi, S., Bashiri Sadr, Z., Niknam, V., & Zarrei, M. (2008). Quantification of valerenic acid and its derivatives in some species of Valeriana L. and Centranthus longiflorus Stev. *Asian Journal of Plant Sciences, 7*(2), 195–200. https://doi.org/10.3923/ajps.2008.195.200
Fakhr Tabatabei, S. M. (1999). Missing points in chain of agricultural production to industrial production of medicinal plants. *Agricultural Economics and Development, 7*(28), 231–251.
Farah, A., & Donangelo, C. M. (2006). Phenolic compounds in coffee. *Brazilian Journal of Plant Physiology, 18*(1), 23–36. https://doi.org/10.1590/S1677-04202006000100003
Feng, J., Zhang, Y., & Yang, T.-Z. (2002). Efect of low-temperature stress on membrane lipid peroxidation and concentration of free-proline in Camptotheca acuminata seedling. *Forestry Research, 15*(2), 197–202.
Gammoudi, N., Zerria, K., Nagaz, K., & Ferchichi, A. (2019). Enhancement of capsaicinoids in vitro production by abiotic elicitors in placenta-derived callus of Capsicum annuum L. Tunisian var. 'Baklouti Medenine'. *Biologia, 74*(6), 725–732. https://doi.org/10.2478/s11756-019-00237-8
Gray, D. E., Pallardy, S. G., Garrett, H. E., & Rottinghaus, G. E. (2003). Acute drought stress and plant age effects on alkamide and phenolic acid content in purple coneflower roots. *Planta Medica, 69*(1), 50–55. https://doi.org/10.1055/s-2003-37026
Guclu-Ustundag, Ö., & Mazza, G. (2007). Saponins: Properties, applications and processing. *Critical Reviews in Food Science and Nutrition, 47*(3), 231–258. https://doi.org/10.1080/10408390600698197
Häkkinen, S. T., Ritala, A., Rischer, H., & Oksman-Caldentey, K.-M. (2012). Medicinal plants, engineering of secondary metabolites in cell cultures. In *Encyclopedia of sustainability science and technology* (pp. 6519–6538). Springer. https://doi.org/10.1007/978-1-4419-0851-3_387
Handa, A. K., Fatima, T., & Mattoo, A. K. (2018). Polyamines: Bio-molecules with diverse functions in plant and human health and disease. *Frontiers in Chemistry, 6*(10), 10. https://doi.org/10.3389/fchem.2018.00010
Hao, D.-C., & Xiao, P.-G. (2015). Genomics and evolution in traditional medicinal plants: Road to a healthier life. *Evolutionary Bioinformatics Online, 11*, 197–212. https://doi.org/10.4137/EBO.S31326
Hartmann, T. (2007). From waste products to ecochemicals: Fifty years research of plant secondary metabolism. *Phytochemistry, 68*(22–24), 2831–2846. https://doi.org/10.1016/j.phytochem.2007.09.017
Hashim, P., Sidek, H., Helan, M. H. M., Sabery, A., Palanisamy, U. D., & Ilham, M. (2011). Triterpene composition and bioactivities of Centella asiatica. *Molecules, 16*, 1310–1322.

Hernández, I., Alegre, L., & Munné-Bosch, S. (2006). Enhanced oxidation of flavan-3-ols and proanthocyanidin accumulation in water-stressed tea plants. *Phytochemistry, 67*(11), 1120–1126. https://doi.org/10.1016/j.phytochem.2006.04.002

Hounsome, N., Hounsome, B., Tomos, D., & Edwards-Jones, G. (2008). Plant metabolites and nutritional quality of vegetables. *Journal of Food Science, 73*(4), 48–65.

Huang, T., & Jander, G. (2017). Abscisic acid-regulated protein degradation causes osmotic stress-induced accumulation of branched-chain amino acids in Arabidopsis thaliana. *Planta, 246*(4), 737–747. https://doi.org/10.1007/s00425-017-2727-3

Hussain, G., Rasul, A., Anwar, H., Aziz, N., Razzaq, A., Wei, W., Ali, M., Li, J., & Li, X. (2018). Role of plant derived alkaloids and their mechanism in neurodegenerative disorders. *International Journal of Biological Sciences, 14*(3), 341–357. https://doi.org/10.7150/ijbs.23247

Hussein, R. A., & El-Anssary, A. A. (2019). Plants secondary metabolites: The key drivers of the pharmacological actions of medicinal plants. In *Herbal medicine*. IntechOpen. https://doi.org/10.5772/intechopen.76139

Iman Santoso, T., Miftahudin, M., Sulistyaningsih, Y. C., & Wiyono, S. (2017). Analysis of secondary metabolites as potential phytoalexins, their secretion sites and proposed resistance markers to vascular streak dieback in Theobroma cacao L. *Edition Pelita Perkebunan, 33*(1), 10–23.

Isah, T. (2019). Stress and defense responses in plant secondary metabolites production. *Biological Research, 52*(1), 39. https://doi.org/10.1186/s40659-019-0246-3

Jamali, N., Mostafavi-Pour, Z., Zal, F., Kasraeian, M., Poordast, T., Ramezani, F., & Zare, R. (2019). Combination effect of caffeine and caffeic acid treatment on the oxidant status of ectopic endometrial cells separated from patients with endometriosis. *Iranian Journal of Medical Sciences, 44*(4), 315–324. https://doi.org/10.30476/ijms.2019.44970

Jamwal, K., Bhattacharya, S., & Puri, S. (2018). Plant growth regulator mediated consequences of secondary metabolites in medicinal plants. *Journal of Applied Research on Medicinal and Aromatic Plants, 9*, 26–38. https://doi.org/10.1016/j.jarmap.2017.12.003

Jazayeri, S. M. (2000). *Effects of drought stress on some metabolites of Eucalyptus camaldulensis*. University of Tehran.

Jazayeri, S. M. (2015). *Characterization of genes related to oil palm (Elaeis guineensis Jacq.) drought stress responses*. National University of Colombia (Universidad Nacional de Colombia).

Jazayeri, S. M., Rivera, Y. D., Camperos-Reyes, J. E., & Romero, H. M. (2015). Physiological effects of water deficit on two oil palm (Elaeis guineensis Jacq.) genotypes. *Agronomía Colombiana, 33*(2), 164–173. https://doi.org/10.15446/agron.colomb.v33n2.49846

Jensen, C. R., Mogensen, V. O., Mortensen, G., Fieldsend, J. K., Milford, G. F. J., Andersen, M. N., & Thage, J. H. (1996). Seed glucosinolate, oil and protein contents of field-grown rape (Brassica napus L.) affected by soil drying and evaporative demand. *Field Crops Research, 47*(2–3), 93–105. https://doi.org/10.1016/0378-4290(96)00026-3

Jeong, H., Lee, S.-H., Yun, H.-S., & Choi, S.-R. (2013). Changes in allicin contents of garlic via light irradiation. *Korean Journal of Food Preservation, 20*(1), 81–87. https://doi.org/10.11002/kjfp.2013.20.1.81

Jiancan, P., Yujie, Z., & Tianzhu, Y. (2002). Effect of low temperature stress on the membrane-lipid peroxidation and the concentration of free proline in Camptotheca acuminata seedling. *Forest Research, 15*(2), 197–202. https://europepmc.org/article/cba/382064

Jurić, S., Sopko Stracenski, K., Król-Kilińska, Ż., Žutić, I., Uher, S. F., Đermić, E., Topolovec-Pintarić, S., & Vinceković, M. (2020). The enhancement of plant secondary metabolites content in Lactuca sativa L. by encapsulated bioactive agents. *Scientific Reports, 10*(1), 1–12. https://doi.org/10.1038/s41598-020-60690-3

Kennedy, B. F., & De Filippis, L. F. (1999). Physiological and oxidative response to NaCl of the salt tolerant Grevillea ilicifolia and the salt sensitive Grevillea arenaria. *Journal of Plant Physiology, 155*(6), 746–754. https://doi.org/10.1016/S0176-1617(99)80092-3

Kim, Y. S., Choi, Y. E., & Sano, H. (2010). Plant vaccination: Stimulation of defense system by caffeine production in planta. *Plant Signaling and Behavior, 5*(5), 489–493. https://doi.org/10.4161/psb.11087

Kleinwächter, M., & Selmar, D. (2015). New insights explain that drought stress enhances the quality of spice and medicinal plants: Potential applications. *Agronomy for Sustainable Development, 35*(1), 121–131. https://doi.org/10.1007/s13593-014-0260-3

Kregiel, D., Berlowska, J., Witonska, I., Antolak, H., Proestos, C., Babic, M., Babic, L., & Zhang, B. (2017). Saponin-based, biological-active surfactants from plants. In *Application and characterization of surfactants*. IntechOpen. https://doi.org/10.5772/68062

Krishnamurthy, R., & Bhagwat, K. A. (1989). Polyamines as modulators of salt tolerance in rice cultivars. *Plant Physiology, 91*(2), 500–504. https://doi.org/10.1104/pp.91.2.500

Ksouri, R., Megdiche, W., Debez, A., Falleh, H., Grignon, C., & Abdelly, C. (2007). Salinity effects on polyphenol content and antioxidant activities in leaves of the halophyte Cakile maritima. *Plant Physiology and Biochemistry, 45*(3–4), 244–249. https://doi.org/10.1016/j.plaphy.2007.02.001

Kubota, N., Mimura, H., & Shimamura, K. (1988). The effects of drought and flooding on the phenolic compounds in peach fruits. *Science Report of the Faculty of Agriculture, Okayama University, 171*, 17–21.

Lahive, F., Hadley, P., & Daymond, A. J. (2019). The physiological responses of cacao to the environment and the implications for climate change resilience. A review. *Agronomy for Sustainable Development, 39*(1), 1–22. https://doi.org/10.1007/s13593-018-0552-0

Laranja, A. T., Manzatto, A. J., & de Campos Bicudo, H. E. M. (2003). Effects of caffeine and used coffee grounds on biological features of Aedes aegypti (Diptera, Culicidae) and their possible use in alternative control. *Genetics and Molecular Biology, 26*(4), 419–429. https://doi.org/10.1590/S1415-47572003000400004

Larson, R. A. (1988). The antioxidants of higher plants. *Phytochemistry, 27*(4), 969–978. https://doi.org/10.1016/0031-9422(88)80254-1

Laxa, M., Liebthal, M., Telman, W., Chibani, K., & Dietz, K. J. (2019). The role of the plant antioxidant system in drought tolerance. *Antioxidants, 8*(4), 94. https://doi.org/10.3390/antiox8040094

Lerner, A. B., Case, J. D., Takahashi, Y., Lee, T. H., & Mori, W. (1958). Isolation of melatonin, the pineal gland factor that lightens melanocytes. *Journal of the American Chemical Society, 80*, 2587.

Liou, G. Y., & Storz, P. (2010). Reactive oxygen species in cancer. *Free Radical Research, 44*(5), 479–496. https://doi.org/10.3109/10715761003667554

Liu, H., Wang, X., Wang, D., Zou, Z., & Liang, Z. (2011). Effect of drought stress on growth and accumulation of active constituents in Salvia miltiorrhiza Bunge. *Industrial Crops and Products, 33*(1), 84–88. https://doi.org/10.1016/j.indcrop.2010.09.006

Liu, Y., Meng, Q., Duan, X., Zhang, Z., & Li, D. (2017). Effects of PEG-induced drought stress on regulation of indole alkaloid biosynthesis in *Catharanthus roseus*. *Journal of Plant Interactions, 12*(1), 87–91. https://doi.org/10.1080/17429145.2017.1293852

Lucci, N., & Mazzafera, P. (2009). Distribution of rutin in fava d'anta (*Dimorphandra mollis*) seedlings under stress. *Journal of Plant Interactions, 4*(3), 203–208. https://doi.org/10.1080/17429140802707035

Luyckx, J., & Baudouin, C. (2011). Trehalose: An intriguing disaccharide with potential for medical application in ophthalmology. *Clinical Ophthalmology, 5*(1), 577–581. https://doi.org/10.2147/OPTH.S18827

M'bo Kacou, A. A., Elain Apshara, S., Hebbar, K. B., Ananda, K. S., Mathias, T. G., & Sévérin, A. (2016). Change in leaf epicuticular wax and biochemical secondary metabolites in cocoa (theobroma cacao l.) under Hydric stress. *International Journal of Current Research, 8*(5), 28988–28999. http://www.journalcra.com/article/change-leaf-epicuticular-wax-and-biochemicalsecondary-metabolites-cocoa-theobroma-cacao-l

Madeo, F., Carmona-Gutierrez, D., Kepp, O., & Kroemer, G. (2018). Spermidine delays aging in humans. *Aging, 10*(8), 2209–2211. https://doi.org/10.18632/aging.101517

Mandal, S. M., Chakraborty, D., & Dey, S. (2010). Phenolic acids act as signaling molecules in plant-microbe symbioses. *Plant Signaling and Behavior, 5*(4), 359–368. https://doi.org/10.4161/psb.5.4.10871

Medina, V., & Laliberte, B. (2017). *A review of research on the effects of drought and temperature stress and increased CO2 on Theobroma cacao L., and the role of genetic diversity to address climate change*. Bioversity International. https://core.ac.uk/download/pdf/132697857.pdf

Moser, G., Leuschner, C., Hertel, D., Hölscher, D., Köhler, M., Leitner, D., Michalzik, B., Prihastanti, E., Tjitrosemito, S., & Schwendenmann, L. (2010). Response of cocoa trees (Theobroma cacao) to a 13-month desiccation period in Sulawesi, Indonesia. *Agroforestry Systems, 79*(2), 171–187. https://doi.org/10.1007/s10457-010-9303-1

Motamayor, J. C., Risterucci, A. M., Lopez, P. A., Ortiz, C. F., Moreno, A., & Lanaud, C. (2002). Cocoa domestication I: The origin of the cocoa cultivated by the Mayas. *Heredity, 89*(5), 380–386. https://doi.org/10.1038/sj.hdy.6800156

Najafi, A., Adutwum, E., Yari, A., Salehi, E., Mikaeili, S., Dashtestani, F., Abolhassani, F., Rashki, L., Shiasi, S., & Asadi, E. (2018). Melatonin affects membrane integrity, intracellular reactive oxygen species, caspase3 activity and AKT phosphorylation in frozen thawed human sperm. *Cell and Tissue Research, 372*(1), 149–159. https://doi.org/10.1007/s00441-017-2743-4

Nasrollahi, V., Mirzaie-Asl, A., Piri, K., Nazeri, S., & Mehrabi, R. (2014). The effect of drought stress on the expression of key genes involved in the biosynthesis of triterpenoid saponins in liquorice (Glycyrrhiza glabra). *Phytochemistry, 103*, 32–37. https://doi.org/10.1016/j.phytochem.2014.03.004

Nisa, M. U., Huang, Y., Benhamed, M., & Raynaud, C. (2019). The plant DNA damage response: Signaling pathways leading to growth inhibition and putative role in response to stress conditions. *Frontiers in Plant Science, 10*, 653. https://doi.org/10.3389/fpls.2019.00653

Nogués, S., Allen, D. J., Morison, J. I. L., & Baker, N. R. (1998). Ultraviolet-B radiation effects on water relations, leaf development, and photosynthesis in droughted pea plants. *Plant Physiology, 117*(1), 173–181. https://doi.org/10.1104/pp.117.1.173

Okogbenin, E., Ekanayake, I. J., & Porto, M. C. M. (2003). Genotypic variability in adaptation responses of selected clones of cassava to drought stress in the Sudan Savanna Zone of Nigeria. *Journal of Agronomy and Crop Science, 189*(6), 376–389. https://doi.org/10.1046/j.1439-037X.2003.00050.x

Pagare, S., Bhatia, M., Tripathi, N., Pagare, S., & Bansal, Y. K. (2015). Secondary metabolites of plants and their role: Overview. *Current Trends in Biotechnology and Pharmacy, 9*(3), 293–304.

Panche, A. N., Diwan, A. D., & Chandra, S. R. (2016). Flavonoids: An overview. *Journal of Nutritional Science, 5*, 1–15. https://doi.org/10.1017/jns.2016.41

Parida, A. K., & Das, A. B. (2005). Salt tolerance and salinity effects on plants: A review. *Ecotoxicology and Environmental Safety, 60*(3), 324–349. https://doi.org/10.1016/j.ecoenv.2004.06.010

Paul, A. L., Liu, L., McClung, S., Laughner, B., Chen, S., & Ferl, R. J. (2009). Comparative interactomics: Analysis of Arabidopsis 14-3-3 complexes reveals highly conserved 14-3-3 interactions between humans and plants. *Journal of Proteome Research, 8*(4), 1913–1924. https://doi.org/10.1021/pr8008644

Pedranzani, H., Racagni, G., Alemano, S., Miersch, O., Ramírez, I., Peña-Cortés, H., Taleisnik, E., Machado-Domenech, E., & Abdala, G. (2003). Salt tolerant tomato plants show increased levels of jasmonic acid. *Plant Growth Regulation, 41*(2), 149–158. https://doi.org/10.1023/A:1027311319940

Phankaen, Y., Manaprasertsak, A., Pluempanupat, W., Koul, O., Kainoh, Y., & Bullangpoti, V. (2017). Toxicity and repellent action of Coffea arabica against Tribolium castaneum (Herbst) adults under laboratory conditions. *Journal of Stored Products Research, 71*, 112–118. https://doi.org/10.1016/j.jspr.2017.01.006

Pipitone, L. (2019). The state and future of the cocoa & coffee markets. United Nations Conference on Trade and Development 11th Multi-year Expert Meeting on Commodities and Development 15-16 April 2019, Geneva.

Plengmuankhae, W., & Tantitadapitak, C. (2015). Low temperature and water dehydration increase the levels of asiaticoside and madecassoside in Centella asiatica (L.) Urban. *South African Journal of Botany, 97*, 196–203. https://doi.org/10.1016/j.sajb.2015.01.013

Ramakrishna, A., & Ravishankar, G. A. (2011). Influence of abiotic stress signals on secondary metabolites in plants. *Plant Signaling and Behavior, 6*(11), 1720–1731. https://doi.org/10.4161/psb.6.11.17613

Rao, S., Usha, K., & Shetty, A. (2015). Production of secondary metabolites from callus cultures of Centella asiatica (L.) Urban. *Annals of Phytomedicine, 4*(1), 74–78.

Resende, M. L. V., Flood, J., Ramsden, J. D., Rowan, M. G., Beale, M. H., & Cooper, R. M. (1996). Novel phytoalexins including elemental sulphur in the resistance of cocoa (Theobroma cacao L.) to Verticillium wilt (Verticillium dahliae Kleb.). *Physiological and Molecular Plant Pathology, 48*(5), 347–359. https://agris.fao.org/agris-search/search.do?recordID=GB9632731

Royal Botanic Gardens Kew. (2017). *The state of the world's plants report 2017 | Kew*. Royal Botanic Gardens Kew.

Rustaiee, A., Sefidkon, F., Tabatabaei, S. M. F., Omidbaigi, R., & Mirahmadi, S. F. (2011). Chemical polymorphism of essential oils from five populations of thymus daenensis Celak. subsp. daenensis endemic to Iran. *Journal of Essential Oil Research, 23*(3), 6–11. https://doi.org/10.1080/10412905.2011.9700450

Samanta, A., Das, G., & Das, S. K. (2011). Roles of flavonoids in plants. *International Journal of Pharmaceutical Science and Technology, 6*(1), 12–35. https://www.researchgate.net/publication/279499208_Roles_of_flavonoids_in_Plants

Saravanan, R., Khristi, S., Gajbhiye, N. A., & Maiti, S. (2009). Effect of plant population and soil moisture stress on herbage yield and andrographolide content in Andrographis paniculata. *Indian Journal of Horticulture, 66*(1), 120–125. https://www.researchgate.net/publication/265972772_Effect_of_plant_population_and_soil_moisture_stress_on_herbage_yield_and_andrographolide_content_in_Andrographis_paniculata

Selmar, D. (2008). Potential of salt and drought stress to increase pharmaceutical significant secondary compounds in plants. *Landbauforsch Völk, 58*, 139–144.

Srivastava, S., & Srivastava, M. (2014). Morphological changes and antioxidant activity of *Stevia rebaudiana* under water stress. *American Journal of Plant Sciences, 5*(22), 3417–3422. https://doi.org/10.4236/ajps.2014.522357

Sun, S.-Q., & Tan, X.-F. (2008). Effects of nitrogen forms on camptothecin content and its metabolism-related enzymes activities in Camptotheca acuminata seedlings. *Zhongguo Zhong Yao Za Zhi, 33*(13), 1519–1523.

Szabó, B., Tyihák, E., Szabó, L. G., & Botz, L. (2003). Mycotoxin and drought stress induced change of alkaloid content of papaver somniferum plantlets. *Acta Botanica Hungarica, 45*(3–4), 409–417. https://doi.org/10.1556/ABot.45.2003.3-4.15

Tari, I., Kiss, G., Deer, A. K., Csiszar, J., Erdei, L., Galle, A., Gemes, K., Horvath, F., Poor, P., Szepesi, A., & Simon, L. M. (2010). Salicylic acid increased aldose reductase activity and sorbitol accumulation in tomato plants under salt stress. *Biologia Plantarum, 54*(4), 677–683. https://doi.org/10.1007/s10535-010-0120-1

Tholl, D. (2015). Biosynthesis and biological functions of terpenoids in plants. *Advances in Biochemical Engineering/Biotechnology, 148*, 63–106. https://doi.org/10.1007/10_2014_295

Thoma, F., Somborn-Schulz, A., Schlehuber, D., Keuter, V., & Deerberg, G. (2020). Effects of light on secondary metabolites in selected leafy greens: A review. *Frontiers in Plant Science, 11*, 497. https://doi.org/10.3389/fpls.2020.00497

Vakili, B., Karimi, F., Sharifi, M., & Behmanesh, M. (2012). Chromium-induced tropane alkaloid production and H6H gene expression in Atropa belladonna L. (Solanaceae) in vitro-propagated plantlets. *Plant Physiology and Biochemistry, 52*, 98–103. https://doi.org/10.1016/j.plaphy.2011.12.003

Varshney, K. A., & Gangwar, L. P. (1988). Choline and betaine accumulation in Trifolium alexandrinum L. during salt stress. *Egyptian Journal of Botany, 31*(1–3), 81–86.
Vashisth, D., Kumar, R., Rastogi, S., Patel, V. K., Kalra, A., Gupta, M. M., Gupta, A. K., & Shasany, A. K. (2018). Transcriptome changes induced by abiotic stresses in Artemisia annua. *Scientific Reports, 8*(1), 1–14. https://doi.org/10.1038/s41598-018-21598-1
Vaughan, M. M., Christensen, S., Schmelz, E. A., Huffaker, A., Mcauslane, H. J., Alborn, H. T., Romero, M., Allen, L. H., & Teal, P. E. A. (2015). Accumulation of terpenoid phytoalexins in maize roots is associated with drought tolerance. *Plant Cell and Environment, 38*(11), 2195–2207. https://doi.org/10.1111/pce.12482
Villamar-Torres, R., Jazayeri, S. M., Liuba-Delfini, G., García Cruzatty, L. C., & Viot, C.-R. (2018). Volatile organic compounds: Plant natural defense mechanisms against herbivorous arthropods and an opportunity for plant breeding of cotton. *Scientia Agropecuaria, 9*(2), 287–297. https://doi.org/10.17268/sci.agropecu.2018.02.14
Vincken, J. P., Heng, L., de Groot, A., & Gruppen, H. (2007). Saponins, classification and occurrence in the plant kingdom. *Phytochemistry, 68*(3), 275–297. https://doi.org/10.1016/j.phytochem.2006.10.008
Wang, D. H., Du, F., Liu, H. Y., & Liang, Z. S. (2010). Drought stress increases iridoid glycosides biosynthesis in the roots of Scrophularia ningpoensis seedlings. *Journal of Medicinal Plants Research, 4*(24), 2691–2699. http://www.academicjournals.org/JMPR
Wang, H., Oo Khor, T., Shu, L., Su, Z.-Y., Fuentes, F., Lee, J.-H., & Tony Kong, A.-N. (2012). Plants vs. cancer: A review on natural phytochemicals in preventing and treating cancers and their druggability. *Anti-Cancer Agents in Medicinal Chemistry, 12*(10), 1281–1305. https://doi.org/10.2174/1871520128038333026
Wang, L., Nägele, T., Doerfler, H., Fragner, L., Chaturvedi, P., Nukarinen, E., Bellaire, A., Huber, W., Weiszmann, J., Engelmeier, D., Ramsak, Z., Gruden, K., & Weckwerth, W. (2016). System level analysis of cacao seed ripening reveals a sequential interplay of primary and secondary metabolism leading to polyphenol accumulation and preparation of stress resistance. *The Plant Journal, 87*(3), 318–332. https://doi.org/10.1111/tpj.13201
Wang, S., Alseekh, S., Fernie, A. R., & Luo, J. (2019). The structure and function of major plant metabolite modifications. *Molecular Plant, 12*, 899–919.
Wei, N., & Deng, X. W. (1999). Making sense of the COP9 signalosome—A regulatory protein complex conserved from Arabidopsis to human. *Trends in Genetics, 15*(3), 98–103. https://doi.org/10.1016/S0168-9525(98)01670-9
Wink, M. (1988). Plant breeding: Importance of plant secondary metabolites for protection against pathogens and herbivores. *Theoretical and Applied Genetics, 75*(2), 225–233. https://doi.org/10.1007/BF00303957
Wink, M. (2018). Plant secondary metabolites modulate insect behavior-steps toward addiction? *Frontiers in Physiology, 9*, 364. https://doi.org/10.3389/fphys.2018.00364
Woodrow, I. E., Slocum, D. J., & Gleadow, R. M. (2002). Influence of water stress on cyanogenic capacity in Eucalyptus cladocalyx. *Functional Plant Biology, 29*(1), 103. https://doi.org/10.1071/PP01116
Xu, D. P., Li, Y., Meng, X., Zhou, T., Zhou, Y., Zheng, J., Zhang, J. J., & Li, H. B. (2017). Natural antioxidants in foods and medicinal plants: Extraction, assessment and resources. *International Journal of Molecular Sciences, 18*(1), 96. https://doi.org/10.3390/ijms18010096
Yang, L., Zhao, Y., Zhang, Q., Cheng, L., Han, M., Ren, Y., & Yang, L. (2019). Effects of drought-re-watering-drought on the photosynthesis physiology and secondary metabolite production of Bupleurum chinense DC. *Plant Cell Reports, 38*(9), 1181–1197. https://doi.org/10.1007/s00299-019-02436-8
Zeng, J., Chen, A., Li, D., Yi, B., & Wu, W. (2013). Effects of salt stress on the growth, physiological responses, and glycoside contents of stevia rebaudiana bertoni. *Journal of Agricultural and Food Chemistry, 61*(24), 5720–5726. https://doi.org/10.1021/jf401237x
Zhang, X., Niu, M., Teixeira da Silva, J. A., Zhang, Y., Yuan, Y., Jia, Y., Xiao, Y., Li, Y., Fang, L., Zeng, S., & Ma, G. (2019). Identification and functional characterization of three new terpene

synthase genes involved in chemical defense and abiotic stresses in Santalum album. *BMC Plant Biology, 19*(1), 115. https://doi.org/10.1186/s12870-019-1720-3

Zhu, Z., Liang, Z., Han, R., & Wang, X. (2009). Impact of fertilization on drought response in the medicinal herb Bupleurum chinense DC.: Growth and saikosaponin production. *Industrial Crops and Products, 29*(2–3), 629–633. https://doi.org/10.1016/j.indcrop.2008.08.002

Diversity and Ecology of Arbuscular Mycorrhization Fungi

Liliana Lara-Capistrán, Luis Guillermo Hernádez-Montiel, Juan José Reyes-Pérez, Ramón Zulueta-Rodríguez, Seyed Mehdi Jazayeri, and Ronald Oswaldo Villamar-Torres

Abbreviation

AM	Arbuscular mycorrhiza
AMF	Arbuscular mycorrhizal fungi

L. Lara-Capistrán
Centro de Investigaciones Atmosféricas y de Ecología, perteneciente a la Universidad Popular Autónoma Veracruzana (UPAV), Veracruz, México

Facultad de Ciencias Agrícolas, Universidad Veracruzana, Veracruz, México

L. G. Hernádez-Montiel
Centro de Investigaciones Biológicas del Noroeste, S.C., Calle Instituto Politécnico Nacional No. 195, La Paz, Baja California Sur, México
e-mail: lhernandez@cibnor.mx

J. J. Reyes-Pérez
Universidad Técnica Estatal de Quevedo, Facultad de Ciencias Agropecuarias, Quevedo, Los Ríos, Ecuador
e-mail: jreyes@uteq.edu.ec

R. Zulueta-Rodríguez
Facultad de Ciencias Agrícolas, Universidad Veracruzana, Veracruz, México

S. M. Jazayeri
Departamento de Biología, Facultad de Ciencias, Universidad Nacional de Bogotá, Bogotá, Colombia
e-mail: smjazayeri@unal.edu.co

R. O. Villamar-Torres (✉)
Universidad Técnica Estatal de Quevedo, Facultad de Ciencias Agropecuarias, Quevedo, Los Ríos, Ecuador

Instituto Superior Tecnológico "Ciudad de Valencia"–Tecnología en Producción Agrícola y Tecnología en Procesamiento de Alimentos, Quevedo, Los Ríos, Ecuador
e-mail: rvillamart@uteq.edu.ec

N. R. Maddela, L. C. García (eds.), *Innovations in Biotechnology for a Sustainable Future*, https://doi.org/10.1007/978-3-030-80108-3_10

1 Introduction

Arbuscular mycorrhiza (AM) is the mutualistic symbiosis established between fungi of the *Phylum Glomeromycota* and most vascular plants (Smith & Read, 2008; Parniske, 2008) including species of great value in agricultural systems (Giovannini et al., 2020; Gao et al., 2020), whose ability to increase the absorption of water and low-mobility nutrients is usually transcendent in its growth and development (Cardoso et al., 2017; Ma et al., 2019).

In addition, AM confers other benefits on plants such as resistance to attack by pests and diseases (Jung et al., 2012; Cameron et al., 2013), tolerance under saline stress (Frosi et al., 2017; Chang et al., 2018; Wang et al., 2019), drought stress (Silva et al., 2015; Chitarra et al., 2016; Fernández-Lizarazo & Moreno-Fonseca, 2016; Wu & Zou, 2017), and presence of heavy metals (Miransari, 2017; Kumar & Singh, 2019; Kumar & Saxena, 2019). These do not underestimate the improvement of physical-chemical properties of the soil like texture, structure (Thirkell et al., 2017; Atakan et al., 2018), and the release of nutrients (Ortaş et al., 2017).

On the other hand, the trend of the current scientific classification of arbuscular mycorrhizal fungi (AMF) is based on the combined application of conventional identification methods by discrete characters of the subcellular structure of spores such as size, color, and supporting hypha (Morton & Bentivenga, 1994), as well as by the phylogenetic-molecular analyzes of the partial region of the β-tubulin gene and primers that flank the SSU rRNA-ITS-LSU region (Oehl et al., 2011a) known as a polyphasic taxonomy (Tindall et al., 2010).

In this context, Schüßler (2006) had proposed the existence of ten AMF genera with around 193 described species, but recently Oehl et al. (2011a) reported 29 genera with 230 species arranged in 14 families.

Despite the fact that information on AMF diversity in natural and managed ecosystems is scarce, there are evidences that this could be higher than estimated (Opik et al., 2008). Thus, and in this manner, Börstler et al. (2006) considered that in the world there could be 1250 species of mycorrhizal fungi. Although the controversy about how their composition changes in natural ecosystems that were transformed into agroecosystems still prevents a standardized criterion among specialists (Belay et al., 2015) especially if the mycorrhizal spore population is forged in response to alterations in the plant community due to the obligate nature of the mycobionts (Paliocha, 2017; Sharma et al., 2018).

As far as agricultural systems are concerned, many the practices are carried out that in one way or another tend to negatively affect the abundance, diversity, and functioning of AMF (Schalamuk et al., 2006, 2013; Gómez et al., 2007; Lovera & Cuenca, 2007; Alguacil et al., 2008). Therefore, it is essential that the focus of research related to the use and application of mycorrhizal inoculum is specified to the development of mycotechnologies aimed at the formation of sustainable agroecosystems where assess the colonizing capacity of native mycobionts and promote the permanence and proliferation of members of the mycorrhizal fungal community with desirable agronomic traits (Rillig et al., 2016; Mukherjee et al., 2018).

Therefore, if the richness and composition of AMF species present in the soil varies according to the abiotic factors and predominant host plants in a given area (Cuenca, 2015), then understanding the effect of agronomic practices could contribute to the identification of management strategies that optimize the benefits of AM in the production of various crops.

2 Arbuscular Mycorrhizal Fungi

Perhaps, the most widespread type of fungus-root symbiosis in nature is the arbuscular mycorrhiza, formed by certain zygomycetes, which do not develop Harting's network and intracellularly colonize the root cortex by means of specialized structures called arbuscules. Arbuscules act as organs of bidirectional exchange of nutrients where carbon flows toward the host and these moves toward the plant (Andrade-Torres, 2010; Schalamuk et al., 2013; Ortiz-Acevedo et al., 2015).

Some genera of mycorrhizal fungi also develop protrusions that are lined by the plasma membrane, called vesicles, composed mainly of mycelial lipid bodies and therefore are considered a reservoir of nutrients for the survival and proliferation of the mycobiont (Rich et al., 2017; Müller et al., 2017).

The presence of both arbuscules and vesicles led to the fact that at first the symbiosis was known as vesicular-arbuscular (VA) mycorrhizas, but since not all species of fungi form vesicles, today the association is known as AM (Smith & Read, 2008; Aguilera Gómez et al., 2008).

3 Systematics (Taxonomy) of Arbuscular Mycorrhizal Fungi

The systematics of AMF were restructured considering the characters of greatest phenological importance in their life cycle and taking into account characters that could be interpreted evolutionarily, which means grouping current taxa that are related by their descendants from a common ancestor in order to organize a natural classification.

From this point of view, Morton and Benny (1990) proposed placing these fungi in the order Glomales; two new suborders: Glomineae and Gigasporineae and two new families: Acaulosporaceae and Gigasporaceae, in addition to the Glomaceae family proposed by Pirozynski and Dalpé (1989); thus the taxonomic location of these symbionts was as follows: Eumycota division, Zygomycotina subdivision, Zygomycetes class, Glomales order, two suborders Glomineae with two families (Glomaceae, with the genus *Glomus*; and Acaulosporaceae, with the genera

Acaulosporineapora and *Entrophospoe*) and Gigasporineapoe with a single family (Gigasporaceae, with two genera: *Gigaspora* and *Scutellospora*).

Recent taxonomic reviews and the use of novel techniques such as DNA sequencing (Schüßler et al., 2001; Redecker, 2002), fatty acid profiling (Olsson et al., 1995; Nakano-Hylander & Olsson, 2007), and immunological reactions to specific monoclonal antibodies (Ligrone et al., 2007) have been crucial in the advances reported to date for the hierarchical and systematic ordering of these microorganisms (Sun & Guo, 2012).

In this sense, Table 1 shows the current taxonomic classification of these symbionts, considering the new families and genera that have so far been reported by internationally renowned researchers such as Oehl and Sieverding (2004), Walker and Schüßler (2004), Spain et al. (2006), Palenzuela et al. (2008), Goto et al. (2012), Oehl et al. (2008, 2011b, c, d); Sieverding et al. (2015), Błaszkowski et al. (2017, 2018), and Symanczik et al. (2018), among others.

To define the main structure used in the morphological identification and correct naming of AMF, the term glomerospore was proposed for any fungus that forms arbuscular-type mycorrhizal symbiosis, and whose function lies in a microscopic and multinucleated asexual structure that is formed for survival in adverse conditions (dormancy or quiescence) (Goto & Maia, 2006; Souza, 2015a) and colonization of any root of a host whose genetic compatibility is related to specify its association (Chaudhary et al., 2020).

Said allochthonous propagule (Vályi et al., 2016) develops in the terminal part of a sporogeneous linear or bulbous hypha, laterally or intercalated from a sporiferous saccule, with germination from an orb or germination shield, or simply from a germ tube (Goto & Maia, 2006; Souza, 2015a).

One of the characteristics for the diagnosis and sensible typing of these mycobionts is to consider the mode of spore formation (sporogenesis) and, accordingly, the morphological variations recognized by Schüßler and Walker (2010) in the ontogeny of survival and resistance structures are five morphotypes (acaulosporoid, entrophosporoid, gigasporoid, glomoid, and glomoid-radial, Fig. 1) that can be depicted as follows:

- **Acaulosporoid morphotype:** They are distinguished by developing laterally from an ephemeral structure called the sporiferous saccule (Fig. 1a), whose content is transferred to the spore during its maturation (Stürmer & Morton, 1999).

  Thus, at the time of its detachment, a scar remains as a result of the contact point between the spore and the saccule. This formation is typical of the Diversisporales and Archaeosporales order, and of the *Acaulospora*, *Ambispora*, *Archaeospora*, and *Otospora* genera (Walker, 1983; Redecker et al., 2013; Souza, 2015b).
- **Entrophosporoid morphotype:** Similar to the previous type, however it differs because the morphotypes are only found in the genus *Entrophospora* and develop within the neck of the sporiferous saccule (Wu et al., 1995; Kaonongbua et al.,

Table 1 Current systematics of the *Phylum Glomeromycota*

Class (3)	Order (5)	Family (16)	Genus (41)
Archaeosporomycetes	Archaeosporales	Archaeosporaceae	*Archaeospora*
			Intraspora
			Palaeospora
		Ambisporaceae	*Ambispora*
		Geosiphonaceae	*Geosiphon*
Glomeromycetes	Glomerales	Glomeraceae	*Glomus*
			Dominikia
			Funneliformis
			Kamienskia
			Rhizoglomus
			Oehlia
			Septoglomus
			Sclerocystis
			Simiglomus
		Entrophosporaceae	*Entrophospora*
			Albahypha
			Claroideoglomus
			Viscospora
	Diversisporales	Diversisporaceae	*Diversispora*
			Desertispora
			Otospora
			Tricispora
			Redeckera
			Corymbiglomus
		Sacculosporaceae	*Sacculospora*
		Pacisporaceae	*Pacispora*
		Acaulosporaceae	*Acaulospora*
	Gigasporales	Gigasporaceae	*Gigaspora*
		Scutellosporaceae	*Scutellospora*
			Bulbospora
			Orbispora
		Racocetraceae	*Racocetra*
			Cetraspora
		Dentiscutataceae	*Dentiscutata*
			Fuscutata
			Quatunica
		Intraornatosporaceae	*Intraornatospora*
			Paradentiscutata
Paraglomeromycetes	Paraglomerales	Paraglomeraceae	*Paraglomus*
			Innospora
		Pervetustaceae	*Pervetustus*

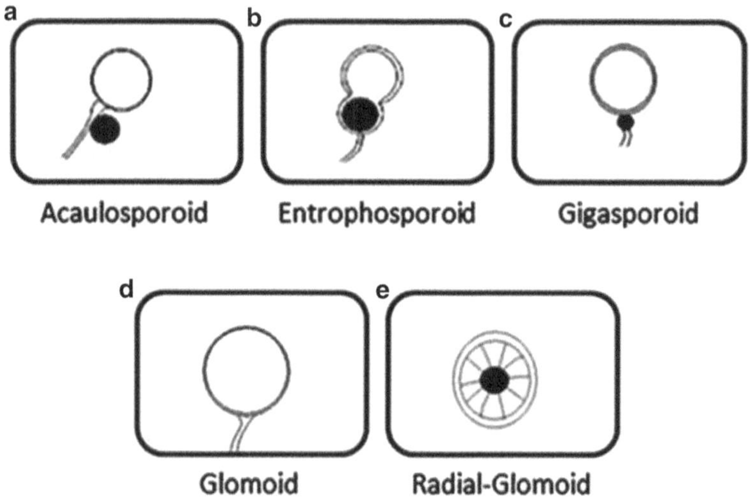

Fig. 1 (**a–e**) Sporogenesis of arbuscular mycorrhizal fungi. (Taken from Souza, 2015b)

2010; Souza, 2015a) (Fig. 1b), very similar to the form of development of the members of the Sacculosporaceae family (Oehl et al., 2011d).

- **Gigasporoid morphotype:** The spores are formed in the terminal part of a bulbous suspensory hypha (Walker, 1983; Redecker et al., 2013) (Fig. 1c) and, as unique characteristics of this morphotype, the size, color and non-formation of vesicles within the roots stand out but, instead, form extra-radical structures adorned or not with spines, called accessory cells (Bentivenga & Morton, 1995).[1]

  The genera within this morphological arrangement are *Bulbospora*, *Cetraspora*, *Dentiscutata*, *Fuscutata*, *Gigaspora*, *Intraornatospora*, *Orbispora*, *Paradentiscutata*, *Quatunica*, *Racocetra* and *Scutellospora* (Oehl et al., 2011c; Goto et al., 2012; Souza, 2015a).
- **Glomoid morphotype:** Morphotype formed in the terminal part of the reproductive hypha and made up of some members of the genus Acaulospora (Fig. 1d). However, in the case of *Ambispora* and *Archaeospora* the spores can be both glomoid and acaulosporid; while all the constituents of *Diversispora*, *Geosiphon*, *Glomus*, *Pacispora* and *Paraglomus* show it (Oehl & Sieverding, 2004; Schüßler & Walker, 2010; Oehl et al., 2011a; Souza, 2015a).
- **Glomoid-radial morphotype:** spores are grouped from a central plexus and covered by a peridium; some species of the *Glomus* and *Sclerocystis* genera are found in this morphology (Fig. 1e) (Redecker et al., 2000; Oehl et al., 2011a).

  On the other hand, subcellular structures such as sporiferous sacculum, scar, holding hypha, septa and spore walls are other preserved and persistent typing

[1]Salmerón-Santiago et al. (2012) mention that Gigasporoid with germination shield can be called Scutelosporoid (see Figure 3 in the original article of the cited publication).

characteristics within the same genus that contribute to the morphological identification of spores.

Each of such structure is briefly described below:

- **Sporiferous sacculus:** inflated sac, formed by a reproductive hypha (Fig. 1a), from which the spore will form either inside (*Entrophospora* and *Sacculospora*) or to one side (*Acaulospora*, *Ambispora* and *Archaeospora*) (Oehl et al., 2008; Kaonongbua et al., 2010).
- **Scar:** It is a mark derived from the point of contact between the neck of the sporiferous saccule and the glomerospore. Thus, once the spore has matured, the saccule collapses and leaves this scar on its way, located in the structural layer "2" of the *Acaulospora*, *Ambispora*, *Archaeospora*, *Entrophospora* and *Sacculospora* genera (Oehl et al., 2008, 2011b; Kaonongbua et al., 2010) (Fig. 1a).
- **Clamping hypha:** It occurs with different number of layers, shapes and occlusions at the base of the glomerospora in the genera *Ambispora*, *Claroideoglomus*, *Diversispora*, *Desertispora*, *Dominikia*, *Innospora*, *Kamienskia*, *Funneliformis*, *Glomus*, *Pacispora*, *Paraglomus*, *Pervetustus*, *Rhizoglomus*, *Redeckera*, *Sclerocystis*, *Septoglomus*, and *Simiglomus* (Fig. 1) (Oehl & Sieverding, 2004; Schüßler & Walker, 2010; Oehl et al., 2011b; Błaszkowski et al., 2015, 2017; Symanczik et al., 2018).
- **Septum:** Structure that separates the glomerospore from the subjection hypha with a septum; the septum can be located at the base or the middle part of the hypha (Fig. 1d) and, according to its position, makes it belong to one of the genera mentioned above (Oehl et al., 2011d; Souza, 2015b).
- **Spore walls:** Once sporogenesis has been recognized, the walls are counted, identified, and described, defined as the number of structural and germ walls that make up the glomerospore. These layers are of utmost importance since they are morphologically equal and highly conserved despite any biotic or abiotic condition (Souza, 2015b).

At present, and after consulting different specialized sources (Walker, 1983, 1986; Morton et al., 1995; Spain et al., 2006; Sieverding & Oehl, 2006; Oehl et al., 2008), the layers differ by the position they have when the spores are broken on a slide:

- *External or structural* wall(s) to the group of layers (2–4) originated from a fertile hypha and, as it grows, the layers can be differentiated, since each wall has color, rigidity, thickness, and particular reaction to Melzer's reagent.
- *Middle or germinal* wall refers to the group of colorless walls (2–3), considered upon germination, which are formed from the outermost to the inner layer without having physical contact with the structural ones and, as the name indicates, their function is predispose the spore to germinate, that is why all AMF have them.
- *Internal wall or germination structure* is the one that initiates germination and, in correspondence with the genus, will be its morphology. Thus, for example, the germ tube is exclusive to the genus *Gigaspora*, emerging from a

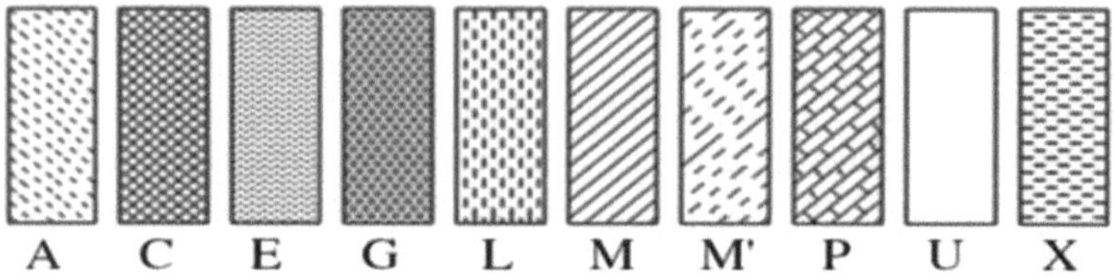

Fig. 2 Murograph for schematizing wall morphological types in endogonaceous spores proposed by Walker (1983). A: amorphous, C: coriaceous, E: evanescent, G: germinal, L: laminated, M: membranous, M′: germinal membranous, P: peridium, U: unitary, X: expansive. (Taken from Souza, 2015b)

specialized ornate layer located in the inner layer, but once it germinates, it can branch directly into soil. They present the structure known as the *germination orb*, which is hyaline-ovoid formed by a hypha positioned in the innermost germ layer, representative of the *Acaulospora* genus. Germination shield, located on the inner wall, with lobed to toothed morphologies, being visible to the naked eye due to its color and rarely disappearing when ripe, being a very important characteristic to differentiate between genera.

After the first publication where, different layers that make up the wall of AMF spores were described (Fig. 2), others have been proposed as instrumental criteria and tools are taken into consideration for the description and identification of new species.

The basic morphological characteristics of the sporal walls proposed by Walker (1983) and Souza (2015b) are mentioned below:

- **Amorphous wall:** thick hyaline layer that tends to deform when pressure is applied in polyvinyl alcohol and lactic acid glycerol (PVLG), whereas with Melzer's reagent it takes on a strong purple color. It is located in the inner part of the spore.
- **Leathery wall:** Colorless, flexible, and resistant layer when mounting, with a leather appearance, being difficult to differentiate from a membranous wall due to its similarity.
- **Evanescent wall:** outer layer that, when the spore matures, degrades, and detaches.
- **Expansive wall:** Layer that when mounted expands and creates slight striations.
- **Germ wall:** single layer of the genus *Gigaspora*, similar to the laminated layer. However, it differs by presenting ornamentation with protrusions and continuing to specialize in germination.
- **Laminate wall:** Composed by the union of several layers that accumulate as the spore matures, providing rigidity.
- **Membranous wall:** thin, flexible and colorless layer that, when mounted, wrinkles, but does not present breaks.

- **Unitary wall:** rigid, uniform, permanent and external layer, colorless or pigmented, adorned—or not—with projections.
- **Peridium:** Network of hyphae that surround the spore or sporocarp, present only in some genera (*Glomus*, *Rhizoglomus*, *Sclerocystis* and *Sacculospora*).

4 Diversity and Ecology of Arbuscular Mycorrhizal Fungi

AMF are essential rhizospheric microorganisms of the natural soil microflora in natural ecosystems and in those managed by human (Berruti et al., 2014; Giraldo et al., 2019). Consequently, it has been speculated that these can colonize a wide range of host plants (Jung et al., 2012; Kokkoris & Hart, 2019). Therefore, aspects related to nutrition, growth, and development have special importance of plants (Begum et al., 2019; Mohammad, 2019), as well as their participation in the biotic regulations of soils, which also influence the stability of the plant communities that make up a certain ecosystem or agroecosystem (Isbell et al., 2017; Bhale et al., 2018).

However, the abundance and wide range of affinity of AMF with their hosts does not necessarily reflect their functional significance because the combinations between the symbionts have different nuances that derive from mutual interaction (Lee et al., 2013). And, although the diversity of these microorganisms is vast and not yet well studied, numerous authors have argued that the mycorrhizal association is crucial in the biogeochemical cycling of nutrients and in the productivity of plants (Pal et al., 2017; Yadav et al., 2017; Banerjee et al., 2018, 2019; Pickles et al., 2020).

For this reason, it is necessary to learn more about the identity of the organisms responsible for this key process both in natural ecosystems and in anthropically modified systems where the sustainability of agri-food production is intended.

AMF not only govern plant biodiversity and ecosystem productivity at different altitudes and latitudes (Sanders et al., 1996; Lugo et al., 2008) where the prevailing biophysical conditions allow them to interact with their hosts (Nouri et al., 2014; Berruti et al., 2016), but these mycobionts establish symbiosis with plants appraised by human. They are subject to various technical management strategies where recent progress in research has demonstrated their ability to overcome the loss of biological fertility of soils and thereby reduce the need for chemical inputs in agro-productive sectors (Ganugi et al., 2019) such as horticultural crops, ornamental species (Perner et al., 2007), medicinal species (Zeng et al., 2013; Tchiechoua et al., 2020), fruit trees (Aseri et al., 2008) and forestry plants (Liang et al., 2007; Xueming et al., 2007), among others.

Most studies consider that mycorrhizal propagules (spores and mycorrhizal roots) are concentrated near the soil surface (in the first 0–20 cm) (Cardoso et al., 2003) and their presence decreases as the penetration of roots are lower in deeper soil layers (Muleta et al., 2008; Shukla et al., 2013). However, Oehl et al. (2005) mention that in corn crops there is a great diversity of AMF spores between 50 and 70 cm, which can be equally elemental in agroforestry systems where trees spread their roots in deeper soil layers (Cardoso et al., 2003).

Like the topsoil, mycorrhizal fungi are also susceptible to agents responsible for erosion such as water and air/wind (Harner et al., 2011; Egan et al., 2014). Then, the dragging of the sediments not only entails problems of physical loss of soil, but also of transport and deposition of mycorrhizal propagules (Bueno & Moora, 2019). Although these erosion agents are not a dominant mode for the dispersal of AM fungal spores (Egan et al., 2014), they can also be moved to different and distant sites (Davison et al., 2015, 2018).

In summary, it has been shown that the distribution, dynamics, and survival of AMF in their different habitats are influenced by several factors, among which intervene the physicochemical properties (Jamiołkowska et al., 2018; Silva-Flores et al., 2019; Han et al., 2019), pH (Melo et al., 2014, 2017, 2019; Gupta et al., 2018), nutrient availability (Tahat & Sijam, 2012; Gosling et al., 2013; Xing et al., 2018) and the geographic and topographic diversity of soils (Jamiołkowska et al., 2018), anthropogenic activities and agrotechnical factors related to intensive agricultural management practices (Tchabi et al., 2009; Säle et al., 2015; Giller et al., 2015; Jamiołkowska et al., 2018; Rillig et al., 2019) and the variability/seasonality of climatic conditions (Wang et al., 2015; Jamiołkowska et al., 2018; Xing et al., 2018; Silva-Flores et al., 2019), among others.

Therefore, the functional traits of plants and the variation in mycorrhizal mutualism not only influence the dynamics of plant populations in different terrestrial ecosystems, but also their ecological distribution (Maherali, 2020).

Acknowledgments The authors are grateful to Tancredo Souza, Universidade Federal de Santa Catarina, Florianopolis, Brazil (ORCID ID https://orcid.org/0000-0001-8729-5478) for allowing the use of two key elements of the chapter: Figs. 1 and 2. Similarly, the authors express special thanks to Editors: Dr. Naga Raju Maddela and Dr. Luz Cecilia García for guidance and accepting our request to write this chapter.

References

Aguilera Gómez, L. I., Olalde Portugal, V., Arriaga, M. R., & Contreras Alonso, R. (2008). Micorrizas arbusculares. *Cien. Ergo Sum, 14*, 300–306.

Alguacil, M. M., Lumini, E., Roldán, A., Salinas-García, J. R., Bonfante, P., & Bianciotto, V. (2008). The impact of tillage practices on arbuscular mycorrhizal fungal diversity in subtropical crops. *Ecological Applications, 18*, 527–536.

Andrade-Torres, A. (2010). Micorrizas: Antigua interacción entre plantas y hongos. *Universidad del Zulia, 61*, 84–90.

Aseri, G. K., Jain, N., Panwar, J., Rao, A. V., & Meghwal, P. R. (2008). Biofertilizers improve plant growth, fruit yield, nutrition, metabolism and rhizosphere enzyme activities of Pomegranate (*Punica granatum* L.) in Indian Thar Desert. *Scientia Horticulturae, 117*, 130–135.

Atakan, A., Özkaya, H. Ö., & Erdoğan, O. (2018). Effects of arbuscular mycorrhizal fungi (AMF) on heavy metal and salt stress. *Turkish Journal of Agriculture - Food Science and Technology, 6*, 1569–1574.

Banerjee, S., Schlaeppi, K., & van der Heijden, M. G. A. (2018). Keystone taxa as drivers of microbiome structure and functioning. *Nature Reviews. Microbiology, 16*, 567–576.

Banerjee, S., Walder, F., Buchi, L., Meyer, M., Held, A. Y., Gattinger, A., Keller, T., Charles, R., & van der Heijden, M. G. A. (2019). Agricultural intensification reduces microbial network complexity and the abundance of keystone taxa in roots. *The ISME Journal, 13*, 1722–1736.

Begum, N., Qin, C., Ahanger, M. A., Raza, S., Khan, M. I., Ashraf, M., Ahmed, N., & Zhang, L. (2019). Role of arbuscular mycorrhizal fungi in plant growth regulation: Implications in abiotic stress tolerance. *Frontiers in Plant Science, 10*, 1068. https://doi.org/10.3389/fpls.2019.01068

Belay, Z., Vestberg, M., & Assefa, F. (2015). Diversity and abundance of arbuscular mycorrhizal fungi across different land use types in a humid low land area of Ethiopia. *Tropical and Subtropical Agroecosystems, 18*, 47–69.

Bentivenga, S. P., & Morton, J. B. (1995). A monograph of the genus *Gigaspora*, incorporating developmental patterns of morphological characters. *Mycologia, 87*, 719–731.

Berruti, A., Borrielo, R., Orgiazzi, A., Barbera, A. C., Lumini, E., & Bianciotto, V. (2014). Arbuscular mycorrhizal fungi and their value for ecosystem management. In O. Grillo (Ed.), *The dynamic balance of the planet* (pp. 159–191). Intech Open.

Berruti, A., Lumini, E., Balestrini, R., & Bianciotto, V. (2016). Arbuscular mycorrhizal fungi as natural biofertilizers: Let's benefit from past successes. *Frontiers in Microbiology, 6*, 1559. https://doi.org/10.3389/fmicb.2015.01559

Bhale, U. N., Bansode, S. A., & Singh, S. (2018). Multifactorial role of arbuscular mycorrhizae in agroecosystem. In P. Gehiot & J. Singh (Eds.), *Fungi and their role in sustainable development: Current perspectives* (pp. 205–220). Springer Nature.

Błaszkowski, J., Chwat, G., Góralska, A., Ryszka, P., & Kovács, G. M. (2015). Two new genera, *Dominikia* and *Kamienska*, and *D. disticha* sp. nov. in Glomeromycota. *Nova Hedwigia, 100*, 225–238.

Błaszkowski, J., Kozłowska, A., Crossay, T., Symanczik, S., & Al-Yahya'ei, M. N. (2017). A new family, Pervetustaceae with a new genus, *Pervetustus*, and *P. simplex* sp. nov. (Paraglomerales), and a new genus, *Innospora* with *I. majewskii* comb. nov. (Paraglomeraceae) in the Glomeromycotina. *Nova Hedwigia, 105*, 397–410.

Błaszkowski, J., Kozłowska, A., Niezgoda, P., Goto, B., & Dalpé, Y. (2018). A new genus, *Oehlia* with *Oehlia diaphana* comb. nov. and an emended description of *Rhizoglomus vesiculiferum* comb. nov. in the Glomeromycotina. *Nova Hedwigia, 107*, 501–518.

Börstler, B., Renker, C., Kahmen, A., & Buscot, F. (2006). Species composition of arbuscular mycorrhizal fungi in two mountain meadows with differing management types and levels of plant biodiversity. *Biology and Fertility of Soils, 42*, 286–298.

Bueno, C. G., & Moora, M. (2019). How do arbuscular mycorrhizal fungi travel? *The New Phytologist, 222*, 645–647.

Cameron, D. D., Neal, A. L., van Wees, S. C. M., & Ton, J. (2013). Mycorrhiza-induced resistance: More than the sum of its parts? *Trends in Plant Science, 18*, 539–545.

Cardoso, I. M., Boddington, C., Janssen, B. H., Oenema, O., & Kuyper, T. W. (2003). Distribution of mycorrhizal fungal spores in soils under agroforestry and monocultural coffee systems in Brazil. *Agroforestry Systems, 58*, 33–43.

Cardoso, E. J. B. N., Nogueira, M. A., & Zangaro, W. (2017). Importance of mycorrhizae in tropical soils. In J. L. de Azevedo & M. C. Quecine (Eds.), *Diversity and benefits of microorganisms from the tropics* (pp. 245–267). Springer International Publishing AG.

Chang, W., Sui, X., Fan, X.-X., Jia, T.-T., & Song, F.-Q. (2018). Arbuscular mycorrhizal symbiosis modulates antioxidant response and ion distribution in salt-stressed *Elaeagnus angustifolia* seedlings. *Frontiers in Microbiology, 9*, 652. https://doi.org/10.3389/fmicb.2018.00652

Chaudhary, V. B., Nolimal, S., Sosa-Hernández, M. A., Egan, C., & Kastens, J. (2020). Trait-based aerial dispersal of arbuscular mycorrhizal fungi. *The New Phytologist, 228*, 238–252.

Chitarra, W., Pagliarani, C., Maserti, B., Lumini, E., Siciliano, I., Cascone, P., Schubert, A., Gambino, G., Balestrini, R., & Guerrieri, E. (2016). Insights on the impact of arbuscular mycorrhizal symbiosis on tomato tolerance to water stress. *Plant Physiology, 171*, 1009–1023.

Cuenca, G. (2015). *Las micorrizas arbusculares: Aspectos teóricos y aplicados. Ediciones IVIC*. Instituto Venezolano de Investigaciones Científicas (IVIC). 432 p.

Davison, J., Moora, M., Öpik, M., Adholeya, A., Ainsaar, L., Bâ, A., Burla, S., Diedhiou, A. G., Hiiesalu, I., Jairus, T., Johnson, N. C., Kane, A., Koorem, K., Kochar, M., Ndiaye, C., Pärtel, M., Reier, Ü., Saks, Ü., Singh, R., … Zobel, M. (2015). Global assessment of arbuscular mycorrhizal fungus diversity reveals very low endemism. *Science, 349*, 970–973.

Davison, J., Moora, M., Öpik, M., Ainsaar, L., Ducousso, M., Hiiesalu, I., Jairus, T., Johnson, N., Jourand, P., Kalamees, R., Koorem, K., Meyer, J.-Y., Püssa, K., Reier, Ü., Pärtel, M., Semchenko, M., Traveset, A., Vasar, M., & Zobel, M. (2018). Microbial island biogeography: Isolation shapes the life history characteristics but not diversity of root-symbiotic fungal communities. *The ISME Journal, 12*, 2211–2224.

Egan, C., Li, D.-W., & Klironomos, J. (2014). Detection of arbuscular mycorrhizal fungal spores in the air across different biomes and ecoregions. *Fungal Ecology, 12*, 26–31.

Fernández-Lizarazo, J. C., & Moreno-Fonseca, L. P. (2016). Mechanisms for tolerance to water-deficit stress in plants inoculated with arbuscular mycorrhizal fungi: A review. *The Agronomia Colombiana, 34*, 179–189.

Frosi, G., Barros, V. A., Oliveira, M. T., Santos, M., Ramos, D. G., Maia, L. C., & Santos, M. G. (2017). Arbuscular mycorrhizal fungi and foliar phosphorus inorganic supply alleviate salt stress effects in physiological attributes, but only arbuscular mycorrhizal fungi increase biomass in woody species of a semiarid environment. *Tree Physiology, 38*, 25–36.

Ganugi, P., Masoni, A., Pietramellara, G., & Benedettelli, S. (2019). A review of studies from the last twenty years on plant–arbuscular mycorrhizal fungi associations and their uses for wheat crops. *Agronomy, 9*, 840. https://doi.org/10.3390/agronomy9120840

Gao, X., Guo, H., Zhang, Q., Guo, H., Zhang, L., Zhang, C., Gou, Z., Liu, Y., Wei, J., Chen, A., Chu, Z., & Zeng, F. (2020). Arbuscular mycorrhizal fungi (AMF) enhanced the growth, yield, fiber quality and phosphorus regulation in upland cotton (*Gossypium hirsutum* L.). *Scientific Reports, 10*, 2084. https://doi.org/10.1038/s41598-020-59180-3

Giller, K. E., Andersson, J. A., Corbeels, M., Kirkegaard, J., Mortensen, D., Erenstein, O., & Vanlauwe, B. (2015). Beyond conservation agriculture. *Frontiers in Plant Science, 6*, 870. https://doi.org/10.3389/fpls.2015.00870

Giovannini, L., Palla, M., Agnolucci, M., Avio, L., Sbrana, C., Turrini, A., & Giovannetti, M. (2020). Arbuscular mycorrhizal fungi and associated microbiota as plant biostimulants: Research strategies for the selection of the best performing inocula. *Agronomy, 10*, 106. https://doi.org/10.3390/agronomy10010106

Giraldo, K. J. R., Correa, M. I. M., Jaramillo, P. H., Gutiérrez, L. A., & Guzmán, L. P. M. (2019). Caracterización de hongos micorrízicos arbusculares de suelos ganaderos del trópico alto y trópico bajo en Antioquia, Colombia. *IDESIA, 37*, 35–44.

Gómez, L. E. D., Sánchez de Prager, M., El-Sharkawi, M., & Cadavid, L. F. (2007). Micorriza arbuscular en agroecosistemas. Caso 1: Algunos indicadores de actividad biológica en el suelo y su relación con el manejo agronómico de la yuca (*Manihot esculenta* Crantz) en la costa norte de Coilombia1. In *Sánchez de Prager, M. (coordinador.). Las endomicorrizas: Expresión bioedáfica de importancia en el trópico Universidad Nacional de Colombia-Sede Palmira, Colombia* (pp. 179–193).

Gosling, P., Mead, A., Proctor, M., Hammond, J. P., & Bending, G. D. (2013). Contrasting arbuscular mycorrhizal communities colonizing different host plants show a similar response to a soil phosphorus concentration gradient. *The New Phytologist, 198*, 546–556.

Goto, B. T., & Maia, L. C. (2006). Glomerospores: A new denomination for the spores of Glomeromycota, a group molecularly distinct from Zygomycota. *Mycotaxon, 96*, 129–132.

Goto, B. T., Silva, G. A., Assis, D. M. A., Silva, D. K. A., Souza, R. G., Ferreira, A. C. A., Jobim, K., Mello, C. M. A., Vieira, H. E. E., Maia, L. C., & Oehl, F. (2012). *Intraornatosporaceae* (*Gigasporales*), a new family with two new genera and two new species. *Mycotaxon, 119*, 117–132.

Gupta, M. M., Gupta, A., & Kumar, P. (2018). Urbanization and biodiversity of arbuscular mycorrhizal fungi: The case study of Delhi, India. *Revista de Biología Tropical, 66*, 1547–1558.
Han, X., Xu, C., Wang, Y., Huang, D., Fan, Q., Xin, G., & Müller, C. (2019). Dynamics of arbuscular mycorrhizal fungi in relation to root colonization, spore density, and soil properties among different spreading stages of the exotic plant threeflower beggarweed (*Desmodium triflorum*) in a *Zoysia tenuifolia* Lawn. *Weed Science, 67*, 689–701.
Harner, M. J., Opitz, N., Geluso, K., Tockner, K., & Rillig, M. C. (2011). Arbuscular mycorrhizal fungi on developing islands within a dynamic river floodplain: An investigation across successional gradients and soil depth. *Aquatic Sciences, 73*, 35–42.
Isbell, F., Adler, P. R., Eisehauer, N., Fornara, D. A., Kimmel, K., Kremen, C., Letourneau, D. K., Liebman, M., Polley, H. W., Quijas, S., & Scherer-Lorenzen, M. (2017). Benefits of increasing plant diversity in sustainable agroecosystems. *Journal of Ecology, 105*, 871–879.
Jamiołkowska, A., Księżniak, A., Gałązka, A., Hetman, B., Kopacki, M., & Skwaryło-Bednarz, B. (2018). Impact of abiotic factors on development of the community of arbuscular mycorrhizal fungi in the soil: A review. *International Agrophysics, 32*, 133–140.
Jung, S. C., Martínez-Medina, A., Lopez-Raez, J. A., & Pozo, M. J. (2012). Mycorrhiza-induced resistance and priming of plant defenses. *Journal of Chemical Ecology, 38*, 651–664.
Kaonongbua, W., Morton, J. B., & Bever, J. D. (2010). Taxonomic revision transferring species in *Kuklospora* to *Acaulospora* (Glomeromycota) and a description of *Acaulospora colliculosa* sp. nov. from field collected spores. *Mycologia, 102*, 1497–1509.
Kokkoris, V., & Hart, M. (2019). *In vitro* propagation of arbuscular mycorrhizal fungi may drive fungal evolution. *Frontiers in Microbiology, 10*, 2420. https://doi.org/10.3389/fmicb.2019.02420
Kumar, S., & Saxena, S. (2019). Arbuscular mycorrhizal fungi (AMF) from heavy metal-contaminated soils: Molecular approach and application in phytoremediation. In B. Giri, R. Prasad, Q.-S. Wu, & A. Varma (Eds.), *Biofertilizers for sustainable agriculture and environment* (pp. 489–500). Springer Nature.
Kumar, S., & Singh, J. (2019). Impact of arbuscular mycorrhizal fungi (AMF) in global sustainable environments. In A. N. Yadav, A. N. Mishra, S. Singh, & A. Gupta (Eds.), *Recent advancement in white biotechnology through fungi* (pp. 419–436). Springer Nature.
Lee, E.-H., Eo, J.-K., Ka, K.-H., & Eom, A.-H. (2013). Diversity of arbuscular mycorrhizal fungi and their roles in ecosystems. *Microbiology, 41*, 121–125.
Liang, Y., Guo, L.-D., Du, X.-J., & Ma, K. P. (2007). Spatial structure and diversity of woody plants and ectomycorrhizal fungus sporocarps in a natural subtropical forest. *Mycorrhiza, 17*, 271–278.
Ligrone, R., Carafa, A., Lumini, E., Bianciotto, V., Bonfante, P., & Duckett, J. G. (2007). Glomeromycotean associations in liverworts: A molecular cellular and taxonomic analysis. *American Journal of Botany, 94*, 1756–1777.
Lovera, M., & Cuenca, G. (2007). Diversidad de hongos micorrízicos arbusculares (HMA) y potencial micorrízico del suelo de una sabana natural y una sabana perturbada de La Gran Sabana, Venezuela. *Interciencia, 32*, 108–114.
Lugo, M. A., Ferrero, M., Menoyo, E., Estévez, M. C., Siñeriz, F., & Anton, A. (2008). Arbuscular mycorrhizal fungi and rhizospheric bacteria diversity along an altitudinal gradient in South American Puna grassland. *Microbial Ecology, 55*, 705–713.
Ma, X., Luo, W., Li, J., & Wu, F. (2019). Arbuscular mycorrhizal fungi increase both concentrations and bioavailability of Zn in wheat (*Triticum aestivum* L.) grain on Zn-spiked soils. *Applied Soil Ecology, 135*, 91–97.
Maherali, H. (2020). Mutualism as a plant functional trait: Linking variation in the mycorrhizal symbiosis to climatic tolerance, geographic range, and population dynamics. *International Journal of Plant Sciences, 181*, 9–19.
Melo, C. D., Walker, C., Rodríguez-Echeverría, S., Borges, P. A. V., & Freitas, H. (2014). Species composition of arbuscular mycorrhizal fungi differ in semi-natural and intensively managed pastures in an isolated oceanic island (Terceira, Azores). *Symbiosis, 64*, 73–85.

Melo, C. D., Luna, S., Krüger, C., Walker, C., Mendonça, D., Fonseca, H. M. A. C., Jaizme-Vega, M., & Machado, A. C. (2017). Arbuscular mycorrhizal fungal community composition associated with *Juniperus brevifolia* in native Azorean forest. *Acta Oecologica, 79*, 48–61.

Melo, C. D., Walker, C., Krüger, C., Borges, P. A. V., Luna, S., Mendonça, D., Fonseca, H. M. A. C., & Machado, A. C. (2019). Environmental factors driving arbuscular mycorrhizal fungal communities associated with endemic woody plant *Picconia azorica* on native forest of Azorean. *Annales de Microbiologie, 69*, 1309–1327.

Miransari, M. (2017). Arbuscular mycorrhizal fungi and heavy metal tolerance in plants. In Q.-S. Wu (Ed.), *Arbuscular mycorrhizas and stress tolerance of plants* (pp. 147–161). Springer Nature.

Mohammad, I. (2019). Mycorrhizae's role in plant nutrition and protection from pathogens. *Current Investigations in Agriculture and Current Research, 8*, 1037–1045.

Morton, J. B., & Benny, G. L. (1990). Revised classification of arbuscular mycorrizhal fungi (Zygomycetes): A new order, Glomales, two new suborders, Glomineae and Gigasporineae, and two new families, Acaulosporaceae and Gigaporaceae, with an emendation of Glomaceae. *Mycotaxon, 37*, 471–491.

Morton, J. B., & Bentivenga, S. P. (1994). Levels of diversity in endomycorrhizal fungi (Glomales, Zygomycetes) and their role in defining taxonomic and non-taxonomic groups. *Plant and Soil, 159*, 47–59.

Morton, J. B., Bentivenga, S. P., & Bever, J. D. (1995). Discovery, measurement, and interpretation of diversity in arbuscular endomycorrhizal fungi (Glomales, Zygomycetes). *Canadian Journal of Botany, 73*, 25–32.

Mukherjee, G., Dey, P., & Dhiman, S. (2018). Agricultural important microorganisms: From rhizosphere to bioformulation as biological control weapons for sustainable agriculture. In P. Gehlot & J. Singh (Eds.), *Fungi and their role in sustainable development: Current perspectives* (pp. 147–158). Springer Nature.

Muleta, D., Assefa, F., Nemomissa, S., & Granghall, U. (2008). Distribution of arbuscular mycorrhizal fungi spores in soils of smallholder agroforestry and monocultural coffee systems in southwestern Ethiopia. *Biology and Fertility of Soils, 44*, 653–659.

Müller, A., Ngwene, B., Peiter, E., & George, E. (2017). Quantity and distribution of arbuscular mycorrhizal fungal storage organs within dead roots. *Mycorrhiza, 27*, 201–210.

Nakano-Hylander, A., & Olsson, P. A. (2007). Carbon allocation in mycelia of arbuscular mycorrhizal fungi during colonization of plant seedlings. *Soil Biology and Biochemistry, 39*, 1450–1458.

Nouri, E., Breuillin-Sessoms, F., Feller, U., & Reinhardt, D. (2014). Correction: Phosphorus and nitrogen regulate arbuscular mycorrhizal symbiosis in *Petunia hybrida*. *PLoS One, 9*, e90841. https://doi.org/10.1371/journal.pone.0090841

Oehl, F., & Sieverding, E. (2004). *Pacispora*, a new vesicular arbuscular mycorrhizal fungal genus in the Glomeromycetes. *Journal of Applied Botany, 78*, 72–82.

Oehl, F., Sieverding, E., Ineichen, K., Ris, E.-A., Boller, T., & Wiemken, A. (2005). Community structure of arbuscular mycorrhizal fungi at different soil depths in extensively and intensively managed agroecosystems. *The New Phytologist, 165*, 273–283.

Oehl, F., de Souza, F. A., & Sieverding, E. (2008). Revision of *Scutellospora* and description of five new genera and three new families in the arbuscular mycorrhiza-forming Glomeromycetes. *Mycotaxon, 106*, 311–360.

Oehl, F., Silva, G. A., Goto, B. T., & Sieverding, E. (2011a). Glomeromycota: Three new genera, and glomoid species reorganized. *Mycotaxon, 116*, 75–120.

Oehl, F., Silva, G. A., Goto, B. T., Maia, L. C., & Sieverding, E. (2011b). Glomeromycota: Two new classes and a new order. *Mycotaxon, 116*, 365–379.

Oehl, F., Silva, D. K. A., Maia, L. C., Ferreira de Souza, N. M., Vieira, H. E. E., & da Silva, G. A. (2011c). *Orbispora* gen. nov., ancestral in the *Scutellosporaceae* (*Glomeromycetes*). *Mycotaxon, 116*, 161–169.

Oehl, F., Sieverding, E., Palenzuela, J., Ineichen, K., & Alves da Silva, G. (2011d). Advances in *Glomeromycota* taxonomy and classification. *IMA Fungus, 2*, 191–199.

Olsson, P. A., Bååth, E., Jakobsen, I., & Söderström, B. (1995). The use of phospholipid and neutral lipid fatty-acids to estimate biomass of arbuscular mycorrhizal fungi in soil. *Mycological Research, 99*, 623–629.

Opik, M., Moora, M., Zobel, M., Saks, U., Wheatley, R., Wright, F., & Daniell, T. (2008). High diversity of arbuscular mycorrhizal fungi in a boreal herb-rich coniferous forest. *The New Phytologist, 179*, 867–876.

Ortaş, I., Rafique, M., & Ahmed, I. A. M. (2017). Application of arbuscular mycorrhizal fungi into agriculture. In Q.-S. Wu (Ed.), *Arbuscular mycorrhizas and stress tolerance of plants* (pp. 305–327). Springer Nature.

Ortiz-Acevedo, A., Osorio-Vega, N. W., Echeverri-Gómez, J., González-Murillo, O. A., & Medina-Sierra, M. (2015). Fisiología de los hongos formadores de micorrizas arbusculares. *LRRD (Livestock Research for Rural Development), 27*, 188. http://www.lrrd.org/lrrd27/9/orti27188.html

Pal, S., Singh, H. B., Rai, A., & Farooqui, A. (2017). Diversity of arbuscular mycorrhiza associated with long term wastewater irrigation in the peri-urban soil of Varanasi. *International Journal of Agriculture Environment and Biotechnology, 10*, 779–784.

Palenzuela, J., Ferrol, N., Boller, T., Azcón-Aguilar, C., & Oehl, F. (2008). *Otospora bareai*, a new fungal species in the Glomeromycetes from a dolomitic shrub land in Sierra de Baza National Park (Granada, Spain). *Mycologia, 100*, 296–305.

Paliocha, M. (2017). Functional trait and life-history variation of arbuscular mycorrhizal fungi during secondary succession. *NMBU Student Journal of Life Sciences, 7*, 1–8.

Parniske, M. (2008). Arbuscular mycorrhiza: The mother of plant root endosymbioses. ***Nature Reviews. Microbiology, 6***, 763–775.

Perner, H., Schwarz, D., Bruns, C., Mäder, P., & George, E. (2007). Effect of arbuscular mycorrhizal colonization and two levels of compost supply on nutrient uptake and flowering of pelargonium plants. *Mycorrhiza, 17*, 469–474.

Pickles, B. J., Truong, C., Watts-Williams, S. J., & Bueno, C. G. (2020). Mycorrhizas for a sustainable world. *The New Phytologist, 225*, 1065–1069.

Pirozynski, K. A., & Dalpé, Y. (1989). Geological history of the Glomaceae with particular reference to mycorrhizal symbiosis. *Symbiosis, 7*, 1–36.

Redecker, D. (2002). Molecular identification and phylogeny of arbuscular mycorrhizal fungi. *Plant and Soil, 244*, 67–73.

Redecker, D., Morton, J. B., & Bruns, T. D. (2000). Ancestral lineages of arbuscular mycorrhizal fungi (Glomales). *Molecular Phylogenetics and Evolution, 14*, 276–284.

Redecker, D., Schüßler, A., Stockinger, H., Stürmer, S. L., Morton, J. B., & Walker, C. (2013). An evidence-based consensus for the classification of arbuscular mycorrhizal fungi (Glomeromycota). *Mycorrhiza, 23*, 515–531.

Rich, M. K., Nouri, E., Courty, P.-E., & Reinhardt, D. (2017). Diet of arbuscular mycorrhizal fungi–bread & butter? *Trends in Plant Science, 22*, 652–660.

Rillig, M. C., Sosa-Hernández, M. A., Roy, J., Aguilar-Trigueros, C. A., Vályi, K., & Lehmann, A. (2016). Towards an integrated mycorrhizal technology: Harnessing mycorrhiza for sustainable intensification in agriculture. *Frontiers in Plant Science, 7*, 1625. https://doi.org/10.3389/fpls.2016.01625

Rillig, M. C., Aguilar-Trigueros, C. A., Camenzind, T., Cavagnaro, T. R., Degrune, F., Hohmann, P., Lammel, D. R., Mansour, I., Roy, J., van der Heijden, M. G. A., & Yang, G. (2019). Why farmers should manage the arbuscular mycorrhizal symbiosis. *The New Phytologist, 222*, 1171–1175.

Säle, V., Aguilera, P., Laczko, E., Mäder, P., Berner, A., Zihlmann, U., van der Heijden, M. G. A., & Oehl, F. (2015). Impact of conservation tillage and organic farming on the diversity of arbuscular mycorrhizal fungi. *Soil Biology and Biochemistry, 84*, 38–52.

Salmerón-Santiago, I. A., Pedraza-Santos, M. E., Mendoza-Oviedo, L. S., & Chávez-Bárcenas, A. T. (2012). Cronología de la taxonomía y cladística de los glomeromicetos. *Revista Fitotecnia Mexicana, 38*, 153–163.

Sanders, I. R., Clapp, J. P., & Wiemken, A. (1996). The genetic diversity of arbuscular mycorrhizal fungi in natural ecosystems - A key to understanding the ecology and functioning of the mycorrhizal symbiosis. *The New Phytologist, 133*, 123–134.

Schalamuk, S., Velazquez, S., Chidichimo, H., & Cabello, M. (2006). Fungal spore diversity of arbuscular mycorrhizal fungi associated with spring wheat: Effects of tillage. *Mycologia, 98*, 16–22.

Schalamuk, S., Druille, M., & Cabello, M. (2013). Hongos formadores de micorrizas arbusculares: Influencia de las prácticas agronómicas sobre la diversidad y dinámica de colonización. In I. E. V. García de Salomone, S. Vázquez, C. Penna, & F. Cassán (Eds.), *Rizósfera, Biodiversidad y Agricultura Sustentable* (pp. 47–71). Asociación Argentina de Microbiología.

Schüßler, A. (2006). Phylogeny and taxonomy of Glomeromycota ('arbuscular mycorrhizal [AM] and related fungi'). http://www.amf-phylogeny.com

Schüßler, A., & Walker, C. (2010). *The Glomeromycota: A species list with new families and new genera.* The Royal Botanic Garden/The Royal Botanic Garden Kew/Botanische Staatssammlung Munich/Oregon State University. Createspace Independent Publication. 58 p. (Copy of the formal publication: October 17 2011).

Schüßler, A., Schwarzott, D., & Walker, C. (2001). A new fungal phylum, the *Glomeromycota*: Phylogeny and evolution. *Mycological Research, 105*, 1413–1421.

Sharma, S., Sharma, S., Thakur, A., & Singh, M. J. (2018). Periodic changes in arbuscular mycorrhizal fungi in the rhizospheric soil of fruit plants. *Agricultural Research Journal, 55*, 685–690.

Shukla, A., Vyas, D., & Jha, A. (2013). Soil depth: An overriding factor for distribution of arbuscular mycorrhizal fungi. *Journal of Soil Science and Plant Nutrition, 13*, 23–33.

Sieverding, E., & Oehl, F. (*2006*). *Revision of Entrophospora and description of Kuklospora and Intraspora, two new genera in the arbuscular mycorrhizal Glomeromycetes. Journal of Applied Botany and Food Quality, 80, 69–81.*

Sieverding, E., da Silva, G. A., Berndt, R., & Oehl, F. (2015). *Rhizoglomus*, a new genus of the *Glomeraceae*. *Mycotaxon, 129*, 373–386.

Silva, E. M., Maia, L. C., Menezes, K. M. S., Braga, M. B., Melo, N. F., & Yano-Melo, A. M. (2015). Water availability and formation of propagules of arbuscular mycorrhizal fungi associated with sorghum. *Applied Soil Ecology, 94*, 15–20.

Silva-Flores, P., Bueno, C. G., Neira, J., & Palfner, G. (2019). Factors affecting arbuscular mycorrhizal fungi spore density in the Chilean Mediterranean-type ecosystem. *Journal of Soil Science and Plant Nutrition, 19*, 42–50.

Smith, S. E., & Read, D. J. (2008). *Mycorrhizal symbiosis* (3rd ed.). Elsevier. 800 p.

Souza, T. (2015a). Glomeromycota classification. In T. Souza (Ed.), *Handbook of arbuscular mycorrhizal fungi* (pp. 87–128). Springer International Publishing AG.

Souza, T. (2015b). Spores: A special tool to survive. In T. Souza (Ed.), *Handbook of arbuscular mycorrhizal fungi* (pp. 65–86). Springer International Publishing AG.

Spain, J. L., Sieverding, E., & Oehl, F. (2006). *Appendicispora*: A new genus in the arbuscular mycorrhiza-forming Glomeromycetes, with discussion of the genus *Archaeospora*. *Mycotaxon, 97*, 163–182.

Stürmer, S. L., & Morton, J. B. (1999). Taxonomic reinterpretation of morphological characters in Acaulosporaceae based on developmental patterns. *Mycologia, 91*, 849–857.

Sun, X., & Guo, L.-D. (2012). Endophytic fungal diversity: Review of traditional and molecular techniques. *Micology, 3*, 65–76.

Symanczik, S., Al-Yahya'ei, M. N., Kozłowska, A., Ryszka, P., & Błaszkowski, J. (2018). A new genus, *Desertispora*, and a new species, *Diversispora sabulosa*, in the family Diversisporaceae (order Diversisporales, subphylum Glomeromycotina). *Mycological Progress, 17*, 437–449.

Tahat, M. M., & Sijam, K. (2012). Mycorrhizal fungi and abiotic environmental conditions relationship. *Research Journal of Environmental Sciences, 6*, 125–133.

Tchabi, A., Hountondji, F., Laouwin, L., Coyne, D., & Oehl, F. (2009). *Racocetra beninensis* from sub-Saharan savannas: A new species in the Glomeromycetes with ornamented spores. *Mycotaxon, 110*, 199–209.

Tchiechoua, Y. H., Kinyua, J., Ngumi, V. W., & Odee, D. W. (2020). Effect of indigenous and introduced arbuscular mycorrhizal fungi on growth and phytochemical content of vegetatively propagated *Prunus africana* (Hook. f.) Kalkman provenances. *Plants, 9*, 37. https://doi.org/10.3390/plants9010037

Thirkell, T. J., Charters, M. D., Elliott, A. J., Sait, S. M., & Field, K. J. (2017). Are mycorrhizal fungi our sustainable saviours? considerations for achieving food security. *Journal of Ecology, 105*, 921–929.

Tindall, B. J., Rosselló-Móra, R., Busse, H. J., Ludwig, W., & Kampfer, P. (2010). Notes on the characterization of prokaryote strains for taxonomic purposes. *International Journal of Systematic and Evolutionary Microbiology, 60*, 249–266.

Vályi, K., Mardiah, U., Rillig, M. C., & Hempel, S. (2016). Community assembly and coexistence in communities of arbuscular mycorrhizal fungi. *The ISME Journal, 10*, 2341–2351.

Walker, C. (1983). Taxonomic concepts in the Endogonaceae: Spore wall characteristics in species descriptions. *Mycotaxon, 18*, 443–455.

Walker, C. (1986). Taxonomic concepts in the Endogonaceae. II. A fifth morphological wall type in endogonaceous spores. *Mycotaxon, 25*, 95–99.

Walker, C., & Schüßler, A. (2004). Nomenclatural clarifications and new taxa in the Glomeromycota. *Mycological Research, 108*, 981–982.

Wang, Y., Li, T., Li, Y., Björn, L. O., Rosendahl, S., Olsson, P. A., Li, S., & Fu, X. (2015). Community dynamics of arbuscular mycorrhizal fungi in high-input and intensively irrigated rice cultivation systems. *Applied and Environmental Microbiology, 81*, 2958–2965.

Wang, J., Fu, Z., Ren, Q., Zhu, L., Lin, J., Zhang, J., Cheng, X., Ma, J., & Yue, J. (2019). Effects of arbuscular mycorrhizal fungi on growth, photosynthesis, and nutrient uptake of *Zelkova serrata* (Thunb.) makino seedlings under salt stress. *Forests, 10*, 186. https://doi.org/10.3390/f10020186

Wu, Q.-S., & Zou, Y.-N. (2017). Arbuscular mycorrhizal fungi and tolerance of drought stress in plants. In Q.-S. Wu (Ed.), *Arbuscular mycorrhizas and stress tolerance of plants* (pp. 25–41). Springer Nature.

Wu, C.-G., Liu, Y. S., Huang, Y. L., Wang, Y. P., & Chao, C. C. (1995). Glomales of Taiwan: V. *Glomus chimonobambusae* and *Entrophospora kentinensis*, spp. nov. *Mycotaxon, 53*, 283–294.

Xing, D., Wang, Z., Xiao, J., Han, S., Luo, C., Zhang, A., Song, L., & Gao, X. (2018). The composition and diversity of arbuscular mycorrhizal fungi in karst soils and roots collected from mulberry of different ages. *Ciência Rural, 48*(10), e20180361. cottlple:/c/dtexd.dforoi.morgm/1u0lb.1e5rr9y0o/0f1d0i3ff-e8r4e7n8tcarg2e0s1. 80361.

Xueming, Z., Qin, P., Wan, S., Zhao, F., Guang, W., Yan, D., & Zhou, J. (2007). Effects of arbuscular mycorrhizal fungi on the rooting and growth of beach plum (*Prunus maritima*) cuttings. *The Journal of Horticultural Science and Biotechnology, 82*, 863–866.

Yadav, R. S., Mahatma, M. K., Thirumalaisamy, P. P., Meena, H. N., Bhaduri, D., Arora, S., & Panwar, J. (2017). Arbuscular mycorrhizal fungi (AMF) for sustainable soil and plant health in salt-affected soils. In S. Arora, A. K. Singh, & Y. P. Singh (Eds.), *Bioremediation of salt affected soils: An Indian perspective* (pp. 133–156). Springer International Publishing AG.

Zeng, Y., Guo, L.-P., Chen, B.-D., Hao, Z.-P., Wang, J.-Y., Huang, L.-Q., Yang, G., Cui, X.-M., Yang, L., Wu, Z.-X., Chen, M.-L., & Zhang, Y. (2013). Arbuscular mycorrhizal symbiosis for sustainable cultivation of Chinese medicinal plants: A promising research direction. *Aperito Journal of Cellular and Molecular Biology, 41*, 1199–1221.

Part IV
Environmental Biotechnology

Microbial Reductive Dehalogenation and Its Role in Bioremediation

Srinivasan Kameswaran, Bellemkonda Ramesh, Gopi Krishna Pitchika, M. Subhosh Chandra, Swapna B., and M. Srinivasulu

1 Introduction

An extraordinarily significant and numerous groups of environmental chemicals are halogenated compounds. Microbiological researches on the biodegradation of halogenated compounds have focused primarily on the physiological techniques charged for their mineralization and on the carbon-halogen bond cleavage enzymes concerned. Due to the fact that their physiological residences and substratum variety can decide the process conditions that need to be used and the variety of transformations that can be obtained in functional treatment systems, the characterization of dehalogenating species is important for their industrial application. The use of a halogenated compound as a source of carbon and an oxidizable substrate with oxygen may be preferred for degradation on the basis of one of four processes or nitrate as an electron acceptor; fermentative metabolism, in which a dehalogenated intermediate acts as an electron acceptor; the use of a halogenated compound as an electron acceptor with halide release; and co-metabolic transformation. The final

S. Kameswaran (✉) · Swapna B.
Department of Botany, Vikrama Simhapuri University PG Centre, Kavali, Andhra Pradesh, India

B. Ramesh
Department of Food Technology, Vikrama Simhapuri University, Nellore, Andhra Pradesh, India

G. K. Pitchika
Department of Zoology, Vikrama Simhapuri University PG Centre, Kavali, Andhra Pradesh, India

M. S. Chandra
Department of Microbiology, Yogi Vemana University, Kadapa, Andhra Pradesh, India

M. Srinivasulu
Department of Biotechnolgy, Yogi Vemana University, Kadapa, Andhra Pradesh, India

N. R. Maddela, L. C. García (eds.), *Innovations in Biotechnology for a Sustainable Future*, https://doi.org/10.1007/978-3-030-80108-3_11

phase occurs when an enzyme with some other physiological feature lacks specificity. True dehalogenation is the main response of halogenated compounds to microbial degradation. The halogen substituent, typically responsible for the poisonous and xenobiotic individual of the compound, is altered at some point in this level, often via hydrogen or a hydroxyl institution. During subsequent metabolic steps, halogen removal decreases both recalcitrance to biodegradation and the possibility of forming toxic intermediates. Where the presence of halogen substituents can contribute to the development of acylhalides or 2-haloaldehydes, which are reactive products due to their electrophilicity and can cause cellular damage, the latter can occur during oxidative conversion. Reductive dehalogenation, which offers new insights into our understanding of the role of halo-organic compounds in microbial physiology and ecology, is one of the most significant biological procedures for this (Smidt & de Vos, 2004). The most important feature of microbial reductive dehalogenation is the use of halogenated compounds as terminal electron acceptors for the transport of anaerobic respiratory electrons, i.e., the metabolic mechanism known as dehalorespiration, halorespiration, or chlororespiration. The majority of halo-organic compounds have long been thought to be xenobiotics, and the occurrence of these compounds in nature is due to contamination from anthropogenic sources. Accumulated scientific knowledge of reductive dehalogenation and dehalorespiring bacteria, however, indicates that organohalides are important as physiological substrates for the production and survival of these specific and anthropogenic microorganisms that can be developed in natural environments (Ahn et al., 2003). Since more than 3800 organohalogen species are estimated to be produced by living organisms and through natural abiogenic processes (Gribble, 2003), much more attention should be paid to the geochemical cycle of halogenated compounds in which important ecological roles may be played by anaerobic dehalogenating microorganisms.

Another important feature of reductive dehalogenation is relevant to its application to environmental bioremediation (Wackett, 1994). Most congeners of polychlorinated biphenyls (PCBs), polychlorinated dibenzo-*p*-dioxins/furans (PCDD/Fs), and other haloaromatics, as well as aliphatic halo-organic compounds, are highly toxic and recalcitrant contaminants whose potential risk to human health and wild life should be taken into account (Ross, 2004). Thus, environmental science and technology are central to the issue of how to remedy organohalogen emissions. Scientifically sound and cost-effective bioremediation procedures can be provided by harnessing microbial reductive dehalogenation since this anaerobic method may function more effectively in extracting halogen atoms from halogen atoms than aerobic biodegradation (poly) halogenated compounds (Hiraishi et al., 2005; Nealson, 2003; Scow & Hicks, 2005).

We are discussing important recent results in the field of microbial dehalogenation in this chapter. Much work has been done on the identifying and characterizing enzymes to remove halogen, both from aerobic and anaerobic organisms. Several biochemical and biophysical studies have been conducted on new dehalogenases, leading to great insight into their evolution, diversity, and mechanisms of catalysis.

2 Reductive Dehalogenation

Reductive dehalogenation is an important process which while aerobic reductive dehalogenation has been reported, usually occurs under anaerobic conditions. Reductive dehalogenation is a common mechanism by which bacteria attack halogenated organic compounds. Many pesticides, herbicides, and solvents are halogenated hydrocarbons, as are the insulating oils, polychlorinated biphenyls (PCBs). The first step in the biodegradation of these compounds is often the removal by anaerobic bacteria of one or more chlorine atoms.

To achieve halogen elimination, two general mechanisms are employed. The displacement of the halogen atom with a proton is hydrogenolysis. With alkyl and aryl compounds, this process can occur; however, bacteria show substrate specificity, with less substrate-specific and more widespread aryl dehalogenators. Alternatively, two electrons can be substituted for the halogen atoms on neighboring carbons, resulting in a double bond between carbon atoms. This mechanism is referred to as vicinal reduction.

The degree to which chlorine atoms can be extracted by reductive dehalogenation is not certain and depends on the composition and behavior of the microbial community and on environmental conditions. Since the isolation of pure strains has proven to be very difficult, most research on reductive dehalogenation has been carried out using mixed microbial populations. In codependent or syntrophic relationships, populations of bacteria that can dehalogenate chlorinated compounds reductively can occur, making the survival of one community member contingent on the survival and operation of others. It is commonly thought that polychlorinated compounds cannot be fully dehalogenated by individual strains, but rather that the removal of several chlorine atoms is the result of many bacterial species' concerted action.

The metabolic benefit of reductive dehalogenation is not well characterized; however, in at least one case, dehalogenation is thought to be coupled to ATP synthesis though a proton ATPase. In anaerobic bacteria, including methanogens, sulfate reducers, and denitrifiers, reductive dehalogenation is most frequently observed. Aryl dehalogenation in certain aerobic bacteria and facultative anaerobes has also been reported. Further biodegradation of the compound is enabled by the removal of chlorine or other halogens from organic compounds. The less halogenated a compound is in general, the more likely it is to be biodegraded further. Dehalogenating bacteria therefore play a significant role in the initiation of halogenated compound biodegradation.

Methanogens dehalogenate TCE via vinyl chloride to ethane. Polychlorinated biphenyls are often dehalogenated reductively to biphenyls with a lower chlorination amount. Dehalogenation makes aerobic biodegradation more prone to PCBs. Often reductively dehalogenated are chlorobenzenes, chlorophenoxyacetates, pentachlorophenol, and other halogenated aromatic hydrocarbons.

3 Reductive Forms of Dehalogenation Reaction

There are four forms of response in which reductive processes are involved. The first reaction, the hydrogenolysis, replaces a halogen substituent by hydrogen. Hydrogenolysis is the reductive dehalogenation reaction most commonly observed and is found with aliphatic and aromatic compounds. The second type is a dihalo-elimination. In this reaction two halogens are eliminated from the target compound at the same time and a double bond is formed. It has been found with cyclic and noncyclic aliphatic halocompounds and is restricted to compounds with more than one carbon atom. The third type is a coupling and can occur when free radicals are involved. Only as side products were products found which resulted from a coupling reaction. The fourth type is a hydrolytic reduction. The products of this type of reaction are oxygenated and formed by the reduction of a polyhalogenated carbon to a carbenoid by two electrons, followed by hydrolysis. This reaction could explain the formation of CO_2 from tetrachloromethane or acetate from 1,1,1-trichloroethane (Criddle, 1989).

4 The Biological Sense of Reductive Dehalogenation

Reductive dehalogenation is defined as the removal of a halogen substituent from a molecule with the concomitant addition of two electrons (Dolfing, 2003). Hydrogenolysis and dichloroelimination are two modes of reductive dehalogenation, but the biological phase takes place mainly as a hydrogenolytic reaction. Co-metabolic reductive dehalogenation, which is of no benefit to the catalyzing microorganisms, happens with enzymes that normally catalyze other reactions. On the other hand, in metabolic reductive dehalogenation, i.e., dehalorespiration, energy is conserved via an anaerobic respiratory process with a halogenated compound as the terminal electron acceptor and reductive dehalogenase (RDase) as the key enzyme involved. During dehalorespiration, dehalogenated end products from polyhalogenated compounds differ among species and strains, as can be seen in the reductive dechlorination of tetrachloroethene (PCE). In the case of dehalorespiration with PBDs and other polychlorinated aromatics, their monochlorinated or dichlorinated congeners are the end products in general, and no complete dechlorination has so far been reported.

The redox potential is relatively high, ranging from 260 to 570 mV, for halo-organic compounds including PBDs and chloroethenes (tetrachloroethene [PCE], trichloroethene [TCE], dichloroethene [DCE], and vinyl chloride [VC]) (Dolfing, 2003). For example, if 1,2,3,4-tetrachlorodibenzo-*p*-dioxin (TCDD) to 2-monochlorodibenzo-*p*-dioxin (MCDD) is dehalogenated through the 1,2,3-trichlorodibenzo-*p*-dioxin (TrCDD) and 2,3-dichlorodibenzo-*p*-dioxin (TD) pathways with H_2 as the electron donor, H_2 as the electron donor is dehalogenated as the intermediate (DCDD) pathway, the total reaction and $\Delta G^{\circ\prime}$ for this is given by:

$$1,2,3,4-\mathrm{TCDD}(\mathrm{C_{12}H_4Cl_4O_2})+6\mathrm{H}^+ + 6e^- \rightarrow 2-\mathrm{MCDD}(\mathrm{C_{12}H_7ClO_2})3\mathrm{HCl}$$

$$\Delta G^{\circ\prime} = -469\,\mathrm{kJ\,mol^{-1}}$$

As described above and previously (Dolfing, 2003), It is clear that the hydrogenotrophic redox process with PBDs can provide ample energy for the growth and survival of dehalogenating microorganisms as terminal electron acceptors.

Although both mandatory and facultative anaerobes have been identified as microorganisms capable of metabolic or co-metabolic reductive dehalogenation, all of the dehalorespiring bacteria so far described are strictly anaerobic, with the exception of *Anaeromyxobacter*. As mentioned above, however, the redox potentials for organohalides are much higher than that for the SO_4^{2}/H_2S redox couple (E_0 = −217 mV) and are comparable to the value for the NO^-/NO^- couple (E'_0 = 433 mV). Therefore, the reductive dehalogenation process itself may not always require strictly anoxic, low-potential conditions (Dolfing, 2003). In fact, populations of "Dehalococcoides" and their phylogenetic relatives within the *Chloroflexi* phylum were identified at a broad Eh gradient of 5 to −75 mV along with ubiquinone containing aerobic bacteria at the lake sediment surface (Hiraishi et al., 2005). Moreover, it has been suggested that reductive dehalogenating activity of "*Dehalococcoides*" can be triggered in anaerobic environments located in close spatial proximity to aerobic environments in wetland systems (Kassenga et al., 2004).

5 Aerobic Growth on Halogenated Aliphatic Compounds

Several different species have been identified that are able to use halogenated compounds as a substrate for development. 1,3-dichloropropylene and 1,2-dibromoethane are relatively recent examples of compounds considered to be degradable in pure bacterial cultures. In both cases, initial dehalogenation starts with a reaction catalyzed by a hydrolytic haloalkane dehalogenase (Poelarends et al., 2000). The haloalkane dehalogenases that degrade these compounds are homologous to the classical enzyme from *Xanthobacter* autotrophicus (Verschueren et al., 1993). Even though sequence similarities are low, the overall structures are very similar and the catalytic residues are conserved (Newman et al., 1999). LinB from *Sphingomonas paucimobilis* is a new example of a recently studied hydrolytic dehalogenase, an enzyme involved in tetrachlorocyclohexadiene conversion and an intermediate in the γ-hexachlorocyclohexane degradation pathway (Marek et al., 2000).

Both these hydrolytic dehalogenases belong to the superfamily of the alpha/β-hydrolase fold and contain a main domain and a domain of the cap. The active site is located between these two domains and catalysis proceeds through a covalent enzyme–substrate intermediate. The enzymes have a catalytic triad formed by the principal domain residues, with the halogen substrate being displaced by an aspartate as the nucleophile. In addition, these enzymes have a retained residue of tryptophan in the main domain and variable residues in the cap domain (phenylalanine or asparagine) that lead to halide binding (Newman et al., 1999; Marek et al., 2000).

Haloacid dehalogenases, which can be classified into at least two classes, known as groups I and II, are other hydrolytic dehalogenases that have recently been studied in more detail, according to Hill et al. (1999). The group II enzymes use covalent catalysis, much like the haloalkane dehalogenases, and extensive insight into the reaction mechanism has been obtained. Both the chloropropionic acid dehalogenase (L-DEX) X-ray structures of *Pseudomonas* sp. YL have been solved (Li et al., 1998) and chloroacetate dehalogenase (DhlB) of *X. autotrophicus* GJ10 (Ridder et al., 1999). Typically, group II enzymes transform L-2-chloropropionic acid, but not D-2-chloropropionic acid, with a configuration inversion at the chiral core.

The structure of group II dehalogenases is somewhat different from that of haloalkane dehalogenases, but they are similar to many phosphatases, the catalytic subunits of P-type ATPases and several other enzymes of the haloacid dehalogenase superfamily. Group II dehalogenases share a conserved residue of nucleophilic aspartate that is found near the N terminus and is involved in covalent intermediate formation. By interaction with a conserved lysine, the aspartate is located. Arginine, asparagine, and a trace of phenylalanine are also retained and are involved in halogen/halide binding. Therefore, from those in alpha/β-hydrolase fold enzymes, both the nucleophilic site and the leaving group site are distinct.

Although no structural information is available as yet, biochemical studies have indicated that group I haloacid dehalogenases (Ridder et al., 1999; Nardi Dei et al., 1999), which include a remarkable enzyme selective for D-2-chloropropionic acid and an enzyme that converts trichloroacetate to carbon monoxide (Stringfellow et al., 1997), do not use a covalent mechanism for catalysis. The use of ^{18}O-labeled water and DL-DEX haloacid dehalogenase from *Pseudomonas* sp. 113 in Nardi Dei et al. (1999) indicated that a water molecule was activated directly by dehalogenase to strike the alpha carbon of 2-haloalkanoic acid, thus displacing the halogen atom. Which amino acids are involved in dehalogenation has not yet been determined, although mutagenesis studies have highlighted a number of candidates.

6 Aerobic Growth on Halogenated Aromatics

The dehalogenation of aromatic substrates may take place before, during, or after aromatic ring cleavage. Genetic and biochemical studies have recently provided fresh insight into these mechanisms. In the metabolism of polyhalogenated compounds such as tetrachlorobenzene, cleavage of carbon-chlorine bonds before ring cleavage appears to be an important reaction. This reaction, catalyzed by dioxygenase, was first involved in fluorocatechol dehalogenation, but appears to be much wider in scope. For example, TecA dioxygenase from *Burkholderia* PS12, which contains the Rieske group of type 2Fe–2S, can fuse two oxygen atoms into 1,2,4,5-tetrachlorobenzene. This is accompanied by dehydrogenation, which means that catechol is formed immediately without the participation of dehydrogenase, which would be necessary for the production of dihydrodiol. The alpha subunit (TecA1) appears to have basic halogen removal determinants and more precisely, Beil et al. (1999) that the non-mega-amino acid at position 220 is required for reaction with tetrachlorobenzene and dioxygen.

Tetrachlorohydroquinone reductive dehalogenase (TCD), an enzyme involved in pentachlorophenol metabolism in *Sphingomonas chlorophenolica*, is used as a second mechanism for dehalogenation before ring cleavage. The reductive removal of chlorine from a ring structure by a glutathione-dependent reaction is catalyzed by this enzyme; the enzyme is identical in sequence to glutathione transferases. Anandarajah et al. (2000) recently indicated that this dehalogenase, an enzyme that isomerizes a cis double bond to a trans double bond, has evolved from a maleylacetoacetate isomerase. This isomerase is involved in tyrosine metabolism, and the isomerase appears to be a eukaryotic organism that shares a large sequence similarity with *S. chlorophenolica* and *S. paucimobilis* reduced dehalogenases that are involved in the degradation of chlorophenols and lindane (-HCH), respectively. A third mechanism is used to remove the halogen before the ring cleavage with the coenzyme 4-chlorobenzoyl A (CoA) dehalogenase. In this case, the cleavage of the carbon and halogen bond is hydrolytic; it takes activation of the 4-chlorobenzoate substrate to grow into the CoA derivative, a process found in *Pseudomonas* CBS3. The dehalogenase is evolutionarily related to a 2-enoyl-CoA hydratase (Xiang et al., 1999).

Dechlorination has long been thought to be a non-realistic choice during ring cleavage of catechol. Indeed, it has been suggested that *meta* cleavage of an aromatic ring between a carbon atom substituted by hydroxy and halogen is forbidden because it would create an unstable ring and toxic acyl halide. Surprisingly, the epithelization enzyme, 3-chlorocatechol dioxygenase from the degrading organism of chlorobenzene *Pseudomonas putida* GJ31, rapidly cleaves and removes 3-chlorocatechol to produce 2-hydroxymuconic acid. Our studies with hybrid enzymes have shown that this activity is linked to the C-terminal part of the ring cleavage dioxygenase (Mars et al., 1999). This enzyme is either immune to inactivation by possible reactive intermediates or by efficient dechlorination, prevents their creation. How widely distributed dehalogenation is by *meta* cleavage remains to be known.

7 On Halogenated Substrates, Anaerobic Growth: Halorespiration

During the oxidation of an electron rich compound, such as hydrogen or an organic substrate, a halogenated compound acts as a terminal electron acceptor during halorespiration. The biochemistry of anaerobic reductive dehalogenation is diverse. Three separate reductive dehalogenases were identified and cloned and sequenced for their genes. These are the trichloroethene reductive dehalogenase from *Dehalococcoides ethenogenes* (Magnuson et al., 1999), tetrachloroethene reductive dehalogenase from *Dehalospirillum multivorans* and orthochlorophenol reductive dehalogenase derived from *Desulfitobacterium dehalogenans* (Van de Pas et al., 1999; Smidt et al., 2000). In three consecutive steps, the trichloroethene reductive dehalogenase from Dehalococcoides ethenogenes dehalogenates trichloroethene to ethene and is also active with 1,2-dichloroethane and 1,2-dibromoethane. Using an initial reductive dehalogenase for the conversion of tetrachloroethene or trichloroethene and another reductive enzyme for the formation of ethene, *D. ethenogenes* will completely dechlorinate perchloroethene to ethene. However, genome sequencing has shown that in *D. ethenogenes*, some more reductive dehalogenases may be present. Remarkably, chlorinated compounds are the only known functional electron acceptors, raising concerns about the evolutionary history of *D. ethenogenes* and their natural ecological niche (Maymo-Gatell et al., 1999). The existence of a signal sequence of twin arginine means that extra-cytoplasmic enzymes are the reductive dehalogenases. Electrons should thus flow to electron transport components that generate a proton motive force for ATP synthesis. Reductive dehalogenases have a 50–65 kDa molecular weight and contain a cobalamin cofactor and clusters of iron-sulfur. The similarity of the sequence with other proteins is limited to the C-terminal portion of the protein, which contain two binding forms of [4Fe–4S] groups, each of which consists of four cysteines. Furthermore, genetic research has shown that a small protein encoded by a gene is closely related to the dehalogenase gene; it has been speculated that this small protein is involved in either membrane interaction or transmission of electrons from the dehalogenase to the components of the membrane (Magnuson et al., 2000; Van de Pas et al., 1999).

8 Factors Affecting the Dechlorination of PCBs

Factors affecting chlorine removal in PCB were hypothesized (Bedard et al., 1996, 1997; Wu et al., 1996; Van Dort et al., 1997; Quensen et al., 1990; Wu & Wiegel, 1997) that a variety of microorganisms with distinct halogen removal enzymes, each showing a unique pattern of selectivity of congeners, are responsible for the different patterns of PCB removal observed in vitro and in the environment. Environmental factors and conditions affect the growth and the variety of metabolic activities of different microorganisms differently (Maddela et al., 2015, 2017) and hence influence divergently the extent and rate of the various PCB-dechlorinating activities. Consequently, to gain an understanding of the diversity of PCB dehalogenation and the circumstances under which a particular pattern of PCB dehalogenation may or may not occur, it is necessary to better understand whether and to what extent individual environmental factors can affect PCB dechlorination. This knowledge will help to predict the potential in a given environment for PCB dechlorination and will help in the creation of bioremediation schemes. Below, environmental factors are discussed which have been shown to influence PCB dechlorination and include especially temperature, pH, available carbon source, H_2 as electron donor and the presence or absence of electron acceptors other than PCBs. No detailed and systematic studies have been carried out so far on the reductive dehalogenation in the absence of different soil/sediment forms (including with different PCB adsorption properties).

8.1 Temperature

Besides the effect of temperature on the bioavailability as well as on transport between sites (due to the influence of temperature on the heats of surface-air exchanges) (Simcik et al., 1999; Halsall et al., 1999), including the absorption and enzymatic dehalogenation of PCB congeners, temperature has an important impact on growth and physiological activity. However, most laboratory studies of microbial PCB dechlorination reported to date have been performed at room temperature, i.e., around 25 °C. PCB-contaminated sediments usually undergo a much wider range of temperatures in the environment. The temperature range depends on the atmosphere and on the water depth and the sediment itself. The influence of temperature is multifaceted. The effects involve improvements in the kinetics of adsorption and desorption of PCBs from soil particles, and hence both hydrolytic (abiotic) dehalogenation and microbial transformation availability of PCBs. However in contrast to the influence of temperature on the growth of microorganisms and the catalytic activity of enzymes, these effects are possibly small.

8.2 pH

Sediments are mostly well buffered systems, but anaerobic microbial processes, unlike purely aerobic processes, may lead to an increase in acidic fermentation products and thus cause local pH changes. The impact of pH on PCB dechlorination in sediments, like the effects of temperature and carbon sources, is complex because of numerous potential interactions between the different dehalogenating microorganisms. In addition, pH influences the balance between dissolved PCBs and those that are adsorbed to organic matter and thus affects the bioavailability of soil PCBs (Jota & Hassett, 1991). At pH values between 5.0 and 8.0 and at incubation temperatures where changes in the dechlorination processes have been observed, dechlorination of 2346-CB applied as primer and residual PCBs in Woods Pond sediment has been studied, i.e., 15, 18, 25, and 34 °C (Chang, 1995). The pH of each slurry was balanced and maintained by adding sterile anaerobic 2 N NaOH or HCl regularly (2 ~ 10 days). Except for 34 °C at pH 5.0, at all temperatures, some PCB dechlorination was observed at all pH values examined. The optimum pH was about 7.0–7.5 for total chlorine elimination.

8.3 Supplementation of Carbon Sources

Anaerobic PCB dechlorination is a reduction process that presumably uses but does not cleave the rings with PCBs as electron acceptors. PCB dechlorinators should therefore need other compounds as sources of carbon and electrons for growth. Until very recently, in the presence of sediment which could provide a variety of organic matter, all research on anaerobic PCB dechlorination was conducted. In addition, all experiments were conducted with both PCB dechlorinators and non-PCB dechlorinators containing primary microcosms or enrichment cultures. Thus, adding a specific carbon source to a farm can enhance PCB chlorine removal by providing a desired carbon source and electron source PCB to dechlorination devices or for non-dechlorinated bacteria that can provide growth stimuli as the most suitable electron donors or the vitamins or carbon sources to the microorganisms that remove the chlorine from the PCBs. The addition may also stimulate the use of PCB dechlorination-inhibiting substances. On the other hand, such additions could also inhibit PCB dechlorination by supplying a carbon source to non-PCB dechlorinators. These could then out-compete the PCB dechlorinators for electron donors or whose products would be more preferred electron or acceptors than PCBs to the dechlorinators and thus inhibit the dehalogenation. Obviously, both effects can occur simultaneously for different halogen removal groups. All these possibilities, which are difficult to distinguish in a mixed culture, complicate the interpretation of the results obtained when supplementing the carbon sources.

Alder et al. (1993) showed that repeated addition of fatty acids (acetate, propionate, butyrate, and hexanoic acid) (500 mg l^{-1} initially and 250 mg l^{-1} monthly) induced dechlorination of added PCBs in carbon-limited sediment slurries, but not in sediment slurries with greater organic carbon content. The addition to sediment slurries of 0.1% (v/v) thioglycolate medium with beef extract or acetate improved PCB dechlorination by shortening or removing the lag time for PCB dechlorination (Abramowicz et al., 1993) and increasing the overall dechlorination rate observed (Tiedje et al., 1991). In Aroclor 1248-contaminated soils, the addition of 0.06% pyruvate and malate significantly increased the extent of PCB dechlorination (Klasson et al., 1996), while in Aroclor 1260-contaminated Woods Pond sediment, malate applied together with a primer only shortened the lag period before the onset of reductive dehalogenation of PCBs (Chang, 1995; Bedard & Van Dort, 1998; Bedard et al., 1998).

8.4 Supplementation of H_2 as Electron Donor

Reductive dechlorination is a two electron transfer reaction in which H_2 is assumed to be directly or indirectly the electron donor (Dalton, 1990; DeWeerd et al., 1991; Zhang & Wiegel, 1990) and water the proton source (Griffith et al., 1992; Nie & Vogel, 1991). Lake sediments contain H_2 producers and usually a variety of competing H_2 utilizers with different affinities for H_2. The successful competition for H_2 by a microorganism depends not only on the partial pressure (i.e., availability) of H_2 and the affinity of the microorganism for H_2 but also on the presence of utilizable carbon sources and electron acceptors. H_2, depending on the partial pressure, can stimulate or inhibit this microbial dechlorination process (DeWeerd et al., 1991; Zhang & Wiegel, 1990; Linkfield & Tiedje, 1990; Madsen & Aamand, 1991). Wu et al. (1996) incubated, under shaking conditions in O_2-free N_2 gas, Woods Pond's Aroclor 1260-contaminated sediment (amended to 2346-CB) at 15, 25, or 34 °C and at pH 6.2 or 7.2 under a regularly replenished gas atmosphere of 0, 1, or 10% H_2 (v/v). In general, there were no major variations in the rate or degree of dechlorination between samples incubated with or without 1% H_2 gas in the gas atmosphere. Assumingly this quantity did not significantly change the available amount of H_2 compared to the microbially produced H_2.

8.5 Electron Acceptors

For anaerobic populations, electron acceptors are also the limiting ability and a significant determinant of the species composition of anaerobic communities. Therefore, electron acceptors can affect the presence of reductively dehalogenating organisms in a culture. In addition, the availability of electron acceptors may be expected to influence the electron flow (i.e., the availability of reductants) needed for reductive dehalogenation. This effect could occur through electron intracellular channeling or through interspecific competition for donors of electrons. Accordingly, evidence which indicates that electron acceptors do affect dehalogenation activity in anaerobic communities is accumulating. This relationship tends, however to be complex.

Reductive dehalogenation of haloaromatic compounds can lead to energy conservation (Mohn & Tiedje, 1992 and literature cited therein; Mackiewicz & Wiegel, 1998). Some results support the proposal (Brown et al., 1987; Quensen et al., 1988; Brown & Wagner, 1990) that PCBs are used by the dechlorinating microorganisms as electron acceptors. Kim and Rhee (1997) reported that the addition of 300 ppm of Aroclor 1248 to anaerobic sediment PCB enrichment cultures resulted in a 188-fold increase in the number of PCB dechlorinators (from 2.5×10^5 to 4.6×10^7 cells per gram of sediment). Conversely, the number decreased by 93% from 4.6×10^5 to 3.1×10^4 initial value in samples without addition of Aroclor 1248. They concluded that the growth of chlorine removal devices PCB required the presence of PCBs.

Recently, we observed (Wu et al., 1999) that the number of microorganisms in Woods Pond sediment capable of dehalogenating 26-BB and PCBs increased nearly 1000-fold (from 3–4.9 × 10^5 to 2–5.8 × 10^8 cells per gram of sediment (dry weight)) after preparation with 26-BB (1050 WM) plus 10 mM malate. These results demonstrate that halogenated PCBs lead PCB to remove chlorine mainly by stimulating the growth of the microorganisms that act on removing chlorine from the PCB. It is possible to further stimulate halogen removal by induction of halogen removal enzymes, but is currently considered to be of less importance.

In the presence of typical electron acceptors for anaerobic microorganisms, several investigations of PCB dechlorination have been performed. PCB dechlorination was usually observed under methanogenic conditions (Alder et al., 1993; May et al., 1992; Morris et al., 1992; Rhee et al., 1993), and the addition of bromoethane sulfonic acid (BESA), an inhibitor of methanogenesis, inhibited dechlorination of certain PCB congeners (Rhee et al., 1993; Williams, 1994) or dechlorination processes (Morris et al., 1992). However, ethanol-treated, pasteurized cultures obtained from Hudson River exhibited *meta* dechlorination of Aroclor 1242 (Ye et al., 1992) and the addition of BESA did not inhibit meta dechlorination of 2346-CB by a 2346-CB enrichment culture (Chang, 1995), indicating the methanogens may not carry out the dechlorination but influence the availability of electron donors in these cultures. The addition of sulfate (10–30 mM), an electron acceptor used by sulfate reducing bacteria, completely inhibited dechlorination or favored one dechlorination process over others (Kuipers et al., 1999; Rhee et al., 1993; Morris et al., 1992; May et al., 1992).

9 Regulation of Dehalogenase Gene Expression

A dechlorination reaction often needs only one protein that can recognize and convert a xenobiotic substrate, but a second protein is necessary for controlled expression by binding a halogenated substrate and interaction with the transcription machinery. Therefore, if halogen removal enzyme synthesis is downregulated, the pathway must be more developed than in the case of constitutive protein expression (Poelarends et al., 2000).

In dehalogenating species, a number of regulatory genes that regulate dehalogenase expression have been characterized. Classical haloacid dehalogenases are generally limited because they are natural compounds, which is not surprising. In several cases, transcription is mediated by an alternative RNA polymerase containing the σ54 factor. For chloroalkane and dichloromethane metabolism, where a negative regulator regulates transcription, controlled gene expression is also detected. Surprisingly, even the synthesis of dehalogenases for xenobiotic compounds such as tetrachloroethene and hexachlorocyclohexane (LinD) seems to be regulated (Nagata et al., 1999).

In species that use certain xenobiotic haloalkanes, examples of primitive gene expression can be found. In *Pseudomonas pavonaceae* 170, which uses

1,3-dichloropropene, *Mycobacterium* sp. strain GP1, which uses 1,2-dibromoethane, and *X. autotrophicus* GJ10, which uses 1,2-dichloroethane, and the expression of elemental dihalogenase is constitutive. Plasmid-localized haloalkane halogenase (dhaA) genes present in various subspecies of *Rhodococcus erythropolis*, which are presumed to be the source of dehalogenase genes now also present in *P. pavonaceae* and *Mycobacterium* sp., is upregulated by the adjacent dhaR product of the gene in parental *Rhodococci* (Poelarends et al., 2000).

The DhaR protein is part of the transcriptional repressor type regulator TetR family and responds to 1-chlorobutane and several other 1-halo-*n*-alkanes, but this protein is not manufactured in *Pseudomonas pavonaceae* 170 and is inactivated by a short deletion of *Mycobacterium* sp. GP1 strain-this removal was probably a necessary evolutionary step because the repressor cannot be inactivated by the respective substrates.

10 Genomic Diversity of Dehalorespiration

The entire genome structure of "*Dehalococcoides ethenogenes*" strain 195 (Seshadri et al., 2005), "*Dehalococcoides*" sp. Strain CBDB1 (Kube et al., 2005), and *Desulfitobacterium hafniense* strain Y51 (Nonaka et al., 2006) has been published in the literature. Additionally, genomic information about several other organisms that cause air clearance including the anaerobic *Anaeromyxobacter dehalogenans* strain 2CP-1^T, *Desulfitobacterium hafniense* strain DCB2^T, "*Dehalococcoides*" sp. strain BAV1, and "*Geobacter lovleyi*" strain SZ became available from databases. Strain 195 "*Dehalococcoides ethenogenes*" chromosome is 1.46 Mb and harbors 1591 expected coding sequences (CDSs). The *Dehalococcoides* sp. there is a 1.39 Mb chromosome in Strain CBDB1 on which 1458 predicted CDSs are known. Of those of the free-living prokaryotes so far identified, the CBDB1 strain chromosome is the smallest. *Desulfitobacterium hafniense* strain Y51, on the other hand, has a much larger chromosome (5.7-Mbp) that harbors 5060 predicted CDSs.

It is noteworthy that the chromosomes of strain 195 "*Dehalococcoides ethenogenes*" (Seshadri et al., 2005) and "*Dehalococcoides*" sp., CBDB1 (Kube et al., 2005) contains at least 18 and 32 RDase homologous genes, respectively, although the functions of these genes have not been fully elucidated yet. Such multienzyme systems for reductive dehalogenation in the "*Dehalococcoides*" strains indicate their strict tendency to use dehalorespiration with a number of organohalides as terminal electron acceptors. This physiological trait contrasts with those of *Desulfitobacterium hafniense* strain Y51, which contains only two *rdh* genes but is more physiologically versatile with utilization of a larger set of specialized electron donors and acceptors for anaerobic respiration (Nonaka et al., 2006). Another strain of *Desulfitobacterium hafniense*, DCB2^T, studied for whole genome sequences, contains at least nine RDase homologous genes. Most of the respiration related genes found in the *Desulfitobacterium hafniense* strains are absent in the "*Dehalococcoides*" strains.

In "*Dehalococcoides*" sp., 12 of the 18 rdhAB pairs present in the 195 "*Dehalococcoides ethenogenes*" strain are orthologous strains of CBDB1 (Kube et al., 2005; Seshadri et al., 2005). The amino acid sequence identities of the orthologs are 86–95% between the two, although the position and orientation are conserved only in six of the orthologs. Two *rdhA* genes functionally identified in strain 195, *pceA* and *tceA*, encode PceA-RDase and TceA-RDase, respectively, absent in the CBDB1 strain. The PCE-RDase gene is also involved in the production of 2,3-dichlorophenol-RDase in strain 195, indicating that naturally occurring chlorophenols compounds are candidates for native substrates of PceA (Fung et al., 2007). One of the 32 RDase homologous genes present in the genome of strain CBDB1 is the cbdbA84 gene (*cbrA*), encoding chlorobenzene-RDase (Adrian et al., 2007). Neither in strains 195 and BAV1 nor among the other "*Dehalococcoides*" cultures studied so far, orthologs of cbdbA84 have been identified. In addition to RDase genes, a variety of genes involved in the transport of respiratory electrons are shared by the "Dehalococcoides" strains, including those required to encode various hydrogenase multisubunit complexes.

Genome analyses of halorespiring bacteria have revealed the presence of RDase-encoding regions of potentially foreign origin. In "*Dehalococcoides ethenogenes*" strain 195, the majority of reductive dehalogenase genes, including *tceA*, were probably acquired by several gene transfer events (Regeard et al., 2005). An intra-chromosomal or inter-chromosomal transfer of *tceAB* between "*Dehalococcoides*" strains has also been proposed by studying the environmental distribution of the gene (Krajmalnik Brown et al., 2007). The VC-RDase genes, vcrA and bvcA, are particularly unusual in that, despite the absence of any tRNAs matching codons that end in T, the third position of codons in the genes is biased toward the nucleotide T (McMurdie et al., 2007). This abnormality in the codon usage of VC-RDase genes suggests that the former genes differ from most other "*Dehalococcoides*" genes in evolutionary history, possibly being acquired by lateral transfer.

11 PCB-Dehalogenating Bacteria and Consortia

The reduced microbial halogen removal of halogenated biphenyls as well as other halogenated organic compounds has been extensively studied using anaerobic sediment cultures and miniature worlds (Bedard & Van Dort, 1998; Boyle et al., 1993; Chang et al., 2001; Cutter et al., 1998; Kuo et al., 1999; Morris et al., 1992; Pakdeesusuk et al., 2005; Palekar et al., 2003; Watts et al., 2001; Wu et al., 1996; Wu & Wiegel, 1997). In terms of quantity and consistency, most of the earlier studies offered little or no conclusive knowledge regarding dehalogenating microorganisms in cultures. Since the discovery of "*Dehalococcoides*", culture-independent molecular approaches have been regularly made for the phylogenetic identification and characterization of microorganisms involved in the dechlorination. The dechlorination of PCBs by axenic cultures has so far been reported only for "*Dehalococcoides ethenogenes*" strain 195 (Fennell et al., 2004).

A pioneering study by Holoman et al. (1998) showed that 16S RNA gene clones corresponding to the DLG bacteria, designated as RFLP-17, were present in a PCB-dechlorinating community. The involvement of DLG bacteria in the dechlorination of PCBs was clearly demonstrated, as already described above by the analysis of highly enriched cultures containing the representative DLG bacteria DF-1 (Wu et al., 2002) and o-17 (Cutter et al., 2001). Watts et al. (2005) developed specific PCR primers optimized for the detection of o-17 and DF-1 and other closely related bacteria. Using these PCR primer sets, they detected the o-17/DF-1 bacteria as the main dechlorinating organisms in sediment microcosms exhibiting active dechlorination of PCBs. The addition of various bicarbonate concentrations had a profound effect on the dechlorination of 2,3,4,5-CB in DLG species as putative dechlorinators in sedimentary cultures (Yan et al., 2006).

Bedard et al. (2006), by contrast, showed based on the analysis of a 16S rRNA gene clone library that "*Dehalococcoides*" organisms but not the DLG bacteria were present in river sediment enrichment cultures dechlorinating a commercially produced PCB mixture, Aroclor 1260. Further work using PCR with group-specific prefixes resulted in the exclusion of any involvement of known chlorine removers other than "dehalococcoides" in dechlorination (Bedard et al., 2007). The existence of *Desulfitobacterium* as the most common dechlorinator and *Dehalobacter* as the second most common form was shown by community specific PCR analyses of anaerobic enrichment cultures with a PCB mixture from uncontaminated soil; neither "*Dehalococcoides*" nor the DLG was detected in any culture (Baba et al., 2007). Microcosm studies of PCB dechlorination by other authors found that there was a small range of PCB congener specificities within the phylum *Chloroflexi*, including "*Dehalococcoides*" and the DLG bacteria (Fagervold et al., 2007, 2005), suggesting the importance of synergistic interactions of different species of microorganisms for enhanced dechlorination.

12 Acquisition and Distribution of Dehalogenase Genes

During the evolution of novel catabolic pathways, gene transfer is a significant process. The acquisition of DNA alien by horizontal transfer of genes requires integration into a well-maintained replica in the recipient microorganism. Gene transfer and integration are key mechanisms for the formation of new stable constructs, and various gene integration processes have been proposed. For example, Ravatn et al. (1998) suggested that the clc component, which encodes the chlorocatechol degradation genes, integrates itself into the Pseudomonas B13 chromosome in a site-specific manner via the intB13 gene product encoded with gene located near to the of clc genes. Site-specific integrations are found in phages. Integration occurs in both systems at the 3′ end of the target tRNA gene, but the original functional tRNA sequence is retained due to the identity of the sequences involved in integration (attB and attP).

Another way to distribute dehalogenase genes is to acquire a functional replica of a new dehalogenase gene. Integrase will mediate this, both in *P. pavonaceae* 170 and in *Mycobacterium* sp. A gene encoding a putative site-specific recombinase (intP in strain 170 and intM in strain GP1) is located directly upstream of a haloalkane dehalogenase gene that is likely to have been recruited from a chlorobutane-degradable (Gram-positive) *Rhodococcus* (Poelarends et al., 2000).

The genes of intP and intM products contribute to important sequence similarities with members of the site-specific recombinase Integrase family. The incorporation of the dehalogenase genes into the bacterial genome may have been mediated by these putative integrase proteins. We anticipate that such acquisition of integrase-mediated genes will occur in a similar way to the acquisition of markers of antibiotic resistance. Genetic elements (integral) that bear a gene for a site-specific DNA integrase, which can catalyze the integration of one or more foreign genes into a site directly upstream of the integrase gene, are found in the case of these markers.

13 Conclusions and Future Perspectives

In our understanding of microbial dehalogenation, there are still many differences. First, thermodynamic analysis shows that the range of physiological processes which at the expense of halogenated substrates should enable microbial growth is much broader than experimentally discovered. This implies that there are biochemical limitations, which might be overcome by further genetic adaptation. This has been the case for major organohalogenic aerobic degradation, such as 1,2-dichloropropane, 1,2,3-trichloropropane, and dichloroethanes.

Second, little attention has been paid so far to the role of auxiliary enzymes involved in the metabolism of halogenated hydrocarbons. Recently, it has been shown that a functional DNA polymerase I, which is involved in DNA repair, is necessary, in addition to a functional catabolic pathway, for the utilization of dichloromethane (Kayser et al., 2000). The requirement for additional detoxification mechanisms may be a more general phenomenon, which would explain the failure of various experiments aimed at the genetic construction of organisms with new catabolic pathways.

Third, unraveling new mechanisms of dehalogenation, such as those of haloalcohol dehalogenases, chloroacrylic acid dehalogenases, and hexachlorocyclohexane dehydrochlorinase, remains a challenge. Recently, a mechanism based on elimination of a 1,2-biaxial HCl pair has been proposed for hexachlorocyclohexane dehydrochlorinase, which has been predicted to share similarities with hydrates and isomerases (Nagata et al., 1999; Trantirek et al., 2001).

Fourth, there is still a daunting area of research regarding the origin and distribution of dehalogenase genes. In particular, the mutations (including, for example, recombinations) that are associated with the recent adaption process to anthropogenic pollutants deserve more attention.

Finally, recent advances in techniques of guided evolution will promote the creation of enzymes and species with degradation capacities that are not readily obtainable through classical adaptation and enrichment.

References

Abramowicz, D. A., Brennan, M. J., Van Dort, H. M., et al. (1993). Factors influencing the rate of polychlorinated biphenyl dechlorination in Hudson River sediments. *Environmental Science & Technology, 27*, 1125–1131. https://doi.org/10.1023/A:1008319306757

Adrian, L., Rahnenfuhrer, J., Gobom, J., et al. (2007). Identification of a chlorobenzene reductive dehalogenase in *Dehalococcoides* sp. strain CBDB1. *Applied and Environmental Microbiology, 73*, 7717–7724. https://doi.org/10.1128/AEM.01649-07

Ahn, Y. B., Rhee, S. K., Fennell, D. E., et al. (2003). Reductive dehalogenation of brominated phenolic compounds by microorganisms associated with the marine sponge Aplysina aerophoba. *Applied and Environmental Microbiology, 69*, 4159–4166. https://doi.org/10.1128/aem.69.7.4159-4166.2003

Alder, A. C., Haggblom, M. M., Oppenheimer, S., et al. (1993). Reductive dechlorination of polychlorinated biphenyls in anaerobic sediments. *Environmental Science & Technology, 27*, 530–538. https://doi.org/10.1021/es00040a012

Anandarajah, K., Kiefer, P. M., Jr., Donohoe, B. S., et al. (2000). Recruitment of a double bond isomerase to serve as a reductive dehalogenase during biodegradation of pentachlorophenol. *Biochemistry, 39*, 5303–5311. https://doi.org/10.1021/bi9923813

Baba, D., Yasuta, T., Yoshida, N., et al. (2007). Anaerobic biodegradation of polychlorinated biphenyls by a microbial consortium originated from uncontaminated paddy soil. *World Journal of Microbiology and Biotechnology, 23*, 1627–1636. https://doi.org/10.1007/s11274-007-9409-4

Bedard, D. L., & Van Dort, H. (1998). Complete reductive dehalogenation of brominated biphenyls by anaerobic microorganisms in sediment. *Applied and Environmental Microbiology, 64*, 940–947. https://doi.org/10.1128/AEM.64.3.940-947.1998

Bedard, D. L., Bunnell, S. C., & Smullen, L. A. (1996). Stimulation of microbial para-dechlorination of polychlorinated biphenyls that have persisted in Housatonic River sediment for decades. *Environmental Science & Technology, 30*, 687–694. https://doi.org/10.1016/S0045-6535(96)00360-8

Bedard, D. L., Van Dort, H. M., May, R. J., et al. (1997). Enrichment of microorganisms that sequentially meta, para-dechlorinate the residue of Aroclor 1260 in Housatonic River sediment. *Environmental Science & Technology, 31*, 3308–3313. https://doi.org/10.1021/es9703483

Bedard, D. L., Van Dort, H. M., & DeWeerd, K. A. (1998). Brominated biphenyls prime extensive microbial reductive dehalogenation of Aroclor 1260 in Housatonic River sediment. *Applied and Environmental Microbiology, 64*, 1786–1795. https://doi.org/10.1128/AEM.64.5.1786-1795.1998

Bedard, D. L., Baily, J. J., Reiss, B. L., et al. (2006). Development and characterization of stable sediment-free anaerobic bacterial enrichment cultures that dechlorinate Aroclor 1260. *Applied and Environmental Microbiology, 72*, 2460–2470. https://doi.org/10.1128/AEM.72.4.2460-2470.2006

Bedard, D. L., Ritalahti, K. M., & Loffler, F. E. (2007). The *Dehalococcoides* population in sediment free mixed cultures metabolically dechlorinates the commercial polychlorinated biphenyl mixture Aroclor 1260. *Applied and Environmental Microbiology, 73*, 2513–2521. https://doi.org/10.1128/AEM.02909-06

Beil, S., Timmis, K. N., & Pieper, D. H. (1999). Genetic and biochemical analyses of the tec operon suggest a route for evolution of chlorobenzene degradation genes. *Journal of Bacteriology, 181*, 341–346. https://doi.org/10.1128/JB.181.1.341-346.1999

Boyle, A. W., Blake, C. K., Price, W. A., et al. (1993). Effects of polychlorinated biphenyl congener concentration and sediment supplementation on rates of methanogenesis and 2,3,6-trichlorobiphenyl dechlorination in an anaerobic enrichment. *Applied and Environmental Microbiology, 59*, 3027–3031. https://doi.org/10.1128/AEM.59.9.3027-3031.1993

Brown, J. F., Jr., Bedard, D. L., & Brennan, M. J. (1987). Polychlorinated biphenyl dechlorination in aquatic sediments. *Science, 236*, 709–712. https://doi.org/10.1126/science.236.4802.709

Brown, J. F., Jr., & Wagner, R. E. (1990). PCB movement, dechlorination, and detoxication in the Acushnet estuary. *Environ. Toxicol. Chem. 9*, 1215–1233.

Chang, K. S. (1995). M.S. Thesis. University of Georgia, Athens, GA.

Chang, B. V., Liu, W. G., & Yuan, S. Y. (2001). Microbial dechlorination of three PCB congeners in river sediment. *Chemosphere, 45*, 849–856. https://doi.org/10.1016/s0045-6535(01)00106-0

Criddle, C. S. (1989). Ph.D. Thesis. Stanford University, Stanford, CA.

Cutter, L., Sowers, K. R., & May, H. D. (1998). Microbial dechlorination of 2,3,5,6-tetrachlorobiphenyl under anaerobic conditions in the absence of soil or sediment. *Applied and Environmental Microbiology, 64*, 2966–2969. https://doi.org/10.1128/AEM.64.8.2966-2969.1998

Cutter, L. A., Watts, J. E., Sowers, K. R., et al. (2001). Identification of a microorganism that links its growth to the reductive dechlorination of 2,3,5,6-chlorobiphenyl. *Environmental Microbiology, 3*, 699–709. https://doi.org/10.1046/j.1462-2920.2001.00246.x

Dalton, D. D. (1990). M.S. Thesis. University of Georgia, Athens, GA.

DeWeerd, K. A., Concannon, F., & Sulfita, J. M. (1991). Relationship between hydrogen consumption, dehalogenation, and reduction of sulfur oxyanions by *Desulfomonile tiedjei*. *Applied and Environmental Microbiology, 57*, 1929–1934.

Dolfing, J. (2003). Thermodynamic consideration for dehalogenation. *In* Dehalogenation: Microbial Processes and Environmental Applications, Haggblom, M. M., and I. D. Bosser. (eds.), Kluwervironmental applications (pp. 89–114). Kluwer, Dordrecht, pp. 89–114.

Fagervold, S. K., Watts, J. E., May, H. D., et al. (2005). Sequential reductive dechlorination of meta-chlorinated poly-chlorinated biphenyl congeners in sediment microcosms by two different *Chloroflexi* phylotypes. *Applied and Environmental Microbiology, 71*, 8085–8090. https://doi.org/10.1128/AEM.71.12.8085-8090.2005

Fagervold, S. K., May, H. D., & Sowers, K. R. (2007). Microbial reductive dechlorination of Aroclor 1260 in Baltimore harbor sediment microcosms is catalyzed by three phylotypes within the phylum *Chloroflexi*. *Applied and Environmental Microbiology, 73*, 3009–3018. https://doi.org/10.1128/AEM.02958-06

Fennell, D. E., Nijenhuis, I., Wilson, S. F., et al. (2004). *Dehalococcoides ethenogenes* strain 195 reductively dechlorinates diverse chlorinated aromatic pollutants. *Environmental Science & Technology, 38*, 2075–2081. https://doi.org/10.1021/es034989b

Fung, J. M., Morris, R. M., Adrian, L., et al. (2007). Expression of reductive dehalogenase genes in *Dehalococcoides etheno-genes* strain 195 growing on tetrachloroethene, trichloroethene, or 2,3-dichlorophenol. *Applied and Environmental Microbiology, 73*, 4439–4445. https://doi.org/10.1128/AEM.00215-07

Gribble, G. W. (2003). The diversity of naturally produced organo-halogens. *Chemosphere, 52*, 289–297. https://doi.org/10.1016/S0045-6535(03)00207-8

Griffith, G. D., Cole, J. R., Quensen, J. F., et al. (1992). Specific deuteration of dichlorobenzoate during reductive dehalogenation by *Desulfomonile tiedjei* in D_2O. *Applied and Environmental Microbiology, 58*, 408–411. https://doi.org/10.1128/AEM.58.1.409-411.1992

Halsall, C. J., Gevao, B., Howsam, M., et al. (1999). Temperature dependence of PCBs in the UK atmosphere. *Atmospheric Environment, 33*, 541–552. https://doi.org/10.1021/es049081f

Hill, K. E., Marchesi, J. R., & Weightman, A. J. (1999). Investigation of two evolutionarily unrelated halocarboxylic acid dehalogenase gene families. *Journal of Bacteriology, 181*, 2535–2547. https://doi.org/10.1128/JB.181.8.2535-2547.1999

Hiraishi, A., Sakamaki, N., & Miyakoda, H. (2005). Estimation of "*Dehalococcoides*" populations in lake sediment contaminated with low levels of polychlorinated dioxins. *Microbes and Environments, 20*, 216–226. https://doi.org/10.1264/jsme2.20.216

Holoman, T. R. P., Elberson, M. A., & Cutter, L. A. (1998). Characterization of a defined 2,3,5,6-tetrachlorobiphenyl-ortho-dechlorinating microbial community by comparative sequence analysis of genes coding for 16S rRNA. *Applied and Environmental Microbiology, 64*, 3359–3367. https://doi.org/10.1128/AEM.64.9.3359-3367.1998

Jota, M. A. T., & Hassett, J. P. (1991). Effects of environmental variables on binding of a PCB congener by dissolved humic substances. *Environmental Toxicology and Chemistry, 10*, 483–491. https://doi.org/10.1002/etc.5620100408

Kassenga, G., Pardue, J. H., & Moe, W. M. (2004). Hydrogen thresholds as indicators of dehalorespiration in constructed treatment wetlands. *Environmental Science & Technology, 38*, 1024–1030. https://doi.org/10.1021/es0348391

Kayser, M. F., Stumpp, M. T., & Vuilleumier, S. (2000). DNA polymerase I is essential for growth of Methylobacterium dichloromethanicum DM4 with dichloromethane. *Journal of Bacteriology, 182*, 5433–5439. https://doi.org/10.1128/jb.182.19.5433-5439.2000

Kim, J., & Rhee, G. Y. (1997). Population dynamics of polychlorinated biphenyl-dechlorinating microorganisms in contaminated sediments. *Applied and Environmental Microbiology, 63*, 1771–1776. https://doi.org/10.1128/AEM.63.5.1771-1776.1997

Klasson, K. T., Barton, J. W., Evans, B. S., et al. (1996). Reductive microbial dechlorination of indigenous polychlorinated biphenyls in soil using a sediment-free inoculum. *Biotechnology Progress, 12*, 310–315. https://doi.org/10.1021/bp960019z

Krajmalnik Brown, R., Sung, Y., Ritalahti, K. M., et al. (2007). Environmental distribution of the trichloroethene reductive dehalogenase gene (*tceA*) suggests lateral gene transfer among *Dehalococcoides*. *FEMS Microbiology Ecology, 59*, 206–214. https://doi.org/10.1111/j.1574-6941.2006.00243.x

Kube, M., Beck, A., Zinder, S. H., et al. (2005). Genome sequence of the chlorinated compound-respiring bacterium *Dehalococcoides* species strain CBDB1. *Nature Biotechnology, 23*, 1269–1273. https://doi.org/10.1038/nbt1131

Kuipers, B., Cullen, W. R., & Mohn, W. W. (1999). Reductive dechlorination of nonachlorobiphenyls and selected octachlorobiphenyls by microbial enrichments cultures. *Environmental Science & Technology, 33*, 3577–3583. https://doi.org/10.1021/es9900712

Kuo, C. E., Liu, S. M., & Liu, C. (1999). Biodegradation of coplanar polychlorinated biphenyls by anaerobic microorganisms from estuarine sediments. *Chemosphere, 39*, 1445–1458. https://doi.org/10.1016/s0045-6535(99)00046-6

Li, Y. F., Hata, Y., Fujii, T., et al. (1998). Crystal structures of reaction intermediates of L-2-haloacid dehalogenase and implications for the reaction mechanism. *The Journal of Biological Chemistry, 273*, 15035–15044. https://doi.org/10.1074/jbc.273.24.15035

Linkfield, T. G., & Tiedje, J. M. (1990). Characterization of the requirements and substrates for reductive dehalogenation by strain DCB-1. *Journal of Industrial Microbiology, 5*, 9–15. https://doi.org/10.1007/BF01569601

Mackiewicz, M., & Wiegel, J. (1998). Comparison of energy and growth yields for *Desulfitobacterium dehalogenans* during utilization of chlorophenyl and various traditional electron acceptors. *Applied and Environmental Microbiology, 64*, 352–355. https://doi.org/10.1128/AEM.64.1.352-355.1998

Maddela, N. R., Scalvenzi, L., Pérez, M., Montero, C., & Gooty, J. M. (2015). Efficiency of indigenous filamentous fungi for biodegradation of petroleum hydrocarbons in medium and soil: Laboratory study from Ecuador. *Bulletin of Environmental Contamination and Toxicology, 95*(3), 385–394.

Maddela, N. R., Rodriguez, L., Sanaguano, S. H., Morán, R. E. B., Venkateswarlu, K., & Scalvenzi, L. (2017). Biodegradation of diesel, crude oil and spent lubricating oil by soil isolates of Bacillus spp. *Bulletin of Environmental Contamination and Toxicology, 98*, 698–705.

Madsen, T., & Aamand, J. (1991). Effects of sulfuroxy anions on degradation of pentachlorophenol by a methanogenic enrichment culture. *Applied and Environmental Microbiology, 57*, 2453–2458. https://doi.org/10.1128/AEM.57.9.2453-2458.1991

Magnuson, J. K., Stern, R. V., & Gossett, J. M. (1999). Reductive dechlorination of tetrachloroethene to ethene by a two-component enzyme pathway. *Applied and Environmental Microbiology, 64*, 1270–1275. https://doi.org/10.1128/AEM.64.4.1270-1275.1998

Magnuson, J. K., Romine, M. F., Burris, D. R., et al. (2000). Trichloroethene reductive dehalogenase from *Dehalococcoides ethenogenes*. sequence of *tceA* and substrate range characterization. *Applied and Environmental Microbiology, 66*, 5141–5147. https://doi.org/10.1128/aem.66.12.5141-5147.2000

Marek, J., Vevodova, J., & Smatanova, I. K. (2000). Crystal structure of the haloalkane dehalogenase from Sphingomonas paucimobilis UT26. *Biochemistry, 39*, 14082–14086. https://doi.org/10.1021/bi001539c

Mars, A. E., Kingma, J., Kaschabek, S. R., et al. (1999). Conversion of 3-chlorocatechol by various catechol 2,3-dioxygenases and sequence analysis of the chlorocatechol dioxygenase region of Pseudomonas putida GJ31. *Journal of Bacteriology, 181*, 1309–1318. https://doi.org/10.1128/JB.181.4.1309-1318.1999

May, H. D., Boyle, A. W., & Price, W. A. (1992). Subculturing of a polychlorinated biphenyl-dechlorinating anaerobic enrichment on solid media. *Applied and Environmental Microbiology, 58*, 4051–4054. https://doi.org/10.1128/AEM.58.12.4051-4054.1992

Maymo-Gatell, X., Anguish, T., & Zinder, S. H. (1999). Reductive dechlorination of chlorinated ethenes and 1,2-dichloroethane by 'Dehalococcoides ethenogenes' 195. *Applied and Environmental Microbiology, 65*, 3108–3113. https://doi.org/10.1128/AEM.65.7.3108-3113.1999

McMurdie, P. J., Behrens, S. F., Holmes, S., et al. (2007). Unusual codon bias in vinyl chloride reductase genes of *Dehalococcoides* species. *Applied and Environmental Microbiology, 73*, 2744–2747.

Mohn, W. W., & Tiedje, J. M. (1992). Microbial reductive dehalogenation. *Microbiol. Rev. 56*, 482–507.

Morris, P. J., Mohn, W. W., & Quensen, I. I. I. (1992). Establishment of a polychlorinated biphenyl-degrading enrichment culture with predominantly *meta* dechlorination. *Applied and Environmental Microbiology, 58*, 3088–3094. https://doi.org/10.1128/AEM.58.9.3088-3094.1992

Nagata, Y., Futamura, A., & Miyauchi, K. (1999). Two different types of dehalogenases, LinA and LinB, involved in (γ)-hexachlorocyclohexane degradation in Sphingomonas paucimobilis UT26 are localized in the periplasmic space without molecular processing. *Journal of Bacteriology, 181*, 5409–5413. https://doi.org/10.1128/JB.181.17.5409-5413.1999

Nardi Dei, V., Kurihara, T., Park, C., et al. (1999). DL-2-Haloacid dehalogenase from Pseudomonas sp. 113 is a new class of dehalogenase catalyzing hydrolytic dehalogenation not involving enzyme–substrate ester intermediate. *The Journal of Biological Chemistry, 274*, 20977–20981. https://doi.org/10.1074/jbc.274.30.20977

Nealson, K. (2003). Harnessing microbial appetites for remediation. *Nature Biotechnology, 21*, 243–244. https://doi.org/10.1038/nbt0303-243

Newman, J., Pea, T. S., Richard, R., et al. (1999). Haloalkane dehalogenases: Structure of a Rhodococcus enzyme. *Biochemistry, 38*, 16105–16114. https://doi.org/10.1021/bi9913855

Nie, L., & Vogel, T. M. (1991). Identification of the proton source for the microbial reductive dechlorination of 2,3,4,5,6-pentachlorobiphenyl. *Applied and Environmental Microbiology, 57*, 2771–2774. https://doi.org/10.1128/AEM.57.9.2771-2774.1991

Nonaka, H., Keresztes, G., & Shinoda, Y. (2006). Complete genome sequence of the dehalorespiring bacterium *Desulfitobacterium hafniense* Y51 and comparison with *Dehalococcoides ethenogenes* 195. *Journal of Bacteriology, 188*, 2262–2274. https://doi.org/10.1128/JB.188.6.2262-2274.2006

Pakdeesusuk, U., Lee, C. M., & Coates, J. T. (2005). Assessment of natural attenuation via in situ reductive dechlorination of polychlorinated biphenyls in sediments of the Twelve Mile Creek arm of Lake Hartwell, SC. *Environmental Science & Technology, 39*, 945–952. https://doi.org/10.1021/es0491228

Palekar, L. D., Maruya, K. A., & Kostka, J. E. (2003). Dehalogenation of 2,6-dibromobiphenyl and 2,3,4,5,6-pentachlorobiphenyl in contaminated estuarine sediment. *Chemosphere, 53*, 593–600. https://doi.org/10.1016/S0045-6535(03)00444-2

Poelarends, G. J., Kulakov, L. A., & Larkin, M. J. (2000). Roles of horizontal gene transfer and gene integration in evolution of 1,3-dichloropropene and 1,2-dibromoethane-degradative pathways. *Journal of Bacteriology, 182*, 2191–2199. https://doi.org/10.1128/JB.182.8.2191-2199.2000

Quensen, J. F., III, Boyd, S. A., & Tiedje, J. M. (1990). Dechlorination of four commercial polychlorinated biphenyl mixtures (Aroclors) by anaerobic microorganisms from sediments. *Applied and Environmental Microbiology, 56*, 2360–2369. https://doi.org/10.1128/AEM.56.8.2360-2369.1990

Quensen, J. F., III, Tiedje, J. M., & Boyd, S. A. (1988). Reductive dechlorination of polychlorinated biphenyls by anaerobic microorganisms from sediments. *Science, 242*, 752–754.

Ravatn, R., Studer, S., Zehnder, A. J., et al. (1998). Int-B13, an unusual site-specific recombinase of the bacteriophage P4 integrase family, is responsible for chromosomal insertion of the 105-kilobase clc element of Pseudomonas sp. strain B13. *Journal of Bacteriology, 180*, 5505–5514. https://doi.org/10.1128/JB.180.21.5505-5514.1998

Regeard, C., Maillard, J., & Dufraigne, C. (2005). Indications for acquisition of reductive dehalogenase genes through horizontal gene transfer by *Dehalococcoides etheno-genes* strain 195. *Applied and Environmental Microbiology, 71*, 2955–2961. https://doi.org/10.1128/AEM.71.6.2955-2961.2005

Rhee, G. Y., Bush, B., Bethoney, C. M., et al. (1993). Anaerobic dechlorination of Aroclor 1242 as effected by some environmental conditions. *Environmental Toxicology and Chemistry, 12*, 1033–1039.

Ridder, I. S., Rozeboom, H. J., Kalk, K. H., et al. (1999). Crystal structures of intermediates in the dehalogenation of haloalkanoates by L-2-haloacid dehalogenase. *The Journal of Biological Chemistry, 274*, 30672–30678. https://doi.org/10.1074/jbc.274.43.30672

Ross, G. (2004). The public health implications of polychlorinated biphenyls (PCBs) in the environment. *Ecotoxicology and Environmental Safety, 59*, 275–291. https://doi.org/10.1074/jbc.274.43.30672

Scow, K. M., & Hicks, K. A. (2005). Natural attenuation and enhanced bioremediation of organic contaminants in groundwater. *Current Opinion in Biotechnology, 16*, 246–253. https://doi.org/10.1016/j.copbio.2005.03.009

Seshadri, R., Adrian, L., & Fouts, D. E. (2005). Genome sequence of the PCE-dechlorinating bacterium *Dehalococcoides ethenogenes*. *Science, 307*, 105–108. https://doi.org/10.1126/science.1102226

Simcik, M. E., Basu, I., Sweet, C. W., et al. (1999). Temperature dependence and temporal trends of polychlorinated biphenyl congeners in the Great Lakes atmosphere. *Environmental Science & Technology, 33*, 1991–1995. https://doi.org/10.1021/es9811896

Smidt, H., & de Vos, W. M. (2004). Anaerobic microbial dehalogenation. *Annual Review of Microbiology, 58*, 43–73. https://doi.org/10.1146/annurev.micro.58.030603.123600

Smidt, H., van Leest, M., & van Der Oost, J. (2000). Transcriptional regulation of the *cpr* gene cluster in ortho-chlorophenol-respiring *Desulfitobacterium dehalogenans*. *Journal of Bacteriology, 182*, 5683–5691. https://doi.org/10.1128/JB.182.20.5683-5691.2000

Stringfellow, J. M., Cairns, S. S., & Cornish, A. (1997). Haloalkanoate dehalogenase II (DehE) of a *Rhizobium sp.* molecular analysis of the gene and formation of carbon monoxide from trihaloacetate by the enzyme. *European Journal of Biochemistry, 250*, 789–793. https://doi.org/10.1111/j.1432-1033.1997.00789.x

Tiedje, J. M., Quensen, J. F., & Mohn, W. W. (1991). Reductive dechlorination of chlorinated aromatic pollutants. In R. W. Rushmore (Ed.), *Biodeterioration and biodegradation* (pp. 293–307) https://www.jstor.org/stable/4251148

Trantirek, L., Hynkova, K., & Nagata, Y. (2001). Reaction mechanism and stereochemistry of γ-hexachlorocyclohexane dehydrochlorinase LinA. *The Journal of Biological Chemistry, 276*, 7734–7740. https://doi.org/10.1074/jbc.M007452200

Van de Pas, B. A., Smidt, H., & Hagen, W. R. (1999). Purification and molecular characterization of ortho-chlorophenol reductive dehalogenase, a key enzyme of halorespiration in *Desulfitobacterium dehalogenans*. *The Journal of Biological Chemistry, 274*, 20287–20292. https://doi.org/10.1074/jbc.274.29.20287

Van Dort, H. M., Smullen, L. A., & May, R. J. (1997). Priming microbial meta-dechlorination of polychlorinated biphenyls that have persisted in Housatonic River sediments for decades. *Environmental Science & Technology, 31*, 3300–3307. https://doi.org/10.1021/es970347a

Verschueren, K. H., Seljee, F., & Rozeboom, H. J. (1993). Crystallographic analysis of the catalytic mechanism of haloalkane dehalogenase. *Nature, 363*, 693–698. https://doi.org/10.1038/363693a0

Wackett, L. P. (1994). Dehalogenation in environmental biotechnology. *Current Opinion in Biotechnology, 5*, 260–265. https://doi.org/10.1016/0958-1669(94)90027-2

Watts, J. E., Wu, Q., & Schreier, S. B. (2001). Comparative analysis of polychlorinated biphenyl-dechlorinating communities in enrichment cultures using three different molecular screening techniques. *Environmental Microbiology, 3*, 710–719. https://doi.org/10.1046/j.1462-2920.2001.00247.x

Watts, J. E., Fagervold, S. K., & May, H. D. (2005). A PCR-based specific assay reveals a population of bacteria within the *Chloroflexi* associated with the reductive dehalogenation of poly-chlorinated biphenyls. *Microbiology, 151*, 2039–2046. https://doi.org/10.1099/mic.0.27819-0

Williams, W. A. (1994). Microbial reductive dechlorination of polychlorinated biphenyls in heat treated and bromoethane sulfonate treated anaerobic sediment slurries. *Environmental Science & Technology, 28*, 2269–2284. https://doi.org/10.1016/0045-6535(94)90191-0

Wu, Q., & Wiegel, J. (1997). Two anaerobic polychlorinated biphenyl-dehalogenating enrichments that exhibit different para-dechlorination specificities. *Applied and Environmental Microbiology, 63*, 4826–4832. https://doi.org/10.1128/AEM.63.12.4826-4832.1997

Wu, Q., Bedard, D. L., & Wiegel, J. (1996). Influence of incubation temperature on the microbial reductive dechlorination of 2,3,4,6-tetrachlorobiphenyl in two freshwater sediments. *Applied and Environmental Microbiology, 62*, 4174–4179. https://doi.org/10.1128/AEM.62.11.4174-4179.1996

Wu, Q., Bedard, D. L., & Wiegel, J. (1999). 2,6-Di bromobiphenyl primes extensive dechlorination of Aroclor 1260 in contaminated sediment at 8-30EC by stimulating growth of PCB-dehalogenating microorganisms. *Environmental Science & Technology, 33*, 595–602. https://doi.org/10.1021/es980741o

Wu, Q., Watts, J. E. M., & Sowers, K. R. (2002). Identification of a bacterium that specifically catalyzes the reductive dechlorination of polychlorinated biphenyls with doubly flanked chlorines. *Applied and Environmental Microbiology, 68*, 807–812. https://doi.org/10.1128/aem.68.2.807-812.2002

Xiang, H., Luo, L., & Taylor, K. L. (1999). Interchange of catalytic activity within the 2-enoyl-coenzyme A hydratase/isomerase superfamily based on a common active site template. *Biochemistry, 38*, 7638–7652. https://doi.org/10.1021/bi9901432

Yan, T., Lapara, T. M., & Novak, P. J. (2006). The effect of varying levels of sodium bicarbonate on polychlorinated biphenyl dechlorination in Hudson River sediment cultures. *Environmental Microbiology, 8*, 1288–1298. https://doi.org/10.1111/j.1462-2920.2006.01037.x

Ye, D., Quensen, J. F., III, & Tiedje, J. M. (1992). Anaerobic dechlorination of polychlorobiphenyls (Aroclor 1242) by pasteurized and ethanol-treated microorganisms from sediment. *Applied and Environmental Microbiology, 58*, 1110–1114. https://doi.org/10.1128/AEM.58.4.1110-1114.1992

Zhang, X., & Wiegel, J. (1990). Sequential anaerobic degradation of 2,4-dichlorophenol in freshwater sediments. *Applied and Environmental Microbiology, 56*, 1119–1127. https://doi.org/10.1128/AEM.56.4.1119-1127.1990

Microbial Capacities for Utilization of Nitroaromatics

Bellemkonda Ramesh, Srinivasan Kameswaran, Ch. Venkatrayulu, M. Subhosh Chandra, G. Vidya Sagar Reddy, and M. Ramakrishna

1 Introduction

Nitrobenzene, *p*-nitrophenol (PNP), nitrotoluenes (TNT, DNT, NT), are critical environmental pollutants that have been reported due to their toxicity to many living organisms (Kumari et al., 2017; Saha et al., 2017, 2014). The nitro substitutes present in NACs have an electron-withdrawing character, causing resistance to biodegradation. Therefore, the oxidative attack by bacterial oxygenases becomes difficult. However, many powerful bacteria have adapted to take advantage of NACs with the use of oxygenate enzyme. The increase in the number of nitro groups and electron-drawing alternatives on the aromatic ring increase the rebellious character, forcing the nitroaromatics to be used by the partial reduction mechanism.

Various microorganisms had been characterized for their capability to degrade NACs (Zheng et al., 2007; Nishino & Spain, 1993, 1995; Torres et al., 1996; Park et al., 1999; Park & Kim, 2000). Microorganisms use oxidative and reductive pathways to degrade to convert NACs entirely to CO_2 and H_2O or partially to an organic compound. It is the genetic mechanism found in microbes that directs the

B. Ramesh
Department of Food Technology, Vikrama Simhapuri University, Nellore, Andhra Pradesh, India

S. Kameswaran (✉) · M. Ramakrishna
Department of Botany, Vikrama Simhapuri University PG Centre, Kavali, Andhra Pradesh, India

C. Venkatrayulu
Department of Marine Biology, Vikrama Simhapuri University, Nellore, Andhra Pradesh, India

M. S. Chandra
Department of Microbiology, Yogi Vemana University, Kadapa, Andhra Pradesh, India

G. V. S. Reddy
Department of Biotechnology, Vikrama Simhapuri University, Nellore, Andhra Pradesh, India

N. R. Maddela, L. C. García (eds.), *Innovations in Biotechnology for a Sustainable Future*, https://doi.org/10.1007/978-3-030-80108-3_12

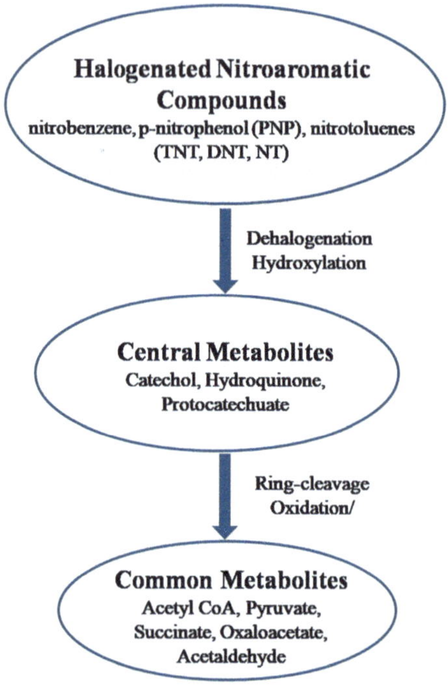

Fig. 1 General microbial pathways of halogenated aromatics degradations. Three major catabolic pathways (upper, middle and lower pathways) are involved in mass biodegradation

conversion of simple products from NACs. While aerobic bacteria use both aerobic and partial catabolic pathways systems, anaerobic bacteria were only able to use only the reduction mechanism to use NACs (Fig. 1).

Substituted NACs and their derivatives have been reported to degrade efficiently by microorganisms, but contaminants are still persist in the environment because microbial strains have been reported to be useful as biological treatment agents in controlled laboratory conditions (Qureshi et al., 2009; Ghosh et al., 2017; Singh et al., 2015). The researchers report that bacteria can only grow only under the influence of pH, temperature, oxygen, humidity, the appropriate level of nutrients, the bioavailability of pollutants, and the presence of other toxic compounds (Yadav et al., 2015; Saha et al., 2017).

Thus, exploiting the maximum potential of bacteria to degrade pollutants in the contaminated environment remains a challenge today, which could be accomplished in the future by understanding the microbial capacities of utilization based on "-omics" analysis.

1.1 Anaerobic Biodegradation of Nitroaromatics

In this process, the nitro group is reduced to nitroso derivative, hydroxylamines, or amines by the action of nitroreductases. The degradation of most of the (poly)nitroaromatic compounds occurs only under anaerobic conditions (Kulkarni & Chaudhari, 2007; Nishino et al., 2002; Zhang & Bennett, 2005). The complete mineralization of nitroaromatic by a single anaerobic strain is very rare (Razo-Flores et al., 1997). There are several reports showing that the initial step during the degradation of mono-, di-, and trinitroaromatic compounds is the reduction of nitro groups to amino groups (Razo-Flores et al., 1997; McCormick et al., 1976; Donlon et al., 1996; Peres et al., 1999; Rieger & Knackmuss, 1995).

1.2 Aerobic Biodegradation of Nitroaromatic Compounds

Mono- and bilateral aromatics mainly undergo aerobic biodegradation and achieve to complete mineralization. In this stage, the nitroaromatics act as a source of carbon, nitrogen, and energy for the microbes. During the past few decades, numerous reports have isolated the mineralized microbes of the various nitroaromatics and their degradation pathway. Few of them have been extensively studied and distinguished. There are different strategies in nitro-mechanical aerobic degradation (Nishino et al., 2000), which is used in nature.

1.3 Nitroaromatic Degradational Pathway Relevance to Research

In order to reduce pollution levels in the atmosphere, a recent survey reported that nitro-/amino-/chloroaromatic containing industrial wastewaters must be non-polluted before being discharged into the environment (Chatterjee et al., 2017; Purohit et al., 2016). Therefore, knowledge of the biological capabilities of bacteria to degrade bacteria at genetic levels and their regulatory pathways becomes crucial. NAC Biodegradation Reports indicate that these reports are microbes that may act as biostimulant degradation (Begum & Arundhati, 2016; Thangaraj et al., 2008).

1.4 Challenges of Nitroaromatics in Biodegradation

One of the main problems and drivers of the growth of the world economy is environmental sustainability. Globally responsible development and use are one of the priorities for sustainable urbanization and industrialization processes, but the discharge and disposal of hazardous waste containing organic and inorganic contaminants remains a major problem.

With the listing in the United States of nitroaromatic compounds, the hazardous group of the Environmental Protection Agency includes the protection of the atmosphere from certain hazardous chemicals. It was not possible to limit the production of nitroaromatics derived from anthropogenic sources, such as the pharmaceutical, defense, and agricultural sectors. Instead, it is important to suggest and practice ways to mitigate their influence in the world. The use of microbes is one of the strongest methods involved in the biodegradation of NACs. From the researchers' inputs into the biodegradation of such nitroaromatics, i.e., by the proper use of microbes and their field applications (bioremediation), it is envisaged that promiscuity may be found in mitigating nitroaromatics. It is important to understand different biodegrading mechanisms, capacities, and genetics of bacteria responsible for the biodegradation of a broad range of nitroaromatics to fill this gap, insights of which have been mentioned in this chapter.

Genomics, proteomics, metabolomics, and their combinations are currently recorded as important emerging methods for understanding microbial capacities and evaluating aromatic biodegradation and substituted aromatic compounds (Tikariha et al., 2016; Kapley & Purohit, 2009; Qureshi et al., 2009).

2 Molecular Processes of Biodegradative Pathways of Aromatic and Nitroaromatics

2.1 Aromatic Biodegradative Pathways

The typical biodegradation pathway of aromatics and substituted aromatic compounds such as NACs. In general, there are two phases of aerobic biodegradation: first, ring cleavage of an aromatic compound occurs through a particular ring modification reaction, leading to the production of a dihydroxylated benzene ring or a replacement dihydroxylated benzene ring. In the second step of degradation, fission of this aromatic ring with a subsequent reaction leads to central carbon intermediates. There are different dioxygenases that are responsible for catalyzing the ring fission stage. It can be classified as either the ortho- or beta-ketoadipate pathway depending on the form of cleavage (because beta-ketoadipate is a critical ortho-cleavage intermediate) when the cleavage of the ring occurs between the hydroxyl groups (intradiol cleavage) or when it is cleaved adjacent to one of the hydroxyls, it is known as the meta pathway (estradiol cleavage).

Significant pathways recorded for aromatic compound catabolism in bacteria have shown that various enzymes have carried out initial conversion steps, but the compounds are converted into a small number of derived metabolites, such as protocatechuates and catechols. Via various enzymes in the ring cleavage pathway, these intermediate metabolites are channeled into central metabolic routes. This generalized scheme of catabolic pathways for aromatic compounds indicated that by evolving "peripheral" enzymes that were able to convert initial substrates into one of the central intermediates such as catechol, resorcinol, and benzoquinones, microorganisms have increased their substrate spectrum and capacities.

2.2 *Biodegradative Pathways for NACs*

Several researchers have documented degrading pathways and intermediates of nitroaromatic pollutants (Tiwari et al., 2017; Ghosh et al., 2017; Min et al., 2017; Qureshi & Purohit, 2002). Studies have shown the bacterial degradation of toxic substances, namely 4-chloro-3-nitrophenol (4C3NP) and 4-chloro-sorcinal intermediate formation in the 4C3NP degradation pathway, as well as the degradation of 2-chloro-4-aminophenol. *Pseudomonas* sp., 4C3NP, bacterial mineralization, from sewage collected from a chemically contaminated area.

2.3 *Nitrobenzene Pathway*

Nitrobenzene has been identified here as one of the representative NACs to understand the biodegradation pathways that various bacteria such as *Pseudomonas pseudoalcaligene*, *Comamonas* sp., *Pseudomonas mendocina*, and *Pseudomonas putida* have adopted.

2.4 *Nitrophenol Pathway*

Several bacteria have been reported to be used as growth substrates for other contaminants, such as 4-nitrobenzoate, 2-nitrobenzoate, 2-nitrotoluene, nitrobenzene, *p*-nitroaniline, and PNP (Yanzhen et al., 2016; Tikariha et al., 2016; Qureshi et al., 2007). In bacteria such as *Pseudomonas* species, para-Nitrophenol has been shown as a representative model compound for nitroaromatic contaminants to understand biodegrading pathways.

2.5 *Pathways for Nitrobenzoate and Nitrobenzaldehyde Catabolism*

Nitrobenzoate and nitrobenzaldehyde compounds have been documented for the understanding of biodegradation pathways of various bacteria such as *Comamonas acidovorans* (Groenewegen et al., 1992), *Ralstonia pickettii* (Yabannavar & Zylstra, 1995), *Ralstonia* sp. (Samanta et al., 2000), *Pseudomonas* sp. (Haigler & Spain, 1993), and *Pseudomonas putida* (James et al., 2000; Rhys-Williams et al., 1993) all use a reductive pathway that results in protocatechuate being the main intermediate in the catabolism of 4-nitrobenzoate.

2.6 Pathways to Nitrotoluene Catabolism

The bacteria strains of *Acidovorax* sp., *Comamonas* sp., *Pseudomonas putida*, *Mycobacterium* sp., *Pseudomonas pseudoalcaligenes*, *Burkholderia cepacia*, *Escherichia coli*, and *Hydrogenophaga palleronii* are capable of using 2-nitrotoluene as the sole carbon, nitrogen and energy source for growth (Haigler et al., 1994). In JS42, a dioxygenase oxidizes the 2 and 3 positions of 2-nitrotoluene to form an unstable nitrohydrodiol, which spontaneously converts to 3-methylcatechol with the release of nitrite.

2.7 Degradation of Trinitrotoluene (TNT)

Trinitrotoluene (TNT) is very difficult to degrade (Nishino & Spain, 2001). The three nitro groups with a nucleophilic aromatic ring structure make TNT vulnerable to reductive attack but resistant to oxygenase attack from aerobic organisms (Lenke et al., 2000). In most current reports, the reduction mechanism predominates in TNT degradation. New evidence indicates that TNT can be reduced by carbon monoxide dehydrogenation from *Clostridium thermoaceticum* (Huang et al., 2000) and by the manganese peroxidase (MnP) from the white mold *Phlebia radiata* (Van Aken et al., 1999). Based on the discovery of the discovery of the reduction of pentaerythritoltetranitrate (PETN) from *Enterobacter cloacae* PB2, French et al. (1998) found that this strain could grow slowly in 2,4,6-TNT under aerobic conditions as the sole source of nitrogen without the producing of dinitrotoluene as an intermediate and catalytic conversion of TNT via the hydride–Meisenheimer complex with nitro group chest in the form of nitrite.

2.8 Pathways for Chloronitrobenzene Catabolism

Only four strains have been identified that can use chloronitrobenzenes as the sole source of carbon and energy for development. Pseudomonas stutzeri has been isolated due to its ability to grow on 2-chloronitrobenzene and has been reported to release chloride and nitrite from this substrate (Liu et al., 2005). However, no further characterization of its degradation pathway has been published. *Comamonas* sp. strain (Wu et al., 2006), *Pseudomonas putida* (Zhen et al., 2006), and *Comamonas* sp. strain LW1 (Katsivela et al., 1999) each use nitroreductase to reduce 4-chloronitrobenzene to 1-chloro-4-hydroxylaminobenzene, which is further transformed to 2-amino-5-chlorophenol by a hydroxylaminobenzene mutase or by Bamberger rearrangement. Ring cleavage of 2-aminophenol 1,6-dioxygenase produces 2-amino-5-chloromuconate, which is converted to an intermediate TCA cycle after additional enzymatic steps (Wu et al., 2006). Recently, the mutant forms of nitrobenzene dioxygenase formed by *Comamonas* sp. (Ju & Parales, 2006) were used for the chlorobenzene-degrading strain *Ralstonia* sp. JS705 to be grown on all three isomers of chloronitrobenzene (Ju & Parales, 2009).

2.9 Pathways for Biologically Generated Nitroaromatic Compounds Catabolism

There is very little understanding of how biologically synthesized nitroaromatic compounds are degraded in the natural world. Despite the extensive use of chloramphenicol in hospitals and research laboratories around the world for more than 50 years, microbial degradation pathways have not yet been understood. There is increasing interest in understanding the metabolic fate of naturally occurring nitroaromatic compounds, but besides 5-nitroanthranilate (Qu & Spain, 2010), 3-nitrotyrosine is currently the only biogenic nitroaromatic compound that bacterial strains have been documented to develop.

Isolated from soils obtained from Cape Cod, MA, *Burkholderia* sp. The JS165 strain and the JS171 paradox *Variovorax* are capable of using 3-nitrotyrosine as the sole carbon, nitrogen, and energy source for development (Nishino & Spain, 2006). 3-Nitrotyrosine is converted to 4-hydroxy-3-nitro-phenylacetate using deaminase-dependent ketoglutarate. NADH-dependent denitratase then removes the nitro group to generate homoprotocatechuate, which is metabolized by the tyrosine rescue pathway. The gene encoding denitratase as described and the characterization of the purified protein had shown that previously uncharacterized flavoprotein monooxygenase appears to be widely distributed in a variety of bacterial genera (Payne et al., 2007).

In rat cells, 3-nitrotyrosine is converted to 4-hydroxy-3-nitrophenylacetate by the formation of 3-nitrotyramine and 4-hydroxy-3-nitrophenylacetaldehyde intermediates (Blanchard-Fillion et al., 2006). *Escherichia coli* MG1655 can use 3-nitrotyramine as the sole source of growth nitrogen but cannot use 3-nitrotyrosine (Rankin et al., 2008). Similar to mammalian cells, MG1655 uses an amine oxidase (TynA) to extract the terminal amino group from 3-nitrotyramine to create 4-hydroxy-3-nitrophenylacetaldehyde, which is then oxidized with phenylacetaldehyde dehydrogenase (FeaB). 4-Hydroxy-3-phenylacetate is a dead-end metabolite in MG1655 because the strain tends to lack the enzymes present in *Burkholderia* sp. JS165 strain and strain of *Variovorax paradoxus* JS171 that complete the metabolism of compounds entering the TCA cycle. Interestingly, the expression of both tynA and feaB was under the regulatory control of the nitric oxide-sensitive repressor (NsrR), further strengthening the relationship between the production of nitric oxide and the nitration of tyrosine residues in proteins (Rankin et al., 2008).

3 Microbes Associated with Biodegradation of Nitroaromatics in the Environment and Its Adaptive Features

Several species of bacteria, including *Pseudomonas, Burkholderia, Rhodococcus, Roseivirga,* and *Shewanella oneidensis*, have been shown to be potent candidates for NAC degraders (Tikariha et al., 2016; Sengupta et al., 2015; Yanzhen et al., 2016; Xu et al., 2016; Selvaratnam et al., 2016; Liu et al., 2017; Min et al., 2016; Ghosh et al., 2017). Phyla Proteobacteria and Gammaproteobacteria, viz., *Pseudomonas, Burkholderia, Bradyrhizobium, Cupriviadus, Shewanella, Raoultella*, are the majority of these bacteria. There are only a few exceptions, such as *Roseivirga* belonging to Actinobacteria and Phyla Bacteriodetes belonging to Rhodococcus. Thus those groups of bacteria that are mostly observed in the soil and often associated with the degradation of a variety of aromatic compounds fall into nitroaromatic-using genera.

3.1 NAC-Utilizing Microbes

To open the aromatic ring, which then goes down to the central aromatic pathway, NAC-degrading bacterial genomes harbor unique genes, such as oxygenases. Moreover, the genetic mapping of the bacteria mentioned above has not revealed any highly specialized feature that could be solely present in nitroaromatic utilizing bacteria. However, they mostly have a wide collection of genes for the use of other aromatic compounds, including other aromatic-degrading bacteria (Tikariha et al., 2016). Many heavy metal stress-resistant genes, antibiotic tolerance, and genes for oxidative stress have also been shown through a closer look at the different types of genes found in aromatic compounds-degrading bacteria. It seems unlikely that bacteria can thrive at higher NAC concentrations, as these compounds act as uncouplers of oxidative phosphorylation and have a toxic effect on bacterial cells (Yanzhen et al., 2016). In addition, the concentration of nitroaromatics found in bodies of water or soil is very low (in most cases, less than 1 ppm) and can be easily degraded by bacteria identified to date that have demonstrated the ability to use up to 50 ppm.

To open the aromatic ring, which then goes down to the central aromatic pathway, NAC-degrading bacterial genomes harbor unique genes, such as oxygenases. Moreover, the genetic mapping of the bacteria mentioned above has not revealed any highly specialized feature that could be solely present in nitroaromatic utilizing bacteria. However, they mostly have a wide collection of genes for the use of other aromatic compounds, including other aromatic-degrading bacteria (Tikariha et al., 2016). Many heavy metal stress-resistant genes, antibiotic tolerance, and genes for oxidative stress have also been shown through a closer look at the different types of genes found in aromatic compounds-degrading bacteria. It seems unlikely that bacteria can thrive at higher NAC concentrations, as these

Table 1 Provided by Genbank Researchers NCBI, the complete genome data showed a nitroaromatic compound using potency in bacteria

S. No.	Bacteria	Operons	Nitroaromatic compound used by bacteria	References
1	*Pseudomonas putida* SF1	*Pnp* gene cluster	*p*-Nitrophenol	Tikariha et al. (2016)
2	*Shewanella oneidensis* MR1	*cymA* and *nfnB*	2,6-Ditrotoluene	Liu et al. (2017)
3	*Rhodococcus* sp. WB1	*Tod* gene cluster	Nitrotoluene and toluene	Xu et al. (2016)
4	*Pseudomonas fragi* PI21	*lysR* gene regulator	Aniline, nitrobenzene	Yanzhen et al. (2016)
5	*Roseivirga* sp.	*lysR* gene regulator	2,4,6-Trinitrotoluene	Selvaratnam et al. (2016)
6	*Raoultella arnithinolytica* TNT	Nitroreductase A, B, and NEM reductase genes	Trinitrotoluene	Thijs et al. (2014)
7	*Rhodococcus imtechensis* RKJ300	*mnp* gene cluster	2-Chloro-4-nitrophenol	Min et al. (2016, 2017)
8	*Paraburkholderia xenovorans* LB400	*amn* gene cluster	2-Aminophenol	Chirino et al. (2013)

compounds act as uncouplers of oxidative phosphorylation and have a toxic effect on bacterial cells (Yanzhen et al., 2016). In addition, the concentration of nitroaromatics found in bodies of water or soil is very low (in most cases, less than 1 ppm) and can be easily degraded by bacteria identified to date that have demonstrated the ability to use up to 50 ppm (Table 1).

Bacteria sense environmental signal by two-component signal transduction system. Some two-component signal regulatory systems are known to control the expression of catabolic pathways and control the global cellular process.

LysR is a regulatory protein that is involved in the control expression of NAC degradation pathways such as 2-nitrotoluene (in species of Acidovorax), where 2,4- and 2,6-dinitrotoluene serve as inducers, and 4-nitrophenol, where it exists in tetrameric forms.

XylR/NtrC-type transcriptional regulators were also found in the regulation of catabolic pathways of nitroaromatics. These regulators activate the alternative sigma factor rho54-holding RNA polymerase (RNAP). Rho54 RNAP holoenzyme forms a stable complex with -12 and -24 promoters but is unable to start transcription without further activators such as NACs. Studies, however, demonstrate that without mediating activation, 3-nitrotoluene can also bind XylR. It thus demonstrates that diversity in control/regulatory mechanisms operates as a source of carbon and energy throughout the use of compounds and leads to the development of microbe utilization capacities.

The pathway of degradation of aromatic compounds is also frequently associated with a number of regulatory genes, such as LTTRs. For example, LTTRs regulate a single target operon such as catR that controls catBCA expression for catechol metabolism in *Pseudomonas putida*.

3.2 Genomes of NACs Degrading Bacteria

Whole-genome sequence (WASs) of many of the nitroaromatic-degrading bacteria since the past decade showcases their genetic capacities toward utilization of NACs.

4 Genome Plasticity: Nitroaromatic Degrading Bacterial Genome Diversification

The innate ability of bacteria in their genome to evolve through mutations, rearrangements, or horizontal gene transfer (HGT) (Juhas et al., 2009). The genome has a collection of genes known as core genes for carrying out important metabolic functions and another set of genes known as HGT-acquired accessory genes that are beneficial under particular environmental conditions. HGT's gene transfer plays a crucial role in the diversification and adaptation of microorganisms in different stress environments, representing a more vivid genome plasticity scenario.

In addition, because of their role in the transmission of different genes, such as antibiotic resistance and virulence genes, which could contribute to the generation of new catabolic genes, genomic islands may also be responsible for the evolution of a wide spectrum of bacteria. This new gene may then be mixed with the current machinery, or the creation of entirely new metabolic pathways takes place.

5 Evolution of New Biodegradative Pathways and Networking of Genes/Operons

The networking of evolving microbiota in a niche has been conceptualized by ecosystem biology. It assumes that the different microbes in the community together form a societal pathway for the use of a complex metabolite. In such a case the entire cascade of nitroaromatic biodegradation enzymes are not found in a single microbe and thus involve the creation of consortia to mitigate the challenges. The implication of a "-omics" based approach to assessing microbial capacities could become an evolving tool for future fields of research and could serve as a scope toward pollutant mitigation.

6 Conclusion

Nitroaromatics are important environmental contaminants that are released and have toxic effects on ecosystems. Based on knowledge encoded on their genomes, microbial utilization capacities for nitroaromatics under aerobic conditions have been addressed in the present study. Bacterial strains such as *Pseudomonas*, *Burkholderia*, *Rhodococcus*, *Roseivirga*, and *Shewanella oneidensis*, based on the sharing of degradative pathway enzymes and genes, have the ability to degrade different NACs. The need for the triggering/inducing factor to express the degrading genes remains a critical element in optimizing the utilization potential of microbes in the case of NACs.

References

Begum, S. S., & Arundhati, A. (2016). A study of bioremediation of methyl parathion in vitro using potential Pseudomonas sp. isolated from agricultural soil, Visakhapatnam. India. *International Journal of Current Microbiology and Applied Sciences, 5*, 464–474. https://doi.org/10.20546/ijcmas.2016.502.052

Blanchard-Fillion, B., Prou, D., Polydoro, M., Spielberg, D., Tsika, E., Wang, Z., Hazen, S. L., Koval, M., Przedborski, S., & Ischiropoulos, H. (2006). Metabolism of 3-nitrotyrosine induces apoptotic death in dopaminergic cells. *The Journal of Neuroscience, 26*, 6124–6130. https://doi.org/10.1523/JNEUROSCI.1038-06.2006

Chatterjee, S., Deb, U., Datta, S., Walther, C., & Gupta, D. K. (2017). Common explosives (TNT, RDX, HMX) and their fate in the environment: Emphasizing bioremediation. *Chemosphere, 184*, 438–451. https://doi.org/10.1016/j.chemosphere.2017.06.008

Chirino, B., Strahsburger, E., Agulló, L., González, M., & Seeger, M. (2013). Genomic and functional analyses of the 2-aminophenol catabolic pathway and partial conversion of its substrate into picolinic acid in *Burkholderia xenovorans* LB400. *PLoS One, 8*, e75746. https://doi.org/10.1371/journal.pone.0075746

Donlon, B. A., Razo-Flores, E., Lettinga, G., & Field, J. A. (1996). Continuous detoxification, transformation and degradation of nitrophenols in upflow anaerobic sludge blanket (USAB) reactors. *Biotechnology and Bioengineering, 51*, 439–449. https://doi.org/10.1002/(SICI)1097-0290(19960820)51:4<439::AID-BIT7>3.0.CO;2-J

French, C. E., Nicklin, S., & Bruce, N. C. (1998). Aerobic degradation of 2,4,6-trinitrotoluene by Enterobacter cloacae PB2 and by pentaerythritol tetranitrate reductase. *Applied and Environmental Microbiology, 64*(8), 2864–2868. https://doi.org/10.1128/AEM.64.8.2864-2868

Ghosh, S., Sanval, P., & Kumar, R. (2017). Evolution of C4 plants and controlling factors: Insight from n-alkane isotopic values of NW Indian Siwalik paleosols. *Organic Geochemistry, 110*, 110–121. https://doi.org/10.1016/j.orggeochem.2017.04.009

Groenewegen, P. E., Breeuwer, P., van Helvoort, J. M., Langenhoff, A. A., de Vries, F. P., & deBont, J. A. (1992). Novel degradative pathway of 4-nitrobenzoate in *Comamonas acidovorans* NBA-10. *Journal of General Microbiology, 138*, 1599–1605. https://doi.org/10.1128/AEM.00739-16

Haigler, B. E., & Spain, J. C. (1993). Biodegradation of 4-nitrotoluene by Pseudomonas sp. strain 4NT. *Applied and Environmental Microbiology, 59*(7), 2239–2243. https://doi.org/10.1128/AEM.59.7.2239-2243.1993

Haigler, B. E., Wallace, W. H., & Spain, J. C. (1994). Biodegradation of 2-nitrotoluene by *Pseudomonas* sp. strain JS42. *Applied and Environmental Microbiology, 60*, 3466–3469. https://doi.org/10.1128/AEM.60.9.3466-3469.1994

Huang, S., Lindahl, P. A., Wang, C., Bennett, G. N., Rudolph, F. B., & Hughes, J. B. (2000). 2,4,6-Trinitrotoluene reduction by carbon monoxide dehydrogenase from *Clostridium thermoaceticum*. *Applied and Environmental Microbiology, 66*, 1474. https://doi.org/10.1128/aem.66.4.1474-1478.2000

James, K. D., Hughes, M. A., & Williams, P. A. (2000). Cloning and expression of *ntnD*, encoding a novel NAD $(P)^+$-independent 4-nitrobenzyl alcohol dehydrogenase from *Pseudomonas* sp. strain TW3. *Journal of Bacteriology, 182*, 3136–3141. https://doi.org/10.1128/jb.182.11.3136-3141.2000

Ju, K. S., & Parales, R. E. (2006). Control of substrate specificity by active site residues in nitrobenzene 1,2-dioxygenase. *Applied and Environmental Microbiology, 72*, 1817–1824. https://doi.org/10.1128/AEM.72.3.1817-1824.2006

Ju, K. S., & Parales, R. E. (2009). Application of nitroarene dioxygenases in the design of novel strains that degrade chloronitrobenzenes. *Microbial Biotechnology, 2*, 241–252. https://doi.org/10.1111/j.1751-7915.2008.00083.x

Juhas, M., Van der Meer, J. R., Gaillard, M., Harding, R. M., Hood, D., & Crook, D. W. (2009). Genomic islands: Tools of bacterial horizontal gene transfer and evolution. *FEMS Microbiology Reviews, 33*, 376–393. https://doi.org/10.1111/j.1574-6976.2008.00136.x

Kapley, A., & Purohit, H. (2009). Genomic tools in bioremediation. *Indian Journal of Microbiology, 49*, 108–113. https://doi.org/10.1007/s12088-009-0012-2

Katsivela, E., Wray, V., Pieper, D. H., & Wittich, R. M. (1999). Initial reactions in the biodegradation of 1-chloro-4-nitrobenzene by a newly isolated bacterium, strain LW1. *Applied and Environmental Microbiology, 65*, 1405–1412. https://doi.org/10.1128/AEM.65.4.1405-1412.1999

Kulkarni, M., & Chaudhari, A. (2007). Microbial remediation of nitro-aromatic compounds: An overview. *Journal of Environmental Management, 85*, 496–512. https://doi.org/10.1016/j.jenvman.2007.06.009

Kumari, S., Regar, R. K., Bajaj, A., Ratnasekhar, C., Satyanarayana, G. N. V., Mudiam, M. K. R., & Manickam, N. (2017). Simultaneous biodegradation of polyaromatic hydrocarbons by a *Stenotrophomonas* sp: Characterization of *nid* genes and effect of surfactants on degradation. *Indian Journal of Microbiology, 57*, 60–67. https://doi.org/10.1007/s12088-016-0612-6

Lenke, H., Achtnich, C., & Knackmuss, H. J. (2000). Perspeptives of bioelimination of polynitroaromatic compounds. In J. C. Spain, J. B. Highes, & H. J. Knackmuss (Eds.), *Biodegradation of nitroaromatic compounds and explosives* (pp. 91–126). Lewis Publishers. https://doi.org/10.1023/B:WIBI.0000021720.03712.12

Liu, H., Wang, S. J., & Zhou, N. Y. (2005). Anew isolate of *Pseudomonas stutzeri* that degrades 2-chloronitrobenzene. *Biotechnology Letters, 27*, 275–278. https://doi.org/10.1007/s10529-004-8293-3

Liu, D. F., Min, D., Cheng, L., Zhang, F., Li, D. B., Xiao, X., Sheng, G. P., & Yu, H. Q. (2017). Anaerobic reduction of 2,6-Dinitrotoluene by Shewanella oneidensis MR-1: Roles of Mtr respiratory pathway and nfnB. *Biotechnology and Bioengineering, 114*, 761–768. https://doi.org/10.1002/bit.26212

McCormick, N., Feeherry, F. E., & Levinson, H. S. (1976). Microbial transformation of 2, 4, 6-trinitrotoluene and other nitroaromatic compounds. *Applied and Environmental Microbiology, 31*, 949–958. https://doi.org/10.1128/AEM.31.6.949-958.1976

Min, J., Zhang, J. J., & Zhou, N. Y. (2016). A two-component para-nitrophenol monooxygenase initiates a novel 2-chloro-4-nitrophenol catabolic pathway in *Rhodococcus imtechensis* RKJ300. *Applied and Environmental Microbiology, 82*, 714–723. https://doi.org/10.1128/AEM.03042-15

Min, J., Chen, W., Wang, J., & Hu, X. (2017). Genetic and biochemical characterization of 2-chloro-5-nitrophenol degradation in a newly isolated bacterium, *Cupriavidus* sp. Strain CNP-8. *Frontiers in Microbiology, 8*, 1778. https://doi.org/10.3389/fmicb.2017.01778

Nishino, S. F., & Spain, J. C. (1993). Degradation of nitrobenzene by a *Pseudomonas pseudoalcaligenes*. *Applied and Environmental Microbiology, 59*(8), 2520–2525. https://doi.org/10.1128/aem.59.8.2520-2525.1993

Nishino, S. F., & Spain, J. C. (1995). Oxidative pathway for the biodegradation of nitrobenzene by Comamonas sp. strain JS765. *Applied and Environmental Microbiology, 61*(6), 2308–2313. https://doi.org/10.1128/AEM.61.6.2308-2313.1995

Nishino, S. F., & Spain, J. C. (2001). Technology status review: Bioremediation of dinitrotoluene (DNT).

Nishino, S., & Spain, J. C. (2006). Biodegradation of 3-nitrotyrosine by *Burkholderia* sp. strain JS165 and *Variovorax paradoxus* JS171. *Applied and Environmental Microbiology, 72*, 1040–1044. https://doi.org/10.1128/AEM.72.2.1040-1044.2006

Nishino, S. F., Spain, J. C., & He, Z. (2000). Strategies for aerobic degradation of nitroaromatic compounds by bacteria: Process discovery to field application. In J. C. Spain, J. B. Hughes, & H. J. Knackmuss (Eds.), *Biodegradation of nitroaromatic compounds and explosives* (pp. 7–61). Lewis Publishers. https://doi.org/10.1023/B:WIBI.0000021720.03712.12

Nishino, S. F., Spain, J. C., & He, Z. (2002). Biodegradation, transformation and bioremediation of nitroaromatic compounds. In C. J. Hurst, R. L. Crawford, G. R. Knudsen, M. J. McInerey, & L. D. Stetzenbach (Eds.), *Manual of environmental microbiology* (2nd ed., pp. 987–996). ASM Press. ISBN-10: 953-51-2179-0, ISBN-13: 978-953-51-2179-4.

Park, H. S., & Kim, H. S. (2000). Identification and characterization of the nitrobenzene catabolic plasmids pNB1 and pNB2 in *Pseudomonas putida* HS12. *Journal of Bacteriology, 182*(3), 573–580. https://doi.org/10.1128/jb.182.3.573-580.2000

Park, H. S., Lim, S. J., Chang, Y. K., Livingston, A. G., & Kim, H. S. (1999). Degradation of chloronitrobenzenes by a coculture of *Pseudomonas putida* and a *Rhodococcus* sp. *Applied and Environmental Microbiology, 65*(3), 1083–1091.

Payne, R. B., Qu, Y., Nishino, S. F., & Spain, J. C. (2007). Abstract. In *General Meeting of the American Society for Microbiology, Toronto, Canada, 21 to 25 May 2007*. abstr. Q-013. https://doi.org/10.1128/MMBR.00006-10

Peres, C. M., Van Aken, B., Naveau, H., & Agathos, S. N. (1999). Continuous degradation of mixtures of 4-nitrobenzoate and 4-aminobenzoate by immobilized cells of *Burkholderia cepacia* strain PB4. *Applied Microbiology and Biotechnology, 52*, 440–445. https://doi.org/10.1007/s002530051544

Purohit, H. J., Kapley, A., Khardenavis, A., Qureshi, A., & Dafale, N. A. (2016). Chapter three-insights in waste management bioprocesses using genomic tools. *Advances in Applied Microbiology, 97*, 121–170. https://doi.org/10.1016/bs.aambs.2016.09.002

Qu, Y., & Spain, J. C. (2010). Biodegradation of 5-nitroanthranilic acid by *Bradyrhizobium* sp. strain JS329. *Applied and Environmental Microbiology, 76*, 1417–1422. https://doi.org/10.1128/AEM.02816-09

Qureshi, A., & Purohit, H. J. (2002). Isolation of bacterial consortia for degradation of p-nitrophenol from agricultural soil. *The Annals of Applied Biology, 140*, 159–162. https://doi.org/10.1111/j.1744-7348.2002.tb00168.x

Qureshi, A., Verma, V., Kapley, A., & Purohit, H. J. (2007). Degradation of 4-nitroaniline by *Stenotrophomonas* strain HPC 135. *International Biodeterioration and Biodegradation, 60*, 215–218. https://doi.org/10.1016/j.ibiod.2007.03.004

Qureshi, A., Mohan, M., Kanade, G. S., Kapley, A., & Purohit, H. J. (2009). In situ bioremediation of organochlorine-pesticide-contaminated microcosm soil and evaluation by gene probe. *Pest Management Science, 65*(7), 798–804. https://doi.org/10.1002/ps.1757

Rankin, L. D., Bodenmiller, D. M., Partridge, J. D., Nishino, S. F., Spain, J. C., & Spiro, S. (2008). *Escherichia coli* NsrR regulates a pathway for the oxidation of 3-nitrotyramine to 4-hydroxy-3-

nitrophenylacetate. *Journal of Bacteriology, 190*, 6170–6177. https://doi.org/10.1128/JB.00508-08

Razo-Flores, E., Donlon, B., Lettinga, G., & Field, J. A. (1997). Biotransformation and biodegradation of N-substituted aromatics in methanogenic granular sludge. *FEMS Microbiology Reviews, 20*, 525–538. https://doi.org/10.1111/j.1574-6976.1997.tb00335.x

Rhys-Williams, W., Taylor, S. C., & Williams, P. A. (1993). A novel pathway for the catabolism of 4-nitrotoluene by *Pseudomonas*. *Journal of General Microbiology, 139*, 1967–1972. https://doi.org/10.1099/00221287-139-9-1967

Rieger, P. G., & Knackmuss, H. J. (1995). Basic knowledge and perspectives on biodegradation of 2,4,6 trinitrotoluene and related nitroaromatic compounds in contaminated soil. In J. C. Spain (Ed.), *Biodegradation of nitroaromatic compounds* (pp. 1–18). Plenum Press. https://doi.org/10.1007/978-3-662-05794-0_4

Saha, S. P., Banik, S. P., Majumder, A., Noor, A., Biswas, K., Hasan, N., Banerjee, N., Saha, P., Das, R., Halder, S., & Parveen, S. (2014). Bioremediation of methyl parathion by bacterial strains isolated from fresh vegetables. *Journal of Environment and Sociobiology, 11*, 43–56.

Saha, S., Badh, N., Pal, S., Biswas, R., & Nandy, T. (2017). Carbon and nutrient-limiting conditions stimulate biodegradation of low concentration of phenol. *Biochemical Engineering Journal, 126*, 40–49. dx.doi.org. https://doi.org/10.1590/0104-6632.20170341s20150388

Samanta, S. K., Bhushan, B., Chauhan, A., & Jain, R. K. (2000). Chemotaxis of a *Ralstonia* sp. SJ98 toward different nitroaromatic compounds and their degradation. *Biochemical and Biophysical Research Communications, 269*, 117–123. https://doi.org/10.1006/bbrc.2000.2204

Selvaratnam, C., Thevarajoo, S., Ee, R., Chan, K. G., Bennett, J. P., Goh, K. M., & Chong, C. S. (2016). Genome sequence of *Roseivirga* sp. Strain D-25 and its potential applications from the genomic aspect. *Marine Genomics, 28*, 29–31. https://doi.org/10.1016/j.margen.2016.04.004

Sengupta, K., Maiti, T. K., & Saha, P. (2015). Degradation of 4-nitrophenol in presence of heavy metals by a halotolerant *Bacillus* sp. Strain BUPNP2, having plant growth promoting traits. *Symbiosis, 65*(3), 157–163. https://doi.org/10.1007/s13199-015-0327-1

Singh, D., Mishra, K., & Ramanthan, G. (2015). Bioremediation of nitroaromatic compounds. In *Wastewater treatment engineering*. Intech Open. https://doi.org/10.5772/61253

Thangaraj, K., Kapley, A., & Purohit, H. J. (2008). Characterization of diverse *Acinetobacter* isolates for utilization of multiple aromatic compounds. *Bioresource Technology, 99*(7), 2488–2494. https://doi.org/10.1016/j.biortech.2007.04.053

Thijs, S., Van Hamme, J., Gkorezis, P., Rineau, F., Weyens, N., & Vangronsveld, J. (2014). Draft genome sequence of Raoultella ornithinolytica TNT, a trinitrotoluene-denitrating and plant growth-promoting strain isolated from explosive-contaminated soil. *Genome Announcements, 2*(3), e00491–e00414. https://doi.org/10.1128/genomeA.00491-14

Tikariha, H., Pal, R. R., Qureshi, A., Kapley, A., & Purohit, H. J. (2016). In silico analysis for prediction of degradative capacity of *Pseudomonas putida* SF1. *Gene, 591*(2), 382–392. https://doi.org/10.1016/j.gene.2016.06.028

Tiwari, J., Naoghare, P., Sivanesan, S., & Bafana, A. (2017). Biodegradation and detoxification of chloronitroaromatic pollutant by *Cupriavidus*. *Bioresource Technology, 223*, 184–191. https://doi.org/10.1016/j.biortech.2016.10.043

Torres, R. M., Grosset, C., Steiman, R., & Alary, J. (1996). Liquid chromatography study of degradation and metabolism of pentachloronitrobenzene by four soil micromycetes. *Chemosphere, 33*(4), 683–692. https://doi.org/10.1016/S1002-0160(10)60076-8

Van Aken, B., Hofrichter, M., Scheibner, K., Hatakka, A., Naveau, H., & Agathos, S. N. (1999). Transformation and mineralization of 2,4,6-trinitrotoluene (TNT) by manganese peroxidase from the white-rot basidiomycete *Phlebia radiata*. *Biodegradation, 10*, 83–91. https://doi.org/10.1023/A:1008371209913

Wu, J. F., Jiang, C. Y., Wang, B. J., Ma, Y. F., Liu, Z. P., & Liu, S. J. (2006). Novel partial reductive pathway for 4-chloronitrobenzene and nitrobenzene degradation in *Comamonas* sp. strain CNB-1. *Applied and Environmental Microbiology, 72*, 1759–1765. https://doi.org/10.1128/AEM.72.3.1759-1765.2006

Xu, Y., Yu, M., & Shen, A. (2016). Complete genome sequence of the polychlorinated biphenyl degrader *Rhodococcus* sp. WB1. *Genome Announcements, 4*(5), e00996–e00916. https://doi.org/10.1128/genomeA.00996-16

Yabannavar, A. V., & Zylstra, G. J. (1995). Cloning and characterization of the genes for p-nitrobenzoate degradation from *Pseudomonas pickettii* YH105. *Applied and Environmental Microbiology, 61*, 4284–4290. https://doi.org/10.1128/AEM.61.12.4284-4290.1995

Yadav, T. C., Pal, R. R., Shastri, S., Jadeja, N. B., & Kapley, A. (2015). Comparative metagenomics demonstrating different degradative capacity of activated biomass treating hydrocarbon contaminated wastewater. *Bioresource Technology, 188*, 24–32. https://doi.org/10.1016/j.biortech.2015.01.141

Yanzhen, M., Yang, L., Xiangting, X., & Wei, H. (2016). Complete genome sequence of a bacterium *Pseudomonas fragi* P121, a strain with degradation of toxic compounds. *Journal of Biotechnology, 224*, 68–69. https://doi.org/10.1016/j.jbiotec.2016.03.019

Zhang, C., & Bennett, G. N. (2005). Biodegradation of xenobiotics by anaerobic bacteria. *Applied and Environmental Microbiology, 67*, 600–618. https://doi.org/10.1007/s00253-004-1864-3

Zhen, D., Liu, H., Wang, S. J., Zhang, J. J., Zhao, F., & Zhou, N. Y. (2006). Plasmid-mediated degradation of 4-chloronitrobenzene by newly isolated *Pseudomonas putida* strain ZWL73. *Applied Microbiology and Biotechnology, 72*, 797–803. https://doi.org/10.1007/s00253-006-0345-2

Zheng, C. L., Zhou, J. T., Zhao, L. H., Lu, H., Qu, B. C., & Wang, J. (2007). Isolation and characterization of a nitrobenzene degrading Streptomyces strain from activated sludge. *Bulletin of Environmental Contamination and Toxicology, 78*(2), 163–167. https://doi.org/10.1007/s00128-007-9031-z

Microbial Interaction with Metals and Metalloids

Bellemkonda Ramesh, Srinivasan Kameswaran, Ch. Venkatrayulu, Somavarapu Silpa, M. Subhosh Chandra, G. Vidya Sagar Reddy, and K. Naveen Kumar

1 Introduction

Microorganisms encounter metals and metalloids of different types in the environment, and thus, it is not surprising that they interact with them, sometimes for their benefit, and other times to their expense. Base metals like vanadium, chromium, manganese, iron, cobalt, nickel, copper, zinc, molybdenum, silver, cadmium, and lead are of particular practical interest; precious metals such as gold and silver; and the metalloids arsenic, selenium, and antimony. In nature, these minerals and metals are mostly found as cations, or oxyanions, or both in an aqueous solution, and mostly as salts or oxides in crystalline (metallic) form or amorphous precipitates in an insoluble form. A few of them, such as copper, iron, and gold, may also be present in the metallic state in nature, but the first two elements are seldom present. Both microbes employ metal species for structural functions and/or catalytic functions, whether prokaryotic or eukaryotic. Structural as well as catalytic functions serve the

B. Ramesh (✉) · S. Silpa
Department of Food Technology, Vikrama Simhapuri University, Nellore, Andhra Pradesh, India

S. Kameswaran
Department of Botany, Vikrama Simhapuri University PG Centre, Kavali, Andhra Pradesh, India

C. Venkatrayulu
Department of Marine Biology, Vikrama Simhapuri University, Nellore, Andhra Pradesh, India

M. S. Chandra
Department of Microbiology, Yogi Vemana University, Kadapa, Andhra Pradesh, India

G. V. S. Reddy
Department of Biotechnology, Vikrama Simhapuri University, Nellore, Andhra Pradesh, India

K. N. Kumar
Department of Chemistry, Vikrama Simhapuri University, Nellore, Andhra Pradesh, India

N. R. Maddela, L. C. García (eds.), *Innovations in Biotechnology for a Sustainable Future*, https://doi.org/10.1007/978-3-030-80108-3_13

alkali metals Ca and Mg. In catalytic functions, metals V, Cr, Mn Fe, Co, Ni, Cu, Zn, Mo, and W, and metalloid Se can participate. For such uses, low environmental concentrations are sufficient.

Microbes change their physical and chemical state so that they can interact with the natural and synthetic environment of the heavy metals and metalloids present at the contamination site. Metals and minerals are responsible for the growth, activity and survival of microorganisms. The microbes can be considered as the geo-active agents. Minerals of biogenic origin have global geological and industrial importance. Besides, it is also the source of many important structural components of many organisms (Ehrlich, 1996a, b; Gadd & Raven, 2010).

Bioremediation is a sustainable strategy that uses the metabolic activity of microorganisms and plants to clean up polluted sites. Bioremediation is a cost-effective and environmentally friendly way to cleans nature with the help of nature (Kaur & Maddela, 2021). Most significant aspect of biological removal is the microbiological aspect. The approach adopted at a contaminated site for bioremediation of pollutants depends on the form and process of contact between the metals and the microbes (Maddela et al., 2016, 2017).

The areas of heavy metal pollution are often inhabited by microorganisms, which are prevalent in nature. They can turn heavy metals into their nonlethal forms effortlessly. In the bioremediation process, microorganisms produce metabolic intermediates by mineralizing organic pollutants or some final products such as carbon dioxide and water, which can be used for cell growth as primary substrates. Via a two-way protection, which includes enzymes that can degrade the target contaminants and also resist the necessary heavy metal, the microorganism can function. Bioremediation includes techniques as diverse as bioaccumulation, bio-absorption, biomineralization, interactions between minerals and microbes, biological bleaching, and biotransformation. Covalent bonding, electrostatic interactions, redox potential, extracellular precipitation, and van der Waals forces are the forces by which metal ions may bind to a cell's surface or a mixture of all these processes (Blanco, 2000). The chemistry behind this association is that metal cations are adsorbed by the negatively charged groups (hydroxyl, phosphoryl and carboxyl) present on the cell wall of the microbes. The nucleation of the metal then limits these (Wase & Forster, 1997).

2 Microbial Interaction with Metal and Metalloids

Metals such as Na, Mn, K, Fe, Cu, Zn, Mg, Ca, and Co are important for life but can become toxic if present in excess of required amount. Whereas metals such as Pb, Hg, Cs, Cd, and Al do not have any known significant metabolic role in living organisms, when accumulated in the atmosphere, they can still cause toxicity. They toxicize the atmosphere at a very high degree (Gadd, 2010).

3 Types of Interaction

The process of interacting a microorganism with minerals depends in part on whether the organism is prokaryotic or eukaryotic. Both the types of microorganisms have the capacity to associate to metal ions present in the external surface of the cell or to transport to the cell to initiate different metabolisms within the cells. Alternatively, only prokaryotic organisms (eubacteria and archaea) are capable to oxidize Mn(II), Fe(II), Co(II), Cu(I), AsO_2^-), Se^0 or SeO_3^{2-}, or reduce Mn(IV), Fe(III), Co (III), AsO_4^{2-}, SeO_4^{2-}, or SeO_3^{2-} widely and energy conservation by uptaking metals ions through transfer the electrons between the celluar interior and extracellular metals (Ehrlich, 1996a, b). Some microbes may reduce metal ions such as Hg^{2+} or Ag^+ to Hg^0 and Ag^0 respectively, but do not conserve energy from these reactions (Summers & Sugarman, 1974). Some prokaryotes and eukaryotes may form metabolic products, like acids or bonds, which dissolve the base minerals found in minerals, such as iron, copper, zinc, nickel, cobalt, and others. Others may form anions, such as sulfides or carbonates, that precipitate dissolved metal ions (see Ehrlich, 1996a, b). Some prokaryotes may methylate some metal and metalloid compounds, producing corresponding volatile metal derivatives (Summers & Silver, 1978; Beveridge & Doyle, 1989; Ehrlich, 1996a, b).

4 Levels of Interaction

4.1 Metabolic/Enzymatic

Uptake of trace metals and their subsequent incorporation into metalloenzymes or utilization in enzyme activation occurs in all microbes (Wackett et al., 1989). Some examples of mineral enzymes are nitrogenase (Mo/Fe or sometimes V/Fe, or Fe only) (Orme-Johnson, 1992; Robson et al., 1986), cytochromes (Fe) and cytochrome oxidase *aa3* (Fe, Cu) (Wackett et al., 1989), superoxide dismutases (Fe, Mn, Cu or Zn) (Fridovich, 1978), bacterial chlorophyll (Mg) (Scheer, 1991), iron and sulfur proteins (Wackett et al., 1989), CO dehydrogenase with Mo in aerobic bacteria (Ferry, 1995), NADP-dependent dehydrogenase (W/Se/Fe) (Yamamoto et al., 1983), and hydrogenformate propensity H (Mo/Se/Fe) (Boyington et al., 1997). For absorption, these minerals must be in ionic form. Uptake may require genetically determined and controlled transport mechanisms (Silver & Walderhaug, 1992). In some cases, uptake may be fast, nonspecific, and constitutive, as for instance with the CorA Mg^{2+} exchange system and the Mgta and Mgtb Mg^{2+} uptake systems in *Salmonella typhimurium* (Snavely et al., 1989). Since ferric iron in the environment, around a neutral pH, is mainly present in a water insoluble form, its absorption under aerobic conditions requires microbial formation of bonds, called iron carriers, to make ferric iron soluble (Neilands, 1974).

The number of microorganisms that are capable to utilize few metals or metalloids as electron donors or acceptors in energy metabolism. They include eubacteria and archaea (Ehrlich, 1996a, b). Depending on the element, the mineral species may be in simple ionic form or in oxyanions form. As sources of energy, oxidizable minerals or metalloids may fully satisfy the energy demand of the organism (chemolithotrophs). For example, the eubacteria *Leptospirillum ferrooxidans* and *Thiobacillus ferrooxidans* and the archaea *Sulfolobus acidocaldarius* and *Acidianus brierleyi* are able to obtain all their energy for growth from the oxidation (FeII) to Fe (III) (Ehrlich, 1996a, b); *Stibiobacter senarmontii* from the oxidation of Sb_2O_3 to Sb_2O_5 (Lyalikova et al., 1976); and *Pseudomonas arsenitoxidans* from the oxidation of AsO^{2-} to $AsO_4{}^{3-}$ (Ilyaletdinov & Abdrashitova, 1981). Some oxidized metal species may serve as terminal electron acceptors in anaerobic respiration by heterotrophs and, depending on the organism, this may enable them to mineralize the organic carbon that serves as reductant. Some autotrophic anaerobic hydrogen-oxidizing bacteria also use types of oxidizing minerals as the final electron acceptors in their respiration. Examples of anaerobic respiration in which an oxidized metal or metalloid species serves as terminal electron acceptor include Fe(III) reduction to Fe^{2+}, Fe_3O_4, or $FeCO_3$ (Lovley & Phillips, 1988; Coleman et al., 1993) and MnO_2 reduction to Mn^{2+} or $MnCO_3$ with acetate by the eubacterium *Geobacter metallireducens* (Lovley & Phillips, 1988), and $SeO_4{}^{2-}$ and $SeO_3{}^{2-}$ reduction to Se^0 by *Thauera selenatis* with nitrate (Rech & Macy, 1992; DeMoll-Decker & Macy, 1993). The archeon *Sulfolobus* sp. has been shown to reduce $MoO_4{}^{2-}$ to a lower oxidation state (Brierley & Brierley, 1982). Aerobic reduction of MnO_2 to Mn^{2+} by a few marine eubacteria and of $CrO_4{}^{2-}$ to Cr(III) by the eubacterium *Pseudomonas fluorescens* LB300 as part of respiration has also been observed (Ehrlich, 1996a, b; Wang & Shen, 1995). The utilization of metals or metalloids as electron donors or acceptors in energy metabolism in eukaryotes is unknown.

Enzymatic microbial detoxification from harmful metals or metalloids is the third type of reaction. In this process, a toxic metal species can be transformed into a less toxic or nontoxic entity by enzymatic oxidation or reduction. The bacterial oxidation of AsO^{2-} to $AsO_4{}^{3-}$ by a strain of *Alcaligenes faecalis*, and reduction of $CrO_4{}^{2-}$ to $Cr(OH)_3$ by *P. fluorescens* LB300 or *Enterobacter cloacae* are examples of such redox reactions (Ehrlich, 1996a, b; Wang & Shen, 1995). The detoxification in the previous two examples is part of the organism's respiration process of the organisms. In some other instances, detoxification may be by enzymatic reduction that is not part of the respiratory process, as in mercury detoxification (Robinson & Tuovinen, 1984). In general, mercury-resistant bacteria produce the enzyme mercuric reductase, which catalyzes the conversion of Hg^{2+} to volatile Hg^0. Mercuric reductase formation is induced by Hg^{2+} in all organisms tested except in *Thiobacillus ferrooxidans*, in which the enzyme is constitutive (Robinson & Tuovinen, 1984). Still other detoxification processes include enzymatic or nonenzymatic methylation of minerals and metals such as Sn, Hg, Pb, As, and Se (Frankenberger & Karlson, 1992; Trevors, 1992; Hallas et al., 1982; Guard et al., 1981; Summers & Silver, 1978; Chau et al., 1976; Wong et al., 1975). When microbes cannot detoxify harmful metals, they often have other genetically determined defenses against them (Ji &

Silver, 1995; Silver, 1992). These defenses include modification or elimination of membrane transport systems into the cell for the harmful metal or metalloid species, or efflux systems (molecular pumps) for their removal from the cell interior if taken up.

Enzymatically induced anaerobic metal biocorrosion is another example of metal/microbial reaction. In the original concept, as formulated by Von Wolzogen Kuehr and Van der Vlugt (1934), sulfate-reducing bacteria promote biocorrosion of cast-iron metal surfaces anaerobically through cathodic depolarization. In this embodiment, the iron surface exposed to aqueous moisture undergoes spontaneous reaction:

$$Fe^0 + 2H_2O \rightarrow Fe^{2+} + 2OH^- + H_2 \quad (1)$$

with the reaction

$$Fe^0 \rightarrow Fe^{2+} + 2e^- \quad (2)$$

at anodic regions, and the reaction

$$2H_2O + 2e \rightarrow 2OH^- + H_2 \quad (3)$$

at cathodic regions. It is believed that H_2 generated in the cathodic region accumulates on the iron surface where it is formed, causing passivation (polarization) of the surface; i.e., stops their accumulation from further erosion. Sulfate-reducing bacteria, when this hydrogen is used to reduce sulfate, as shown by the reaction

$$4H_2 + SO_4{}^{2-} + H^+ \rightarrow HS^- + 4H_2O \quad (4)$$

were thought to depolarize the surface, thereby promoting continuation of the corrosion process.

The sulfide they generate can react with Fe^{2+} produced in the anodic areas, which may also help promoting corrosion if iron sulfide does not precipitate on the iron surface as a uniform film that would the iron surface as long as the film is not disturbed. Although some previous experiments seemed to support this model, the general opinion now is that anaerobic biological erosion is the result of several different microbiological interactions. Minerals surfaces are often colonized by biofilms and their activity must be taken into account. These biofilms consist of the union of bacteria, with specific sites in the biofilm. The metabolic products released by one member of the bond into the biofilm and not consumed by any of the other organs may be corrosive to the metal or act as chemical depolarizers in conjunction with the H_2-consuming activity of the sulfate-reducing bacteria in the lower part of the biofilm (Videla, 1995).

4.2 Metabolic/Nonenzymatic

Prokaryotic and eukaryotic microbes are capable accumulating metals by binding them as cations to the cell surface in a passive process (Beveridge & Doyle, 1989; Gadd, 1993). Even dead cells can bind metal ions. Depending on conditions, this linkage may be selective or nonselective. In some cases, if the cell surface becomes saturated with a metal species, the cell may later act as a nucleus in the formation of a mineral containing the metal (Macaskie et al., 1987, 1992; Schultze-Lam et al., 1996).

Some bacteria and fungi could enhance selective and nonselective leaching of one or more mineral components from ore or other rock with metabolic products such as acids and/or the bonds they produce (Ehrlich, 1996a, b). The acids may be organic or inorganic. Groudev and Groudeva (1986) were able to filter aluminum from clays with the oxalic and citric acids formed by fungus *Aspergillus niger*. Alibhai et al. (1991) were able to leach nickel selectively from low-grade Greek laterites with citric acid produced by various species of *Aspergillus* and *Penicillium*. The process discriminated against iron, probably because of a higher affinity of citric acid for nickel than for iron.

Microorganisms may release inorganic metabolic products like sulfide, carbonate, or phosphate ions into the respiratory metabolism and with them toxic metal ions precipitate as a form of nonenzymatic detoxification (Macaskie et al., 1987; Ehrlich, 1996a, b). To be effective, precipitation must reduce the concentration of dissolved mineral species below their inhibitory level. Under some conditions, microorganisms can affect nonenzymatic corrosion of metals such as aluminum or iron or some metal alloys through formation and release of corrosive metabolic products (Edyvean, 1995). These products are chiefly organic and inorganic acids.

5 Natural Occurrences of Metal/Microbe Interactions

In nature, a marked microbial interaction with minerals is often demonstrated through mineral freezing or packing (Ferris et al., 1989; Ghiorse & Ehrlich, 1992; Ehrlich, 1996a, b). Mineral fixation may be through cell isolation and accumulation, or through extracellular sedimentation. Metal mobilization results from dissolution of insoluble metal-containing phases. Bioleaching of metals from ores is a practical example (Ehrlich & Brierley, 1990). These processes are essential for controlling the bioavailability of minerals in soil, sediments, and water.

Extracellular mineral accumulations that result from the metabolism of the microbial respiratory system or through mineral attachment to the surfaces of microbial cells include some iron and manganese oxide deposits, some iron and manganese carbonate deposits, a few mineral sulfide deposits (most of which have a hydrothermal origin), and some gold deposits (Schultze-Lam et al., 1996; Ghiorse & Ehrlich, 1992; Doyle, 1991; Ferris, 1991; Beveridge et al., 1982; Beveridge & Koval, 1981).

6 Metal–Microbes Interaction

Microbes use a range of mechanisms with minerals and metals found in both synthetic and natural environments. The change in the physical and chemical state helps to convert heavy metals into nontoxic forms. Minerals and metals affect the microbial growth, activity, and survival. Microbes are predominant in heavy metal-contaminated soil and operate though a two-way defense that includes production of the enzyme that degrades target pollutants and appropriate heavy metal resistance (Dixit et al., 2015). There are various forms of bioremediation which are bio-absorption, bio-mineralization, biological bleaching, interaction between minerals and microbes, biotransformations, and bioaccumulation. Depending on the mechanism of interaction of minerals with microorganisms at the contaminated site, the bioremediation strategy is adopted. Microbes can dissolve the metal and oxidize or reduce transition metals. The interaction of metals with microbes occurs by oxidizing, binding, volatilizing, reducing, stabilizing, and transformation of heavy metals.

7 Bioremediation by Adsorption

The forces through which the metal ions can bind to the surface of cell are electrostatic interactions, covalent bonding, van der Waals forces, redox interactions and extracellular precipitation or a combination of all these processes (Blanco, 2000). The chemistry behind this binding is that the metal cations are adsorbed by the negatively charged groups (carboxyl, hydroxyl and phosphoryl) present on the cell wall of the microbes. Then it is bounded by metallic nucleation (Wase & Forster, 1997). Binding of minerals to the surfaces outside the cell immobilizes the metal and thus prevents its entry into the cell. The phosphoryl and phospholipids groups of bacterial lipopolysaccharides in the outer membrane strongly interact with the cationic minerals.

8 Biosorption

Biosorption is a process that involves a bio-absorbent material with a higher affinity for sorbates (metal ions) and it is extended until equilibrium occurs between the two components (Dixit et al., 2015; Maddela et al., 2015). The extent of bio-absorption varies with the level of the mineral and the microorganisms. The metals that can be extracted through this technique are U, Mn, Ni, Hg, Au, Zn, Pb, Cd, Cu, Th, Cs, Ag, and Sn.

9 Natural Occurrences of Metal–Microbe Interaction

Microbes are closely correlated with the biogeochemical cycle of metals and metalloids and depending upon the interaction and mechanism involved in an environment, metals are either immobilized or mobilized (Gadd, 2010). Metal immobilization, in nature, can occur through extracellular precipitation or cellular sequestration and accumulation. Dissolution of insoluble metal-containing phases results in metal mobilization. One of the practical examples is the bioleaching of metals from the ores (Ehrlich & Brierley, 1990). These methods act as an essential part for controlling biological availability of metals in soils, sediments, and water.

10 Metal Mobilization

Metals can be mobilized by protonolysis, Fe^{3+}-binding siderophores, redox reactions, methylation and indirect Fe^{3+} attack (Gadd, 2010). This can result in volatilization of metals. Microbes can assemble metals and through redox processes can outbreak the mineral surfaces (Ehrlich, 1996a, b; Lloyd & Lovley, 2001). The solubility of Fe^{3+} and Mn^{4+} is raised by reduction to Fe^{2+} and Mn^{2+} respectively. As a result of methylation, methylated derivatives of some metals and metalloids are formed that increases the mobility of these elements. Various microbes like methanogens, sulfate-reducing bacteria and clostridia act under anaerobic state and fungi such as *Alternaria* and *Penicillium* spp. act under aerobic state. Such microbes could mediate methylation of Pb, Hg, and Sn and the metalloids Te, Se, and As (Gadd, 2010). The methyl derivatives of these elements differ in their volatility, toxicity, and solubility. The methylated substances of a volatile nature are often lost from the soil. However, methylation of some heavy metals may not treat soil. For example, bacteria and fungi can methylate the mercury ion (Hg^{2+}) to a more toxic compound, methylmercury [$(CH_3)Hg^+$] (Barkay & Wagner-Dobler, 2005). But methylmercury can be methylated to dimethyl mercury, by naturally volatile bacteria. Likewise, phenylmercury can be converted to volatile diphenyl mercury by microbial reactions.

11 Metal Immobilization

This technique is used to reduce the mobilization of minerals by changing their physical and chemical states. This can be done in two ways: off-site and on-site installation, in order to treat contaminated soil. In off-site technology, the contaminated soil is removed from the original place but its storage is risky (for example, in the case of radionuclides). An on-site technique is applied to non-excavated soils.

The primary role of stabilization, which is performed by the bio-absorption method, is to change the original soil minerals into more geochemically stable stages. Each and every microbial material can act as an effective biosorbent for the metals, excluding the alkali metal that are cations like K^+ and Na^+, and which can be a crucial passive method in dead and living organisms (Gadd, 2010). Absorption plays a significant role as a biological treatment method by influencing bioavailability and thus is beneficial in interaction between microbes and minerals. Heavy metal microbes could reduce the redox condition and this can decrease the transport and toxicity of many elements. The reduction of U^{4+} to U^{6+} forms the base for removal of uranium from the contaminated leachates and waters and also the uranium ores are formed such as uraninite (UO_2) (Lovley & Coates, 1997; Landa, 2005). For reductive precipitation of metals like U^{6+}, Cr^{6+}, Tc^{7+} and Pd^{2+}, sulfur- and sulfate-reducing bacteria are important (Gadd, 2010). The microbial reduction of gold species and ionic silver results in formation of gold Au (0) and elemental silver Ag (0) (Kierans et al., 1991; Southam et al., 2009). Many vital organic and inorganic minerals such as phosphates, oxalates, oxides, sulfides, and carbonates are formed by microbes, resulting in mineral freezing (Gadd et al., 2007). Iron-containing minerals in rocks, sediments and soils are weathered partly by chemical activity and partly by fungi and bacteria. Ferrous iron ablation may precipitate Fe^{3+}. Aqueous iron oxides formed by microbes can accumulate minerals by co-sedimentation or adsorption, in the aqueous environment. Reduction of iron oxides or acidification can result in remobilization of adsorbed metals (Ehrlich & Newman, 2009).

12 Mechanisms of Metal Tolerance by Microbes

Heavy metals are different to remove from environments because they do not degrade easily and thus persist in the environment. Heavy metal toxicity generally occurs due to the interaction of metals with enzymes (proteins) which leads to inhibition of metabolic actions. If the presence of these minerals in the environment exceeds the threshold value, they are toxic to all life (Kumar et al., 2014). If a bacterial strain could grow in a contaminated area containing a high concentration of heavy metal, it would provide for the metal to be tolerated by the microbe present. There are some microorganisms that can tolerance metals. The reason behind this is the early exposure of the microbe to heavy metals. But in some microbes, this trait is thought to have evolved due to genetic changes after exposure to heavy metals in past years. To remove the toxic metals from polluted area, new technologies are being implemented. Biosorption is one of the important methods which are based on the ability of different materials to bind to a metal. In the bioremediation process, microorganisms mineralize organic pollutants into specific metabolic intermediates or end products such as carbon dioxide and water, which can be used for cell growth as primary substrates. Microorganism can act through a two-way defense, which includes enzymes that can break down target pollutants as well as resist the appropriate heavy metal (Dixit et al., 2015). Due to the presence of the mineral in the environment, microorganisms have developed methods for metal resistance (for example, the use of a specific plasmid for a specific mineral) and detoxification (Gomathy & Sabarinathan, 2010). The metal resistance mechanism can be broadly classified:

1. General mechanism of metal resistance
2. Metal dependent mechanism of metal resistance

13 General Mechanism of Metal Resistance by Microbes

Binding minerals to additional cellular materials immobilizes minerals and prevents them from entering the cell. Binding of minerals to microbial cells is both environmentally and practically important. Ecological cell surface attachment plays a large role in mineral distribution, especially in the aquatic environment. In practical terms, the ability of microbes to absorb minerals has been exploited for the purpose of bioremediation i.e., removing of mineral contaminants from nature. Four phenomena were observed that contribute to the general mechanism of metal resistance.

13.1 Exopolymer Binding

Extracellular polymeric materials (EPMs) or outer polymers are diffuse in nature and offer defense against dehydration, phagocytosis and parasitism. External polymers include carbohydrates, sugars, and sometimes fatty acids, DNA and fats responsible for extracellular binding (Schiewer & Volesky, 2000). Many microorganisms produce EPMs resulting in a strong metal-bonding. EPSs prevent binding the toxic metals to enter the cell by mobilizing or immobilizing them and thus play a crucial role in metal cycling (Gomathy & Sabarinathan, 2010). These interactions can efficiently bind lead, cadmium, and uranium. The presence of negatively charged groups on exopolymer such as hydroxyl, succinyl, phosphate, amine, amide, and uronic acids, contribute to binding of metals (as metals are positively charged). It results in immobilization of metals and thus prevents their entry into the cell.

13.1.1 Siderophores

These compounds that belong to the biggest well-known compounds can connect and transfer or shuttle Fe. They are highly specific Fe^{3+} ligands. Their main function is to increase the concentration of iron in the areas having very low concentrations of iron and then transfer it into the cell. Fe^{2+} along with Fe^{3+} is mediated into the cell by the siderophore. Siderophores interact with metals having similar chemical structure like iron (e.g., aluminum, gallium, chromium, etc.) that is, forms similar size of trivalent ions as irons. The metal bioavailability is reduced when siderophore binds to the metals and thus results in reduction of metal toxicity. For example, iron acid reduces copper toxicity in cyanobacteria (Roane & Pepper, 2000). The organisms have developed certain methods that can ensure Fe demand is completed either by attachment to solid iron mineral (for example, Fe oxides or by production of species-specific siderophores) (Gomathy & Sabarinathan, 2010).

13.1.2 Biosurfactant Complexation

Some compounds produced by the microbes are excreted out. This class of compounds is classified as biosurfactant. They can form complex with the metals such as lead and cadmium. It increases the mobility of the resultant complex and thus increases the solubility. These complexes are nontoxic to the cells. Various researches have concluded that metal-contaminated sites provide a better site for the isolation of biosurfactant producting microorganism than the uncontaminated sites (Gomathy & Sabarinathan, 2010).

13.1.3 Precipitation

It causes immobilization of metals or heavy metals, which results the soluble metals to become insoluble in nature. It may be dependent or independent on the cellular metabolism. In case of dependent precipitation, the metal removed from the solution is involved with the dynamic defense system of the microbes while independent precipitation results from the chemical interplay between metal and the cell surface.

14 Metal Dependent Mechanism of Metal Resistance

14.1 Metallothioneins

These are cysteine-rich proteins which have low molecular weight. They are divided into three different types according to their cysteine structure and function i.e., Cys-Cys, Cys-X-Cys and Cys-X-X-Cys. These motifs are characteristic and invariant for metallothioneins. Metallothioneins are classified into class-I metallothioneins (MTs) and they include all which are found in animals and class-II MTs includes those which are present in plants and other microbes (Fowler et al., 1987). The thiol group for the mercaptide bonds is obtained from the invariant alignment of the cysteine group (cys) and an arrangement of metal-thiolate clusters is made. In cadmium (Cd) and zinc (Zn), the alpha domain in carboxy terminal region is a metal-cys cluster.

14.2 Methylation of Metals

In this mechanism of interaction, only certain metals are involved. Hence it is considered a resistance mechanism based on metal. Methyl generally increases mineral toxicity due to increased lipid susceptibility and thus increased permeability through the cell membrane (Gomathy & Sabarinathan, 2010). But with the help of volatilization of minerals, it diffuses easily away from the cell and the toxicity of the minerals is reduced. This phenomenon of metal volatilization is observed in lead (Pb), selenium (Se), mercury (Hg), tin (Sn), and arsenic (As). For example, mercury (Hg^{2+}) is oxidized to methylmercury and dimethylmercury, which are both volatile and highly toxic forms of mercury and can rapidly diffuse away from the cell (Roane & Pepper, 2000). Minerals can be removed from highly contaminated surface water by methylation. In Gram-negative and Gram-positive bacteria, mercury resistance may involve the reduction of Hg^{2+} to Hg^0 (elemental form of mercury).

14.3 Biosorption

The process of biosorption involves a biosorbent having high affinity toward sorbate (metal ions) and the interaction is extended until equilibrium is achieved between both the components (Dixit et al., 2015). The extent of bio-absorption varies with the level of the mineral and microorganism. The mechanism of bio-absorption can be divided into two classes based on the dependence of the cell's metabolism: Metabolism dependent and non-metabolically dependent.

Metabolism dependent biosorption occur due to intracellular accumulation of the metal when it is transported across the cell membrane. Non-metabolism dependent biosorption occurs when the uptake of the metal is due to the physicochemical interaction between the metal and the functional group present on the surface of the microbe. This could be observed from the bio-absorption of U, Cu, Ni, Zn, Pb, Cd, Hg, Th, Cs, Au, Ag, Sn, and Mn. Successful remediation of Zn^{2+} and Cd^{2+} can be done by this method through the ion exchange mechanism.

14.4 Efflux System

Plasmid-encoded energy dependent metal efflux systems are used by certain microbes to remove the metals from the cell. They include chemically contrasting ion/proton pumps and ATPases system associated with resistance to Cadmium (Cd), Chromium (Cr), and Arsenic (Ar).

15 Interaction at Molecular Level

Mostly, efflux forms the basis for the resistance system to metals by microbes. Two groups of efflux are known: P-type ATPases (e.g., the Cu^{2+}, Cd^{2+}) and Zn^{2+} ATPases of gram-negative bacteria. Analog chemical pumps, for example, the three-component divalent cation arrangement of czc, ncc, and cnr, of *R. metallidurans* CH34 (Taghavi et al., 1997).

15.1 Lead

The lead resistant bacteria *Ralstonia metallidurans* CH34, contains the active lead resistance factor *pbr*. The unique property of this trigger is that it blends the functions involved in absorption, flow, and accumulation of Pb^{2+} (Borremans et al., 2001). Metallothioneins (MTs) are encrypted by the *smt* locus which consists of two individually transcribed genes *smtb* and *smtA*. The elementary role of MTs is zinc homeostasis, but Pb^{2+} is also competent of switching on the expression of *smtA*. Efflux of Pb^{2+} is mediated mostly via P-type ATPases from PIB family. Some PIB pumps are: CadA from *S. aureus*, ZntA from *E. coli*, CadA2 from *P. putida* KT2440, and PbrA from *C. metallidurans* (Jarosławiecka & Piotrowska-Seget, 2014).

15.2 Arsenic

Three or Five membered operons involved in arsenate and arsenite resistance contains both *ars1* and *ars2*. These operons were recognized in either the chromosomal DNA or Plasmid of several bacteria including *Corynebacterium glutamicum*, *Achromobacter xyloxidan*, *Staphylococcus aureus*, *Pseudomonas putida*, *Bacillus* sp., etc. These operons are placed at certain distance from each other, in the bacterial chromosome (Sarkar et al., 2016; Roychowdhury et al., 2002). Both of them contain genes encoding a regulatory protein *arsR*, an arsenite reductase *arsC1′* and an arsenite permease *arsB*. Arsenate reductase genes and arsenite permease and (*arsC4* and *arsB3*) were also identified scattered on the chromosome. Another type of periplasmic dissimilatory reductase, *arr* is involved in reduction of arsenate and use it in respiratory metabolism. Arsenite oxidation is facilitated by a periplasmic enzyme Aio.

15.3 Zinc

Zinc resistance is conferred by czrC operon. This gene was identified in methicillin-resistant *Staphylococcus hyicus* (Slifierz et al., 2014). From several studies, it was found that czrC gene conferred widespread zinc resistance in several microorganisms (Cavaco et al., 2010).

15.4 Copper

cop is the copper-resistance operon. Three protein products of *cop* operon were characterized which provides a better understanding of copper-resistant mechanism. The *cop* proteins are *copA* (72 kDa), *copB* (39 kDa), and *copC* (12 kDa). *copA* and *copC* are periplasmic proteins and *copB* is an outer membrane protein. The *cop* proteins serve in the copper-resistant mechanism by mediating the sequestration of copper out of the cytoplasm (Cha & Cooksey, 1991).

15.5 Nickel and Cobalt

rcnA (*yohM*) gene is responsible for nickel and cobalt resistance. Different studies were conducted and it was inferred that membrane bound polypeptide is encoded by the gene *yohM* that shows increased nickel and cobalt resistance in *E. coli* (Rodrigue et al., 2005). This gene was specifically induced by Co and Ni only, not by other metals like Cu, Zn or Cd. *rcnA* is proposed as the new denomination to *yohM*.

15.6 Chromium

Several chromium resistance species that belong to different genera have been isolated with five or seven member operon. One of such strain, *Ochrobactrum tritici* strain 5bvl1 was found to contain transposon-located (TnOtChr) chromate resistance operon with five numbers of genes *chrB*, *chrA*, *chrC*, and *chrF*. The *chrA* and *chrB* contributed to high resistance but this was not found in *chrC* or *chrF* genes (Morais et al., 2011).

15.7 Cadmium

cadA and *cadC* genes conferred cadmium resistance. These genes seem to be organized in an operon and their transcription occurs in vivo and it is cadmium dependent. $cadC_{st}$ and $cadA_{st}$ were the two genes located on the chromosome of *Streptococcus thermophilus* 4134 that constituted a cadmium resistance cassette (Viti et al., 2014).

15.8 Mercury

The reduction of Hg^{2+} to Hg (0) is mediated by mercuric reductase (*MerA*). The diversity of *MerA* is not much known. From places such as brine bacteria in sea ice, freshwater and high arctic snow were isolated and seven markers of *merA* were identified (Moller et al., 2014). Two classes of mercury resistance are: narrow spectrum defines resistance to inorganic mercury and broad spectrum defines resistance to organic mercury, which is encoded by the *merB* gene.

15.9 Iron

Iron acid contains the largest subset of compounds known that can bind and transfer or mix iron. *Escherichia coli* contain six specific ferric receptors (*Cir*, *Fiu*, *FecA*, *FepA*, *FhuE*, *FhuA*) which provide specificity to many loops. Many bacteria can absorb anaerobic iron via *FeoB*. Besides, *Escherichia coli* contain three iron storage proteins (*FtnA*, *FtnB*, and *Bfr*). Among these *FtnA* plays a major storage role (Sarkar et al., 2016).

16 The Role of Microbes in Disseminating Metals and Metalloids into the Environment

All minerals/metalloids have been present, either in elemental or mineral form, in the subsurface crust of the Earth, for a long period of time. Metal erosion, sediment resuspension, atmospheric deposition, mineral evaporation from water resources into soil and groundwater, soil erosion of mineral ions and heavy metal leaching leads to pollution (Pacyna & Nriagu, 1988). Moreover, there are reports revealing that natural phenomena such as weathering and volcanic eruptions can also cause heavy metal contamination (Jung, 2008; He et al., 2005; Shallari et al., 1998; Pacyna & Nriagu, 1988). The role of microbes in mineral bioremediation is well known and is the most popular approach. At the same time, these microbes play lead role in metal/metalloids dissemination from subsurface earth crust. The organisms concerned with rock decomposition are bacteria, fungi, and other soil microbes.

Microbial metabolisms directs the decomposition of minerals, including oxidation of various metal as oxides/hydroxides which leads to the generation of acid mine discharge, which in turn leads to heavy metal contamination during mining activities. Moreover, microbial metabolism helps form various minerals over geological time. In the presence of moisture some microbes secrete carbonic acid or various other acids that erode rocks. For example, the microbial transformation of As where bacteria could lead to a reduction of As methylation could lead to the formation of gaseous arsines, which could cause mineralization of an organic As a compound into inorganic As or could lead to volatilization of As. Transitions like these give rise to cycling and accumulation in the soil. Arsenic accumulation in soil leads to toxicity and contamination of the groundwater thus becomes an important issue to be looked upon. Also, arsines are the highly toxic forms of As thus its evaluation in contaminated environment is of great concern (Turpeinen et al., 2002).

17 Biological Remediation

Biological remediation or bioremediation proves out to be more economical over physical and chemical remediation techniques, since the pollutants here can be treated on-site. Moreover the effluent volumes generated by bioremediation are smaller to a great extent and thus it can be disposed of easily. Moreover since these practice is based upon natural processes (makes use of microbes to neutralize contaminants at a site), it is highly acceptable. As a result the hazardous substances are broken down to less toxic or nontoxic substances. Certain microorganisms require heavy metals, as essential micronutrient, in different amount to facilitate their growth and development. Microorganisms are regarded as metal accumulators due to presence of distinctive original property of remediation of toxic metals in the soil. Genetic engineering of such metal accumulators can help in expressing a missing trait and thus resulting in a differentially expressed gene.

17.1 Zinc

Saccharomyces cerevisiae, through ion exchange mechanism, acts as a biosorbent and removes Zn^{2+}. Ectomycorrhizal fungi (*Paxillus involutus* and *Suillus granulates*) are able to deliver elements from wood and apatite (K, Pb, Ca, Mn, and Ti) and gather them in the mycelia (Wallander et al., 2003). Ectomycorrhizal fungi and Ericoid mycorrhizal have the potential to dissolve a variety of zinc-bearing minerals (Leyval & Joner, 2001). *Synechococcus* sp. (cynobacterial strains) has been reported with the expression of the smtA gene and production of metal-binding protein (Gadd, 2010). *Penicillium chrysogenum* and *Aspergillus niger* provide zinc resistance and are also included in leaching.

17.2 Copper

Due to the increased uptake of metal, *Saccharomyces cerevisiae* mutants (pmr1D) are highly receptive to heavy metals. By combining biosorption with continuous metabolic uptake after physical adsorption, the mutants are able to remove Cu^{2+} from synthetic effluents. The capability of *Citrobacter* spp. to generate phosphate enzymatically results in copper precipitation (Gomathy & Sabarinathan, 2010). Biosorption of copper, by *Z. ramigera* and *C. vulgaris*, occurs through both formation of coordination bonds between metals and amino and carboxyl groups of cell wall polysaccharides and adsorption.

17.3 Cobalt

Saccharomyces cerevisiae mutants (pmr1D) have the ability to remove Co^{+} from synthetic effluents by a combination of biosorption and continuous metabolic uptake after physical adsorption. *Zooglea* spp. is involved in cobalt metal uptake.

17.4 Nickel

Phormidium valderianum helps immobilize nickel. *Penicillium* and *Aspergillus* spp. helps resist nickel resistance. Hydrogen cyanide forming bacteria, for example, *Chromobacterium violaceum* and *Pseudomonas fluorescens* are capable of galvanizing nickel as different cyanide compounds and complexes (Gadd, 2010). Phytochelatin synthase (PCS) is an expressing gene present in *P. fluorescens* 4F39 that provides an effective method for nickel bioremediation (Dixit et al., 2015).

17.5 Arsenic

Alcaligenes faecalis aids in the bacterial oxidation of AsO_2^{-} to AsO_4^{3-}. *Escherichia coli*, *Staphylococcus aureus*, *Pseudomonas putida*, *Bacillus subtilis* are associated with arsenic resistance. These are related to plasmid-encoded energy dependent flow systems and include ions/chemical pumps and ATPases (Rajendran et al., 2003).

17.6 Lead

Soil contaminated with Pb^{2+} can be degraded by bio-absorption method by using fungal species such as *Cephalosporium aphidicola* and *Aspergillus parasitica* (Dixit et al., 2015). Ectomycorrhizal fungi (*Paxillus involutus* and *Suillus granulatus*) can deliver elements of wood ash and apatite (Pb, K, Mn, Ca, Ti,) and synthesize them in the mycelia (Wallander et al., 2003). Fungi and oriental rootstock have the ability to dissolve a variety of lead-bearing minerals (Leyval & Joner, 2001). Lead fixation has been observed in several bacterial species, including *Azotobacer* spp., *Staphylococcus aureus*, *Citrobacter* spp., and *Micrococcus luteus*, which can enzymatically produce phosphate and lead to lead deposition (Gomathy & Sabarinathan, 2010).

17.7 Chromium

Enterobacter cloacae or *Pseudomonas fluorescens* are involved in reducing CrO_4^{2-} to Cr $(OH)_3$. *Pseudomonas putida*, *Staphylococcus aureus*, *Escherichia coli*, *Bacillus subtilis* are associated with chromium resistance. These are related to plasmid-encoded energy dependent flow systems and include ions/chemical pumps and ATPases (Rajendran et al., 2003).

17.8 Cadmium

Ion-exchange mechanism is utilized *Saccharomyces cerevisiae* for the removal of Cd^{2+} and thus acts as biosorbent. Genetically modified *Ralstonia eutropha* is used to decrease the toxic effect of Cd^{2+} by expressing mouse metallothionein on the cell surface. Ectomycorrhizal fungi (*Suillus granulates* and *Paxillus involutus*) are able to release elements of wood ash and apatite (K, Ti, Pb, Mn, Ca) and then synthesize them into mycelia (Wallander et al., 2003). External root mycelium and mycorrhizal fungi have the ability to dissolve a variety of cadmium-bearing minerals (Leyval & Joner, 2001). *Deinococcus radiodurans* (radiation resistant bacterium) is genetically engineered to naturally reduces Cr^{4+} to Cr^{3+}. It has been done for complete toluene (fuel hydrocarbon) degradation by cloned genes of *xyl* and *tod* operons of *Pseudomonas putida* (Brim et al., 2006). *Staphylococcus aureus*, *Bacillus subtilis*, *Escherichia coli*, *Pseudomonas putida*, are associated with cadmium resistance. These are related to plasmid-encoded energy dependent flow systems and ions/chemical pumps and ATPases (Rajendran et al., 2003).

17.9 Iron

Iron can be oxidized enzymatically by specific bacteria, for example acidic like *Acidithiobacillus ferrooxidans*, *Leptospirillum ferrooxidans Acidianus brierleyi*, *Sulfobacillus thermosulfidooxidans*, and *Sulfolobus* spp. (Gadd, 2010). *Leptothrix* spp., *Aspergillus*, *Cladosporium*, *Gallionella* spp. and *Alternaria*, also exhibit iron resistance (Ehrlich & Newman, 2009).

17.10 Mercury

Mercury resistant fungi *Verticillum terrestre*, *Neocosmospora vasinfecta*, and *Hymenoscyphus ericae*, convert Hg^{2+} to a nontoxic state, from a toxic state. Bacterium *Deinococcus geothermalis* was genetically engineered and mer operon from *E. coli* was added that coded for Hg^{2+} reduction. Reports showed a decrease in mercury at elevated temperatures (Brim et al., 2003). The mercury-resistant bacteria *Cupriavidus metallidurans* MSR33 was genetically modified microorganism by transformed the pTP6 plasmid. It provided genes (*merB* and *merG*) that aid in the regulation of mercury biodegradation along with the synthesis of mercuric reductase (MerA) and organic lysine protein (MerB) (Dixit et al., 2015). In general, there are two different methods for the decomposition of mercury by bacteria like *Klebsiella pneumonia* M426: volatilization of mercury by reduction of Hg^{2+} to Hg (0) and precipitation of mercury in the form of insoluble mercury. The plant peptide Phytochelatin 20 activation on the cell surface of *Escherichia coli* and *Moraxella* sp. Mercury collect 25 times extra than wild-type strains (Bae et al., 2001).

17.11 Uranium

Geobacter species can change the state of Uranium from its soluble state U^{6+} to insoluble state U^{4+} and immobilize them. *Citrobacter* spp. and *Glomus intraradices* also impart resistance to uranium in contaminated soil (Gadd, 2010).

18 Conclusions

Heavy metals are an essential and important trace element, but as these heavy metals increase in concentration due to natural or industrial activities, they become toxic to many microbes. On the other hand, microbes have adapted to tolerate minerals or can even use them for growth. Hence this interaction between microbes and minerals on environmental matrices is an essential part of the Earth's biogeochemical cycle. This type of activity has both negative and positive effects on the environment. Processes, packing and fixation, disintegration and binding are subject to microbial metabolism/ catabolism. These activities are the basis of microbial biological therapy. Bioremediation with microbes has shown an excellent mineral-microbial interaction effect which will be most promising when genetic engineering comes into play.

References

Adriano, D. C. (2001). *Trace elements in the terrestrial environment: Biogeochemistry, bioavailability and risks of metals* (2nd ed.). Springer. https://doi.org/10.1007/978-0-387-21510-5

Ahuja, R. (2016). Characterization of heavy metals in contaminated agricultural soil. *International Journal of Theoretical and Applied Sciences, 8*(1), 38–40.

Ali, M., & Bhat, A. K. (2014). Soil microbiological indices of polluted soils of industrial belts of Jammu, India. *International Journal of Current Microbiology and Applied Sciences, 3*(1), 559–576.

Alibhai, K., Leak, D. J., Dudeney, A. W. L., Agatzini, S., & Tzeferis, P. (1991). Microbial leaching of nickel from low grade Greek laterites. In R. W. Smith & M. Misra (Eds.), *Mineral bioprocessing* (pp. 191–205). The Minerals, Metals, and Materials Society.

Andrews, D., & Walker, B. (2016). U.S. EPA, Occurrence data for the unregulated contaminant monitoring rule. www.epa.gov/dwucmr/occurrence-data-unregulated-contaminant-monitoring-rule#3

Bae, W., Mehra, R. K., Mulchandani, A., et al. (2001). Genetic engineering of Escherichia coli for enhanced uptake and bioaccumulation of mercury. *Applied and Environmental Microbiology, 67*(11), 5335–5338. https://doi.org/10.1128/AEM.67.11.5335-5338.2001

Baldi, F., Filippelli, M., & Olson, G. J. (1989). Biotransformation of mercury by bacteria isolated from a river collecting cinnabar mine waters. *Microbial Ecology, 17*, 263–274. https://doi.org/10.1007/BF02012839

Banerjee, S., Das, B., Umlong, I. M., et al. (2011). Heavy metal contaminants of underground water in Indo Bangla Border Districts of Tripura, India. *International Journal of ChemTech Research, 3*(1), 516–522.

Banerjee, S., Kumar, A., Maiti, S. K., et al. (2016). Seasonal variation in heavy metal contaminations in water and sediments of Jamshedpur stretch of Subarnarekha River, India. *Environment and Earth Science, 75*(3), 1–12. https://doi.org/10.1007/s12665-015-4990-6

Banerjee, S., Pramanik, A., Sengupta, S., et al. (2017). Distribution and source identification of heavy metal concentration in Chilika Lake, Odisha India: An assessment over salinity gradient. *Current Science, 112*(1), 87. https://doi.org/10.18520/cs/v112/i01/87-94

Banfield, J. F., Cervini-Silva, J., & Nealson, K. H. (Eds.). (2005). *Molecular geomicrobiology, Reviews in mineralogy and geochemistry* (Vol. 59). Mineralogical Society of America.

Barkay, T., & Wagner-Dobler, I. (2005). Microbial transformations of mercury: Potentials, challenges, and achievements in controlling mercury toxicity in the environment. *Advances in Applied Microbiology, 57*, 1–52. https://doi.org/10.1016/S0065-2164(05)57001-1

Bell, N., & Quan, L. (1997). The application of Bactech (Australia) Ltd. technology for processing refractory gold ores at Youanmi Gold Mine. In *Conference proceedings. International Bio-hydrometallurgy Symposium IBS97 BIOMINE 97. Australian Mineral Foundation, Glenside, South Australia* (pp. M2.3.1–M2.3.9).

Beveridge, T. J., & Doyle, R. (1989). *Metal ions and bacteria*. Wiley.

Beveridge, T. J., & Koval, S. F. (1981). Binding of metals to cell envelopes of *Escherichia coli* K-12. *Applied and Environmental Microbiology, 42*, 325–335.

Beveridge, T. J., Forsberg, C. W., & Doyle, R. C. (1982). Major sites of metal binding in *Bacillus licheniformis* walls. *Journal of Bacteriology, 150*, 1438–1448. https://doi.org/10.1128/JB.150.3.1438-1448.1982

Blanco, A. (2000). Immobilization of nonviable cyanobacteria and their use for heavy metal adsorption from water. In E. J. Oluguin, G. Sanchez, & E. Hernandez (Eds.), *Environmental biotechnology and cleaner bioprocesses* (p. 135). Taylor and Francis.

Borremans, B., Hobman, J. L., Provoost, A., et al. (2001). Cloning and functional analysis of the pbr lead resistance determinant of *Ralstonia metallidurans* CH34. *Journal of Bacteriology, 183*(19), 5651–5658. https://doi.org/10.1128/JB.183.19.5651-5658.2001

Bosecker, K. (1993). Biosorption of heavy metals by filamentous fungi. In A. E. Torma, M. L. Apel, & C. L. Brierley (Eds.), *Biohydro-metallurgical technologies* (Vol. II, pp. 55–64). The Minerals, Metals and Materials Society.

Boyington, J. C., Gladyshev, V. N., Kangulov, S. V., et al. (1997). Crystal structure of formate dehydrogenase H: Catalysis involving Mo, molybdopterin, selenocysteine, and an Fe_4S_4 cluster. *Science, 275*, 1305–1308. https://doi.org/10.1126/science.275.5304.1305

Brierley, C. L., & Brierley, J. A. (1982). Anaerobic reduction of molybdenum by *Sulfolobus* species. *Zentralblatt für Bakteriologie Mikrobiologie und Hygiene: I. Abt. Originale C: Allgemeine, angewandte und ökologische Mikrobiologie, 3*, 289–294.

Brierley, J. A., Brierley, C. L., & Goyak, G. M. (1986). AMT-BIOCLAIM: A new wastewater treatment and metal recovery technology. In R. W. Lawrence, R. M. R. Branion, & H. G. Ebner (Eds.), *Fundamental and applied biohydrometallurgy* (pp. 291–308). Elsevier.

Briggs, A. P., & Millard, M. (1997). Cobalt recovery using bacterial leaching at the Kasese Project, Uganda. In *Conference proceedings. International Biohydrometallurgy Symposium IBS97 BIOMINE 97* (pp. M2.4.1–M2.4.11). Australian Mineral Foundation.

Brim, H., Venkateswaran, A., Kostandarithes, H. M., et al. (2003). Engineering Deinococcus geothermalis for bioremediation of high-temperature radioactive waste environments. *Applied and Environmental Microbiology, 69*(8), 4575–4582. https://doi.org/10.1128/AEM.69.8.4575-4582.2003

Brim, H., Osborne, J. P., Kostandarithes, H. M., et al. (2006). Deinococcus radiodurans engineered for complete toluene degradation facilities Cr(IV) reduction. *Microbiology, 152*, 2469–2477. https://doi.org/10.1099/mic.0.29009-0

Brus, D. J., Li, Z., Song, J., et al. (2009). Predictions of spatially averaged cadmium contents in rice grains in the Fuyang Valley, PR China. *Journal of Environmental Quality, 38*(3), 1126–1136. https://doi.org/10.2134/jeq2008.0228

Cantafio, A., Hagen, K. D., Lewis, G. E., et al. (1996). Pilot-scale selenium bioremediation of San Joaquin Drainage water with *Thauera selenatis*. *Applied and Environmental Microbiology, 62*, 3298–3303. https://doi.org/10.1128/aem.62.9.3298-3303.1996

Cavaco, L. M., Hasman, H., Stegger, M., et al. (2010). Cloning and occurrence of czrC, a gene conferring cadmium and zinc resistance in methicillin-resistant Staphylococcus aureus CC398 isolates. *Antimicrobial Agents and Chemotherapy, 54*(9), 3605–3608. https://doi.org/10.1128/AAC.00058-10

Cha, J. S., & Cooksey, D. A. (1991). Copper resistance in pseudomonas syringae mediated by periplasmic and outer membrane proteins. *Proceedings of the National Academy of Sciences, 88*(20), 8915–8919. https://doi.org/10.1073/pnas.88.20.8915

Chaitanya, I., Satyaprakash, M., & Reddy, T. B. (2016). Bioaccumulation of heavy metals in marine fish samples at Visakhapatnam and Bheemili region, north east coast of Andhra Pradesh, India. *International Journal of Science, Environment and Technology, 5*, 1718–1729.

Charan, P. D., Singh, M., Rakhecha, P., et al. (2015). Study of heavy metals concentration in ground water samples collected from Bikaner city, Rajasthan. *International Journal of Engineering Research and Management Technology, 2*(5), 16–17.

Chau, Y. K., Wong, P. T. S., Silverberg, B. A., et al. (1976). Methylation of selenium in the aquatic environment. *Science, 192*, 1130–1131. https://doi.org/10.1126/science.192.4244.1130

Coleman, M. L. L., Hedrick, D. B., Lovley, D. R., et al. (1993). Reduction of Fe(III) in sediments by sulfate-reducing bacteria. *Nature (London), 361*, 436–438. https://doi.org/10.1038/361436a0

Das, P., Kumar, M., & Sarma, K. P. (2015). Speciation of heavy metals in surface sediment of the Brahmaputra River, Assam, India. *Journal of Environmental Research And Development, 9* (3A), 944–952.

Dash, A., Das, H. K., & Mishra, B. (2016). Heavy metals contamination of ground water in and around Joda of Keonjhar district, Odisha, India. *IOSR Journal of Environmental Science, Toxicology and Food Technology, 10*(10), 44–50. https://doi.org/10.9790/2402-1010024450

DeLeo, P. C., & Ehrlich, H. L. (1994). Reduction of hexavalent chromium by *Pseudomonas fluorescens* LB300 in batch and continuous culture. *Applied and Environmental Microbiology, 40*, 756–759. https://doi.org/10.1007/BF00173341

DeMoll-Decker, H., & Macy, J. M. (1993). The periplasmic nitrite reductase of *Thauera selenatis* may catalyze the reduction of selenite to elemental selenium. *Archives of Microbiology, 160*, 241–247. https://doi.org/10.1007/BF00249131

Dew, D. W., & Miller, D. M. (1997). The BioNIC process. Bioleaching of minerals sulfide concentrates for recovery of nickel. In *Conference Proceedings. International Bio-hydrometallurgy Symposium IBS97 BIOMINE 97* (pp. M7.1.1–M7.1.9). Australian Mineral Foundation.

Dey, R., & Choudhary, S. K. (2015). Heavy metal in sediments of Kabar Lake, a tropical wetland in Begusarai district of Bihar. *Ecology, Environment & Conservation, 22*(3), 1509–1515.

Dhankher, O. P., Li, Y. J., Rosen, B. P., et al. (2002). Engineering tolerance and hyperaccumulation of arsenic in plants by combining arsenate reductase and gamma-glutamylcysteine synthetase expression. *Nature Biotechnology, 20*, 1140–1145. https://doi.org/10.1038/nbt747

Dheeba, B., & Sampathkumar, P. (2012). Evaluation of heavy metal contamination in surface soil around industrial area, Tamil Nadu, India. *International Journal of ChemTech Research, 4*(3), 1229–1240.

Dixit, R., Malaviya, D., Pandiyan, K., et al. (2015). Bioremediation of heavy metals from soil and aquatic environment: An overview of principles and criteria of fundamental processes. *Sustainability, 7*(2), 2189–2212. https://doi.org/10.3390/su7022189

Doyle, R. J. (1991). How cell walls of gram-positive bacteria interact with metal ions. In T. J. Beveridge & R. J. Doyle (Eds.), *Metal ions and bacteria* (pp. 275–293). Wiley.

Edyvean, R. G. J. (1995). The influence of marine macrofouling on corrosion. In C. C. Gaylarde & H. A. Videla (Eds.), *Bioextraction and biodeterioration of metals* (pp. 169–196). Cambridge University Press.

Ehrlich, H. L. (1996a). *Geomicrobiology*. Dekker. https://doi.org/10.4319/lo.1982.27.5.0984

Ehrlich, H. L. (1996b). Microbes and metals. *Applied Microbiology and Biotechnology, 48*(6), 687–692. https://doi.org/10.1007/s002530051116

Ehrlich, H. L., & Brierley, C. L. (1990). *Microbial mineral recovery*. Mc-Graw-Hill.

Ehrlich, H. L., & Newman, D. K. (2009). *Geomicrobiology* (5th ed.). CRC Press/Taylor & Francis.

Fasinu, P. S., & Orisakwe, O. E. (2013). Heavy metal pollution in sub-Saharan Africa and possible implications in cancer epidemiology. *Asian Pacific Journal of Cancer Prevention, 14*(6), 3393–3402. https://doi.org/10.7314/apjcp.2013.14.6.3393

Fazil, M. I., Iqbalb, M. A., & Abdullah, S. (2012). A study on heavy metal ion contamination of groundwater reserves in Beed City, Maharashtra, India. *Bulletin of Environment, Pharmacology and Life Sciences, 1*(1), 18–21.

Fernandez-Cadena, J. C., Andrade, S., Silva-Coello, C. L., & De la Iglesia, R. (2014). Heavy metal concentration in mangrove surface sediments from the north-west coast of South America. *Marine Pollution Bulletin, 82*(1), 221–226. https://doi.org/10.1016/j.marpolbul.2014.03.016

Ferris, F. G. (1991). Metallic ion interactions with the outer membrane of gram-negative bacteria. In T. J. Beveridge & R. J. Doyle (Eds.), *Metal ions and bacteria* (pp. 295–323). Wiley.

Ferris, F. G., Schultze, S., Witten, T. C., Fyfe, W. S., & Beveridge, T. J. (1989). Metal interactions with microbial biofilms in acidic and neutral pH environments. *Applied and Environmental Microbiology, 55*, 1249–1257. https://doi.org/10.1128/AEM.55.5.1249-1257.1989

Ferry, J. G. (1995). CO dehydrogenases. *Annual Review of Microbiology, 49*, 305–333. https://doi.org/10.1146/annurev.mi.49.100195.001513

Fowler, B. A., Hildebrand, C. E., Kojima, Y., & Webb, M. (1987). Nomenclature of metallothionein. In J. H. R. Kagi & Y. Kojima (Eds.), *Metallothionein II* (pp. 19–22). https://doi.org/10.1007/978-3-0348-6784-9_2

Frankenberger, W. T., Jr., & Karlson, U. (1992). Dissipation of soil selenium by microbial volatilization. In D. C. Adriano (Ed.), *Bio-geochemistry of trace metals* (pp. 365–381). Lewis.

Fridovich, I. (1978). The biology of oxygen radicals. *Science, 201*, 875–880. https://doi.org/10.1126/science.210504

Gadd, G. M. (1993). Interactions of fungi with toxic metals. *The New Phytologist, 124*, 25–60. https://doi.org/10.1007/978-1-4899-0981-7_28

Gadd, G. M. (2010). Metals, minerals and microbes: Geomicrobiology and bioremediation. *Microbiology, 156*(3), 609–643. https://doi.org/10.1099/mic.0.037143-0

Gadd, G. M., & Raven, J. A. (2010). Geomicrobiology of eukaryotic microorganisms. *Geomicrobiology Journal, 27*(6–7), 491–519. https://doi.org/10.1080/01490451003703006

Gadd, G. M., Burford, E. P., Fomina, M., & Melville, K. (2007). Mineral transformation and biogeochemical cycles: A geomycological perspective. In G. M. Gadd, P. Dyer, & S. Watkinson (Eds.), *Fungi in the environment* (pp. 78–111). Cambridge University Press. https://doi.org/10.1017/CBO9780511541797.006

Gaylarde, C. C., & Videla, H. A. (1995). *Bioextraction and biodeterioration of metals*. Cambridge University Press. https://doi.org/10.1017/CBO9780511541797.006

Ghiorse, W. C., & Ehrlich, H. L. (1992). Microbial biomineralization of iron and manganese. In H. C. W. Skinner & R. W. Fitzpatrick (Eds.), *Biomineralization. Processes of iron and manganese. Modern and ancient environments. Catena supplement 21* (pp. 75–99). Catena Cremlingen-Destedt.

Giri, S., Mahato, M. K., Singh, G., et al. (2012). Risk assessment due to intake of heavy metals through the ingestion of groundwater around two proposed uranium mining areas in Jharkhand, India. *Environmental Monitoring and Assessment, 184*(3), 1351–1358. https://doi.org/10.1007/s10661-011-2045-3

Goher, M. E., Farhat, H. I., Abdo, M. H., et al. (2014). Metal pollution assessment in the surface sediment of Lake Nasser, Egypt. *Egyptian Journal of Aquatic Research, 40*(3), 213–224. https://doi.org/10.1016/j.ejar.2014.09.004

Gomathy, M., & Sabarinathan, K. G. (2010). Microbial mechanisms of heavy metal tolerance-A review. *Agricultural Reviews, 31*(2), 133–138.

Govil, P., Reddy, G., & Krishna, A. (2001). Contamination of soil due to heavy metals in the Patancheru industrial development area, Andhra Pradesh, India. *Environmental Geology, 41*(3), 461–469. https://doi.org/10.1007/s002540100415

Groudev, S. N., & Groudeva, V. I. (1986). Biological leaching of aluminum from clays. Workshop on biotechnology for the mining, metal-refining and fossil fuel industries. *Biotechnology and Bioengineering Symposium, 16*, 91–99.

Guard, H. E., Cobet, A. B., & Coleman, W. M., III. (1981). Methylation of trimethyltin compounds by estuarine sediments. *Science, 213*, 770–771. https://doi.org/10.1126/science.213.4509.770

Gupta, S., Bhatnagar, M., & Jain, R. (2003). Physico-chemical characteristics and analysis of Fe and Zn in tubewell water and sewage water of Bikaner City. *Asian Journal of Chemistry, 15*(2), 727.

Hallas, L. E., Means, J. C., & Cooney, J. J. (1982). Methylation of tin by estuarine microorganisms. *Science, 215*, 1505–1507. https://doi.org/10.1126/science.215.4539.1505

Haloi, N., & Sarma, H. P. (2012). Heavy metal contaminations in the groundwater of Brahmaputra flood plain: An assessment of water quality in Barpeta District, Assam (India). *Environmental Monitoring and Assessment, 184*(10), 6229–6237. https://doi.org/10.1007/s10661-011-2415-x

Harikrishnan, N., Suresh Gandhi, M., Chandrasekaran, A., & Ravisankar, R. (2015). Assessment of heavy metal pollution and potential ecological risk of sediments of East Coast of Tamil Nadu by Energy Dispersive X-Ray Fluorescence Spectroscopy (EDXRF) and Sediment Quality

Guidelines (SQGS). *Journal of Heavy Metal Toxicity and Diseases, 3*, 1–7. https://doi.org/10.21767/2473-6457.100003

He, Z. L., Yang, X. E., & Stoffella, P. J. (2005). Trace elements in agroecosystems and impacts on the environment. *Journal of Trace Elements in Medicine and Biology, 19*(2), 125–140. https://doi.org/10.1016/j.jtemb.2005.02.010

Hejabi, A. T., Basavarajappa, H. T., Karbassi, A. R., et al. (2011). Heavy metal pollution in water and sediments in the Kabini River, Karnataka, India. *Environmental Monitoring and Assessment, 182*(1), 1–13. https://doi.org/10.1007/s10661-010-1854-0

Herawati, N., Suzuki, S., Hayashi, K., et al. (2000). Cadmium, copper, and zinc levels in rice and soil of Japan, Indonesia, and China by soil type. *Bulletin of Environmental Contamination and Toxicology, 64*(1), 33–39. https://doi.org/10.1007/s001289910006

Hoffland, E., Kuyper, T. W., et al. (2004). The role of fungi in weathering. *Frontiers in Ecology and the Environment, 2*(5), 258–264.

Holmgren, G. G. S., Meyer, M. W., Chaney, R. L., et al. (1993). Cadmium, lead, zinc, copper, and nickel in agricultural soils of the United States of America. *Journal of Environmental Quality, 22*(2), 335–348. https://doi.org/10.2134/jeq1993.00472425002200020015x

Ilyaletdinov, A. N., & Abdrashitova, S. A. (1981). Autotrophic oxidation of arsenic by a culture of *Pseudomonas arsenitoxidans*. *Mikrobiologiya, 50*, 197–204. PMID: 7242389.

Ingledew, W. J. (1982). *Thiobacillus ferrooxidans*. The bioenergetics of an acidophilic chemolithotroph. *Biochimica et Biophysica Acta, 683*, 89–117. https://doi.org/10.1016/0304-4173(82)90007-6

Jain, C. K., Bandyopadhyay, A., & Bhadra, A. (2010). Assessment of ground water quality for drinking purpose, District Nainital, Uttarakhand, India. *Environmental Monitoring and Assessment, 166*(1), 663–676. https://doi.org/10.1007/s10661-009-1031-5

Jamir, T. T., Devi, W. B., Singh, U. I., et al. (2011). Lead, iron and manganese contamination in spring pond and well water in Nagaland, one of the seven North-Eastern States of India, A future danger. *Journal of Chemical and Pharmaceutical Research, 3*, 403–411.

Jarosławiecka, A., & Piotrowska-Seget, Z. (2014). Lead resistance in microorganisms. *Microbiology, 160*(1), 12–25. https://doi.org/10.1099/mic.0.070284-0

Ji, G., & Silver, S. (1995). Bacterial resistance mechanisms for heavy metals of environmental concern. *Journal of Industrial Microbiology, 14*, 61–75. https://doi.org/10.1007/BF01569887

Jinwal, A., & Dixit, S. (2008). Pre and post monsoon variation in physio-chemical characteristic in groundwater quality in Bhopal, India. *Asian Journal of Experimental Sciences, 22*, 311–316.

Jo, I. S., & Koh, M. H. (2004). Chemical change in agricultural soils of Korea: Date review and suggested countermeasures. *Environmental Geochemistry and Health, 26*, 105–107. https://doi.org/10.1023/b:egah.0000039573.05245.cc

Jung, M. C. (2008). Heavy metal concentrations in soils and factors affecting metal uptake by plants in the vicinity of a Korean Cu-W mine. *Sensors, 8*(4), 2413–2423. https://doi.org/10.3390/s8042413

Kachenko, A. G., & Singh, B. (2005). Heavy metals contamination in vegetables grown in urban and metal smelter contaminated sites in Australia. *Water, Air, and Soil Pollution, 169*(1), 101–123. https://doi.org/10.1007/s11270-006-2027-1

Kashyap, R., & Vera, K. S. (2015). Seasonal variation of certain heavy metals in Kuntbhyog lake of Himachal Pradesh, India. *Journal of Environment, Ecology, Family and Urban Studies, 1*(1), 15–26.

Kaur, J., & Maddela, N. R. (2021). Microbial bioremediation: A cutting-edge technology for xenobiotic removal. In N. R. Maddela, L. C. García Cruzatty, & S. Chakraborty (Eds.), *Advances in the domain of environmental biotechnology. Environmental and microbial biotechnology*. Springer. https://doi.org/10.1007/978-981-15-8999-7_16

Khan, A. G., Kuek, C., Chaudhry, T. M., et al. (2000). Role of plants, mycorrhizae and phytochelators in heavy metal contaminated land remediation. *Chemosphere, 41*(1), 197–207. https://doi.org/10.1016/S0045-6535(99)00412-9

Kierans, M., Staines, A. M., Bennett, H., et al. (1991). Silver tolerance and accumulation in yeasts. *Biology of Metals, 4*, 100–106. https://doi.org/10.1007/BF01135386
Kim, B. H., & Gadd, G. M. (2008). *Bacterial physiology and metabolism*. Cambridge University Press.
Krishna, A. K., & Govil, P. K. (2007). Soil contamination due to heavy metals from an industrial area of Surat, Gujarat, Western India. *Environmental Monitoring and Assessment, 124*, 263–275. https://doi.org/10.1007/s10661-006-9224-7
Kulshreshtha, A., Soni, R. K., & Shinde, C. P. (2015). Quantitative estimation of heavy metals in ground water in Meerut Region in Uttar Pradesh. *Journal of Applied Pharmaceutical Science, 8* (8), 46–49. https://doi.org/10.9790/5736-08824649
Kumar, V., & Chopra, A. K. (2015). Heavy metals accumulation in soil and agricultural crops grown in the province of Asahi India Glass Ltd., Haridwar (Uttarakhand), India. *Advances in Crop Science and Technology, 4*(1), 203. https://doi.org/10.4172/2329-8863.1000203
Kumar, S. D., & Srikantaswamy, S. (2012). Heavy metals pollution assessment in industrial area soil of Mysore city, Karnataka, India. *International Journal of Applied Sciences and Engineering Research, 1*(4), 604–611. https://doi.org/10.6088/ijaser.0020101062
Kumar, S. K., Magesh, N. S., & Chandrasekar, N. (2012). Trace element concentration in groundwater, Tuticorin City, Tamil Nadu, India. *Bulletin of Environmental Contamination and Toxicology, 88*(6), 876–879. https://doi.org/10.1007/s00128-012-0614-y
Kumar, B., Kumari, S., & Flores, L. C. (2014). Plant mediated detoxification of mercury and lead. *Arabian Journal of Chemistry, 10*, S2335. https://doi.org/10.1016/j.arabjc.2013.08.010
Landa, E. R. (2005). Microbial biogeochemistry of uranium mill tailings. *Advances in Applied Microbiology, 57*, 113–130. https://doi.org/10.1016/S0065-2164(05)57004-7
Leyval, C., & Joner, E. J. (2001). Bioavailability of heavy metals in the mycorrhizosphere. In G. R. Gobran, W. W. Wenzel, & E. Lombi (Eds.), *Trace elements in the rhizosphere* (pp. 165–185). CRC Press.
Lloyd, J. R., & Lovley, D. R. (2001). Microbial detoxification of metals and radionuclides. *Current Opinion in Biotechnology, 12*(3), 248–253. https://doi.org/10.1016/s0958-1669(00)00207-x
Lovley, D. R. (1987). Organic matter mineralization with the reduction of ferric iron: A review. *Geomicrobiology Journal, 5*, 375–399. https://doi.org/10.1080/01490458709385975
Lovley, D. R. (1991). Dissimilatory Fe(III) and Mn(IV) reduction. *Microbiological Reviews, 55*, 259–287. https://doi.org/10.1016/S0065-2911(04)49005-5
Lovley, D. R. (1993). Dissimilatory metal reduction. *Annual Review of Microbiology, 47*, 263–290. https://doi.org/10.1146/annurev.mi.47.100193.001403
Lovley, D. R. (1995). Bioremediation of organic and inorganic metal contaminants with dissimilatory metal reduction. *Journal of Industrial Microbiology, 14*, 85–93. https://doi.org/10.1007/BF01569889
Lovley, D. R., & Coates, J. D. (1997). Bioremediation of metal contamination. *Current Opinion in Biotechnology, 8*, 285–289. https://doi.org/10.1016/s0958-1669(97)80005-5
Lovley, D. R., & Phillips, E. J. P. (1988). Novel mode of microbial energy metabolism: Organic carbon oxidation coupled to dissimilatory reduction of iron or manganese. *Applied and Environmental Microbiology, 54*, 1472–1480. https://doi.org/10.1128/AEM.54.6.1472-1480.1988
Lovley, D. R., & Phillips, E. J. P. (1992). Bioremediation of uranium contamination with enzymatic uranium reduction. *Environmental Science & Technology, 26*, 2228–2234. https://doi.org/10.1021/es00035a023
Luo, Y. M., & Teng, Y. (2006). Status of soil pollution degradation and countermeasures in China. *Soil Science, 38*(5), 505–508.
Lyalikova, N. N., Vedenina, I. Y., & Romanova, A. K. (1976). Assimilation of carbon dioxide by a culture of *Stibiobacter senarmontii*. *Mikrobiologiya, 45*, 552–554. (Microbiology NY 45: 476–477). PMID: 1004256.
Macaskie, L. E., Dean, A. C. R., Cheetham, A. K., et al. (1987). Cadmium accumulation by a *Citrobacter* sp.: The chemical nature of the accumulated metal precipitate and its location on the

bacterial cells. *Journal of General Microbiology, 133*, 539–544. https://doi.org/10.1099/00221287-133-3-539

Macaskie, L. E., Empson, R. M., Cheetham, A. K., et al. (1992). Uranium bioaccumulation by a *Citrobacter* sp. as a result of enzymically mediated growth of polycrystalline HUO2-PO4. *Science, 257*, 782–784. https://doi.org/10.1126/science.1496397

Machender, G., Dhakate, R., Mallikharjuna, S. T., et al. (2012). Heavy metal contamination in sediments of Balanagar industrial area, Hyderabad, Andhra Pradesh, India. *Arabian Journal of Geosciences, 7*(2), 513–525. https://doi.org/10.1007/s12517-012-0759-3

Maddela, N. R., Reyes, J. J. M., Viafara, D., & Gooty, J. M. (2015). Biosorption of copper (II) by microorganisms isolated from crude oil contaminated soil. *Soil and Sediment Contamination: An International Journal, 24*(8), 898–908.

Maddela, N. R., Burgos, R., Kadiyala, V., Banganegiri, M., & Carrión, A. R. (2016). Removal of crude oil from soil by using novel microorganisms of Ecuador soils: Solid and slurry phase methods. *International Biodeterioration and Biodegradation, 108*, 85–90.

Maddela, N. R., Scalvenzi, L., & Venkateswarlu, K. (2017). Microbial degradation of total petroleum hydrocarbons in crude oil: A field-scale study at the low-land rainforest of Ecuador. *Environmental Technology, 38*, 2543–2550.

Marg, B. Z. (2011). *Hazardous metals and minerals pollution in India: Sources, toxicity and management. A position paper*. Indian National Science Academy.

Maurya, P. K., & Malik, D. S. (2016). Distribution of heavy metals in water, sediments and fish tissue (*Heteropneustis fossilis*) in Kali River of western U.P. India. *International Journal of Fisheries and Aquatic Studies, 4*(2), 208–215.

McCready, R. G. L., & Gould, W. D. (1990). Bioleaching of uranium. In H. L. Ehrlich & C. L. Brierley (Eds.), *Microbial mineral recovery* (pp. 107–125). McGraw-Hill.

Meitei, L. S., & Rakesh, K. (2013). A comparative study of the ground and surface water quality with reference to heavy metal concentrations in the Imphal valley Manipur, India. *International Journal of Environmental Sciences, 3*(6), 1857.

Mohankumar, K., Hariharan, V., & Rao, N. P. (2016). Heavy metal contamination in groundwater around industrial estate vs residential areas in Coimbatore, India. *Journal of Clinical and Diagnostic Research, 10*(4), BC05. https://doi.org/10.7860/jcdr/2016/15943.7527

Mohsenzadeh, F., & Shahrokhi, F. (2014). Biological removing of Cadmium from contaminated media by fungal biomass of Trichoderma species. *Journal of Environmental Health Science and Engineering, 12*(1), 102. https://doi.org/10.1186/2052-336X-12-102

Moller, A. K., Barkay, T., Hansen, M. A., et al. (2014). Mercuric reductase genes (merA) and mercury resistance plasmids in High Arctic snow, freshwater and sea-ice brine. *FEMS Microbiology Ecology, 87*(1), 52–63. https://doi.org/10.1111/1574-6941.12189

Morais, P. V., Branco, R., & Francisco, R. (2011). Chromium resistance strategies and toxicity: What makes Ochrobactrum tritici 5bvl1 a strain highly resistant. *Biometals, 24*(3), 401–410. https://doi.org/10.1007/s10534-011-9446-1

Nagajyoti, P. C., Lee, K. D., & Sreekanth, T. V. M. (2010). Heavy metals, occurrence and toxicity for plants: A review. *Environmental Chemistry Letters, 8*, 199–216. https://doi.org/10.1007/s10311-010-0297-8

Naik, M. M., Pandey, A., & Dubey, S. K. (2012). Biological characterization of lead enhanced exopolysaccharide produced by a lead resistant Enterobacter cloacae strain P2B. *Biodegradation, 23*, 775–783. https://doi.org/10.1007/s10532-012-9552-y

Nanda, P. (2015). Bioaccumulation of heavy metals and physiological response in anabas testudineus on exposure to paper mill effluent. *Journal of Environmental & Analytical Toxicology, 5*(1), 1. https://doi.org/10.4172/2161-0525.1000244

Nath, T. N. (2013). Heavy metals contamination of tea estates soil in Sivasagar and Dibrugarh district of Assam, India. *International Journal of Advanced Research and Technology, 2*(4), 2278–7763.

Neilands, J. B. (Ed.). (1974). *Microbial iron metabolism*. Academic Press.

Olson, G. J. (1994). Microbial oxidation of gold ore and gold bio-leaching. *FEMS Microbiology Letters, 119*, 1–6. https://doi.org/10.1111/j.1574-6968.1994.tb06858.x

Orme-Johnson, W. H. (1992). Nitrogenase structure: Where to now? *Science, 257*, 1639–1640. https://doi.org/10.1126/science.1529351

Pacyna, J. M., & Nriagu, J. O. (1988). Quantitative assessment of worldwide contamination of air, water and soils by trace metals. *Nature, 333*(6169), 134–139. https://doi.org/10.1038/333134a0

Patel, M., & Manoj, K. (2015). Assessment of heavy metals in potable ground water of Olpad Taluka, Surat, Gujarat, India. *International Journal of Current Microbiology and Applied Sciences, 4*, 124–128.

Patel, K. S., Shrivas, K., & Hoffmann, P. (2006). A survey of lead pollution in Chhattisgarh State, central India. *Environmental Geochemistry and Health, 28*(1), 11–17. https://doi.org/10.1007/s10653-005-9006-0

Pethkar, A. V., & Paknikar, K. M. (1997). Recovery of silver from low-tenor solutions using *Cladosporium cladosporioides* biomass beads. In *Conference Proceedings. International Biohydro-metallurgy Symposium IBS97 BIOMINE 97* (pp. PE8.1–PE8.2). Australian Mineral Foundation.

Phillips, E. J. P., Landa, E. R., & Lovely, D. R. (1995). Remediation of uranium contaminated soils with bicarbonate extraction and microbial U(VI) reduction. *Journal of Industrial Microbiology, 14*, 203–207. https://doi.org/10.1007/BF01569928

Pope, D. H., Duquette, D. J., Johannes, A. H., et al. (1984). Microbially influenced corrosion of industrial alloys. *Materials Performance, 23*, 14–18. https://doi.org/10.5772/intechopen.70735

Prasanth, K. M., Sreekala, P. P., Sandeep, S., et al. (2013). Heavy metals and its fractions in soils of Koratty Region, Kerala. *Research Journal of Recent Sciences, 2*, 171–176.

Raja, I. A., Khan, M. Y., Khan, N. A., et al. (2013). Assessment of some metals in the drinking water of Dal Lake Kashmir. *Nature and Science, 11*(3), 63–64.

Rajendran, P., Muthukrishnan, J., & Gunasekaran, P. (2003). Microbes in heavy metal remediation. *Indian Journal of Experimental Biology, 41*(9), 935–944. PMID: 15242287.

Rech, S., & Macy, J. M. (1992). The terminal reductases for selenate and nitrate respiration *in Thauera selenatis* are two distinct enzymes. *Journal of Bacteriology, 174*, 7361–7320. https://doi.org/10.1128/jb.174.22.7316-7320.1992

Roane, T. M., & Pepper, I. L. (2000). Microorganisms and metal pollutants. In *Environmental Microbiology* (pp. 421–441). https://doi.org/10.1016/B978-0-12-370519-8.00021-3

Robinson, J. B., & Tuovinen, O. H. (1984). Mechanism of microbial resistance and detoxification of mercury and organomercury compounds: Physiological, biochemical, and genetic analyses. *Microbiological Reviews, 48*, 95–124. PMCID: PMC373215.

Robson, R. L., Eady, R. R., Richardson, T. H., et al. (1986). The alternative nitrogenase of *Azotobacter chroococcum* is a vanadium enzyme. *Nature, 322*, 388–390. https://doi.org/10.1038/322388a0

Rodrigue, A., Effantin, G., & Mandrand-Berthelot, M. A. (2005). Identification of rcnA (yohM), a nickel and cobalt resistance gene in Escherichia coli. *Journal of Bacteriology, 187*(8), 2912–2916. https://doi.org/10.1128/JB.187.8.2912-2916.2005

Roychowdhury, T., Uchino, T., Tokunaga, H., et al. (2002). Arsenic and other heavy metals in soils from an arsenic affected area of West Bengal, India. *Chemosphere, 49*(6), 605–618. https://doi.org/10.1016/s0045-6535(02)00309-0

Sand, W., Rohde, K., Sobotke, B., et al. (1992). Evaluation of *Leptospirillum ferrooxidans* for leaching. *Applied and Environmental Microbiology, 58*, 85–92. https://doi.org/10.1128/AEM.58.1.85-92.1992

Sand, W., Gehrke, T., Hallman, R., et al. (1995). Sulfur chemistry, biofilm, and the (in)direct attack mechanism - A critical evaluation of bacterial leaching. *Applied Microbiology and Biotechnology, 43*, 961–966. https://doi.org/10.1007/BF00166909

Sandstroem, A. E., Sundkvist, J. E., & Petersson, S. (1997). Bio-oxidation of a complex zinc sulfide ore: A study performed in continuous bench and pilot scale. In *Conference Proceedings.*

International Biohydrometallurgy Symposium IBS97 BIOMINE 97 (pp. M1.1.1–M1.1.11). Australian Mineral Foundation.

Santos, I. R., Ilho, E. V. S., Schaefer, C., et al. (2005). Heavy metal contamination in coastal sediments and soils near the Brazilian Antarctic Station, King George Island. *Marine Pollution Bulletin, 50*(2), 185–194. https://doi.org/10.1016/j.marpolbul.2004.10.009

Sarkar, A., Paul, D., Kazy, S. K., et al. (2016). Molecular analysis of microbial community in arsenic-rich groundwater of Kolsor, West Bengal. *Journal of Environmental Science and Health, Part A, 51*(3), 229–239. https://doi.org/10.1080/10934529.2015.1094339

Scheer, H. (Ed.). (1991). *Chlorophylls*. CRC.

Schiewer, S., & Volesky, B. (2000). Biosorption processes for heavy metal removal. In *Environmental microbe-metal interactions* (pp. 329–362). American Society of Microbiology.

Schultze-Lam, S., Fortin, D., Davis, B. S., et al. (1996). Mineralization of bacterial surfaces. *Chemical Geology, 132*, 171–181. https://doi.org/10.1016/S0009-2541(96)00053-8

Sekhon, G. S., & Singh, B. (2013). Estimation of heavy metals in the groundwater of Patiala District of Punjab, India. *Earth Resources, 1*(1), 1–4. https://doi.org/10.12966/er.05.01.2013

Shallari, S., Schwartz, C., Hasko, A., et al. (1998). Heavy metals in soils and plants of serpentine and industrial sites of Albania. *Science of the Total Environment, 209*(2–3), 133–142. https://doi.org/10.1016/S0048-9697(98)80104-6

Sharma, A., & Kumar, A. (2016). Assessment of heavy metal contamination in soil sediments of Jaipur and Kota Industrial Areas, Rajasthan, India. *International Journal of Engineering, Management & Sciences, 3*(10), 1–7.

Sharma, P., Dubey, A., & Chatterjee, S. K. (2013). Determination of heavy metals in surface and ground water in an around (Agrang Block) Raipur District, Chhattisgarh, India. *International Journal of Scientific and Engineering Research, 4*, 722–724.

Sharma, M. C., Baxi, S., Sharma, K. K., et al. (2014). Heavy metal ions levels and related physicochemical parameters in soils in the vicinity of a paper industry location in Nahan Area of Himachal Pradesh. *Journal of Environmental & Analytical Toxicology, 4*(6), 1. https://doi.org/10.4172/2161-0525.1000236

Sharma, B. B., Sarma, H. P., & Borah, L. (2015). Chemical speciation of copper and cadmium in Kameng river sediments using sequential extraction procedure. *International Journal of Environmental Sciences, 6*(1), 88–96. https://doi.org/10.6088/ijes.6010

Sheela, A. M., Letha, J., Joseph, S., et al. (2012). Assessment of heavy metal contamination in coastal lake sediments associated with urbanization: Southern Kerala, India. *Lakes & Reservoirs: Research and Management, 17*(2), 97–112. https://doi.org/10.1111/j.1440-1770.2012.00501.x

Shrivastava, V. (2014). Geochemical assessment of heavy metal pollution and toxicity of Kunda River Sediment at Khargone District, Madhya Pradesh, India. *International Journal of Engineering Research & Technology, 3*(2), 329–333.

Silver, S. (1992). Bacterial heavy metal detoxification and resistance systems. In S. Mongkolsuk, P. S. Lovett, & J. Trempy (Eds.), *Biotechnology and environmental science: Molecular approaches* (pp. 109–129). Plenum. https://doi.org/10.1007/b102447

Silver, S., & Walderhaug, M. (1992). Gene regulation of plasmid- and chromosome-determined inorganic ion transport in bacteria. *Microbiological Reviews, 56*, 195–228. PMCID: PMC372861.

Singare, P. U., Trivedi, M. P., & Ravindra, M. (2012). Sediment heavy metal contaminants in Vasai Creek of Mumbai: Pollution impacts. *American Journal of Chemistry, 2*(3), 171180. https://doi.org/10.5923/j.chemistry.20120203.13

Singh, G., & Kamal, R. K. (2017). Heavy metal contamination and its indexing approach for groundwater of Goa mining region, India. *Applied Water Science, 7*(3), 1479–1485. https://doi.org/10.1007/s13201-016-0430-3

Slifierz, M. J., Friendship, R. M., & Weese, J. S. (2014). Methicillin-resistant Staphylococcus aureus in commercial swine herds is associated with disinfectant and zinc usage. *Applied and Environmental Microbiology, 81*(8), 2690–2695. https://doi.org/10.1128/AEM.00036-15

Slifierz, M. J., Friendship, R. M., & Weese, J. S. (2015). Methicillin-resistant Staphylococcus aureus in commercial swine herds is associated with disinfectant and zinc usage. *Applied and Environmental Microbiology, 81*(8), 2690–2695. https://doi.org/10.1128/AEM.00036-15

Snavely, M. D., Florer, J. B., Miller, C. G., et al. (1989). Magnesium transport in *Salmonella typhimurium*: Magnesium-28 ion transport by the CorA Mgta, and Mgtb systems. *Journal of Bacteriology, 171*, 4761–4766. https://doi.org/10.1128/jb.171.9.4761-4766.1989

Sonawane, N. S., Sawant, C. P., & Patil, R. V. (2013). Soil quality assessment and heavy metal contamination in agricultural soil in and around Toranmal (Triable Region) of Maharashtra. *Archives of Applied Science Research, 5*(2), 294–298.

Southam, G., Lengke, M. F., Fairbrother, L., et al. (2009). The biogeochemistry of gold. *Elements, 5*, 303–307. https://doi.org/10.2113/gselements.5.5.303

Srinivas, J., Purushotham, A. V., & Murali Krishna, K. V. S. G. (2013). A study of heavy metals contamination in surface and groundwater of rural and urban areas of Kakinada, East Godavari district, A. P. *International Journal of Civil, Structural, Environmental and Infrastructure Engineering Research and Development, 3*, 231–236.

Summers, A. P., & Silver, S. (1978). Microbial transformations of metals. *Annual Review of Microbiology, 32*, 637–672. https://doi.org/10.1146/annurev.mi.32.100178.003225

Summers, A. P., & Sugarman, L. I. (1974). Cell-free mercury(II) reducing activity in a plasmid-bearing strain of *Escherichia coli*. *Journal of Bacteriology, 119*, 242–249. https://doi.org/10.1128/JB.119.1.242-249.1974

Taghavi, S., Mergeay, M., & Van der Lelie, D. (1997). Genetic and physical maps of the Alcaligenes eutrophusCH34 Megaplasmid pMOL28 and its derivative pMOL50 obtained after temperature-induced mutagenesis and mortality. *Plasmid, 37*(1), 22–34. https://doi.org/10.1006/plas.1996.1274

Talukdar, B., Basumatary, S., Kalita, H. K., et al. (2015). Histopathological alternations in liver and kidney of Tor tor (Ham) inhabited in coal mining affected areas of Simsang River, Garohills; Meghalaya. *National Academy Science Letters, 38*(4), 321–324.

Thakur, B. K., & Gupta, V. (2015). Groundwater arsenic contamination in Bihar: Causes, issues and challenges. *Manthan: Journal of Commerce and Management, 2*(1). https://doi.org/10.17492/manthan.v2i1.6434

Toth, G., Hermann, T., Szatmari, G., et al. (2016). Maps of heavy metals in the soils of the European Union and proposed priority areas for detailed assessment. *Science of the Total Environment, 565*, 1054–1062. https://doi.org/10.1016/j.scitotenv.2016.05.115

Trevors, J. T. (1992). Mercury methylation by bacteria. *Journal of Basic Microbiology, 26*, 499–504. https://doi.org/10.1002/jobm.3620260811

Turpeinen, R., Pantsar-Kallio, M., & Kairesalo, T. (2002). Role of microbes in controlling the speciation of arsenic and production of arsines in contaminated soils. *Science of the Total Environment, 285*(1), 133–145. https://doi.org/10.1016/S0048-9697(01)00903-2

Urmila, Garg, A., & Annu. (2016). Assessment of heavy metal pollution in soil of Jhajjar, Haryana-India. *Journal of Chemical and Pharmaceutical Research, 8*(5), 629–634.

Vanita, C., Piar, C., Avinash, N., et al. (2014). Evaluation of heavy metals contamination and its genotoxicity in agricultural soil of Amritsar, Punjab, India. *International Journal of Research in Chemistry and Environment, 4*(4), 20–28.

Varghese, J., & Jaya, D. S. (2014). Metal pollution of groundwater in the vicinity of Valiathura sewage farm in Kerala, South India. *Bulletin of Environmental Contamination and Toxicology, 93*(6), 694–698. https://doi.org/10.1007/s00128-014-1410-7

Videla, H. A. (1995). Electrochemical aspects of biocorrosion. In C. C. Gaylarde & H. A. Videla (Eds.), *Bioextraction and biodeterioration of metals* (pp. 85–127). Cambridge University Press.

Viti, C., Marchi, E., Decorosi, F., et al. (2014). Molecular mechanisms of Cr (VI) resistance in bacteria and fungi. *FEMS Microbiology Reviews, 38*(4), 633–659. https://doi.org/10.1111/1574-6976.12051

Von Wolzogen Kuehr, C. A. H., & Van der Vlugt, L. S. (1934). Graphitization of cast iron as an electro-biochemical process in anaerobic soils. *Water, 18*, 147–165.

Wackett, L. P., Orme-Johnson, W. H., & Walsh, C. T. (1989). Transition metal enzymes in bacterial metabolism. In T. J. Beveridge & R. J. Doyle (Eds.), *Metal ions and bacteria* (pp. 165–206). Wiley.

Wallander, H., Mahmood, S., Hagerberg, D., et al. (2003). Elemental composition of ectomycorrhizal mycelia identified by PCR-RFLP analysis and grown in contact with apatite or wood ash in forest soil. *FEMS Microbiology Ecology, 44*(1), 57–65. https://doi.org/10.1016/S0168-6496(02)00456-7

Wang, Y. T., & Shen, H. (1995). Bacterial reduction of hexavalent chromium. *Journal of Industrial Microbiology, 14*, 159–163. https://doi.org/10.1007/BF01569898

Warren, L. A., & Haack, E. A. (2001). Biogeochemical controls on metal behaviour in freshwater environments. *Earth-Science Reviews, 54*, 261–320. https://doi.org/10.1016/S0012-8252(01)00032-0

Wase, J., & Forster, C. F. (1997). *Biosorbents for metal ions*. Taylor & Francis. https://doi.org/10.3109/9780203483046

Wong, P. T. S., Chau, Y. K., & Luxon, P. L. (1975). Methylation of lead in the environment. *Nature, 253*, 263–264. https://doi.org/10.1038/253263a0

Yadav, A., Yadav, P. K., & Shukla, D. N. (2013). Investigation of heavy metal status in soil and vegetables grown in urban area of Allahabad, Uttar Pradesh, India. *International Journal of Scientific and Research Publications, 3*(9), 1–7.

Yamamoto, I., Takashi, S., Liu, S.-M., et al. (1983). Purification and properties of NADP-dependent dehydrogenase from *Clostridium thermoaceticum*, a tungsten-selenium-iron protein. *The Journal of Biological Chemistry, 258*, 1826–1832. PMID: 6822536.

Zeng, F., Wei, W., Li, M., et al. (2015). Heavy metal contamination in rice-producing soils of Hunan province, China and potential health risks. *International Journal of Environmental Research and Public Health, 12*(12), 15584–15593. https://doi.org/10.3390/ijerph121215005

Microbial Remediation of Pharmaceuticals and Personal Care Products

M. Srinivasulu, M. Subhosh Chandra, G. Jaffer Mohiddin, A. Madhavi, B. Ramesh, S. Kameswaran, and P. Suresh Yadav

1 Introduction

A wide variety of pharmaceuticals are used to prevent and treat the diseases of humans and animals. These products primarily are the medicines, nutritional supplements, and drugs. The personal care products (PCPs) are mainly used to help and enhance the quality of life by cleaning and adorning our bodies. These PCPs include detergents, shampoos, lotions, moisturizers, cosmetics, insect repellents, pharmaceuticals and antibacterial soaps, odorants, and sunscreens (Yang & Toor, 2015). The human contribution of the pharmaceutical and personal care products (PPCPs) includes fecal matter, washings in the sinks and baths and outside the homes, the contaminants released by the pharmaceutical companies, hospitals, clinics, antibiotics, steroids. The PPCPs are released into the sewage system and wastewater must be treated before released into the environment (Kinney et al., 2006; Aydin & Talini,

M. Srinivasulu (✉)
Department of Biotechnology, Yogi Vemana University, Kadapa, India

M. S. Chandra · P. S. Yadav
Department of Microbiology, Yogi Vemana University, Kadapa, India

G. J. Mohiddin
Ciencia De la vida y Agricultura, Sede Santo Domingo, Universidad de las Fuerzas Armadas – ESPE, Sangolquí, Ecuador

A. Madhavi
Department of Microbiology, Sri Krishnadevaraya University, Anantapuramu, India

B. Ramesh
Department of Food Technology, Vikrama Simhapuri University, Nellore, Andhra Pradesh, India

S. Kameswaran
Department of Botany, Vikrama Simhapuri University PG Centre, Kavali, Andhra Pradesh, India

N. R. Maddela, L. C. García (eds.), *Innovations in Biotechnology for a Sustainable Future*, https://doi.org/10.1007/978-3-030-80108-3_14

2013; Blair et al., 2013a, b; Tewari et al., 2013). Pharmaceuticals and personal care products (PPCPs) are a group of contaminants of emerging concern (CEC) and defined by US EPA as any product used by individuals for the personal health or cosmetic reasons or used by agribusiness community to improve growth and health of the livestock (www.epa.gov/). PPCPs comprise thousands of chemicals which make up fragrances, cosmetics, over-the-counter drugs, and veterinary medicines.

Monitoring the PPCPs often seems to be argumentative when compared to other water problems like eutrophication, salinity algal blooms, geogenic contamination like fluoride, iron, arsenic, and nitrate (Berg et al., 2007; Heisler et al., 2008). The PPCPs might cause adverse effects to the aquatic organisms and human health owing to their high potency to act biologically even at nanogram level (Gerbersdorf et al., 2015). On the other hand, they are not yet monitored in a routine manner and therefore their levels of contamination to water bodies are unknown. Very few studies are available on the acute and chronic effects of these contaminants and the hazard they cause to the environment or to human health as they are released into aquatic environment. The recent investigations showed that the occurrence of PPCPs in water and soil, and the sources of these pollutants are the simple activities like laundry, shaving, cleaning of households, industries and offices, using sunscreen lotions or using of prescribed medication. Furthermore, in today's market, 4000 compounds have been developed for the purpose of human or veterinary pharmaceutical uses, which are biologically active (Monteiro & Boxall, 2010). The PPCPs are extensively used to cure, control, and for the prevention of diseases. The PPCPs include human and veterinary drugs, fragrances, disinfectants, and household chemicals. Annual production of PPCPs is higher than 23,107 tons, and it increases annually because of high demand (Wang & Wang, 2016). In fact, the extensive application of PPCPs led to the environmental pollution because of the presence of PPCPs in aquatic ecosystems that causes adverse effects on aquatic living organisms. A few of the sources of PPCPs are direct application of drugs in aquaculture, agricultural activities, and release from hospitals, and manufacturing sites. The majority of PPCPs are persistent in the environment and lethal to the non-target organisms. Furthermore, these have a tendency to accumulate at different tropic levels in the environment (Ramakrishnan et al., 2020). Apart from the aquatic environments, soil is also one of the sinks for PPCPs because these are easily adsorbed to soil through their various active sites. The schematic representation of sources and contamination of soil and water with PPCPs are presented in Fig. 1. Usually, studies have shown that PPCPs concentrations are in the range of ng/L to μg/L, ng/mg, and μg/kg in the surface water, groundwater, and soil, respectively (Wu et al., 2015; Roberts et al., 2016; Gottschall et al., 2012).

The PPCPs have been concern globally over the past two decades because of their widespread applications in the medicine, industry, livestock farming, aquaculture, and people's daily life and also due to their adverse impacts on wildlife and people (Pan et al., 2009; Evgenidou et al., 2015; Paredes et al., 2016). On the basis of their different purposes, PPCPs are divided into several classes (Wang & Wang, 2016), which include hormones, antibiotics, lipid regulators, non-steroidal anti-inflammatory drugs, β-blockers, anti-depressants, anticonvulsants,

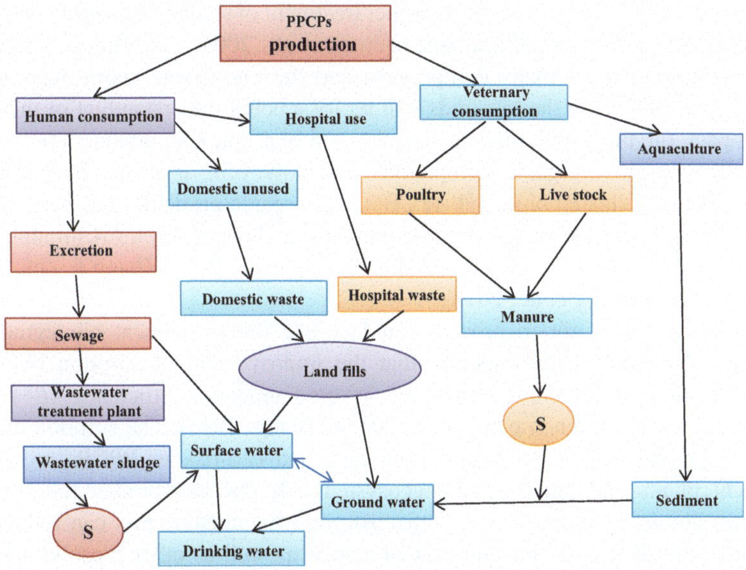

Fig. 1 Schematic representation of sources and contamination of soil and water with PPCPs

antineoplastic, diagnostic contrast media, fragrances, preservatives, disinfectants, and sunscreen agents.

PPCPs are the micropollutants that can be detected ubiquitously in soil, surface water, wastewater, and drinking water; in addition, they also have similar environmental behaviors with persistent organic pollutants (POPs) and are also called pseudo POPs. The wastewater is often recognized as one of the important sinks of PPCPs. Various treatment technologies like fungal biodegradation, activated sludge treatment, nanofiltration, reverse osmosis, and advance oxidation methods have been used for the removal of PPCPs from wastewater (Joss et al., 2004; Jelic et al., 2011; Dialynas & Diamadopoulos, 2012). In addition, the type of biological treatment plays a key role in the removal, transformation, and transportation of the PPCPs' environmental fate and behavior.

2 Biosorption of PPCPs

Sorption of the organic pollutants, which include PPCPs by biosolids or biomass, is recognized as a branch of the biotechnology that effectively decreases the dose of chemicals in wastewater influents, and as an essential step for the subsequent transmembrane transportation and intercellular biodegradation (Bokbolet et al., 1999; Zhang et al., 2018). Moreover, the occurrence of PPCPs and their metabolites in biosolids were measured (Miao et al., 2005; McClellan & Halden, 2010). The micropollutants sorption onto sludge/biosolids represents their major route into the surface water or soil, or whenever improper treatment of sludge takes place or is used as a fertilizer on the agricultural soil (Ternes et al., 2004). Sorption is a removal mechanism applied for many compounds, and the knowledge about the sorption properties of PPCPs on the biosolids will let for a better understanding of their fate and impact on the environment. Sorption of hydrophobic organic compounds especially non-polar organic compounds like polycyclic aromatic hydrocarbons (PAHs), polychlorinated biphenyls (PCBs), and organic pesticides has been widely studied for soil, sewage sludge, microorganisms, and nanoparticles (Carballa et al., 2008; Chen et al., 2010; Lin & Gan, 2011; Zhuang et al., 2011; Zhang & Zhu, 2012; Yu et al., 2013; Zhang et al., 2018).

In addition, the sorption process can facilitate the subsequent biodegradation efficiency. Previous studies specify that the improvement of sorption (K^*d) of pyrene resulted in an enhancement of its biodegradation efficiency (B^*) by a *Klebsiella oxytoca* strain (Zhang et al., 2013a; Zhang, 2013). The sorption process controls the removal, biotransformation, fate, and eco-risk of PPCPs in sewage treatment plants and other aquatic ecosystems. If one can predict the sorption behavior simply by Kd values or other sorption parameters, this can reduce the economic and time costs for analyses of sludge as well as other biosolid samples (Zhang, 2020).

3 Bioremediation

The bioremediation is a biological process that involves the application of microorganisms and their enzymes to convert xenobiotic/recalcitrant pollutants into less toxic forms and, thus, shortens lifetimes in the environment or even their complete mineralization (end products are the carbon dioxide and water) (Boopathy, 2011; Maddela & Scalvenzi, 2018; Kaur & Maddela, 2021). The bioremediation of organic compounds was investigated in more detail in the laboratory level, with metabolic pathways, during the degradation of some pharmaceutical compounds (Kartheek et al., 2011; Gillespie & Philip, 2013). The process of bioremediation is influenced by the physical and chemical factors, namely, the stereochemistry, toxicity, and the dose of contaminant, efficiency of the microbial strain, and environmental conditions during the degradation (e.g., pH and temperature), retention time and the presence of other compounds (Misal et al., 2011; Maddela et al., 2015a, b). A few advantages of bioremediation are as follows: it is accepted as a safe and eco-friendly process, transforms contaminants instead of simply moving them from one medium to another (Mashi, 2013), and presents lesser costs as compared to other technologies (Andra et al., 2010). Though bioremediation proved to be a promising option, research is essential to overcome some disadvantages of the process, that are the incomplete transformations, the limitation to the biodegradable compounds, and the necessity of the selection and the application of different microbes with definite metabolism for the different contaminants. Some of these drawbacks can be overcome by the application of genetically modified microorganisms (GEMs) (Gaylarde

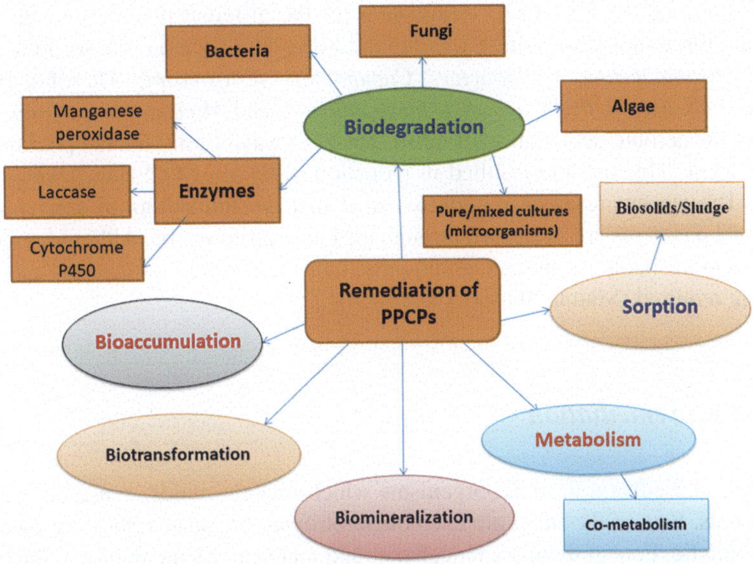

Fig. 2 Illustration of bioremediation of pharmaceuticals and personal care products (PPCPs)

et al., 2005). The overview of bioremediation of pharmaceuticals and personal care products (PPCPs) is presented in Fig. 2.

3.1 Bacterial Remediation

The bacteria play a crucial role in the bioremediation and ultimately assist in attaining the environmental sustainability. Proper usage of samples from the site of interest in the environment, isolation and characterization of the desired strain are necessary to select the suitable microorganism and also their consortium (Mamta Shashi et al., 2020). The bacterial strains of a single or different species associate together to form bacteria consortium and each pure culture used for growing the consortia. Subcultured strains are mixed together in a reactor, then they aggregate and form a biofilm. The bacteria required carbon as well as other nutrients for growth and degradation. Aissaoui et al. (2017) developed bacterial consortia of four strains (*Enterobacter hormaechei*, *Citrobacter youngae*, *Arthrobacter nicotianae*, and *Pseudomonas* sp.) that are excellent biodegraders. Drugs and their metabolites are applied as source of carbon and glucose, respectively. In the absence of glucose, the bacterial consortia eliminated 13.4% and 23.08% of ibuprofen after incubating for 24 h and 48 h, respectively. Das et al. (2012) used *Bacillus megatherium*, Phosphate solubilizing bacteria, *Pseudomonas fluorescens*, *Pseudomonas putida*, *Bacillus subtilis*, *Bacillus pumilus*, *Aspergillus niger*, *Nitrobacter*, *Bacillus licheniformis*, *Nitrosomonas*, and *Rhodococcus* organisms to set up a bacterial consortium. This obtained around 80–90% COD elimination for the different pharmaceutical effluents, and the removal of sulfate was around 80%. The bacterial consortium comprises *Agrobacterium radiobacter*, *Comamonas testosteroni*, *Microbacterium esteraromaticum*, *Methylobacterium mesophilicum*, and *Microbacterium saperdae* used for the aerobic degradation of nZVI (nano zerovalent iron) treated pharmaceutical effluent. This method resulted in reduction of 95.5% COD (Jagadevan et al., 2012). The current research reports revealed that the limitations of conventional biological processes are addressed through the bacterial consortia. Although to make this system available for human community, more in-depth investigations and field trials are required (Mamta Shashi et al., 2020).

3.2 Mycoremediation

The fungi are eukaryotic microorganisms which include molds, yeasts as well as mushrooms. Few of the fungi are chemoheterotrophic organisms; they are parasitic or saprophytic. Few of them are unicellular, and many are filamentous and have cell walls. The kingdom fungi include phyla *Chytridiomycota*, *Zygomycota*, *Ascomycota*, *Basidiomycota*, *Deuteromycota*, and *Glomeromycota*. According to their mode of sexual reproduction or by using molecular data, the classification of fungi was established (Alexopoulos et al., 1996). The water samples polluted with micropollutants were treated with fungi efficiently (Badia-Fabregat et al., 2015; Zhang et al., 2013c), particularly the pharmaceuticals. They have the abilities to transform a wide variety of recalcitrant compounds using nonspecific intracellular as well as extracellular oxidative enzymes (Durairaj et al., 2013; Jebapriya & Gnanadoss, 2013; Morel et al., 2013). Physiology and the colonization strategy of mycelial fungi tolerate them to more easily withstand sudden changes in pH or humidity, as well as to degrade the complex organic compounds more efficiently (Anastasi et al., 2013), even though they are limited by a long growth cycle and spore formation (Spina et al., 2012).

The fungi are one of the most diverse groups of microbes, play important roles in nature as decomposers, mutualists, or pathogens (Schmit & Mueller, 2007). The universal fungal species richness is controversial since most of these groups are not yet described (Schmit & Mueller, 2007; Bass & Richards, 2011; Blackwell, 2011); however, the mainstream of this richness involves terrestrial basidiomycetes and ascomycetes (Kirk et al., 2008). Both phyla contain contaminant degraders (Harms et al., 2011). *Mucoromycotina* (*Zygomycota*), a subdivision of fungi of incertae sedis (undefined relationships), represents a significant group less represented than those mentioned above, but which includes some well-studied species that metabolize xenobiotics/pollutants (Cha et al., 2001; Asha & Vidyavathi, 2009). The fungi have various strategies to counteract with numerous toxic compounds like recalcitrant PAHs (polycyclic aromatic hydrocarbons) (Maddela et al., 2015a, b) and pesticides (Cerniglia, 1997). These approaches comprise nonenzymatic methods like bioadsorption, biomineralization and biotransformation, and also the biodegradation driven by enzymatic process (Harms et al., 2011). The bioadsorption is driven by the definite composition of the cell wall such as chitosan or chitin (Gadd & Pan, 2016). In few fungi, like *Ascomycete Phoma* sp., biosorption into the fungal mycelia has significant function in the removal of bisphenol, triclosan and 17a-ethinylestradiol (Hofmann & Schlosser, 2016). Additionally, fungi can generate biosurfactants, functionally varied amphiphilic surface compounds with hydrophilic and hydrophobic portions, that interact between the phases of various polarities, result in interfacial tension decrease as well as interactions between molecules increases (Cicatiello et al., 2016; Günther, 2017). It is reported that the chemical structure of fungal biosurfactants involve, among others, sophorolipids, glycolipids, protein-lipid/polysaccharide complexes, and glycolipoproteins. These molecules represent an important tool for the purpose of bioremediation (Bhardwaj et al., 2013). Majority of the pharmaceutically active compounds contain aromatic structures, therefore, it might be possible that biosurfactants enhance the mobility and bioavailability of pharmaceutically active compounds, as it observed previously for PAHs (Souza et al., 2014).

The efficacy of two white-rot fungi (WRF), *Trametes versicolor* and *Ganoderma lucidum*, to eliminate 13 pharmaceutical contaminants with concomitant production of biodiesel from the accumulating lipid content after the treatment, was studied. Single and the combined strains exhibited a total removal (100%) of diclofenac, progesterone, gemfibrozil, ranitidine, and ibuprofen. The low removals were obtained for 4-acetamidoantipyrin, clofibric acid, atenolol, caffeine, carbamazepine, sulfamethoxazole, hydrochlorothiazide, and sulpiride, though the combination of both strains enhanced the system's efficacy, with removals ranging from 15% to 41%. The enzymatic and cytochrome P450 test showed that both extracellular (MnP, laccase, LiP) and intracellular oxidation mechanisms involve in the biological elimination of pharmaceuticals. The oxidative biological removal of pharmaceuticals by extracellular laccase and MnP enzymes and the intracellular cytochrome P450 system was clearly proved. Ultimately, fungal biomass remained after each individual biological elimination process was used for the production of biodiesel by one-step in-situ process, converting between 27% and 30% of the dried fungal sludge mass into FAME after 3 h of reaction (Vasiliadou et al., 2016).

3.3 Phycoremediation

The algae include prokaryotic (cyanobacteria) and eukaryotic organisms (all the algae species) that contain chlorophyll and perform oxygenic photosynthesis. Most of the algae are in microscopic size, hence, they are known as microorganisms (microalgae); several forms of macroscopic algae (macroalgae) grow to over 30 m in length. They are either unicellular or colonial. As the cells are arranged end to end, the alga is said to be in filamentous form (Sze, 1998). According to Ruggiero et al. (2015), algae are classified according to the pigments they possess. The algae color differences are mainly because of the presence of different photosynthetic pigments in addition to green chlorophylls. Chlorophyta and Euglenophyta are green as chlorophyll a is dominant. If the carotenoids are dominant they provide them a golden brown color, like Chrysophyta, whose chloroplasts contain chlorophyll a, c1, and c2, fucoxanthin, and carotene, that are carotenoids mainly responsible for the golden brown color. The Dinophyta (dinoflagellates) have a greenish, reddish, or brown appearance owing to chlorophylls a and c2 and carotenoids. The Rhodophyta (red algae) have chlorophylls a and d, phycobiliproteins, and floridean starch as storage products accumulated in the cytoplasm outside the chloroplast. In the Phaeophyta (brown algae), color results from the dominance of fucoxanthin over chlorophylls a, c1, and c2.

The algae are highly adaptive microorganisms and they are able to grow autotrophically, heterotrophically, or mixotrophically. They can grow in cruel environmental conditions, like low nutrient levels, and high pH and temperature, which are an advantage over some species of fungi (Subashchandrabose et al., 2013). Unlike strictly heterotrophic microorganisms, decrease in nutrient concentrations does not limit the growth of algae (Fu et al., 2016). The microalgae are capable of acclimatizing to changes in temperature, salinity, light, and the availability nutrients, which allow the enhancement of their tolerance and the capacity of biodegradation. The adaptation mechanism to extreme conditions is explained by genetic changes caused by the spontaneous mutations or with physiological adaptation (Osundeko et al., 2014; Cho et al., 2016). Xiong et al. (2017) evaluated that the biodegradation capacity of *Chlorella vulgaris* after acclimation with multiple exposures to levofloxacin and an increase in salinity and levofloxacin biodegradation significantly enhanced after acclimation.

The biomass of alga (Scenedesmus obliquus) was modified with alkaline solution and used for the biosorption of tramadol (TRAM) as well as other pharmaceuticals. The adsorption kinetics and isotherms were investigated. The high adsorption capacity of tramadol over the modified algal biomass with removal percentage was 91% after 45 min. The biosorption of tramadol on the modified algal biomass ensues with Freundlich isotherm model with correlation coefficient (0.942) that emphasized uptake of TRAM by modified algal biomass is driven by the chemisorption. Algal biomass has high potential in reusability for the adsorption of pharmaceutical compounds from wastewater. The ecotoxicological test is very essential to determine the efficiency of modified biomass for the adsorption of pharmaceutical mixture. The application of modified algal biomass might be a promising alternative and reusable sorbent for the elimination of pharmaceutical compounds from the wastewater (Mohamed et al., 2018).

The parent compounds and their potential transformation products were analyzed in both the water and algae phases. The results demonstrated that ibuprofen was primarily biotransformed because of synergistic relationships between the algae and bacteria. Ibuprofen biotransformation products are tentatively identified as hydroxy-ibuprofen, carboxy-ibuprofen, and 4-isobutylcatechol from several samples. In almost all the reactors exposed to light, triclosan undergoes both phototransformation and biotransformation. Triclosan biotransformation takes place in *Scenedesmus obliquus*, as shown by the presence of triclosan-O-sulfate in the extracts of algae. Hence, it is suggested that microalgal + bacterial consortia facilitates the transformation of PPCPs in the algae-based passive water treatment process (Larsen et al., 2019).

The microalgae can uptake the contaminants along with nutrients and accumulate and/or metabolize them inside the cell wall. The negative charge at the active site of the cell wall increases at higher pH, which attracts the metallic cations. Therefore, metals are adsorbed on the cell wall and decreasing the solubility leads to precipitation (Suresh et al., 2015). In addition, they produce enzymes as well as polysaccharides that also remediate contaminants. The microalgae have a capacity to adsorb around 0–16.7% of pharmaceuticals (Xiong et al., 2018). The pharmaceutical elimination capacity of algae depends on the different factors like temperature, pH, light, nutrients, hydrolytic retention time, the toxicity of the compounds, etc. The microalgae potential to degrade persistent pollutants was investigated in a range of wastewaters. Delrue et al. (2016) reported the capability of microalgae in treating the pharmaceuticals. The microalgae and their consortia degrade antibiotics, phenolic compounds, endocrine disruptors, and also transform hormones from the heavily contaminated wastewater. Similarly the microalgal consortia comprising *Chlorella* sp., *Scenedesmus* sp., and *C. zofingiensis* reduced 57.01–62.86% COD and 91.16–95.96% phosphorus from the dairy wastewater, which was higher than the removal efficiency of *Chlorella* sp. monoculture (Zhu et al., 2019).

Owing to its widespread use, huge quantity of oxytetracycline is released into the water, which has a harmful impact on the aquatic ecosystem and human health. Though different physicochemical methods are available for the removal of oxytetracycline, there is an increasing interest in the use of bioremediation. The microalga biomass (living) demonstrates high efficacy than the dead biomass with highest sorption capacities of 29.18 mg/g and 4.54 mg/g, respectively. Combination of living biomass as well as photodegradation in the culture eliminated 13.2 mg/L of oxytetracycline in 11 h of the culture and with an initial oxytetracycline dose of 15 mg/L. With an initial oxytetracycline concentration of 2.5 mg/L, 97% of oxytetracycline was eliminated. This elimination was mostly caused by the bioremediation than the photodegradation. It confirms that potential practical application of living *Phaeodactylum tricornutum* biomass for a low-cost and efficient elimination of oxytetracycline from the seawater. The application of living biomass was more effective than the same amount of dead biomass for the removal of oxytetracycline from the seawater solutions (Sergio et al., 2016).

Enrofloxacin is a fluoroquinolone antibiotic gained a large scientific concern owing to its toxic effect on the aquatic microbiota. The toxicity and elimination of enrofloxacin with five microalgal species and with their consortium were investigated for correlating the behavior and interaction of enrofloxacin in natural ecosystems. Single microalgal species (*Scenedesmus obliquus*, *Chlamydomonas mexicana*, *Chlorella vulgaris*, *Ourococcus multisporus*, *Micractinium resseri*) and consortium of them can withstand highest dose of enrofloxacin (1 mg/L). The growth inhibition (68–81%) of the individual microalgae sp. and their consortium was observed in enrofloxacin (100 mg L^{-1}) when compared with control after 11 days of cultivation. The microalgae can recover from the toxicity of higher doses of enrofloxacin while cultivation. The individual microalgae species and their consortium eliminated 18–26% enrofloxacin at 11-day. The interactions between different microalgal sps. during the degradation of organic contaminants required to be studied carefully in future (Jiu-Qiang et al., 2017).

4 Biodegradation of Pharmaceutical Compounds

The detailed biodegradation studies with batch tests and flow through soil columns under unsaturated, aerobic conditions also demonstrated for the biodegradation of pharmaceuticals, like ibuprofen and diclofenac (Tiehm et al., 2011). Carballa et al. (2007) reported 30–60% removal of ibuprofen under anoxic conditions and up to 80% degradation of diclofenac in laboratory scale experiments. The complete removal of ibuprofen and diclofenac was demonstrated with the white-rot fungus *Phanerochaete chrysosporium* under aerobic conditions in the fed-batch bioreactors (Rodarte-Morales et al., 2012). This indicates that there are currently different bioreactors being tested for their removal capacity that are effective for treatment of the organic micropollutants. Thus far, little is known about bacteria that degrade these pharmaceuticals and the involved biodegradation pathways, for example, only one bacterial strain has been described that degrades ibuprofen and uses ibuprofen as carbon and energy source (Murdoch & Hay, 2005). Diclofenac has shown to be biodegradable but the responsible bacteria are not known. Furthermore, the white-rot fungus *Phanerochaete chrysosporium* can completely degrade ibuprofen and diclofenac (Rodarte-Morales et al., 2012).

Triclosan has shown to be biodegradable in sediment and soil to their low water solubility and high *n*-octanol water portion coefficient (Chen & Rosazza, 1994). The bacterium *Nitrosomonas europaea*, and the *betaproteobacterial Methylobacillus* sp. isolated from activated sludge and found to degradation of triclosan (0.5–2 mg/L) in liquid media (Murdoch & Hay, 2005). Meade et al. (2001) reported that two soil bacteria, *Pseudomonas putida TriRY* and *Alcaligenes xylosoxidans* subsp. *denitrificans TR1*, had high resistance to triclosan (0.4 mg L^{-1}) and use it as their sole source carbon.

The gabapentin is an anti-seizure drug used for bipolar disorder globally, which is highly persistent, and usually detected in surface water and sometimes in groundwater. It is most frequently detected pharmaceutical compound in European natural water, at concentration of 10μg L^{-1} (Quintana et al., 2005). It is found in groundwater and noticed in soils as well as in sediments where treated wastewater is used to recharge groundwater and therefore it was suggested as a molecular indicator for anthropogenic contamination of water bodies recharged with reclaimed water or polluted riverine, groundwater as well as coastal environments by sewage. Gabapentin also has been accumulated from year to year in soil (Liu et al., 2009) and other studies also reported that gabapentin is highly recalcitrant to the microbial degradation in soil (Lin et al., 2006). On the other hand, several fungi and bacteria were able to degrade it, such as *C. elegans* ATCC 9245 which transforms it in a liquid culture (Bueno et al., 2012), *T. versicolor* and *G. lucidum* (Quintana et al., 2005).

The biodegradation of pharmaceuticals involves transfer of the parent compound to its metabolites in the presence of microorganisms in the wastewater treatment plants (WWTPs) or both in aerobic or anaerobic environmental conditions (Fent et al., 2006). In aerobic conditions, microorganisms convert the molecules into simple organic intermediates or mineral products (CO_2 and H_2O) during the successive oxidation reactions (Cirja et al., 2008). In the anoxic conditions, drugs are transformed to partially or completely mineralize and produce methane and CO_2 (Velagaleti, 1997). These reactions are catalyzed by several enzymes, for example diclofenac is converted to 4′-hydroxydiclofenac, 5-hydroxydiclofenac, and 4′,5-dihydroxydiclofenac in *Phanerochaete sordida* during the oxidation reactions by cytochrome P450, manganese peroxidase, and laccase that are the source for 90% removal of diclofenac after the period of 6 days incubation (Hata et al., 2010). The genera *Pseudomonas*, *Arthrobacter*, and *Enterobacter* use a wide range of organic compounds such as pharmaceuticals as carbon and energy sources. The *Enterobacter hormaechei* and *Enterobacter cloacea* show a considerable degrading ability towards the pharmaceutical effluent (Nilambari & Dhanashree, 2014). Likewise, several investigations reported that the capacity of the genus *Pseudomonas* and *Arthrobacter* for degrading the polluting compounds like sulfamethoxazole, chlorophenoxy acids, and pentachloronitrobenzene (Jiang et al., 2014; Evangelista et al., 2010; Wang et al., 2015).

Under metabolic conditions, i.e., in the absence of glucose the mixed bacterial culture has the capacity of eliminating 13.4% and 23.08% of ibuprofen (IBU) after 24 h and 48 h of incubation, respectively. But for diclofenac (DCF), only 9.12% of the total concentration was removed after 48 h of incubation. While in co-metabolic conditions, i.e., in the presence of glucose, a complete removal of IBU was observed and for DCF 56% of the total concentration was removed after 48 h of incubation. In both metabolic and co-metabolic conditions, elimination of SMX was not observed (Salima et al., 2017).

The application of drugs mixture in the presence of a consortium of bacterial strains might reflect the actual situation in the environment as well as wastewater treatment plants (WWTPs) where the pharmaceuticals and other contaminants are present as mixtures (Mishra & Anushree, 2014) (6). Kraigher et al. (2008) reported that the harmful effect of such mixtures on the growth of microorganisms is higher than a single drug, where a mixture of five acidic pharmaceuticals such as ibuprofen, ketoprofen, naproxen, diclofenac, and clofibric acid at the final concentration of 50μg L^{-1} caused shifts in the bacterial community composition and also reduced the bacterial diversity. Additionally, the removal of micropollutants by the use of mixed cultures of bacteria and fungi is better than the application of pure strains (Nguyen et al., 2013).

In the environment, bioremediation process usually depends on the cooperation of metabolic activities of the mixed microorganisms; the benefit of this population can be attributed to the important metabolic capabilities and the synergistic effect between the associated members. For instance, some species be able to eliminate the toxic metabolites of the preceding species and others can degrade compounds which are partially degraded by the first species by promoting the co-metabolic processes (Cerqueira et al., 2011). Numerous investigations reported that the application of microbial consortia improved the biodegradation rate of xenobiotics (Mishra & Anushree, 2014; Mikeskova et al., 2012). In this concern, Reis et al. (2014) studied the biodegradation of sulfamethoxazole (SMX) by pure (*Achromobacter*) and the mixed cultures of bacteria. They demonstrated that the mixed cultures of bacteria showed a higher biodegradation when compared with single microbial strain. On the contrary, the removal of SMX by the use of pure *Rhodococcus equi* is more significant than in the presence of mixed bacterial culture (Larcher & Yargeau, 2011). Moreover, these results are in agreement with those obtained by Rodarte-Morales et al. (2011), which reported that three ligninolytic fungi were able to eliminate the mixture of drugs with different removal rates. The complete degradation of SMX, DCF, IBU, citalopram naproxen, and carbamazepine was achieved after 14 days, whereas for fluoxetine and diazepam lesser elimination percentages were obtained (23–57%). In another study by the same authors, biotransformation of a mixture of three anti-inflammatories (DCF, IBU, and naproxen) by the pellets of *Phaerochaete chrysosporium* in fed-batch bioreactors operating in continuous air supply or periodic pulsation of oxygen was investigated. They observed the total elimination of DCF and IBU in both aerated and oxygenated reactors with a rapid oxidation of DCF in presence of oxygen (77% after 2 h). However, in the case of naproxen, it oxidized in the range of 77 up to 99% under both air and oxygen supply (Rodarte-Morales et al., 2012).

Quintana et al. (2005) studied the biodegradation of five acidic pharmaceuticals in the presence and absence of an external carbon source. They noticed the co-metabolic degradation of IBU, naproxen, and bezafibrate. On the contrary, ketoprofen degraded in the absence of extra carbon source after a lag phase for 10 days. *Planococcus* sp. S5 removed 30% of naproxen after 35 days as sole carbon source, whereas in co-metabolic conditions 75.14% and 86.27% of naproxen was eliminated in the presence of glucose and phenol, respectively (Domaradzka et al., 2015).

The white-rot fungi (WRF) mediated treatment of PPCPs is a promising and an eco-friendly technology. Various PPCPs are effectively eliminated by whole-cell WRF as well as crude/purified lignin modifying enzymes (LMEs). The hydrophilic and persistent PPCPs like naproxen, ketoprofen, and carbamazepine are considerably eliminated only in whole-cell WRF treatment because of the synergistic effects of extracellular as well as intercellular enzymes and sorption onto the fungal biomass. The different redox mediators were introduced to enhance the elimination of persistent PPCPs but continuous addition of mediators is an expensive process. Hence, mediator type and concentration should be selected vigilantly and also the techniques to recover the mediators (Muhammad et al., 2017). Various microorganisms used for remediation of pharmaceutical and personal care products (PPCPs) are presented in Table 1.

Table 1 Microorganisms used for remediation of pharmaceutical and personal care products (PPCPs)

S. No.	Microorganisms	Pollutants (PPCPs)	Reference
1.	Bacteria		
	Enterobacter, hormaechei, Arthrobacter nicotianae, Pseudomonas sp. *Citrobacter youngae* (*Bacterial consortia*).	Ibuprofen, diclofenac	Aissaoui et al. (2017)
	Nitrosomonas europaea, betaproteobacterial Methylobacillus sp.	Triclosan	Murdoch and Hay (2005)
	Pseudomonas putida TriRY, Alcaligenes xylosoxidans subsp. *denitrificans TR1*	Triclosan	Meade et al. (2001)
	Pseudomonas sp. *Arthrobacter* sp.	Sulfamethoxazole, chlorophenoxy acids, and pentachloronitrobenzene	Jiang et al. (2014), Evangelista et al. (2010), Wang et al. (2015)
	Rhodococcus equi	Sulfamethoxazole	Larcher and Yargeau (2011)
	Achromobacter denitrificans	Sulfamethoxazole sulfonamides	Reis et al. (2014)
	Streptomyces MIUG	carbamazepine	Popa et al. (2014)
	Pseudomonas sp. I-24	Iopromide	Liu et al. (2013)
2.	Fungi		
	Ascomycete phoma sp.	Triclosan, 17a-ethinylestradiol	Hofmann and Schlosser (2016)
	Trametes versicolor, Ganoderma lucidum (White-rot fungi)	Diclofenac, progesterone, gemfibrozil, ranitidine ibuprofen	Vasiliadou et al. (2016)
	Phanerochaete chrysosporium (White-rot fungi)	Ibuprofen and diclofenac	Rodarte-Morales et al. (2012)
	Cunninghamella elegans ATCC 9245	Gabapentin	Bueno et al. (2012)
	Phanerochaete sordida	Diclofenac	Hata et al. (2010)
	Delftia tsuruhatensis, Pseudomonas aeruginosa, Stenotrophomonas	Paracetamol	Santosa et al. (2012)
3.	Algae		
	Chlorella vulgaris	Levofloxacin	Xiong et al. (2017)
	Scenedesmus obliquus	Triclosan	Larsen et al. (2019)
	Phaeodactylum tricornutum	Oxytetracycline	Sergio et al. (2016)
	Microalgal species		
	Scenedesmus obliquus, Chlamydomonas mexicana, Chlorella vulgaris, Ourococcus multisporus, Micractinium resseri	Enrofloxacin	Jiu-Qiang et al. (2017)

5 Pure Cultures

Many studies reported that pure cultures isolated from the activated sludge, wastewater, or sediment used to remove the commonly detected PPCPs, like carbamazepine (Santosa et al., 2012; Popa et al., 2014), sulfamethoxazole (Jiang et al., 2014; Reis et al., 2014), iopromide (Liu et al., 2013), ibuprofen (Murdoch & Hay, 2005; Almeida et al., 2013), paracetamol (Zhang et al., 2013b), diclofenac (Hata et al., 2010), and triclosan (Zhou et al., 2014). A few pure cultures isolated from the activated sludge show the capacity to degrade various types of PPCPs. For example, *Achromobacter denitrificans* not only degrade the sulfamethoxazole but also the other sulfonamides (Reis et al., 2014). Additionally, for particular PPCPs, many pure cultures use it as sole source of carbon and energy, but with different degradation mechanism (Zhang et al., 2013b; Murdoch & Hay, 2005; Almeida et al., 2013). *Delftia tsuruhatensis*, *Pseudomonas aeruginosa*, and *Stenotrophomonas* are used for the degradation of paracetamol. Biosorption played negligible role in the removal of paracetamol for *Delftia tsuruhatensis* and *Pseudomonas aeruginosa*, whereas biosorption contributed to the removal of paracetamol for *Stenotrophomonas*. This inconsistency might be due to the differences in enzymes involved in the degradation process. For the specific PPCPs, pure cultures can hardly use them as source of carbon and energy. In this case, other substrates can be supplied to provide the carbon and energy for their metabolic purposes. For instance, carbamazepine has steady structure, which results in poor biodegradability. However, two pure cultures, unidentified *basidiomycete* (Santosa et al., 2012) and *Streptomyces* MIUG (Popa et al., 2014) degrade the carbamazepine in the presence of glucose. In sequence with carbamazepine, iopromide only degraded with the extra substrate. Liu et al. (2013) proved that *Pseudomonas* sp. I-24 has the ability to eliminate iopromide with starch as a primary substrate. Diclofenac show high resistance to the biodegradation in the activated sludge. Nevertheless, Hata et al. (2010) reported that white-rot fungi almost completely remove diclofenac and eliminate its toxicity to the organisms in the absence of extra substrate. Enzyme induction of the microbes is a key factor for the PPCPs biodegradation. The PPCPs biodegradation depends on whether microorganism is able to produce the specific enzyme to decompose them. For example, triclosan is able to induce *Nitrosomonas europaea* for the production of ammonia monooxygenase, which decompose triclosan (Roh et al., 2009). The recalcitrant PPCPs like tetracycline, trimethoprim, and ciprofloxacin cannot induce the microorganisms to produce the specific enzyme which led to their poor biodegradability. This clarifies why there is no pure culture isolated capable of degrading tetracycline, trimethoprim, and ciprofloxacin. To enhance the biodegradability of recalcitrant PPCPs, the important step is thus to induce microorganism to generate the desired enzyme.

6 Mixed Cultures

When compared with pure culture, mixed cultures are easier to achieve the target for degrading the PPCPs because in few cases it is very difficult to get the pure culture. Khunjar et al. (2011) studied the removal of PPCPs by the mixed cultures. In line with pure culture, mixed culture also has the capacity to eliminate the PPCPs. In the mixed media, mixed culture of ammonia oxidizing and heterotrophic bacteria has been demonstrated to be able of enhancing the elimination of 17a-ethinylestradiol (Khunjar et al., 2011). Actually, the most commonly used biological treatment process activated sludge in the WWTPs, depends on the synergy effect of the mixed culture to eliminate the PPCPs. In a few cases, activated sludge shows little removal capability to the PPCPs. Therefore, steps have been taken to improve the elimination of PPCPs by activated sludge. Zhou et al. (2014) reported the improved elimination of PPCPs by adding the mixed culture into the activated sludge. Interestingly, the mixed culture showed higher biodegradation rate in eliminating the mixed PPCPs compared to individual PPCPs (Vasiliadou et al., 2013). This may be because of some PPCPs utilized as source of carbon and energy by mixed culture, and in turn promote the decomposition of the other PPCPs. Previous studies revealed that mixed cultures might be a potential option for enhancing the elimination of PPCPs. Various microorganisms such as bacteria, algae, and fungi used to remediate the pharmaceuticals from the wastewater. These microorganisms convert or accumulate the pollutants into less toxic compounds (Kumar et al., 2020; Rana et al., 2017). The biodegradation of these contaminants depends on certain properties of wastewater like dissolved oxygen (DO), chemical oxygen demand (COD), pH, temperature, etc. (Choi et al., 2017).

7 Hazardous Effects of PPCPs

The general exposure to the pharmaceutical pollutants might have harmful effects on the health of ecosystem (Fong et al., 2015; Niemuth et al., 2015). These effects can alter the bacterial communities by inhibiting growth and interfering with their metabolic pathways. In addition, pharmaceutical pollutants like steroidal estrogens, 17β-estradiol and 17α-ethynylestradiol act as endocrine disruptors, inducing fish feminization in the *Pimephales promelas* sp. (Caracciolo et al., 2015; Kramer et al., 1998). Moreover, pharmaceutical pollutants are considered as ecotoxic compounds that are responsible for health-related issues (Xiong et al., 2017), as in the case of diverse amines derived from pharmaceutical pollutants (ranitidine, carbinoxamine, doxylamine, and nizatidine), which are precursors for the making of *N,N*-nitrosodimethylamine, a compound that represents a risk to human health because of its carcinogenic properties (Kramer et al., 1998). The antibiotics like tetracyclines, macrolides, sulfonamides, and quinolones may induce resistance in bacteria, for instance *Acinetobacter baumannii* sp. is resistant to all the recent antibiotics as a result of their chronic adaptation to hospital environment (Dijkshoorn et al., 2007).

8 Conclusions

In these decades, the PPCPs have been using continuously for various health benefits such as to cure the diseases as well as to improve the human life. The use of these PPCPs has become inevitable in order to treat the variety of infections and health complications in humans and animals. These pharmaceuticals are excreted into the environment through urine and feces. These PPCPs, either in parent form or as a metabolite form, are then released in the surface water as well as groundwater, thus polluting them. Veterinary pharmaceuticals excreted into the soil environment by manure. The PPCPs may cause adverse health effects to aquatic organisms even at lower doses. PPCPs enter the environment; they are still active and may cause various ecotoxicological effects on non-target organisms. This might negatively affect the important physiological functions, metabolism as well as reproduction even at lower quantity. Hence, it is very much essential to detoxify or completely eliminate from the environment. Thus, several researchers used different biological techniques, such as bioremediation, biodegradation, and biosorption, by the application of potential microorganisms such as bacteria, fungi, and algae as well as their enzymes to remove these pollutants from the ecosystem. Although different methods are available to treat or eliminate these pharmaceutical pollutants, further in-depth research is required to find out the efficient bioremediation techniques that are useful for the complete removal of contaminants present in the wastewater generated from various sectors such as domestic, hospital, and industrial sources.

Conflicts of Interests The authors declare that there are no conflicts of interests to disclose.

References

Aissaoui, S., Ouled-Haddar, H., Sifour, M., Beggah, C., & Benhamada, F. (2017). Biological removal of the mixed pharmaceuticals: Diclofenac, ibuprofen, and sulfamethoxazole using a bacterial consortium. *Iranian Journal of Biotechnology, 15*, 135–142.

Alexopoulos, C. J., Mims, C. W., & Blackwell, M. (1996). *Introductory mycology* (4th ed.). John Wiley.

Almeida, B., Kjeldal, H., Lolas, I., Knudsen, A. D., Carvalho, G., Nielsen, K. L., Crespo, M. B., Stensballe, A., & Nielsen, J. L. (2013). Quantitative proteomic analysis of ibuprofen-degrading *Patulibacter* sp. strain I11. *Biodegradation, 24*(5), 615–630.

Anastasi, A., Tigini, V., & Varese, G. C. (2013). The bioremediation potential of different ecophysiological groups of fungi. In E. M. Goltapeh, Y. R. Danesh, & A. Varma (Eds.), *Fungi as bioremediators* (Vol. 32, pp. 29–49). Springer.

Andra, J. A., Augusto, F., & Jardim, I. C. S. F. (2010). Biorremediação de solos contaminados por petróleo e seusderivados. *Eclética Química, 35*, 17–43.

Asha, S., & Vidyavathi, M. (2009). Cunninghamella – A microbial model for drug metabolism studies – A review. *Biotechnology Advances, 27*, 16–29.

Aydin, E., & Talini, I. (2013). Analysis, occurrence and fate of commonly used pharmaceuticals and hormones in the Buyukcekmece Watershed, Turkey. *Chemosphere, 90*, 2004–2012.

Badia-Fabregat, M., Lucas, D., Gros, M., Rodríguez-Mozaz, S., Barceló, D., & Caminal, G. (2015). Identification of some factors affecting pharmaceutical active compounds (PhACs) removal in real wastewater. Case study of fungal treatment of reverse osmosis concentrate. *Journal of Hazardous Materials, 283*, 663–671.

Bass, D., & Richards, T. A. (2011). Three reasons to re-evaluate fungal diversity 'on Earth and in the ocean'. *Fungal Biology Reviews, 25*, 159–164.

Berg, M., Stengel, C., Trang, P. T. K., Viet, P. H., Sampson, M. L., & Leng, M. (2007). Magnitude of arsenic pollution in the Mekong and Red River Deltas-Cambodia and Vietnam. *Science of the Total Environment, 372*, 413–425.

Bhardwaj, G., Cameotra, S. S., & Chopra, H. K. (2013). Biosurfactants from fungi: A review. *Journal of Petroleum & Environmental Biotechnology, 4*, 1–6.

Blackwell, M. (2011). The fungi: 1, 2, 3 ... 5.1 million species? *American Journal of Botany, 98*, 426–438.

Blair, B., Crago, J. P., Hedman, C. J., & Klaper, R. D. (2013a). Pharmaceuticals and personal care products in the Great Lakes above concentrations of environmental concern. *Chemosphere, 93*, 2116–2123.

Blair, B. D., Crago, J. P., Hedman, C., Treguer, R. J. F., Magruder, C., & Royer, L. S. (2013b). Evaluation of a model for the removal of pharmaceuticals, personal care products, and hormones from wastewater. *Science of the Total Environment, 444*, 515–521.

Bokbolet, M., Yenigun, O., & Yucel, I. (1999). Sorption studies of 2,4-D on selected soils. *Water, Air, and Soil Pollution, 111*(1), 75–88.

Boopathy, R. (2011). Factors limiting bioremediation technologies. *Bioresource Technology, 74*, 63–67.

Bueno, M. J., Gomez, M. J., Herrera, S., Hernando, M. D., Agüera, A., & Fernández-Alba, A. R. (2012). Occurrence and persistence of organic emerging contaminants and priority pollutants in five sewage treatment plants of Spain: Two years pilot survey monitoring. *Environmental Pollution, 164*, 267–273.

Caracciolo, A. B., Topp, E., & Grenni, P. (2015). Pharmaceuticals in the environment: Biodegradation and effects on natural microbial communities. A review. *Journal of Pharmaceutical and Biomedical Analysis, 106*, 25–36.

Carballa, M., Omil, F., Ternes, T., & Lema, J. M. (2007). Fate of pharmaceutical and personal care products (PPCPs) during anaerobic digestion of sewage sludge. *Water Research, 41*(10), 2139–2150.

Carballa, M., Fink, G., Omil, F., Lema, J. M., & Ternes, T. (2008). Determination of the solidwater distribution coefficient (Kd) for pharmaceuticals, estrogens and musk fragrances in digested sludge. *Water Research, 42*, 287–295.

Cerniglia, C. E. (1997). Fungal metabolism of polycyclic aromatic hydrocarbons: Past, present and future applications in bioremediation. *Journal of Industrial Microbiology & Biotechnology, 19*, 324–333.

Cerqueira, V. S., Hollenbach, E. B., Maboni, F., Vainstein, M. H., Camargo, F. A., & Do Carmo, R. (2011). Biodegradation potential of oily sludge by pure and mixed bacterial cultures. *Bioresource Technology, 102*(23), 11003–11010.

Cha, C. J., Doerge, D. R., & Cerniglia, C. E. (2001). Biotransformation of malachite green by the fungus *Cunninghamella elegans*. *Applied and Environmental Microbiology, 67*, 4358–4360.

Chen, Y., & Rosazza, J. P. N. (1994). Microbial transformation of ibuprofen by a *Nocardia* species. *Applied and Environmental Microbiology, 60*, 1292–1296.

Chen, B. L., Wang, Y. S., & Hu, D. F. (2010). Biosorption and biodegradation of polycyclic aromatic hydrocarbons in aqueous solutions by a consortium of white-rot fungi. *Journal of Hazardous Materials, 179*, 845–851.

Cho, K., Lee, C. H., & Ko, K. (2016). Use of phenol-induced oxidative stress acclimation to stimulate cell growth and biodiesel production by the oceanic microalga *Dunaliella salina*. *Algal Research, 17*, 61–66.
Choi, Y. Y., Baek, S. R., Kim, J. I., Choi, J. W., Hur, J., & Lee, T. U. (2017). Characteristics and biodegradability of wastewater organic matter in municipal wastewater treatment plants collecting domestic wastewater and industrial discharge. *Water, 9*, 409.
Cicatiello, P., Gravagnuolo, A. M., Gnavi, G., Varese, G. C., & Giardina, P. (2016). Marine fungi as source of new hydrophobins. *International Journal of Biological Macromolecules, 92*, 1229–1233.
Cirja, M., Ivashechkin, P., Schaffer, A., & Corvini, P. F. X. (2008). Factors affecting the removal of organic micropollutants from wastewater in conventional treatment plants (CTP) and membrane bioreactors (MBR). *Reviews in Environmental Science and Biotechnology, 7*, 61–78.
Das, M. P., Bashwant, M., Kumar, K., & Das, J. (2012). Control of pharmaceutical effluent parameters through bioremediation. *Journal of Chemical and Pharmaceutical Research, 4*, 1061–1065.
Delrue, F., Álvarez-Díaz, P. D., Fon-Sing, S., Fleury, G., & Sassi, J. F. (2016). The environmental biorefinery: Using microalgae to remediate wastewater, a win-win paradigm. *Energies, 9*, 1–19.
Dialynas, E., & Diamadopoulos, E. (2012). The effect of biomass adsorption on the removal of selected pharmaceutical compounds in an immersed membrane bioreactor system. *Journal of Chemical Technology and Biotechnology, 87*, 232–237.
Dijkshoorn, L., Nemec, A., & Seifert, H. (2007). An increasing threat in hospitals: Multidrug resistant *Acinetobacter baumannii. Nature Reviews. Microbiology, 5*, 939–951.
Domaradzka, D., Guzik, U., Hupert-Kocurek, K., & Wojcieszyńska, D. (2015). Co-metabolic degradation of naproxen by *Planococcus* sp. strain S5. *Water, Air, and Soil Pollution, 226*, 297.
Durairaj, P., Malla, S., Nadarajan, S. P., Lee, P. G., Jung, E., & Park, H. H. (2013). Fungal cytochrome P450 monooxygenases of *Fusarium oxysporum* for the synthesis of hydroxy fatty acids in engineered *Saccharomyces cerevisiae. Microbial Cell Factories, 14*, 45.
Evangelista, S., Cooper, D., & Yargeau, V. (2010). The effect of structure and a secondary carbon source on the microbial degradation of chlorophenoxy acids. *Chemosphere, 79*, 1084–1088.
Evgenidou, E. N., Konstantinou, I. K., & Lambropoulou, D. A. (2015). Occurrence and removal of transformation products of PPCPs and illicit drugs in wastewater: A review. *Science of the Total Environment, 505*, 905–926.
Fent, K., Weston, A., & Caminada, D. (2006). Ecotoxicology of human pharmaceuticals. *Aquatic Toxicology, 76*, 122–159.
Fong, P. P., Bury, T. B., & Dworkin-Brodsky, A. D. (2015). The antidepressants venlafaxine ("Effexor") and fluoxetine ("Prozac") produce different effects on locomotion in two species of marine snail, the oyster drill (Urosalpinx cinerea) and the starsnail (Lithopomaamericanum). *Marine Environmental Research, 103*, 89–94.
Fu, W., Chaiboonchoe, A., Khraiwesh, B., Nelson, D. R., AlKhairy, D., & Mystikou, A. (2016). Algal cell factories: Approaches, applications, and potentials. *Marine Drugs, 14*(225), 1–19.
Gadd, G. M., & Pan, X. (2016). Biomineralization, bioremediation and biorecovery of toxic metals and radionuclides. *Geomicrobiology Journal, 33*, 175–178.
Gaylarde, C. C., Bellinaso, M. L., & Manfilo, G. P. (2005). Biorremediação - aspetosbiológicos e técnicos da biorremediação de xenobióticos. *Biotecnologia Ciencia e Desenvolvimento, 34*, 36–43.
Gerbersdorf, S. U., Cimatoribus, C., Class, H., Engesser, K. H., Helbich, S., & Hollert, H. (2015). Anthropogenic trace compounds (ATCs) in aquatic habitats-research needs on sources, fate, detection and toxicity to ensure timely elimination strategies and risk management. *Environment International, 79*, 85–105.
Gillespie, I. M. M., & Philip, J. C. (2013). Bioremediation, an environmental remediation technology for the bioeconomy. *Trends in Biotechnology, 31*, 329–332.
Gottschall, N., Topp, E., Metcalfe, C., Edwards, M., Payne, M., & Kleywegt, S. (2012). Pharmaceutical and personal care products in groundwater, subsurface drainage, soil, and wheat grain,

following a high single application of municipal biosolids to a field. *Chemosphere, 87*(2), 194–203.

Günther, M. (2017). Fungal glycolipids as biosurfactants. *Current Biotechnology, 5*, 1–13.

Harms, H., Schlosser, D., & Wick, L. Y. (2011). Untapped potential: Exploiting fungi in bioremediation of hazardous chemicals. *Nature Reviews. Microbiology, 9*, 177–192.

Hata, T., Kawai, S., Okamura, H., & Nishida, T. (2010). Removal of diclofenac and mefenamic acid by the white rot fungus *Phanerochaetesordida* YK-624 and identification of their metabolites after fungal transformation. *Biodegradation, 21*(5), 681–689.

Heisler, J., Glibert, P. M., Burkholder, J. M., Anderson, D. M., Cochlan, W., & Dennison, W. C. (2008). Eutrophication and harmful algal blooms: A scientific consensus. *Harmful Algae, 8*, 3–13.

Hofmann, U., & Schlosser, D. (2016). Biochemical and physicochemical processes contributing to the removal of endocrine-disrupting chemicals and pharmaceuticals by the aquatic *Ascomycete Phoma* sp. UHH 5-1-03. *Applied Microbiology and Biotechnology, 100*, 2381–2399.

Jagadevan, S., Jayamurthy, M., Dobson, P., & Thompson, I. P. (2012). A novel hybrid nano zerovalent iron initiated oxidation-Biological degradation approach for remediation of recalcitrant waste metal working fluids. *Water Research, 46*, 2395–2404.

Jebapriya, G. R., & Gnanadoss, J. J. (2013). Bioremediation of textile dye using white-rot fungi: A review. *International Journal of Current Research and Review, 5*, 1–13.

Jelic, A., Gros, M., Ginebreda, A., Cespedes-Sanchez, R., Ventura, F., & Petrovic, M. (2011). Occurrence, partition and removal of pharmaceuticals in sewage water and sludge during wastewater treatment. *Water Research, 45*(3), 1165–1176.

Jiang, B., Li, A., Cui, D., Cai, R., Ma, F., & Wang, Y. (2014). Biodegradation and metabolic pathway of sulfamethoxazole by *Pseudomonas psychrophila* HA-4, a newly isolated cold-adapted sulfamethoxazole-degrading bacterium. *Applied Microbiology and Biotechnology, 98* (10), 4671–4681.

Jiu-Qiang, X., Mayur, B. K., & Byong-Hun, J. (2017). Ecotoxicological effects of enrofloxacin and its removal by monoculture of microalgal species and their consortium. *Environmental Pollution, 226*, 486–493.

Joss, A., Andersen, H., Ternes, T. A., Richle, P. R., & Siegrist, H. (2004). Removal of estrogens in municipal wastewater treatment under aerobic and anaerobic conditions: Consequences for plant optimization. *Environmental Science & Technology, 38*, 3047–3055.

Kartheek, B. R., Maheswaran, R., Kumar, G., & Banu, G. S. (2011). Biodegradation of pharmaceutical wastes using different microbial strains. *International Journal of Pharmaceutical and Biological Archive, 2*, 1401–1404.

Kaur, J., & Maddela, N. R. (2021). Microbial bioremediation: A cutting-edge technology for xenobiotic removal. In N. R. Maddela, L. C. García Cruzatty, & S. Chakraborty (Eds.), *Advances in the domain of environmental biotechnology. Environmental and microbial biotechnology*. Springer. https://doi.org/10.1007/978-981-15-8999-7_16

Khunjar, W., Mackintosh, S., Skotnicka-Pitak, J., Baik, S., Aga, D., & Love, N. (2011). Elucidating the relative roles of ammonia oxidizing and heterotrophic bacteria during the biotransformation of 17a-ethinylestradiol and trimethoprim. *Environmental Science & Technology, 45*(8), 3605–3612.

Kinney, C. A., Furlong, E. T., Werner, S. L., & Cahill, J. D. (2006). Presence and distribution of wastewater-derived pharmaceuticals in soil irrigated with reclaimed water. *Environmental Toxicology and Chemistry, 25*, 317–326.

Kirk, P., Cannon, P., Minter, D., & Stalpers, J. (2008). *Dictionary of the fungi*. CABI. https://doi.org/10.1079/9780851998268.0000

Kraigher, B., Kosjek, T., Heath, E., Kompare, B., & Mandic-Mulec, I. (2008). Influence of pharmaceutical residues on the structure of activated sludge bacterial communities in wastewater treatment bioreactors. *Water Research, 42*, 4578–4588.

Kramer, V. J., Miles-Richardson, S., Pierensa, S. L., & Giesy, J. P. (1998). Reproductive impairment and induction of alkaline-labile phosphate, a biomarker of estrogen exposure, in fathead minnows (*Pimephalespromelas*) exposed to waterborne 17β-estradiol. *Aquatic Toxicology, 40*, 335–360.

Kumar, V., Thakur, I. S., & Shah, M. P. (2020). Bioremediation approaches for treatment of pulp and paper industry wastewater: Recent advances and challenges. In M. P. Shah (Ed.), *Microbial bioremediation & biodegradation*. Springer Nature. https://doi.org/10.1007/978-981-15-1812-6-1

Larcher, S., & Yargeau, V. (2011). Biodegradation of sulfamethoxazole by individual and mixed bacteria. *Applied Microbiology and Biotechnology, 91*, 211–218.

Larsen, C., Yu, Z. H., Flick, R., & Passeport, E. (2019). Mechanisms of pharmaceutical and personal care product removal in algae-based wastewater treatment systems. *Science of the Total Environment, 695*, 133772.

Lin, K. D., & Gan, J. (2011). Sorption and degradation of wastewater-associated non-steroidal anti-inflammatory drugs and antibiotics in soils. *Chemosphere, 83*, 240–246.

Lin, A. Y., Plumlee, M. H., & Reinhard, M. (2006). Natural attenuation of pharmaceuticals and alkylphenol polyethoxylate metabolites during river transport: Photo-chemical and biological transformation. *Environmental Toxicology and Chemistry, 25*, 1458–1464.

Liu, F., Ying, G. G., Tao, R., Zhao, J. L., Yang, J. F., & Zhao, L. F. (2009). Effects of six selected antibiotics on plant growth and soil microbial and enzymatic activities. *Environmental Pollution, 157*, 1636–1642.

Liu, Y., Hu, J., Xu, B., He, J., Gao, P., Liu, K., Xue, G., & Ognier, S. (2013). Isolation and identification of an iopromide-degrading strain and its application in an A2/O system. *Bioresource Technology, 134*, 36–42.

Maddela, N. R., & Scalvenzi, L. (2018). Petroleum degradation: Promising biotechnological tools for bioremediation. In Z. Mansoor (Ed.), *Recent insights in petroleum science and engineering*. InTech Open. https://doi.org/10.5772/intechopen.70109

Maddela, N. R., Reyes, J. J. M., Viafara, D., & Gooty, J. M. (2015a). Biosorption of copper (II) by microorganisms isolated from crude oil contaminated soil. *Soil and Sediment Contamination: An International Journal, 24*(8), 898–908.

Maddela, N. R., Scalvenzi, L., Pérez, M., Montero, C., & Gooty, J. M. (2015b). Efficiency of indigenous filamentous fungi for biodegradation of petroleum hydrocarbons in medium and soil: Laboratory study from Ecuador. *Bulletin of Environmental Contamination and Toxicology, 95*(3), 385–394.

Mamta Shashi, B., Mohit, S. R., Shaon, R., Halis, S., & Sanjeev, K. P. (2020). Algae and bacteria-driven technologies for pharmaceutical remediation in wastewater. Chapter 15 Algae-and bacteria-driven technologies. In *Removal of toxic pollutants through microbiological and tertiary treatment* (pp. 373–408). Elsevier.

Mashi, B. H. (2013). Biorremediation: Issues and challenges. *JORIND, 11*, 1596–8303.

McClellan, K., & Halden, R. U. (2010). Pharmaceuticals and personal care products in archived U.S. biosolids from the 2001 EPA national sewage sludge survey. *Water Research, 44*, 658–668.

Meade, M. J., Rebecca, L. W., & Terrence, M. C. (2001). Soil bacteria *Pseudomonas putida* and *Alkaligenesxylosodans* subsp. denitrificans inactivate triclosan in liquid and solid substrates. *FEMS Microbiology Letters, 204*, 45–58.

Miao, X. S., Yang, J. J., & Metcalfe, C. D. (2005). Carbamazepine and its metabolites in wastewater and in biosolids in a municipal wastewater treatment plant. *Environmental Science & Technology, 39*, 7469–7475.

Mikeskova, H., Novotny, C., & Svobodova, K. (2012). Interspecific interactions in mixed microbial cultures in a biodegradation perspective. *Applied Microbiology and Biotechnology, 95*, 861–870.

Misal, S. A., Lingojwar, D. P., Shinde, R. M., & Gawai, K. R. (2011). Purification and characterization of azoreductase from alkaliphilic strain Bacillus badius. *Process Biochemistry, 46*, 1264–1269.

Mishra, A., & Anushree, M. (2014). Novel fungal consortium for bioremediation of metals and dyes from mixed waste stream. *Bioresource Technology, 171*, 217–226.

Mohamed, E. M. A., Azza, M., El-Aty, A., Mohamed, I. B., & Rizka, K. A. (2018). Removal of pharmaceutical pollutants from synthetic wastewater using chemically modified biomass of green alga Scenedesmus obliquus. *Ecotoxicology and Environmental Safety, 151*, 144–152.

Monteiro, S. C., & Boxall, A. B. A. (2010). Occurrence and fate of human pharmaceuticals in the environment. *Reviews of Environmental Contamination and Toxicology, 202*(1), 53–154.

Morel, M., Meux, E., Mathieu, Y., Thuillier, A., Chibani, K., & Harvengi, L. (2013). Xenomic networks variability and adaptation traits in wood decaying fungi. *Microbial Biotechnology, 6*, 248–263.

Muhammad, B. A., Faisal, I., Singh, H. L., William, E. P., & Long, D. N. (2017). Degradation of pharmaceuticals and personal care products by white-rot fungi-A critical review. *Current Pollution Reports, 3*, 88–103.

Murdoch, R. W., & Hay, A. G. (2005). Formation of catechols via removal of acid side chains from ibuprofen and related aromatic acids. *Applied and Environmental Microbiology, 71*(10), 6121–6125.

Nguyen, L. N., Hai, F. I., Yang, S., Kang, J., Leusch, F. D. L., & Roddick, F. (2013). Removal of trace organic contaminants by an MBR comprising a mixed culture of bacteria and white-rot fungi. *Bioresource Technology, 148*, 234–241.

Niemuth, N. J., Jordan, R., & Crago, J. (2015). Metformin exposure at environmentally relevant concentrations causes potential endocrine disruption in adult male fish. *Environmental Toxicology and Chemistry, 34*, 291–296.

Nilambari, D., & Dhanashree, T. (2014). Isolation and 16s rRNA sequence analysis of beneficial microbes isolated from pharmaceutical effluent. *Bionano Frontier, 7*(2), 243–248.

Osundeko, O., Dean, A. P., Davies, H., & Pittman, J. K. (2014). Acclimation of microalgae to wastewater environments involves increased oxidative stress tolerance activity. *Plant & Cell Physiology, 55*, 1848–1857.

Pan, B., Ning, P., & Xing, B. S. (2009). Part V-sorption of pharmaceuticals and personal care products. *Environmental Science and Pollution Research, 16*, 106–116.

Paredes, L., Fernandez-Fontaina, E., Lema, J. M., Omil, F., & Carballa, M. (2016). Understanding the fate of organic micropollutants in sand and granular activated carbon biofiltration systems. *Science of the Total Environment, 551–552*, 640–648.

Popa, C., Favier, L., Dinica, R., Semrany, S., Djelal, H., Amrane, A., & Bahrim, G. (2014). Potential of newly isolated wild *Streptomyces* strains as agents for the biodegradation of a recalcitrant pharmaceutical, carbamazepine. *Environmental Technology, 35*(24), 3082–3091.

Quintana, J. B., Weiss, S., & Reemtsma, T. (2005). Pathways and metabolites of microbial degradation of selected acidic pharmaceutical and their occurrence in municipal wastewater treated by a membrane bioreactor. *Water Research, 39*, 2654–2664.

Ramakrishnan, B., Maddela, N. R., Venkateswarlu, K., & Megharaj, M. (2020). Organic farming: Does it contribute to contaminant-free produce and ensure food safety? *Science of the Total Environment, 769*, 145079.

Rana, R. S., Singh, P., Kandari, V., Singh, R., Dobhal, R., & Gupta, S. (2017). A review on characterization and bioremediation of pharmaceutical industries' wastewater: An Indian perspective. *Applied Water Science, 7*, 1–12.

Reis, P. J., Reis, A. C., Ricken, B., Kolvenbach, B. A., Manaia, C. M., & Corvini, P. F. (2014). Biodegradation of sulfamethoxazole and other sulfonamides by Achromobacter denitrificans PR1. *Journal of Hazardous Materials, 280*, 741–749.

Roberts, J., Kumar, A., Du, J., Hepplewhite, C., Ellis, D. J., & Christy, A. G. (2016). Pharmaceuticals and personal care products (PPCPs) in Australia's largest inland sewage treatment plant, and its contribution to a major Australian river during high and low flow. *Science of the Total Environment, 541*(16), 25–1637.

Rodarte-Morales, I., Feijoo, G., Moreira, M. T., & Lema, J. (2011). Degradation of selected pharmaceutical and personal care products (PPCPs) by white-rot fungi. *World Journal of Microbiology and Biotechnology, 27*, 1839–1846.

Rodarte-Morales, A. I., Feijoo, G., Moreira, M. T., & Lema, J. M. (2012). Biotransformation of three pharmaceutical active compounds by the fungus *Phanerochaete chrysosporium* in a fed batch stirred reactor under air and oxygen supply. *Biodegradation, 23*, 145–156.

Roh, H., Subramanya, N., Zhao, F., Yu, C. P., Sandt, J., & Chu, K. H. (2009). Biodegradation potential of wastewater micropollutants by ammonia-oxidizing bacteria. *Chemosphere, 77*(8), 1084–1089.

Ruggiero, M. A., Gordon, D. P., Orrell, T. M., Bailly, N., Bourgoin, T., & Brusca, R. C. (2015). Correction: A higher level classification of all living organisms. *PLoS One, 10*, e0130114. https://doi.org/10.1371/journal.pone.0130114

Salima, A., Ouled-Haddar, H., Mohamed, S., Chérifa, B., & Farida, B. (2017). Biological removal of the mixed pharmaceuticals: Diclofenac, ibuprofen, and sulfamethoxazole using a bacterial consortium. *Iranian Journal of Biotechnology, 15*(2), 135–142.

Santosa, I. J., Grossmana, M. J., Sartorattob, A., Ponezib, A. N., & Durranta, L. R. (2012). Degradation of the recalcitrant pharmaceuticals carbamazepine and 17a-ethinylestradiol by ligninolytic fungi. *Chemical Engineer, 27*, 169–174.

Schmit, J. P., & Mueller, G. M. (2007). An estimate of the lower limit of global fungal diversity. *Biodiversity and Conservation, 16*, 99–111.

Sergio, S., Enrique, T., Roi, M., & Julio, A. (2016). Bioremediation of oxytetracycline in seawater by living and dead biomass of the microalga Phaeodactylum tricornutum. *Journal of Hazardous Materials, 320*, 315–325.

Souza, E. C., Vessoni-Penna, T. C., & de Souza, O. (2014). Biosurfactant-enhanced hydrocarbon bioremediation: An overview. *International Biodeterioration and Biodegradation, 89*, 88–94.

Spina, F., Anastasi, A., Prigione, V., Tigini, V., & Varese, G. C. (2012). Biological treatment of industrial wastewaters: A fungal approach. *Chemical Engineering Transactions, 27*, 175–180.

Subashchandrabose, S. R., Ramakrishnam, B., Megharaj, M., Venkateswarlu, K., & Naidu, R. R. (2013). Mixotrophic Cyanobacteria and microalgae as distinctive biological agents for organic pollutant degradation. *Environment International, 51*, 59–72.

Suresh, K. K., Dahms, H. U., Won, E. J., Lee, J. S., & Shin, K. H. (2015). Microalgae-A promising tool for heavy metal remediation. *Ecotoxicology and Environmental Safety, 113*, 329–352.

Sze, P. A. (1998). *Biology of the algae* (3rd ed.). WCB; McGraw-Hill.

Ternes, T. A., Herrmann, N., Bonerz, M., Knacker, T., Siegrist, H., & Joss, A. (2004). A rapid method to measure the solid-water distribution coefficient (Kd) for pharmaceuticals and musk fragrances in sewage sludge. *Water Research, 38*, 4075–4084.

Tewari, S., Jindal, R., Kho, Y. L., Eo, S., & Choi, K. (2013). Major pharmaceutical residues in wastewater treatment plants and receiving waters in Bangkok, Thailand, and associated ecological risks. *Chemosphere, 91*, 697–704.

Tiehm, A., Schmidt, N., Stieber, M., Sacher, F., Wolf, L., & Hoetzl, H. (2011). Biodegradation of pharmaceutical compounds and their occurrence in the Jordan valley. *Water Resources Management, 25*(4), 1195–1203.

Vasiliadou, I. A., Molina, R., Martínez, F., & Melero, J. A. (2013). Biological removal of pharmaceutical and personal care products by a mixed microbial culture: Sorption, desorption and biodegradation. *Biochemical Engineering Journal, 81*, 108–119.

Vasiliadou, I. A., Sanchez-Vazquez, R., Molina, R., Martínez, F., Melero, J. A., & Bautista, L. F. (2016). Morales Biological removal of pharmaceutical compounds using white-rot fungi with concomitant FAME production of the residual biomass. *Journal of Environmental Management, 180*, 228–237.

Velagaleti, R. (1997). Behavior of pharmaceutical drugs (human and animal health) in the environment. *Drug Information Journal, 31*, 715–722.

Wang, J., & Wang, S. (2016). Removal of pharmaceuticals and personal care products (PPCPs) from wastewater: A review. *Journal of Environmental Management, 182*, 620–640.

Wang, Y., Wang, C., Li, A., & Gao, J. (2015). Biodegradation of pentachloronitrobenzene by *Arthrobacter nicotianae* DH19. *Letters in Applied Microbiology, 61*(4), 403–410.

Wu, M., Xiang, J., Que, C., Chen, F., & Xu, G. (2015). Occurrence and fate of psychiatric pharmaceuticals in the urban water system of Shanghai, China. *Chemosphere, 138*, 486–493.

Xiong, J. Q., Kurade, M. B., & Jeon, B. H. (2017). Biodegradation of levofloxacin by an acclimated freshwater alga Chlorella vulgaris. *Chemical Engineering Journal, 313*, 1251–1257.

Xiong, J. Q., Kurade, M. B., & Jeon, B. H. (2018). Can microalgae remove pharmaceutical contaminants from water? *Trends in Biotechnology, 36*(1), 30–44.

Yang, Y. Y., & Toor, G. S. (2015). *Contaminants in the Urban environment: Pharmaceuticals and personal care products (PPCPs)-Part 2*[1] (p. SL420). Department of Soil and Water Sciences, Center for Landscape Conservation and Ecology; UF/IFAS Extension. http://edis.ifas.ufl.edu/ss633

Yu, Y., Liu, Y., & Wu, L. S. (2013). Sorption and degradation of pharmaceuticals and personal care products (PPCPs) in soils. *Environmental Science and Pollution Research, 20*, 4261–4267.

Zhang, D. (2013). *Surfactant controlled bacterial interfacial behaviors of PAHs and its mechanisms*. Ph.D. dissertation. Zhejiang University (in Chinese).

Zhang, D. (2020). The role of microorganisms in the removal of pharmaceutical and personal care products. In M. N. Vara Prasad, M. Vithanage, & K. Atya (Eds.), *Pharmaceuticals and personal care products: Waste management and treatment technology* (pp. 341–382). Elsevier.

Zhang, D., & Zhu, L. Z. (2012). Effects of Tween 80 on the removal, sorption and biodegradation of pyrene by Klebsiella oxytocaPYR-1. *Environmental Pollution, 164*, 169–174.

Zhang, D., Zhu, L. Z., & Li, F. (2013a). Influences and mechanisms of surfactants on pyrene biodegradation based on interactions of surfactant with a Klebsiella oxytoca strain. *Bioresource Technology, 142*, 454–461.

Zhang, L., Hu, J., Zhu, R., Zhou, Q., & Chen, J. (2013b). Degradation of paracetamol by pure bacterial cultures and their microbial consortium. *Applied Microbiology and Biotechnology, 97* (8), 3687–3698.

Zhang, Y., Xie, J., Liu, M., Tian, Z., He, Z., & Van Nostrand, J. D. (2013c). Microbial community functional structure in response to antibiotics in pharmaceutical wastewater treatment systems. *Water Research, 47*, 6298–6308.

Zhang, D., Lu, L., Zhao, H. T., Jin, M. Q., Lü, T., & Lin, J. (2018). Application of *Klebsiella oxytoca* biomass in the biosorptive treatment of PAH-bearing wastewater: Effect of PAH hydrophobicity and implications for prediction. *Water, 10*, 675. https://doi.org/10.3390/w10060675

Zhou, N. A., Lutovsky, A. C., Andaker, G. L., Ferguson, J. F., & Gough, H. L. (2014). Kinetics modeling predicts bioaugmentation with *Sphingomonad* cultures as a viable technology for enhanced pharmaceutical and personal care products removal during wastewater treatment. *Bioresource Technology, 166*, 158–167.

Zhu, S., Huo, S., & Feng, P. (2019). Developing designer microalgal consortia: A suitable approach to sustainable wastewater treatment. In *Microalgae biotechnology for development of biofuel and wastewater treatment* (pp. 569–598). Springer.

Zhuang, Y., Ahn, S., Seyfferth, A. L., Masue-Slowey, Y., Fendorf, S., & Luthy, R. G. (2011). Dehalogenation of polybrominated diphenyl ethers and polychlorinated biphenyl by bimetallic, impregnated, and nanoscale zerovalent iron. *Environmental Science & Technology, 45*(11), 4896–4903.

Detoxification of Heavy Metals Using Marine Metal Resistant Bacteria: A New Method for the Bioremediation of Contaminated Alkaline Environments

A. Madhavi, M. Srinivasulu, M. Subhosh Chandra, and V. Rangaswamy

1 Introduction

Modernization, industrialization, and fertilization are the root cause of contamination in the ecosystem, which is a grave problem in all corners of the globe. Heavy metals pollution has become a major concern worldwide due to their toxicity, intrinsic persistence, nonbiodegradable nature, and accumulative behaviors (Islam et al., 2018). Heavy metal contamination is not only a threat to living organisms but also a global environmental concern (Konate et al., 2017). The water and soil pollution by heavy metals is a major environmental problem and the majority of conventional approaches do not provide suitable solutions, moreover, these solutions are expensive. The environment is composed of the land, the Earth's atmosphere, and the water, where humans, plants, animals, and microorganisms live or work. Marine environment is the huge body of water that comprises 71% of the earth's coverage. The aquatic ecosystem is interconnected with terrestrial environment, thus, changes in one system ultimately have an impact on another. For past few decades, different factors comprising anthropogenic activities have a stress on the coastal as well as marine ecosystems (Richmond, 2015). This stress includes pollution and also physical devastation of the environment. Environmental pollution by heavy metals has become a serious threat to living organisms in an ecosystem (Okolo et al., 2016; Siddiquee et al., 2015). Heavy metals are widely distributed in almost all types of soils (Zhao et al., 2012), sediment (Gati et al., 2016), and water bodies (Tang et al., 2014). Heavy metal pollution mainly comes from paper making,

A. Madhavi (✉) · V. Rangaswamy
Department of Microbiology, Sri Krishnadevaraya University, Anantapuramu, India

M. Srinivasulu
Department of Biotechnology, Yogi Vemana University, Kadapa, India

M. S. Chandra
Department of Microbiology, Yogi Vemana University, Kadapa, India

N. R. Maddela, L. C. García (eds.), *Innovations in Biotechnology for a Sustainable Future*, https://doi.org/10.1007/978-3-030-80108-3_15

smelting, electroplating, and other industrial wastewater and the overuse of pesticide and fertilizer (Azimi et al., 2017). The rapid industrialization has led to a series of ecological and environmental problems (Sawut et al., 2018). The environmental pollutants are the chemical compounds that are present in elevated levels than in any sector of the environment (Masindi & Muedi, 2018). The rapid industrialization, agriculture and human activities lead to spreading of harmful contaminants like heavy metals in the environment (Ye et al., 2017, 2019). The environmental pollution with heavy metals enhanced beyond the recommended limit and is harmful to all the organisms (Dixit et al., 2015). The heavy metals are serious risk to the human health as well as ecosystem integrity (Ogbomida et al., 2018). The heavy metals are commonly defined as metals that required in trace quantities and considered as harmful (Maitra, 2016). The metals densities >5g/cm^3 are considered as the heavy metals have the atomic number >20 and are toxic at lower doses (Abbas et al., 2014). The metals are natural constituents present in the ecosystem, atmosphere, earth crust, water bodies, and also accumulate into the biological organisms, which include plants and animals. Among the 35 natural available metals, twenty three possess high density above 5 g/cm^3 with atomic weight >40.04 and are generally known as heavy metals (Li et al., 2017), which comprise antimony, tin, thallium, tellurium, bismuth, gold, cerium, gallium, arsenic, cadmium, chromium, copper, iron, cobalt, lead, manganese, nickel, mercury, silver, uranium, vanadium, platinum, and zinc (Li et al., 2017). The heavy metals occur in the atmosphere, lithosphere biosphere, and hydrosphere (Krishna & Mohan, 2016). The heavy metals at higher doses damage the human health as well as ecosystems (Ehya & Marbouti, 2016). The heavy metal contamination becomes a major problem throughout the world owing to their toxicity, persistence, nonbiodegradable, and accumulative nature (Islam et al., 2018). The hazardous effect and bioaccumulation of heavy metals in the environment is a severe risk to the health of organisms. Unlike organic contaminants, heavy metals cannot be broken down by chemical or biological processes. The heavy metals are released into the environment from the industrial and domestic wastes, leakage from the dump sites (Javed et al., 2018) and also from the air, water, soil as well as food in different forms. The survival capacity of microorganisms to heavy metals is often determined by minimum inhibitory concentration (MIC) test. MIC is defined as the lowest concentration of metal that inhibits the growth of microorganism (Vipra et al., 2013).

The contamination of water through heavy metals is one of the main types of pollution which affects the biotic community in the aquatic ecosystems. The heavy metals in the aquatic environments received significant attention around the globe due to their extensive availability, long period of incubation, strong concealment, and the environmental toxicity (Soliman et al., 2019; Yang et al., 2019). Marine environment of the Indian Ocean and the Red Sea has many sources of heavy metals because of human activities like port activity, mining (petroleum, metals and gas) activity, discharges of untreated domestic residues and polluted river water with heavy metals where precious metals and rocks are produced (El Nemr et al., 2016; Usman et al., 2013).

The heavy metal contamination is responsible for the cause of various diseases worldwide, like minamata disease (poisoning by mercury), itai-itai disease (poisoning by cadmium), arsenous acid poisoning, and air-pollution related asthma (Matsuo, 2003). The remarkable increase in the use of heavy metals resulted in an imminent rush of metallic substances in both terrestrial and aquatic environments (Gautam et al., 2016). In the aquatic environments, the heavy metal contamination resulted from the atmospheric deposition, geologic weathering or through the discharge of the agricultural, municipal, residential, or industrial waste products as well as by wastewater treatment plants (Garcia et al., 2015; Maier et al., 2014). Usually, metals are categorized as biologically essential and non-essential. The non-essential metals (cadmium (Cd), aluminum (Al), tin (Sn), mercury (Hg), and lead (Pb)) have no proven biological function, and their toxicity increases with rising concentrations (Sfakianakis et al., 2015). On the other hand, the essential metals (copper (Cu), chromium (Cr), zinc (Zn), nickel (Ni), molybdenum (Mo), cobalt (Co), and iron (Fe)) have known biological roles (Abadi et al., 2014) and the toxicity occurs either at metabolic deficiencies or at higher doses (Sivaperumal et al., 2007). The sources of heavy metals comprise activities like mining, smelting, burning fossil fuels, and electroplating. The heavy metals are also present in batteries, paints and metal products (ammunitions and pipes), cosmetics, fertilizers, ceramics, pesticides, electronic equipment, and wood preservatives (Concórdio-Reis & Freitas, 2019).

2 Marine Ecosystem

In the aquatic ecosystems, biotic components interact with abiotic components. The aquatic ecosystems are usually divided into two types, i.e., the marine ecosystem and freshwater ecosystem (Barange et al., 2010). The marine ecosystem is the biggest water ecosystem that covers 70% of the Earth's surface. Marine ecosystem is subdivided into estuaries, coral reefs oceans, and coastal ecosystems. The freshwater ecosystem covers less than 1% of the Earth's surface. The aquatic heavy metal pollution usually represents high levels of Hg, Cr, Pb, Cd, Cu, Zn, Ni, etc. in water system. The marine environment provides a very important sink for several heavy metals as well as their compounds. The toxic effects of heavy metals in marine environment have a most important concern since they constitute a potential risk to the different flora and fauna species, comprising humans, through food chains. The non-essential toxic plant heavy metals include arsenic (As), cobalt (Co), lead (Pb), cadmium (Cd), chromium (Cr), mercury (Hg), nickel (Ni), and vanadium (V), but others are essential, like copper (Cu), manganese (Mn), zinc (Zn), and iron (Fe). The heavy metals cause detrimental effects on animals, plants, and humans as a result of long-term/acute exposure. The heavy metals generated from industrial wastes enter the aquatic ecosystems and cause health affects in plants, animals, human as well as aquatic biotopes (Balasubramanian, 2012) (Fig. 1).

The extracellular polysaccharides (EPS) secreted by different microorganisms were effective in metal sequestration. For example, the EPS synthesized by *Bacillus*

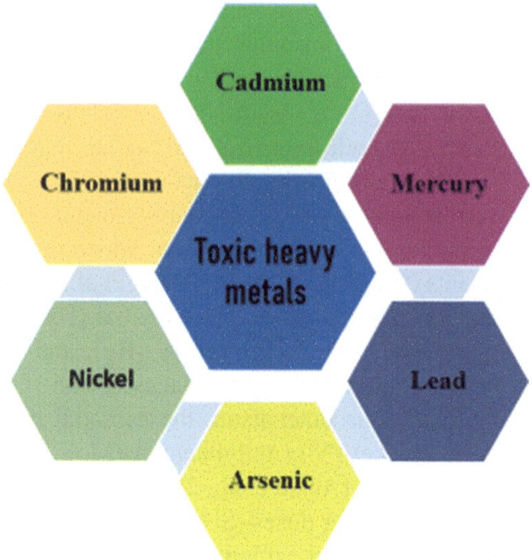

Fig. 1 Toxic heavy metals

firmus (Salehizadeh & Shojaosadati, 2003), *Paenibacillus jamilae* (Morillo Pérez et al., 2008), *Herbaspirillium* sp. (Lin & Harichund, 2012) and *Paenibacillus peoriae* TS7 (Fella-Temzi et al., 2018), *Bacillus licheniformis* KX657843 (Biswas et al., 2020) were capable of removing several heavy metals (e.g., Cd^{2+}, Co^{2+}, Hg^{2+}, Zn^{2+}, Pb^{2+}, Ni^{2+}, Cu^{2+}) from the aqueous systems. The heavy metal resistant bacteria studied extensively to be implemented as bioremediation agents. The biosorption of heavy metal by bacteria had been reported by Bhakta et al. (2012). The negative charge of bacterial cell wall is predicted to bind the cationic heavy metal. Furthermore, it has been suggested that the gram-positive bacteria such as LAB have higher absorption activity than the gram-negative bacteria due to the differences in the structure of cell wall (Gourdon et al., 1990). Metal biosorption usually occurs passively involving the specific binding proteins; however, the dead cells also absorb the heavy metals (Zoghi et al., 2014). The heavy metal, mercury is most toxic to the environment (Shaolin & David, 1997) and causes a serious hazard to animals, plants, and humans (Patra et al., 2004). It is known as a severe environmental contaminant due to its toxicity and ability to enter the biological system even at low doses (Porto et al., 2005). The aquatic ecosystems and the wetlands are more susceptible to mercury contamination (Gilbertson & Carpenter, 2004). Besides its occurrence in the nature, anthropogenic activities are also responsible for the deposition of mercury in the marine environment. The industrial effluents produced by mercury mining, gold smelting, and fuel combustion are the sources for release of heavy metal into the environment (Moreno et al., 2008). A number of techniques like chemical precipitation, reverse osmosis, ultrafiltration, conventional coagulation, magnetic filtration, activated carbon adsorption, ion exchange, and chemical reduction are used for the elimination of toxic heavy metals from the wastewater (Al-Garni, 2005). The bioremediation is used as an alternative technique for the elimination of mercury because it is simple and cost-effective (Zamil et al., 2009). Among the different biosorption techniques, the application of microorganisms plays an important role in the adsorption of heavy metals from the polluted wastewater (Sri Kumaran et al., 2011). Karaca et al. (2010) reported that the application of microorganisms is an effective and easy method for the elimination of heavy metals. The hazardous materials are easily broken down or transformed into simple and non-toxic compounds. The biosorption method received widespread attention because of its efficiency in the elimination of toxic metals from the wastewater (Hoostal et al., 2008). Various types of microorganisms such as bacteria, fungi, and algae are involved in biosorption and elimination of mercury from the environment (Beveridge, 1989). The bacteria have an effective mechanism for the adsorption of metals when compared with other groups of microorganisms. The bacterial communities exposed to mercury for longer periods acquire resistance because of their ability to tolerate heavy metals (Kafilzadeh & Mirzaei, 2008). Many studies are available for the biosorption of heavy metals by marine bacteria like *Bacillus thuringiensis* and *Pseudomonas aeruginosa* (Nithya et al., 2011). Mathe et al. (2012) reported that the bacterial strains, *Rhodococcus erythropolis* and *Pseudomonas corrugate* have the highest tolerance to metals. There are investigations on the biosorption efficiency of *B. thuringiensis* collected from terrestrial environment. It removed 79.4% of Pb, 87.9 of Ni (Oves et al., 2013), and 59.3% of Cr (VI) (Demir & Arisoy, 2007). Several reports revealed that the bacterial strains exposed to mercury for longer periods can develop resistance mechanisms like metal uptake, mineralization, accumulation, oxidation, sorption, etc. which finally enable them to detoxify mercury (Binish et al., 2015). Several species of *Arthrobacter* (Bafana et al., 2010) and *Pseudomonas* (Pepi et al., 2011) isolated from various environments have potential application in the detoxification of mercury. Tsibakhashvili et al. (2010) reported that *Arthrobacter* sp. accumulate mercury only up to a concentration of 1.0 ppm, whereas above that the uptake of mercury was retarded.

3 Sources and Toxic Effects of Heavy Metals

3.1 Cadmium

The toxic metal contamination can seriously threaten the biological sustainability of coastal ecosystems, and become a major problem for the aquatic environment (Liu et al., 2018). The exposure to higher level of metal may negatively affect fish as well as other aquatic organisms and hamper the physiological functions, reproduction and growth rate, or even mortality also increases (Öz et al., 2018). Among the toxic metals, cadmium (Cd) is the most hazardous and usually found in marine environment (Gu et al., 2019). Cadmium (Cd) is one of the main contaminants, nonbiodegradable, non-essential heavy metal and extremely lethal to the organisms even at lower doses and carcinogenic to humans (0.001–0.1 mg/L) (Maddela et al., 2020; Wu et al., 2016). Cadmium is present in almost all environments like air, soil,

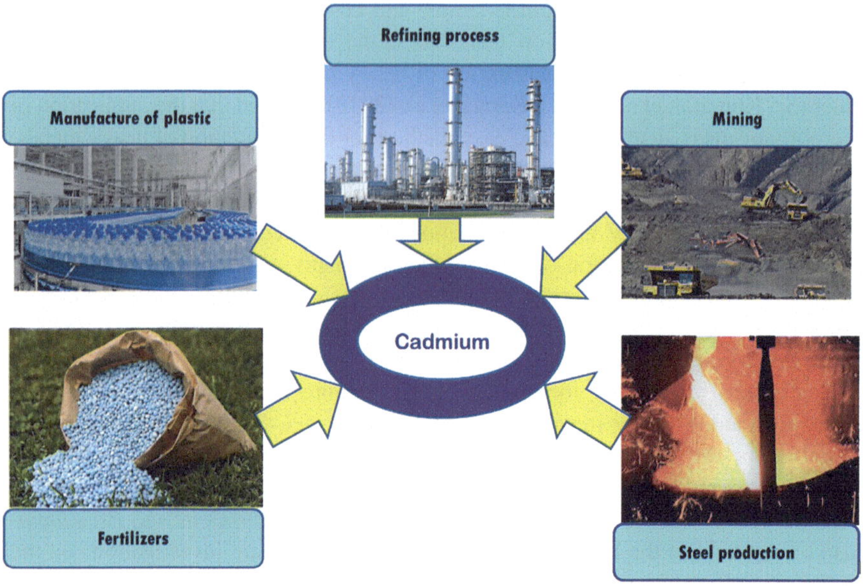

Fig. 2 Sources of cadmium release

water, and food (Hutton, 1983). Cadmium is commonly released into the environment due to the industrial activities, like mining, refining, and manufacture of plastics (Dong et al., 2019). Cadmium usually occurs in the environment in the ionic form Cd^{2+} (CdO_2, $CdCl_2$, or $CdSO_4$) (Castro-González & Méndez-Armenta, 2008). The cadmium is considered as the seventh most toxic non-essential heavy metal (Jaishankar et al., 2014) and released into the environment through natural sources, like volcanism (Hutton & Hutchinson, 1987), and anthropogenic. The anthropogenic activities include mining nonferrous metals, smelting, production of nonferrous metals, iron and steel and the production and disposal of cadmium-containing materials (electroplating, pigments, stabilizers, and Ni-Cd batteries) (Hutton, 1983), use of phosphate fertilizers (Fielder & Dale, 1983) plastic stabilizers, wood preservatives, arsenic pesticides, herbicides, fungicides, and others (Thornton, 1992) (Fig. 2).

Cadmium has a long half-life and is accumulated in organisms through the food chain. Cadmium is persistent and remains in the environment for decades (Rehman et al., 2020).

3.1.1 Effects of Cadmium on Aquatic Life

Accumulation of Cd in the living aquatic animals exerts a broad range of hazardous effects on the different tissues of animals and their human consumers (García-Navarro et al., 2017). The cadmium pollution of the aquatic habitat has significantly increased in the last few decades, resulting in an increase of cadmium deposits in tissues of aquatic organisms in all food chain systems (Giles, 1988). Cadmium is highly toxic for all mammals and fish and it causes a number of structural and pathomorphological changes in various organs of fish. The maximum cadmium levels were identified in the kidney as well as liver of fish (Thophon et al., 2003). While fishes occupy higher trophic level in the food chain, they considered as one of the most common bioindicators for contaminants (Authman et al., 2015; Idriss & Ahmad, 2015). Furthermore, fishes are generally consumed by humans as a main source of protein. Thus, human body is largely vulnerable to enriched heavy metal concentration in fishes (Ali & Khan, 2018).

3.1.2 Impact of Cadmium on Human Health

One of the main routes for cadmium exposure in humans is by the consumption of rice (Shi et al., 2020). No physiological function of cadmium in human cellular metabolism has reported and it is very toxic in minute quantity (Aksoy et al., 2014). In the humans, Cd exposure results in various adverse effects, like renal and hepatic dysfunction, testicular damage, pulmonary edema, osteomalacia, and damage to the hemopoietic system and adrenals (Tinkov et al., 2018). Cadmium causes the cardiovascular, cancer, respiratory and renal, skeletal system in the humans when taken up beyond the threshold limit (Radwan & Salama, 2006) (Fig, 3). Cadmium enters into the animal and human bodies by the food web and causes different diseases (Ali et al., 2019). Cadmium is not involved in any biological function; however, it inhibits the DNA-mediated transformation in microbes, their cellular enzyme functions, and also affects the symbiotic relationship between microbes and plants (Kabata-Pendias & Pendias, 2001). In addition, the bioaccumulation of cadmium in most of the plants may disturb various biochemical functions, which include alteration in mineral uptake, photosynthesis, interfering with the enzymes required for Calvin cycle and metabolism of carbohydrates, alters antioxidant metabolism in plants, and lowers the crop productivity (Feng et al., 2010).

3.1.3 Biodetoxification of Cadmium by Bacteria

The microbes uptake metal by bioaccumulation and/or biosorption (Johncy et al., 2010). Many microbes have developed chromosomally or extra chromosomally controlled detoxification mechanisms to overcome the harmful effects of the heavy metals (Ehrlich, 1997). The biosorption method is emerging as one of the attractive technologies to eliminate heavy metals from the aqueous solution. The biosorption is a process by which living and non-living microbial cells as well as cellular products are used for the removal of heavy metals from aqueous effluents (Demey et al., 2018). In the bacteria, the ability of absorption was higher because of high surface to volume ratios as well as the existence of active chemosorption sites on the surface of cell wall (Ma et al., 2020). Hou et al. (2015) reported cadmium biosorption by

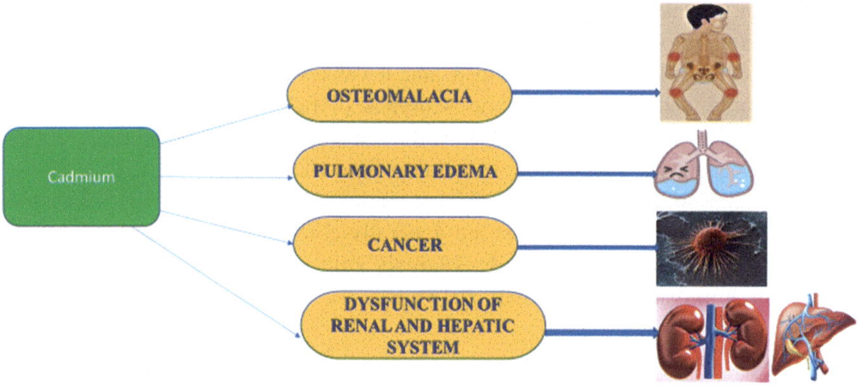

Fig. 3 Cadmium toxicity on human health

Klebsiella sp. at pH 5.0 and 30 °C. The main advantages of biosorption method when compared to other techniques include high efficiency, low cost, minimization of chemical and biological sludge, possibility of metal recovery and regeneration of biosorbent (Lacerda et al., 2019). Some authors reported the high ability of bioaccumulation of heavy metals by gram-negative and heterotrophic marine bacteria (Vogel & Fisher, 2010). *Vibrio* sp. is a potential microorganism for the bioaccumulation of heavy metals (Vogel & Fisher, 2010). *Vibrio harveyi*, a normal inhabitant of the marine environment is reported as potential for bioaccumulation of cadmium up to 23.3 mg Cd^{2+}/g of dry cells (Abd-Elnaby et al., 2011). The marine bacteria also possess the properties of chelation of heavy metals, thus removing them from the contaminated environment by the secretion of exopolysaccharides which have been evident from the reports of *Enterobacter cloaceae*, a marine bacterium. This bacterium has been reported to chelate up to 65% of cadmium, at 100 mg/L of metal concentration (Iyer et al., 2005). The purple non-sulfur marine bacterial isolates such as *Rhodobium marinum* and *Rhodobacter sphaeroides* are more potential for the removal of cadmium, from the contaminated environments through biosorption or biotransformation (Panwichian et al., 2011).

The bacteria eliminate heavy metal ions including Cd^{2+} from the environment either by metabolism-independent adsorption on their cell walls or metabolism-dependent intracellular accumulation (Vargas-García et al., 2012). The hyperaccumulation of Cd^{2+} has been reported to disturb the cell physiology by reactive oxygen species (ROS) production and disruption of bacterial respiratory proteins (Zeng et al., 2012). The bacteria evolved many resistance mechanisms include precipitation, efflux transport, transformation, and intracellular sequestration by metallothionein, glutathione, and other thiol containing compounds to combat the negative effects of heavy metal ions intracellular accumulation (Maynaud et al., 2014). Stanbrough et al. (2013) also isolated a cadmium-resistant bacterium *Achromobacter* sp. strain AO22 from soil and noted its resistance against cadmium of up to 100 mg L^{-1}. The bacterial strains *Acinetobacter brisouii*, *Pseudomonas abietaniphila*, *Exiguobacterium aestuarii*, and *Planococcus rifietoensis* which were isolated from coastal sediments of Vietnam also showed high resistance against cadmium 100, 130, 60, and 400 mg L^{-1}, respectively (Bhakta et al., 2014). Chovanova et al. (2004) studied the cadmium-resistant bacterial community isolated from sewage sludge contaminated by cadmium ions. Hussein et al. (2005) isolated the *Pseudomonas* sp. resistant to cadmium (3–11 mM). The biological elimination of cadmium by *Alcaligenes eutrophus* CH34 was observed that the efficiency for cadmium removal was over 99% (Mahvi & Diels, 2001). The bacterial sps. such as *Alcaligenes xylosoxidans*, *Pseudomonas fluorescens*, *Klebsiella planticola*, *Pseudomonas putida*, *Comamonas testosteroni*, *Serratia liquefaciens*, and *Pseudomonas* sp. showed resistance to Cd between 3 and 11 mM (Nath et al., 2012). Panwichian et al. (2011) stated that *Pseudomonas aeruginosa* strain eliminated more than 75% of the Cd from Cd-amended industrial wastewater under laboratory conditions. The study conducted by Henriques et al. (2015) revealed 80% removal of Cd using *Pseudomonas putida* strain from the culture medium. Bacterial cell walls or the envelope of bacteria are able to absorb metal ions from the medium by electrostatic interactions, and the heavy metal removing mechanism is a nonspecific interaction of heavy metals to the extracellular polysaccharides or cell envelope, proteins, teichoic acids, siderophores, and teichronic acids (Vijayaraghavan & Yun, 2008). The extracellular polysaccharides are also involved in the bioaccumulation of heavy metals. These extracellular polysaccharides have many functional groups, namely amide, carboxyl, imidazole, amino, hydroxyl, phosphate, sulfhydryl, carbonyl, amide, and phosphodiester groups that give very strong negative charge (Samina et al., 2010). Hence metal ions from the medium may be attracted to the cell surface of bacteria. The heavy metals from the environment are also transported through the membrane of the bacteria through the permeation of lipids, carrier-mediated transport, endocytosis, complex permeation, and ion pumps.

3.2 *Mercury*

Mercury (Hg) is a pervasive global pollutant with severe consequences for the integrity of environment and human health. The oceans play a very important role in global mercury (Hg) cycling. Once deposited, Hg is reduced to elemental Hg and volatilized back into the atmosphere, accounting for as much as half of the mercury present in the global atmosphere. The Mediterranean Sea is of special concern since it is affected by multiple Hg pollution sources. The heavy metal, Hg exists mostly in three forms such as metallic elements, inorganic salts, and organic compounds, each of which possesses different toxicity and bioavailability. These forms of Hg are present extensively in water resources like rivers, lakes, and oceans where they get transformed into methyl mercury within the microorganism, eventually undergoing biomagnification causing significant disturbance to aquatic lives. The Hg poisoning is referred to as pink disease or acrodynia. The liberation of mercury into environment is by the activities of number of industries such as pharmaceuticals, paper and pulp preservatives, agriculture industry, and chlorine and caustic soda production industry (Morais et al., 2012). ASGM (artisanal small scale gold mining) is one of the main sources of contamination by anthropogenic mercury (Hg) on the environment because miners use metallic Hg to extract gold from river banks or river beds (Veiga et al., 2006). This Hg can escape to the atmosphere and contaminate soil and aquatic ecosystems of surrounding areas. Natural Hg is derived from tectonic activities and volcanic emissions, while anthropogenic sources of Hg originate from industrial and mining activities. In oceanic waters, mercury mainly occurs in the forms of Hg^0, Hg^{2+}, methyl mercury, and dimethyl mercury and in colloidal form (Morel et al., 1998). Mercury (Hg) is a calamitous environmental pollutant, especially after the environmental disaster at Minamata (Japan) and several other poisoning accidents due to the utilization of mercury pesticides in agriculture (Begam & Sengupta, 2015). Human activities for instance, releases from the coal-fired plants and mining have dramatically increased the mercury concentrations and monomethyl mercury (MeHg) in the environment, marine ecosystems and their population (Lamborg et al., 2014). Hg exists as elemental form, inorganic (iHg) and organic Hg (primarily methylmercury, MeHg). The major cause of Hg into the marine environment is by atmospheric deposition (Driscoll et al., 2013). Among multiple heavy metals, mercury (Hg^{2+}) is the most toxic element for the organisms (Epstein, 2002). The inflow of mercury (Hg) into an aquatic ecosystem occurs naturally, through mineral deposits, forest fires, volcanoes, oceanic emission, and crust degassing. Metal may also be released into the ecosystems by human activities, like smelters processing sulfide ores (e.g., during the production of metals like iron, copper, gold, lead, and zinc), and other industrial activities, such as coal burning (Li & Wang, 2019). Mercury contamination in aquatic ecosystems has received a special focus since the discovery of mercury as the cause of Minimata disease in Japan in the 1950s (Allen et al., 1988).

3.2.1 Toxic Effects of Mercury on Aquatic Organisms

The major source of environmental contamination is industrial waste that has been released into the water bodies and has been transferred into the aquatic system such as fish. Fish are particularly at risk and greatly exposed to pollutants due to feeding and living in aquatic ecosystems, because they cannot escape from undesirable effects of pollutants (Ahmed et al., 2020). Fish are one of the chief organisms in the aquatic food chain ecosystem, having the potential to accumulate mercury at 1000 times exceeding in concentration than their surrounding aquatic environment (Beckers & Rinklebe, 2017). The absorption and accumulation of various kinds of heavy metals constitute a big threat for the fish ecosystem (Zeitoun & Mehana, 2014). The penetration of heavy metals penetrate into the fish is by direct absorption from water through their gills and skin, or by ingestion of contaminated food (Ayyat et al., 2020) and then the metals enter into the bloodstream of the fish and progressively accumulate in their tissues, especially in the liver, where they are bio-transformed and excreted or passed over to consumers through the food chain

(Amini et al., 2013). The heavy metals accumulation in fish tissues is affected by different forms of metal uptake, and environmental parameters as water temperature or intrinsic factors such as size and age (Rajeshkumar & Li, 2018). Kim et al. (2018) stated that the fish-based diets will have increased mercury levels, and also arsenic and cadmium compared to poultry or red meat diets. Whenever Hg enters the water, it is consumed by microbes, which are then consumed by small fish, and these, in turn, by bigger fish. At each step up the food chain, the Hg is retained in the muscle meat of the fish, resulting in the highest concentrations of Hg in large, long-lived predatory fishes. Sub-lethal exposures of fish to methyl mercury might lead to impaired ability to locate, capture, and ingest prey, and to avoid predation (Fjeld et al., 1998).

3.2.2 Mercury Toxicity on Human Health

Seafood is also a source of mercury (Hg) primarily in the form of methylmercury (MeHg), a potent neurotoxin which can harm the human fetus (Sheehan et al., 2014). It is the source of nourishment of number of major nutrients containing high-quality proteins, the marine long-chain omega-3 polyunsaturated fatty acids (n-3 LCPUFAs), eicosapentaenoic acid (EPA), docosahexaenoic acid (DHA), vitamin B12, vitamin D, iodine, selenium, and zinc (Aakre et al., 2019). Humans may be exposed to mercury through food, fish and using or breaking products containing mercury (Norouzi et al., 2012). Bioaccumulated mercury may cause neuronal damage for fish consumers (Kaur et al., 2011). Methyl mercury is the prevalent form of organic mercury in the environment and it is a neurodevelopmental toxicant (Obi et al., 2015), and it is also the most toxic form of mercury (Henriques et al., 2015) and humans exposed to raised levels could have bad health implications. Particularly, methyl mercury has ability to cross the blood–brain barrier (Silbernagel et al., 2011). The methyl and dimethyl mercury (organic mercury) usually originate from biological sources, primarily fresh or saltwater fish (Burger et al., 2011). The target organ for mercury is brain and also it can cause damage to any organ and promote dysfunction of nerves, kidneys, and muscles and it causes disruption to the membrane potential and interrupts intracellular calcium homeostasis (Fig. 4).

Mercury vapors could lead to bronchitis, asthma, and short-term respiratory problems. Mercury causes damage to the tertiary and quaternary protein structure and transforms the cellular function. It also interferes with the transcription and translation processes resulting in the disappearance of ribosomes and destruction of endoplasmic reticulum and the natural killer cells activity. The aspiration of elemental mercury for prolonged period of time induces tremors, gingivitis, and excitability. Ingestion of other common forms of mercury such as Hg^{2+} damages the gastrointestinal tract. Children, when exposed to methyl mercury while they are in the womb would affect on their fine motor skill, language, attention, cognitive thinking, visual spatial skills, and memory (Boylan et al., 2003). Around three thousand lakes in the United States have been shut down to fishing due to mercury impurities and lot of species of ocean fish are also polluted with substantial mercury concentrations

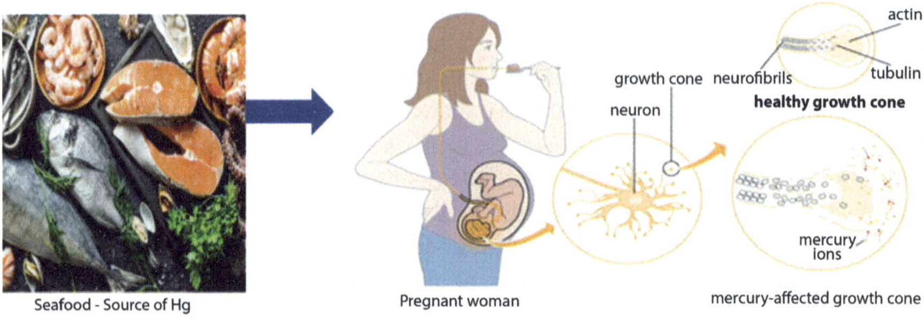

Fig. 4 Toxic effects of mercury on pregnant woman

(Boylan et al., 2003). The consumption of methyl mercury by humans from the seafood is absorbed into the gastrointestinal tract and thereafter transported to the blood. The main Hg toxicity target organs (Silbernagel et al., 2011) are the kidneys (Lund et al., 1993) and brain (Franco et al., 2007). The manifestations from seafood exposure were observed mainly in the central nervous system, involving sensory disorders (Castoldi et al., 2001), ataxia (Harada, 1995), memory lapses, gastrointestinal illness, weakened fine motor coordination, hypertension, muscle and joint pain, thinning of the hair, numbness, heart rate disturbance, fatigue, sleep difficulties (Silbernagel et al., 2011), coma and death (Bose-O'Reilly et al., 2010). Long-standing exposure to methyl mercury induces teratogenic effects (Davidson et al., 2004) and carcinogenicity (Boffetta et al., 1993), self-expression in offspring such as congenital anomalies and tumor formation (Lopez et al., 2009). Mercury shows undesirable effects in fertility, both in men and women (Maeda et al., 2019). The exposure of women to mercury causes infertility and is influenced by a female hormonal system disbalance. Especially, mercury is lethal for the women in labor, who may transmit mercury to the unborn baby through the fetus-feeding organ, lactating mothers (who can pass the Hg to the baby through breast) and young children, whose nervous systems are developing. The elemental form (Hg^0) as well as organic form of mercury crosses the placental barrier and reaches the embryo, leading to developmental disorders (Zheng et al., 2019).

3.2.3 Biodetoxification of Mercury

Mercury-resistant bacteria are now considered as potential approach to biological remediation. Because of their simplicity, lack of secondary contamination, and cost-effectiveness, when compared with other treatment technologies mercury-resistant bacteria are the prominent alternative tool for bioremediation (Singh et al., 2008). The bioremediation strategies (biotransformation, biosorption, and bioprecipitation) of mercurials have been designed and implemented at few cases to remediate mercurials in the environment (Von Canstein et al., 2001). Earlier analysis has outlined the isolation of numerous organic and/or inorganic mercury-resistant bacteria belonging to species of the genera *Bacillus*, *Pseudomonas*, *Staphylococcus*, and

Escherichia from various mercury-contaminated environments (Bafana et al., 2010). There have been reports on several bacteria holding novel genetic mechanisms to convert the toxic form of mercury into less toxic forms to clear away the polluted environments (Chien et al., 2010; Pepi et al., 2013). One of the best-studied bacterial detoxification mechanism is the *mer* operon-mediated mechanism. The isolates of bacteria harboring mer operons carry many functional genes like *merA*, *merB* (optional in broad-spectrum mercury-resistant bacteria), *merT*, *merP*, and *mer F* in addition to operator and promoter regions. Mer operon (positively inducible operon) is responsible for the mercury deintoxification by transforming the organic form (CHg) into inorganic form (Hg^{2+}, ${Hg_2}^{2+}$), and mediated by MerB (organomercurial lyase) followed by conversion of inorganic mercury to elemental/volatile mercury by MerA (mercuric ion reductase) (Dash & Das, 2012). Mercury-resistant determinants have been identified in a vast array of gram-negative and gram-positive bacteria isolated from different environments. *Pseudomonas putida* SP-1 is a biofilm-forming, non-pathogenic marine bacterial isolate having the ability of volatilizing 89% of mercury and it has been confirmed that the mercury-resistant marine bacteria is more efficient in bioremediation (Zhang et al., 2012). Similarly, Chang and Law (1998) developed a detoxification process using *Pseudomonas aeruginosa* PU21 in batch, fed batch, and continuous bioreactor systems. A photosynthetic bacterium, *Rhodopseudomonas palustris* was engineered to accumulate Hg(II) using MerP and MerT, and to sequester it by chelating with phytochelatin (Deng & Jia, 2011). Canstein et al. (1999) proved the removal of mercury from chloralkali electrolysis wastewater by a mercury-resistant *Pseudomonas putida* strain.

3.3 Arsenic

Among the heavy metals, As is attracting media attention owing to its high toxicity. Humans (140 million) in more than 50 countries are exposed to arsenic-contaminated drinking water (Ravenscroft et al., 2009). The word "arsenic" evokes a reaction of fear in most people. This is because arsenic has a long history of being a possible hazard to humans (Hughes et al., 2011). Arsenic is the natural part of the Earth's crust and it is released into the environment from the natural (e.g., volcanic activity, biomethylation, and microbial reduction) and the anthropogenic (e.g., coal-fired power generation, smelting, and vegetation burning) sources (Litynska et al., 2017) and also found in the lakes, groundwater, rivers, warm springs, and seawater. It (As) is a mixture of arsenate and arsenite, in the water with arsenate usually predominating. The contamination of natural waters such as groundwater, seawater, and freshwater with arsenic is an issue of global concern as the rising concentrations have been recorded for water samples in various corners of the world (Rahman et al., 2012). On an international scale, across 200 millions of human beings are at risk of arsenic exposure at levels of concern for public health (Naujokas et al., 2013). Over centuries arsenic has acquired a terrible reputation as "the poison of kings" or "king of poisons" and difficult to detect even after the death since it is colorless, odorless, and tasteless element (Akhtar et al., 2017). In the healthcare service, arsenic has been employed as a therapeutic agent. The arsenic is hazardous in all the four oxidation states (+V, +III, 0, −III), trivalent and pentavalent arsenic is the most common species found in the ecosystems (Rahman et al., 2014). The arsenic enters the atmosphere by means of dust particles coming from volcanic emissions (ashes), wind erosion, low-temperature volatilization from soils, marine aerosols, and pollution and is returned to the Earth's surface (mainly to water bodies) by atmospheric deposition and then, it passes through terrestrial runoff and groundwater discharge to the water bodies. There it binds to or (co)precipitates with suspended particles and tends to sink to the sediments (Sultana et al., 2015). Arsenic is the by-product of the smelting of copper, lead, cobalt, and gold ores (Brininstool, 2017; Li et al., 2017). There have been reports that the toxicity of arsenic compounds follow the order As (III) > As(V) > monomethyl arsonate (MMA) > dimethylarsinic acid (DMA) > organic arsenic species (ATSDR, 2017). Arsenic metabolites exist both in organic and inorganic forms. Inorganic arsenic species, As(III) and As(V) have classified by the International Agency for Research on Cancer as Class I chemicals, carcinogenic to humans; meanwhile, MMA and DMA species are classified as Class IIB chemicals, possibly carcinogenic to humans based on in vitro evidence (Escudero-Lourdes et al., 2012). The inorganic arsenic is a highly poisonous compound, is a cancer-causing agent recognized by the international agency for research on cancer (IARC), and the organic arsenic is less toxic (Styblo et al., 2000).

3.3.1 Toxicity of Arsenic to Aquatic Organisms

Arsenic is highly poisonous to the life forms (Flora, 2015). Panagiotaras and Nikolopoulos (2015) revealed that the typical arsenic leftovers in the marine organisms range from 1 $\mu g\ g^{-1}$ to about 100 $\mu g\ g^{-1}$. This implies that the aquatic organisms may be exposed to arsenic-contaminated water and/or sediments and these organisms accumulate, store and modify arsenic species inside their body. As a result, arsenic can be biomagnified inside the aquatic food web (Khan et al., 2014). Fishes and the marine mammals may be subjected to acute and chronic poisoning of arsenic in the aquatic environment. Acute exposure could occur within a few hours and is generally linked with transformed behavior and hematological shifts. This type of intoxication may also generate destructive effects. The lasting toxicity could adversely affect gonads and raise issues with the development of young specimens (Kumari et al., 2016). Both acute and chronic exposures bring many negative symptoms. Algae, bivalves, mollusks, and fish are particularly vulnerable to toxic effects of arsenic. The accumulation of maximum arsenic concentration was observed in macroalgae due to their strong affinity for trace metals. Meanwhile, the exposure of the fish and marine mammals is ceaseless to arsenic-contaminated food. Metallic materials entering into the stretches of water store up in the tissues of aquatic animals and this accumulation of metals including arsenic influences a variety of physiological systems—fish growth, immune function, reproduction, and enzyme activity (Datta et al., 2009). Besides, fluctuations in water temperatures might possibly disturb the fish metabolism, and also temperatures of water outside the appropriate temperature range have devastating effectson fish (Bagnyukova et al., 2007). The analysis performed by Abdel-Hameid (2009) proved the dramatic expansion in GOT and GPT of Nile Catfish, *Clarias gariepinus*, exposed to arsenic, and higher levels can lead to hepatic impairment due to arsenic poisoning.

3.3.2 Effect of Arsenic on Human Health

Human beings have been affected by the harmful health risks of arsenic (Smedley & Kinniburgh, 2002). Latest epidemiological studies have reported detrimental effects of arsenic on humans due to its high toxicity. The toxicity occurs even at the ppb level (Edwards et al., 2014). In view of the abovementioned reasons, the world health organization (WHO) advocated that the arsenic concentration in potable water must not be more than 10 μg/L (Smedley & Kinniburgh, 2002). Arsenic is a known neurotoxin (Edwards et al., 2014). The reports of several trials proved that the arsenic exposure could lead to neuropathological disorders such as angiogenesis (Messner & Bernhard, 2010), oxidative stress (Engström et al., 2010), inflammation (Vega et al., 2001), and endothelial cell dysfunction (Luo et al., 2009). It also oulines that some of the above disorders are intimately linked to cognitive dysfunction and Alzheimer's disease (Tan et al., 2003). Also human epidemiological studies reported that dermatological disorders and carcinogenesis are the other consequences noted due to the arsenic exposure (Chiou et al., 1995). The greatest problem is that the larger segments of population who were exposed to arsenic-contaminated water in Taiwan, Japan, Bangladesh, West Bengal-India, Chile, and Argentina were riskiest of developing cancer (Mandal & Suzuki, 2002). The more risks were identified for cancer of the skin and then for cancers of lung, bladder, kidney, and liver (Mandal & Suzuki, 2002). The arsenic warfare agents (CWA) belong mostly to the arseno-organic group and are characterized by high affinity for the sulfhydryl groups and high toxicity (Frith, 2013). The CWA irritate the mucous membranes of the eyes, nose, and throat and cause tearing, coughing, sneezing, pain in lungs, and breathing difficulties. The target organ for arsenic poisoning is brain, affecting learning and concentration due to its ability of crossing blood–brain barrier easily (Mundey et al., 2013). Arsenic species are disseminated in all parts of the brain. But, in hypophysis the maximum accumulation was inspected (pituitary gland) (Sanchez-Pena et al., 2010). Development of neurological complications due to acute and chronic exposure of arsenic is rather quick and usually reported as symmetrical sensorimotor neuropathy. Sensory nerves are more susceptible to arsenic than motor nerves and also neurons with long axons are affected more than neurons with short axon. Neuropsychological studies proved that there is serious failure of memory and verbal learning skills following arsenic exposure (Vahidnia et al., 2007). Also, arsenic exposure affects hematopoietic system including bone marrow, spleen, and erythrocytes. When compared to other organs such as heart, liver, and kidney, the increased rate of arsenic accumulation was identified in spleen with two to threefold greater accumulation (Zhang et al., 2014). Arsenic may cause a range of autoimmune diseases including diabetes, atherosclerosis, and non-melanoma skin cancers (Banerjee et al., 2009) and it is clearly evident that the accumulation of arsenic in the pancreas decreases the secretion of insulin besides the cells viability (Lu et al., 2011).

3.3.3 Biodetoxification of Arsenic

In previous decades, the ecology of arsenic was fully examined and numerous arsenic-metabolizing microorganisms isolated from diverse ecosystems have been characterized at the genomic level (Andres & Bertin, 2016). It has been pointed out that among the bacteria the marine bacterium *Marinomonas communis* removes the highest quantity of arsenic by withdrawing 2.290 mg As/g dw from cultures containing 5.0 mg As/l As(V) (Takeuchi et al., 2007). The foremost bacterium that was described capable of metabolizing arsenic is *Herminiimonas arsenicoxydans* (β-proteobacterium) which was isolated from an industrial wastewater treatment plant in Germany was shown to withstand enhanced levels of arsenic and to oxidize arsenite, As (III), into arsenate, As(V) (Muller et al., 2007).

3.4 Lead

Lead is the second most toxic metal after Arsenic (As), comprises 0.002% of Earth's crust (ATSDR, 2017) and its natural level remains to be below 50 mg kg^{-1} (Pais & Jones, 1997). Lead occurs naturally combined with two or more other elements to form lead compounds sources (Flegal, 1986). The most toxic form of lead is organic lead when compared with the inorganic lead due to its lipid soluble nature, leading to expeditious implications (Timbrell, 2008). Lead is the one of the metals which do not have any nutritive value and cause pollution in aquatic environment (Vasanthi et al., 2019). Although earlier literature did not focus on the biological importance of Pb, recent findings suggest that traces of Pb (~29 ng/g diet) is important for enzyme activities and cellular systems, especially during cell development, hematopoiesis, and reproduction (Assi et al., 2016). Lead is broadly dispersed in the environment since it was discovered and utilized by the humans over a very long period of time (Pompeani et al., 2013). Natural lead pollution results from volcanic explosions and forest fire. Non-natural sources were from human activities, mainly referring to the lead emission from the industry and transportation. Lead is usually found in ores, mostly with copper, zinc, and silver. The principal advantage of lead is making rechargeable storage batteries. Trucks, airplanes, automobiles, electric vehicles, tanks, and broadcasting stations all use rechargeable storage batteries as the energy source for light. Dozens of kilograms of lead were employed to manufacture one battery (Crompton, 2000) (Fig. 5).

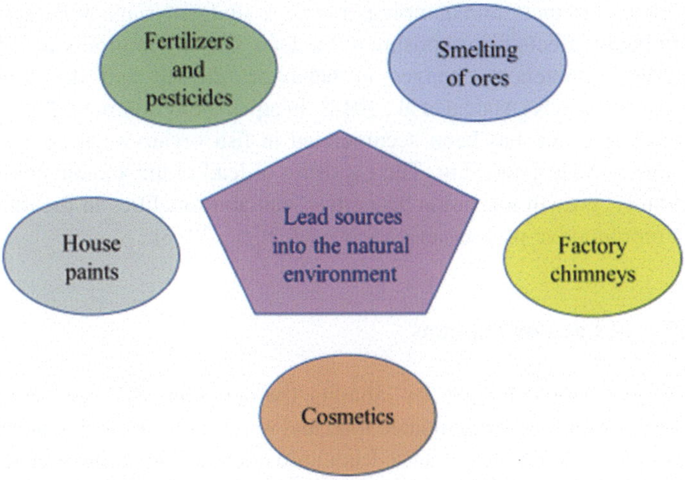

Fig. 5 Multiple sources of lead

Lead is also employed in the fuse wires, house and building roofs, sport instruments, water pipes, various alloys, bearings, and lead crystal glassware (Zietz et al., 2009) and also in prior centuries, lead was often used for covering roofs, window frames, pipes, tableware, jewelry, weights, making glass, shooting balls, printing fonts and for the toy industry for the production of lead soldiers (Wieczorek et al., 2018). Lead poisoning occurs from various forms of human-related activities such as smoking-related activities, leaded petrol, contaminated food, and drinking water, painting of home and smelting and mainly from the manufacturing industries. More than 100–200,000 tons of lead per year gets liberated from the vehicle exhausts in United States, some may be taken by plants, fixation to soil and flow into water bodies. Hence, human exposure of lead in the common population is either due to food or water (Goyer, 1990). Various physiological processes of plants were disturbed due to its extreme toxicity of the lead. A plant with elevated lead concentration fastens the production of reactive oxygen species (ROS), causing lipid membrane damage which eventually leads to chlorophyll damage and photosynthetic processes, thus inhibiting the overall plant growth (Najeeb et al., 2014). Because lead is nonbiodegradable, it is persistent in the environment and accumulates in soils, water bodies and sediments through deposition, leaching, and erosion (Maja-Lena et al., 1999).

3.4.1 Toxicity of Lead on Aquatic Organisms

In few aquatic life forms, the rising lead levels in water could provoke undesirable effects and may alter the blood parameters and neurological system in fish and other animals. The fundamental sources of lead release into the water systems are domestic wastewater, smelting and refining, manufacturing processes like metals, pulp and paper, petroleum products, and dumping of sewage sludge (Nriagu & Pacyna, 1988). Fishes were badly affected by exposure to the lead, causing mortality at lethal dose and promoting impotency, changes in behavior, growth and development at non-fatal concentrations (Afshan et al., 2014). In aquatic environment, Pb^{2+} is highly stable form of lead and has been accumulated in fish organs such as, liver, gills, kidney, scales, muscles, and skin. The exposure of lead in the aquatic environment causes mortality, growth inhibition properties, and abnormalities in the muscle and changes in reproductive performance (Srivastav et al., 2013).

3.4.2 Effect of Lead on Humans

Exposure to lead would happen by inhaling the contaminated dust particles and aerosols or by consuming contaminated food and water. Lead in food or water causes disruption of hemoglobin biosynthesis leading to anemia (Shivakumar et al., 2014). It can further include a rise in blood pressure, kidney damage, miscarriages, subtle abortions, and disruption of nervous system (Nnaji et al., 2007). Bhupander et al. (2011) demonstrated that excessive consumption of lead through food/water might lead to behavioral disturbances in children, brain damage, lower fertility in males, and reduced study skills. In the broader sense, lead harms children more when compared to the adults. Newborn and small kids are particularly vulnerable to lower lead levels than adults. In the developing nations, more than 15 million children were affected with long-lasting neurological disorders due to lead poisoning (HEI, 2004). Mahaffey et al. (1982) observed that the children in rural community contain blood lead concentrations of about 13.9 μg/dl, whereas those from cities with populations less than one million had values of 16.5 μg/g/dl of blood. A number of adverse health effects caused by lead mainly include the renal system, central nervous system, hematopoietic system, and hepatic systems and many health risks linked with chronic exposure to elevated blood lead levels are irreversible, with the nervous system being predominantly important (Flora et al., 2012). The severity of lead toxicity is correlated with higher blood lead levels, but manifestations may vary. The rise in blood lead concentration in children is associated with IQ deficits, attention-related behaviors, and poor progress in education (Centers for Disease Control and Prevention (CDC), 2016, 2018). Exposure to lead in pregnancy is connected with miscarriages and cognitive defects in the child (Centers for Disease Control and Prevention, 2017). The lead impairs the multiple biochemical processes include inhibition of calcium and reacts with proteins. After entering into body, Pb takes the place of calcium and then interacts with biological molecules and interferefs with their normal function. Lead decreases the activities of different enzymes, causes changes in their structure, and inhibits their activity by competing with the necessary cations for binding sites. The oxidative stress caused by lead is the major mechanism responsible for its toxicity, causing changes in the composition of fatty acids in the membranes (affecting processes like exocytosis and endocytosis, and signal transduction processes). Pb can also cause gene expression alterations. Lead poisoning in humans damages the kidneys, liver, heart, brain, skeleton, and the nervous system (Flora et al., 2006). Early signs of poisoning associated with exposure to lead may include memory loss, headache, dullness, and being irritable (Centers for Disease Control and Prevention (CDC), 2002). The International Agency for Research on Cancer (IARC) stated that lead is a possible carcinogenic agent in humans (Jarup, 2003). US EPA reported that the regulatory limit of lead in drinking water is 15 ppb (Martin & Griswold, 2009). The WHO recommended safe level of lead in wastewater and soils employed for agriculture are 0.01 and 0.1 ppm, respectively (Chiroma et al., 2014). Different sources and toxic effects of heavy metals on human health is presented in Table 1.

3.4.3 Biodetoxification of Lead

The lead-resistant microorganisms were isolated from metal polluted soils, industrial wastes, and from plants growing on metal-contaminated soil. Among these isolates, the following gram-positive bacteria which have been identified are *Bacillus cereus*, *Arthrobacter* sp., and *Corynebacterium* sp. and the gram-negative bacteria include *Pseudomonas marginalis*, *Pseudomonas vesicularis*, and *Enterobacter* sp. (Chen & Wang, 2007). In gram-positive bacteria, peptidoglycan together with teichoic and teichuronic acids are responsible for lead binding (Beveridge & Fyfe, 1985). Plenty of microorganisms synthesize extracellular polymers (EPs) that bind cations of toxic metals, thus protecting metal-sensitive and essential cellular components (Bruins et al., 2000). There have been reports of lead(II) binding by EPs about *Paenibacillus jamilae* (Perez et al., 2008), *Bacillus firmus*, *Halomonas* sp. (Amoozegar et al., 2012), *Pseudomonas* sp. (Salehizadeh & Shojaosadati, 2003), and *Cyanobacteria* (Paperi et al., 2006). This effect has been observed in lake sediments (Silverberg et al., 1977). Arctic marine bacteria cultured under polar conditions were capable of converting Pb(II) into trimethyl lead (Me3Pb) (Pongratz & Heumann, 1999) while *Pseudomonas* sp., *Acinetobacter* sp., *Flavobacterium* sp., and *Aeromonas* sp. transformed lead nitrate or trimethyl lead acetate into tetramethyl lead (Me4Pb) (Hughes & Poole, 1989). *Aeromonas* sp., isolated from Lake Ontario, transformed lead acetate into tetramethyl lead (Me4Pb). The precipitation of Pb(II) is used by several microorganisms to lower the concentration of free Pb(II) by sequestering it in the form of phosphate salts outside and inside the cell. The *Citrobacter freundii* precipitates Pb(II) as extracellular phosphate ($PbHPO_4$). *Staphylococcus aureus* precipitates Pb(II) inside the cell as lead phosphate $Pb_3(PO_4)_2$. This kind of protection allows this strain to withstand a 600-fold higher dose of Pb(II) compared to a sensitive strain. The marine bacterium *Vibrio harveyi* is capable of precipitating Pb (II) inside the cell in the form of an unusual phosphate compound $Pb_9(PO_4)_6$. This process is regulated partially by quorum sensing. The exact mechanisms of this regulation are not known, but it has been suggested that quorum sensing controls the availability of inorganic phosphates (Mire et al., 2004). Different types of heavy metal resistant bacteria used for biodetoxification is presented in Table 2.

Table 1 Impact of toxic heavy metals on health of humans

S. No.	Heavy metal	Source	Toxic effect	Reference
1.	Mercury	Volcanoes, geothermal sources, and topsoil enriched in mercury pertains and (from a primary source) re-emission from vegetation, land or water surfaces due to the use of land, biomass burning, artisanal small scale gold mining, nonferrous metals manufacturing, cement production, waste disposal, and caustic soda production (Pacyna et al., 2006)	Targets brain and kidneys. Methyl mercury intoxication causes "Minimata disease" and also methyl mercury causes fetal poisoning in the exposed mother which is called congenital Minamata disease (symptoms include: intellectual impairment and delayed physical development). Mercury vapors can cause bronchitis, asthma, and temporary respiratory problems. It plays a key role in damaging the tertiary and quaternary protein structure and alters the cellular function	(Baraldi et al., 2002; Kondo, 2000; Lund et al., 1993; Mason, 2009; Pirrone et al., 2001)
2	Cadmium	Smelting, mining nonferrous metals, production of nonferrous metals, iron and steel and the production and disposal of cadmium containing materials (electroplating, pigments, stabilizers and Ni-Cd batteries)	Central nervous system, brain, liver, immune system, reproductive system, chronic renal failure, atherosclerosis, cardiovascular diseases, and cancer	(Di Gioacchino et al., 2008; Thompson & Bannigan, 2008; Castro-González & Méndez-Armenta, 2008; Nordberg, 1978; Hutton, 1983; Shi et al., 2018; Messner & Bernhard, 2010; Nawrot et al., 2010)
3	Lead	Smelting, painting, plumbing and printing, battery manufacture and recovery, soldering, lead mining, imported cookware, cosmetics, contaminated tap water, antique toys painted with lead-based paint	Abdominal pain, anorexia, constipation, and convulsions. Impaired neurocognitive and behavioral development in children, structural damage to cells, proteins, nucleic acid, membranes, and lipids	(Weidenhamer & Clement, 2007; Hauptman et al., 2017; Reuben et al., 2017; Centers for Disease Control and Prevention (CDC), 2018; Mathew et al., 2011)
4	Arsenic	Volcanic emissions (ashes), wind erosion, etc.	Various skin diseases, cancers, and cardiovascular diseases	Polya & Middleton, 2017

5	Nickel	Refining operations, metal mining, alloy production, electroplating. chocolates, coffee, teas, legumes, and nuts are the primary sources (having higher nickel levels) and medical devices like dental appliances and joint prostheses Nickel containing foods include hazelnuts; cocoa and dark chocolate; fruits such as almonds, dates, figs, pineapple, plums, raspberries, and grains including bran, buckwheat, millet, whole grain bread, oats, brown rice, sesame seeds, sunflower seeds. seafood (shrimps, mussels, oysters, crab, salmon); vegetables (beans, savoy cabbage, leeks, lettuce, lentils, peas, spinach, cabbage)	Asthma, lung fibrosis, and respiratory tract cancer	(European Chemicals Agency (ECHA), 2018; International Agency for Research on Cancer (IARC), 2017; Chen et al., 2017)

Table 2 Different types of heavy metal resistant bacteria used in biodetoxification

S. No.	Name of the bacteria	Name of the heavy metal	Reference
1.	*Pseudomonas aeruginosa* *Pseudomonas putida*	Reduces toxic Hg^{2+} to volatile Hg^0 Reduces mercury	(Outten et al., 2000; Wagner-Dobler et al., 2000)
2	*Stenotrophomonas maltophilia*	Cadmium and lead	(Congeevaram et al., 2007)
3	*Bacillus cereus* M16	Biosorption of lead	(Paul et al., 2006)
4	*Alcaligenes eutrophus* CH34.	Cadmium	(Nies et al., 1989)
5	*Pseudomonas aeruginosa* RA65 (~ 9.5 kb) *P. aeruginosa* strain JCM 5962 *Pantoea agglomerans*	Plasmid encoded resistance to cadmium Cadmium Cadmium Cd (II)	(Bojorquez et al., 2016; Leonila et al., 2018; Mohamed & Abo-Amer, 2012)

4 Conclusion

Heavy metal pollution is currently a major environmental problem because metal ions persist in the environment due to their nondegradable nature. The toxicity and bioaccumulation tendency of heavy metals in the environment is a serious threat to the health of living organisms. Unlike organic contaminants, heavy metals cannot be broken down by chemical or biological processes. Hence, they can only be transformed into less toxic species. Heavy metals are generally toxic to the body at very low level. The main mechanism of heavy metal toxicity includes the generation of free radicals to cause oxidative stress, damage of biological molecules such as enzymes, proteins, lipids, and nucleic acids, damage of DNA which is key to carcinogenesis as well as neurotoxicity. Microbes have various mechanisms of metal sequestration that hold greater metal biosorption capacities. Several microorganisms like bacteria, fungi, and algae have been used to clean up heavy metal contaminated environments. Further research should be stretched out on the heavy metal biodetoxification that may promote the development of advanced techniques for the heavy metals biodetoxification in the biosphere.

References

Aakre, I., Næss, S., Kjellevold, M., Markhus, M. W., Alvheim, A. R., & Dalane, J. (2019). New data on nutrient composition in large selection of commercially available seafood products and its impact on micronutrient intake. *Food & Nutrition Research, 63*, 3573.

Abadi, D. R. V., Dobaradaran, S., Nabipour, I., Lamani, X., & Ravanipour, M. (2014). Comparative investigation of heavy metal, trace, and macro element contents in commercially valuable fish species harvested off from the Persian Gulf. *Environmental Science and Pollution Research, 22* (9), 6670–6678.

Abbas, S. H., Ismail, I. M., Mostafa, T. M., & Sulaymon, A. H. (2014). Biosorption of heavy metals: a review. *International Journal of Chemical Sciences, 3*(4), 74–102.

Abdel-Hameid, N. A. H. (2009). A protective effect of calcium carbonate against arsenic toxicity of the Nile catfish, Clarias gariepinus. *Turkish Journal of Fisheries and Aquatic Sciences, 9*, 2.

Abd-Elnaby, H., Abou-Elela, G. M., & El-Sersy, N. A. (2011). Cadmium resisting bacteria in Alexandria Eastern Harbor (Egypt) and optimization of cadmium bioaccumulation by Vibrio harveyi. *African Journal of Biotechnology, 10*, 3412–3423.

Afshan, S., Ali, S., Ameen, U. S., Farid, M., Bharwana, S. A., Hannan, F., & Ahmad, R. (2014). Effect of different heavy metal pollution on fish. *Research Journal of Chemical and Environmental Sciences, 2*(1), 74–79.

Ahmed, N. F., Sadek, K. M., Soliman, M. K., Khalil, R. H., Khafaga, A. F., Ajarem, J. S., Maodaa, S. N., & Allam, A. A. (2020). Moringa oleifera leaf extract repairs the oxidative misbalance following sub-chronic exposure to sodium fluoride in nile tilapia Oreochromis niloticus. *Animals, 10*, 626.

Akhtar, A., Wang, S. X., Ghali, L., Bell, C., & Wen, X. (2017). Recent advances in arsenic trioxide encapsulated nanoparticles as drug delivery agents to solid cancers. *Journal of Biomedical Materials Research, 31*(3), 177–188.

Aksoy, E., Salazar, J., & Koiwa, H. (2014). Cadmium determinant 1 is a putative heavy-metal transporter in Arabidopsis thaliana. *The FASEB Journal, 28*(617), 4.

Al-Garni, S. M. (2005). Biosorption of lead by gram-ve capsulated and non-capsulated bacteria. *Watermark, 31*(3), 345–350.

Ali, H., & Khan, E. (2018). Bioaccumulation of non-essential hazardous heavy metals and metalloids in freshwater fish. Risk to human health. *Environmental Chemistry Letters, 16*, 903–917.

Ali, H., Khan, E., & Ilahi, I. (2019). Environmental chemistry and ecotoxicology of hazardous heavy metals: Environmental persistence, toxicity, and bioaccumulation. *Journal of Chemistry, 2019*, 6730305.

Allen, P., Yoke, S., & Keong, W. M. (1988). Acute effects of mercury chloride on intracellular CSH level and mercury distribution in the fish Oreochromis aureus. *Bulletin of Environmental Contamination and Toxicology, 40*, 178–184.

Amini, Z., Pazooki, J., Abtahi, B., & Shokri, M. R. (2013). Bioaccumulation of Zn and Cu in Chasar bathybius (Gobiidae) tissue and its nematode parasite Dichelyne minutus, southeast of the Caspian Sea, Iran. *Indian Journal of Geo-Marine Sciences, 42*, 196–200.

Amoozegar, M. A., Ghazanfari, N., Didari, M. (2012). Lead and cadmium bioremoval by Halomonas sp., an exopolysaccharide producing halophilic bacterium. *Progress in Biological Sciences, 2*, 1–11.

Andres, J., & Bertin, P. (2016). The microbial genomics of arsenic. *FEMS Microbiology Reviews, 40*, 299–322.

Assi, M. A., Hezmee, M. N. M., Haron, A. W., Sabri, M. Y., & Rajion, M. A. (2016). The detrimental effects of lead on human and animal health. *Veterinary World, 9*, 660–671.

ATSDR. (2017). *Case studies in environmental medicine (CSEM) lead toxicity*. Retrieved November 27, 2017, from https://www.atsdr.cdc.gov/csem/lead/docs/CSEM-Lead_toxicity_508.pdf

Authman, M. M., Zaki, M. S., Khallaf, E. A., & Abbas, H. H. (2015). Use of fish as bio-indicator of the effects of heavy metals pollution. *Journal of Aquaculture Research & Development, 6*, 1–13.

Ayyat, M. S., Ayyat, A. M., Naiel, M. A., & Al-Sagheer, A. A. (2020). Reversal effects of some safe dietary supplements on lead contaminated diet induced impaired growth and associated parameters in Nile tilapia. *Aquaculture, 515*, 734580.

Azimi, A., Azari, A., Rezakazemi, M., & Ansarpour, M. (2017). Removal of heavy metals from industrial wastewaters: a review. *ChemBioEng Reviews, 4*(1), 37–59.

Bafana, A., Krishnamurthi, K., Patil, M., & Chakrabarti, T. (2010). Heavy metal resistance in Arthrobacter ramosus strain G2 isolated from mercuric salt-contaminated soil. *Journal of Hazardous Materials, 177*, 481–486.

Bagnyukova, T. V., Lushchak, O. V., Storey, K. B., & Lushchak, V. I. (2007). Oxidative stress and antioxidant defense responses by goldfish tissues to acute change of temperature from 3 to 23 C. *Journal of Thermal Biology, 32*, 227–234.

Balasubramanian, T. (2012). Heavy metal contamination and risk assessment in the marine environment of Arabian Sea, along the southwest coast of India. *American Journal of Chemistry, 2*, 191–208.

Banerjee, N., Banerjee, S., Sen, R., Bandyopadhyay, A., Sarma, N., Majumder, P., Das, J. K., Chatterjee, M., Kabir, S. N., & Giri, A. K. (2009). Chronic arsenic exposure impairs macrophage functions in the exposed individuals. *Journal of Clinical Immunology, 29*, 582–594.

Baraldi, M., Zanoli, P., Tascedda, F., Blom, J. M. C., & Brunello, N. (2002). Cognitive deficits and changes in gene expression of NMDA receptors after prenatal methylmercury exposure. *Environmental Health Perspectives, 110*, 855–858.

Barange, M., Field, J. G., Harris, R. P., Eileen, E., Hofmann, E. E., Perry, R. I., & Werner, F. (2010). *Marine ecosystems and global change* (p. 464). Oxford University Press.

Beckers, F., & Rinklebe, J. (2017). Cycling of mercury in the environment: Sources, fate, and human health implications: A review. *Critical Reviews in Environmental Science and Technology, 47*, 693–794.

Begam, M., & Sengupta, M. (2015). Immunomodulation of intestinal macrophages by mercury involves oxidative damage and rise of pro-in flammatory cytokine release in the fresh water fish Channa punctatus Bloch. *Fish & Shellfish Immunology, 45*, 378–385.

Beveridge, T. J. (1989). Role of cellular design in bacterial metal accumulation and mineralization. *Annual Review of Microbiology, 43*, 147–171.

Beveridge, T. J., & Fyfe, W. S. (1985). Metal fixation by bacterial cell walls. *Canadian Journal of Earth Sciences, 22*, 1892–1898.

Bhakta, J. N., Munekage, Y., Ohnishi, K., Jana, B., & Balcazar, J. (2014). Isolation and characterization of cadmium-and arsenic-absorbing bacteria for bioremediation. *Water, Air, and Soil Pollution, 225*, 1–10.

Bhakta, J. N., Ohnishi, K., Munekage, Y., Iwasaki, K., & Wei, M. Q. (2012). *Journal of Applied Microbiology, 112*, 1193–1206.

Bhupander, K., Mukherjee, D. P., Sanjay, K., Meenu, M., Dev, P., Singh, S. K., & Sharma, C. S. (2011). Bioaccumulation of heavy metals in muscle tissue of fishes from selected aquaculture ponds in East Kolkata Wetlands. *Annals of Biological Research, 2*(5), 125–134.

Binish, M. B., Sruthy, S., & Mahesh, M. (2015). Effect of heavy metals (Pb, Cd, Cu) on the growth of sulphate reduction associated bacterium Clostridium bifermentans isolated from Cochin estuary, Southwest coast of India. *IJMS, 5*, 1–5.

Biswas, J. K., Banerjee, A., Sarkar, B., Sarkar, D., Sarkar, S. K., Rai, M., & Vithanage, M. (2020). Exploration of an extracellular polymeric substance from earthworm gut bacterium (*Bacillus licheniformis*) for bioflocculation and heavy metal removal potential. *Applied Sciences, 10*, 349.

Boffetta, P., Merler, E., & Vainio, H. (1993). Carcinogenicity of mercury and mercury compounds. *Scandinavian Journal of Work, Environment and Health, 19*, 1–7.

Bojorquez, C., Frias Espericueta, M. G., & Voltolina, D. (2016). Removal of cadmium and lead by adapted strains of Pseudomonas aeruginosa and Enterobacter cloacae. *Revista Internacional de Contaminación Ambiental, 32*, 407–412.

Bose-O'Reilly, S., McCarty, K. M., Steckling, N., & Lettmeier, B. (2010). Mercury exposure and children's health. *Current Problems in Pediatric and Adolescent Health Care, 40*, 186–215.

Boylan, H. M., Cain, R. D., & Kingston, H. S. (2003). A new method to assess mercury emissions: A study of three coal-fired electric-generating power station configurations. *Journal of the Air & Waste Management Association, 53*, 1318–1325.

Brininstool, M. (2017). *Mineral commodity summaries 2017*. U.S. Geological Survey, Reston. Retrieved February 9, 2017, from https://minerals.usgs.gov/minerals/pubs/mcs/2017/mcs2017.pdf

Bruins, M. R., Kapil, S., & Oehme, F. W. (2000). Microbial resistance to metals in the environment. *Ecotoxicology and Environmental Safety, 45*, 198–207.

Burger, J., Jeitner, C., & Gochfeld, M. (2011). Locational differences in mercury and selenium levels in 19 species of saltwater fish from New Jersey. *Journal of Toxicology and Environmental Health, 74*, 863–874.

Canstein, V. H., Li, Y., Timmis, K. N., Deckwer, W. D., & Wagner-Dobler, I. (1999). Removal of mercury from chloralkali electrolysis wastewater by a mercury-resistant Pseudomonas putida strain. *Applied and Environmental Microbiology, 65*, 52795284.

Castoldi, A. F., Coccini, T., Ceccatelli, S., & Manzo, L. (2001). Neurotoxicity and molecular effects of methylmercury. *Brain Research Bulletin, 55*, 197–203.

Castro-González, M., & Méndez-Armenta, M. (2008). Heavy metals: Implications associated to fish consumption. *Environmental Toxicology and Pharmacology, 26*, 263–271.

Centers for Disease Control and Prevention. (2017). *What do parents need to know to protect their children?* Retrieved May 17, 2017, from https://www.cdc.gov/nceh/lead/acclpp/blood_lead_levels.htm

Centers for Disease Control and Prevention (CDC). (2002). Managing elevated blood lead levels among young children: Recommendations from the advisory committee on childhood lead poisoning prevention, Atlanta.

Centers for Disease Control and Prevention (CDC). (2016). *Standard surveillance definitions and classifications*. Retrieved November 18, 2016, from https://www.cdc.gov/nceh/lead/data/definitions.htm

Centers for Disease Control and Prevention (CDC). (2018). *Screening young children for lead poisoning: Guidance for state and local public health officials*. Retrieved October 2, 2018, from https://www.cdc.gov/nceh/lead/publications/screening.htm

Chang, J. S., & Law, W. S. (1998). Development of microbial mercury detoxification process using a mercury-hyper resistant strain of *Pseudomonas aeruginosa* PU21. *Biotechnology and Bioengineering, 57*, 462470.

Chen, C., & Wang, J. (2007). Response of *Saccharomyces cerevisiae* to lead ion stress. *Applied Microbiology and Biotechnology, 74*, 683–687.

Chen, Q. Y., Brocato, J., Laulicht, F., & Costa, M. (2017). Mechanisms of nickel carcinogenesis. In A. Mudipalli & J. T. Zelikoff (Eds.), *Essential and non-essential metals. Molecular and integrative toxicology* (pp. 181–197). Springer International Publishing AG.

Chien, M. F., Lin, K. H., Chang, J. E., Huang, C. C., Endo, G., & Suzuki, S. (2010). Interdisciplinary studies on environmental chemistry—biological responses to contaminants. In N. Hamamura, S. Suzuki, S. Mendo, C. M. Barroso, H. Iwata, & S. Tanabe (Eds.), *Distribution of mercury resistance determinants in a highly mercury polluted area in Taiwan* (p. 3136). TERRAPUB.

Chiou, H., Hsueh, Y., & Liaw, K. F. (1995). Incidence of internal cancers and ingested inorganic As: A seven-year follow-up study in Taiwan. *Cancer Research, 55*, 1296–1300.

Chiroma, T. M., Ebewele, R. O., & Hymore, F. K. (2014). Comparative assessment of heavy metal levels in soil, vegetables and urban grey water used for irrigation in Yola and Kano. *International Refereed Journal of Engineering and Science, 3*(2), 1–9.

Chovanova, K., Sladekova, D., & Kmet, V. (2004). Identification and characterization of eight cadmium resistant bacterial. *Biologia, 59*(6), 817–827.

Concórdio-Reis, P., & Freitas, F. (2019). Environmental applications: Biopolymer sorbents for heavy metal removal. In *Encyclopedia of polymer applications* (pp. 1069–1086). CRC Press.

Congeevaram, S., Dhanarani, S., Park, J., Dexilin, M., & Thamaraiselvi, K. (2007). Biosorption of chromium and nickel by heavy metal resistant fungal and bacterial isolates. *Journal of Hazardous Materials, 146*, 270–277.

Crompton, T. R. (2000). *Battery reference book* (pp. 2–4). Newnes.

Dash, H. R., & Das, S. (2012). Bioremediation of mercury and importance of bacterial mer genes. *International Biodeterioration and Biodegradation, 75*, 207213.

Datta, S., Ghosh, D., Saha, D. R., Bhattacharaya, S., & Mazumder, S. (2009). Chronic exposure to low concentration of arsenic is immunotoxic to fish: Role of head kidney macrophages as biomarkers of arsenic toxicity to Clarias batrachus. *Aquatic Toxicology, 92*, 86–94.

Davidson, P. W., Myers, G. J., & Weiss, B. (2004). Mercury exposure and child development outcomes. *Pediatrics, 113*, 1023–1029.

Demey, H., Vincent, T., & Guibal, E. (2018). A novel algal-based sorbent for heavy metal removal. *Chemical Engineering Journal, 332*, 582–595.

Demir, A., & Arisoy, M. (2007). Biological and chemical removal of Cr (VI) from waste water: Cost and benefit analysis. *Journal of Hazardous Materials, 147*, 275–280.

Deng, X., & Jia, P. (2011). Construction and characterization of a photosynthetic bacterium genetically engineered for Hg21 uptake. *Bioresource Technology, 102*, 30833088.

DiGioacchino, M., Petrarca, C., Perrone, A., Farina, M., Sabbioni, E., Hartung T., Martino, S., Esposito, D. L., Lotti, L. V., Mariani-Costantini, R. (2008). Autophagy as an ultrastructural marker of heavy metal toxicity in human cord blood hematopoietic stem cells. *Science of the Total Environment, 392,* 50–58.

Dixit, R., Malaviya, D., Pandiyan, K., Singh, U. B., Sahu, A., Shukla, R., Singh, B. P., Rai, J. P., Sharma, P. K., & Lade, H. (2015). Bioremediation of heavy metals from soil and aquatic environment: An overview of principles and criteria of fundamental processes. *Sustainability, 7*, 2189–2212.

Dong, Q., Fang, J., Huang, F., & Cai, K. (2019). Silicon amendment reduces soil Cd availability and Cd uptake of two pennisetum species. *International Journal of Environmental Research and Public Health, 16*, 1624.

Driscoll, C. T., Mason, R. P., Chan, H. M., Jacob, D. J., & Pirrone, N. (2013). Mercury as a global pollutant: sources, pathways, and effects. *Environmental Science & Technology, 47*, 4967–4983.

Edwards, M., Johnson, L., Mauer, C., Barber, R., & Hall, J. (2014). Regional specific groundwater arsenic levels and neuropsychological functioning: a crosssectional study. *International Journal of Environmental Health Research, 24*, 546–557.

Ehrlich, H. L. (1997). Microbes and metals. *Applied Microbiology and Biotechnology, 48*, 687–692.

Ehya, F., & Marbouti, Z. (2016). Hydrochemistry and contamination of groundwater resources in the Behbahan plain, SW Iran. *Environment and Earth Science, 2016*, 75–45.

El Nemr, A., El-Said, G. F., Khaled, A., & Ragab, S. (2016). Distribution and ecological risk assessment of some heavy metals in coastal surface sediments along the Red Sea, Egypt. *International Journal of Sediment Research, 31*, 164–172.

Engström, K., Vahter, M., & Johansson, G. (2010). Chronic exposure to cadmium and arsenic strongly influences concentrations of 8-oxo-7,8-dihydro-2′- deoxyguanosine in urine. *Free Radical Biology & Medicine, 48*, 1211–1217.

Epstein, E. (2002). Health issues related to beneficial use of biosolids. In: *16th annual residuals and biosolids management conference of the water environment federation* (p. 9)

Escudero-Lourdes, C., Wu, T., Camarillo, J. M., & Gandolfi, A. J. (2012). Interleukin-8 (IL-8) over-production and autocrine cell activation are key factors in monomethylarsonous acid [MMA (III)]- induced malignant transformation of urothelial cells. *Toxicology and Applied Pharmacology, 258*(1), 10–18.

European Chemicals Agency (ECHA). (2018). *Annex 1–background document in support of the committee for risk assessment (RAC) for evaluation of limit values for nickel and its compounds in the workplace; ECHA/RAC/A77-0-0000001412-86-189/F* (pp. 1–211). European Chemicals Agency.

Fella-Temzi, S., Yalaoui-Guellal, D., Rodriguez-Carvajal, M. A., Belhadi, D., Madani, K., & Kaci, Y. (2018). Removal of lead by exopolysaccharides from Paenibacillus peoriae strainTS7 isolated from rhizosphere of durum wheat. *Biocatalysis and Agricultural Biotechnology, 16*, 425–432.

Feng, J., Shi, Q., Wang, X., Wei, M., Yang, F., & Xu, H. (2010). Silicon supplementation ameliorated the inhibition of photosynthesis and nitrate metabolism by cadmium (Cd) toxicity in Cucumis sativus L. *Scientia Horticulturae, 123*, 521–530.

Fielder, R., & Dale, E. (1983). *Cadmium and its compounds*. HM Stationery Office.

Fjeld, E., Haugen, T. O., & Vøllestaed, L. A. (1998). Permanent impairment in the feeding behavior of grayling (*Thymallus thymallus*) exposed to methylmercury during embryogenesis. *Science of the Total Environment, 213*, 247–254.

Flegal, A. R. (1986). Lead in tropical marine systems: A review. *Science of the Total Environment, 58*, 1–8.

Flora, G., Gupta, D., & Tiwari, A. (2012). Toxicity of lead: A review with recent updates. *Interdisciplinary Toxicology, 5*(2), 47–58.

Flora, S. J. S. (2015). Arsenic toxicology. In S. J. S. Flora (Ed.), *Handbook of arsenic toxicology*. Elsevier Inc.

Flora, S. J. S., Flora, G. J. S., & Saxena, G. (2006). Environmental occurrence, health effects and management of lead poisoning. In S. B. Cascas & J. Sordo (Eds.), *Lead: Chemistry, analytical aspects, environmental impacts and health effects* (pp. 158–228). Elsevier Publication.

Franco, J. L., Braga, H., Stringari, J., Missau, F. C., Posser, T., Mendes, B. G., Leal, R., Dos Santos, A. R. S., Dafre, A. L., & Pizzolatti, M. G. (2007). Mercurial-induced hydrogen peroxide generation in mouse brain mitochondria: Protective effects of quercetin. *Chemical Research in Toxicology, 20*, 1919–1926.

Frith, J. (2013). Arsenic – the "poison of kings" and the "saviour of syphilis". *JMVH, 21*(4), 11–17.

Garcia, J. C., Martinez, D. S. T., Alves, O. L., Leonardo, A. F. G., & Barbieri, E. (2015). Ecotoxicological effects of carbofuran and oxidized multiwalled carbon nanotubes on the freshwater fish Nile tilapia: Nanotubes enhance pesticide ecotoxicity. *Ecotoxicology and Environmental Safety, 111*, 131–137.

García-Navarro, J. A., Franco, L., & Romero, D. (2017). Differences in the accumulation and tissue distribution of Pb, Cd, and Cu in Mediterranean mussels (Mytilus galloprovincialis) exposed to single, binary, and ternary metal mixtures. *Environmental Science and Pollution Research, 24*, 6599–6610.

Gati, G., Pop, C., Brudasca, F., Gurzău, A. E., & Spînu, M. (2016). The ecological risk of heavy metals in sediment from the Danube delta. *Ecotoxicology, 25*, 688–696.

Gautam, P. K., Gautam, R. K., Chattopadhyaya, M. C., Banerjee, S., Chattopadhyaya, M. C., & Pandey, J. D. (2016). Heavy metals in the environment: Fate, transport, toxicity and remediation technologies thermodynamic profiling of pollutants view project materials for solid oxide fuel cells view project heavy metals in the environment: Fate, transport, toxicity and rem.

Gilbertson, M., & Carpenter, D. O. (2004). An ecosystem approach to the health effects of mercury in the Great Lakes basin ecosystem. *Environmental Research, 95*(3), 240–246.

Giles, M. A. (1988). Accumulation of cadmium by rainbow trout, salmo gairdneri, during extended exposure. *Canadian Journal of Fisheries and Aquatic Sciences, 45*, 1045–1053.

Gourdon, R., Bhende, S., Rus, E., & Sofer, S. S. (1990). *Biotechnology Letters, 12*, 839e42.

Goyer, R. A. (1990). Lead toxicity: From overt to subclinical to subtle health effects. *Environmental Health Perspectives, 86*, 177–181.

Gu, B. G., Liang, W., Yang, T. Z., Hu, Z. J., & Shen, H. D. (2019). Metallothionein, hemocyte status and superoxide dismutase/aspartate aminotransferase activity are sensitive biomarkers of cadmium stress in Onchidium reevesii. *Aquatic Toxicology, 215*, 105284.

Harada, M. (1995). Minamata disease: Methylmercury poisoning in japan caused by environmental pollution. *Critical Reviews in Toxicology, 25*, 1–24.

Hauptman, M., Bruccoleri, R., & Woolf, A. D. (2017). An update on childhood lead poisoning. *Clinical Pediatric Emergency Medicine, 18*(3), 181–192.

HEI. (2004). *Special report 15, health effects of outdoor air pollution in developing countries of Asia: a literature review*. Health Effects Institute.

Henriques, B., Rocha, L. S., Lopes, C. B., Figueira, P., Monteiro, R. J. R., Duarte, A. C., Pardal, M. A., & Pereira, E. (2015). Study on bioaccumulation and biosorption of mercury by living marine macroalgae: Prospecting for a new remediation biotechnology applied to saline waters. *Chemical Engineering Journal, 281*, 759–770.

Hoostal, M. J., Bidart-Bouzat, M. G., & Bouzat, J. L. (2008). Local adaptation of microbial communities to heavy metal stress in polluted sediments of Lake Erie. *FEMS Microbiology Ecology, 65*, 156–168.

Hou, Y., Cheng, K., & Li, Z. (2015). Biosorption of cadmium and manganese using free cells of *Klebsiella* sp. Isolated from waste water. *PLoS One, 10*(10), 0140962.

Hughes, M. F., Beck, B. D., Chen, Y., Lewis, A. S., & Thomas, D. J. (2011). Arsenic exposure and toxicology: a historical perspective. *Toxicological Sciences, 123*(2), 305–332.

Hughes, M. N., & Poole, R. K. (1989). *Metals and micro-organisms* (p. 277). Chapman & Hall.

Hussein, H., Farag, S., & Kandil, K. (2005). Tolerance and uptake of heavy metals by *Pseudomonads*. *Process Biochemistry, 40*, 955–961.

Hutton, M. (1983). Sources of cadmium in the environment. *Ecotoxicology and Environmental Safety, 7*, 9–24.

Hutton, M. L., & Hutchinson, T. C. (1987). *Lead, mercury, cadmium and arsenic in the environment*. John Wiley & Sons Ltd..

Idriss, A., & Ahmad, A. (2015). Heavy metal concentrations in fishes from Juru River, estimation of the health risk. *Bulletin of Environmental Contamination and Toxicology, 94*, 204–208.

International Agency for Research on Cancer (IARC). (2017). *IARC monographs on the evaluation of carcinogenic risks to humans: Nickel and nickel compounds monograph* (Vol. 100, pp. 169–218). WHO Press.

Islam, M. S., Hossain, M. B., Matin, A., & Sarker, M. S. I. (2018). Assessment of heavy metal pollution, distribution and source apportionment in the sediment from Feni River estuary, Bangladesh. *Chemosphere, 202*, 25–32.

Iyer, A., Mody, K., & Jha, B. (2005). Biosorption of heavy metals by a marine bacterium. *Marine Pollution Bulletin, 50*, 340–343.

Jaishankar, M., Tseten, T., Anbalagan, N., Mathew, B. B., & Beeregowda, K. N. (2014). Toxicity, mechanism and health effects of some heavy metals. *Interdisciplinary Toxicology, 7*, 60–72.

Jarup, L. (2003). Hazards of heavy metal contamination. *British Medical Bulletin, 68*(1), 167–182.

Javed, H., Islam, A., Chauhan, A., Kumar, S., & Sushil, K. (2018). Efficacy of engineered GO Amberlite XAD-16 picolylamine sorbent for the trace determination of Pb (II) and Cu (II) in fishes by solid phase extraction column coupled with inductively coupled plasma optical emission spectrometry. *Scientific Reports, 8*, 17560.

Johncy, M., Hemambika, B., Hemapriya, J., & Rajesh Kannan, V. (2010). Comparative assessment of heavy metal removal by immobilized and dead bacterial cells: A biosorption approach. *African Journal of Environmental Science and Technology, 4*(2), 77–83.

Kabata-Pendias, A., & Pendias, H. (2001). *Trace elements in soils and plants*. CRC Press.

Kafilzadeh, F., & Mirzaei, N. (2008). Growth pattern of Hg resistant bacteria isolated from Kor River in the presence of mercuric chloride. *Pakistan Journal of Biological Sciences, 11*(18), 2243–2248.

Karaca, H., Tay, T., & Kivanac, M. (2010). Kinetics of lead ion bisorption from aqueous solution onto lyophilized Aspergillus niveus. *Water Practice Technology, 5*(1), 1–10.

Kaur, S., Kamli, M. R., Ali, A. (2011). Role of arsenic and its resistance in nature. *Canadian Journal of Microbiology, 57*, 769–774. https://doi.org/10.1139/w11-062

Khan, Z., Kunal, M., Soni, A., & Madamwar, D. (2014). Microaerophilic degradation of sulphonated azo dye Reactive Red 195 by bacterial consortium AR1 through cometabolism. *International Biodeterioration and Biodegradation, 94*, 167–175.

Kim, H. T., Loftus, J. P., Mann, S., & Wakshlag, J. J. (2018). Evaluation of arsenic, cadmium, lead and mercury contamination in over-the-counter available dry dog foods with different animal ingredients (Red Meat, Poultry, and Fish). *Frontiers in Veterinary Science, 5*, 264.

Konate, A., He, X., Zhang, Z., Ma, Y., Zhang, P., Alugongo, G. M., & Rui, Y. (2017). Magnetic (Fe_3O_4) nanoparticles reduce heavy metals uptake and mitigate their toxicity in wheat seedling. *Sustainability, 9*, 790.

Kondo, K. (2000). Congenital minamata disease: Warnings from Japan's experience. *Journal of Child Neurology, 157*, 458–464.

Krishna, A. K., & Mohan, K. R. (2016). Distribution, correlation, ecological and health risk assessment of heavy metal contamination in surfac e soils around an industrial area, Hyderabad, India. *Environment and Earth Science, 75*, 411.

Kumari, B., Kumar, V., Sinha, A. K., Ahsan, J., & Ghosh, A. K. (2016). Toxicology of arsenic in _ sh and aquatic system. *Environmental Chemistry Letters, 15*, 43–64.

Lacerda, E. C. M., dos Passos Galluzzi Baltazar, M., dos Reis, T. A., Côrrea, B., & Gimenes, L. J. (2019). Copper biosorption from an aqueous solution by the dead biomass of *Penicillium ochrochloron*. *Environmental Monitoring and Assessment, 191*, 247.

Lamborg, C. H., Hammerschmidt, C. R., Bowman, K. L., Swarr, G. J., Munson, K. M., Ohnemus, D. C., Lam, P. J., Heimbürger, L. E., Rijkenberg, M. J. A., & Saito, M. A. (2014). A global ocean inventory of anthropogenic mercury based on water column measurements. *Nature, 512*, 65.

Leonila, M. L., Cavalcanti, A. D., Marcos, C., Luna José, C. V., Júnior Rosileide, F. S., Andrade Thayse, A., de Lima Silva Camilo, E., & La Rotta Galba, M. (2018). Campos-Takaki. Cadmium removal from aqueous solutions by strain of *Pantoea agglomerans* UCP1320 isolated from laundry effluent. *Open Microbiology Journal, 12*, 297–307.

Li, F., Qiu, Z. Z., & Zhang, J. D. (2017). Investigation, pollution mapping and simulative leakage health risk assessment for heavy metals and metalloids in groundwater from a typical brownfield, middle China. *International Journal of Environmental Research and Public Health, 14*(7), 768.

Li, W., & Wang, W. X. (2019). Inter-species differences of total mercury and methylmercury in farmed fish in Southern China: Does feed matter? *Science of the Total Environment, 651*, 1857–1866.

Lin, J., & Harichund, C. (2012). Production and characterization of heavy-metal removing bacterial bioflocculants. *African Journal of Biotechnology, 11*, 9619–9629.

Litynska, M., Astrelin, I., & Tolstopalova, N. (2017). Ways of arsenic compounds getting into natural. *MESE, 3*(1), 50–60.

Liu, Q. Q., Wang, F. F., Meng, F. P., Jiang, L., Li, G. J., & Zhou, R. G. (2018). Assessment of metal contamination in estuarine surface sediments from Dongying City, China: Use of a modified ecological risk index. *Marine Pollution Bulletin, 126*, 293–303.

Lopez, M. E. C., Macêdo, G. L., Pereira, S. I., Arrifano, G. D. P., Picanço-Diniz, D. L., Nascimento, J. L. M. D., & Herculano, A. (2009). Mercury and human genotoxicity: Critical considerations and possible molecular mechanisms. *Pharmacological Research, 60*, 212–220.

Lu, T. H., Su, C. C., Chen, Y. W., Yang, C. Y., Wu, C. C., Hung, D. Z., Chen, C. H., Cheng, P. W., Liu, S. H., & Huang, C. F. (2011). Arsenic induces pancreatic beta-cell apoptosis via the oxidative stress regulated mitochondria-dependent and endoplasmic reticulum stress-triggered signaling pathways. *Toxicology Letters, 201*, 15–26.

Lund, B. O., Miller, D. M., & Woods, J. S. (1993). Studies on Hg(II)-induced H_2O_2 formation and oxidative stress in vivo and in vitro in rat kidney mitochondria. *Biochemical Pharmacology, 45*, 2017–2024.

Luo, J., Qiu, Z., Shu, W., Zhang, Y., & Zhang, L. (2009). Effects of arsenic exposure from drinking water on spatial memory, ultra-structures and NMDAR gene expression of hippocampus in rats. *Toxicology Letters, 184*, 121–125.

Ma, H., Li, X., Wei, M., Zeng, G., Hou, S., Li, D., & Xu, H. (2020). Elucidation of the mechanisms into effects of organic acids on soil fertility, cadmium speciation and ecotoxicity in contaminated soil. *Chemosphere, 239*, 124706.

Maddela, N. R., Kakarla, D., Garcia, L. C., Chakraborty, S., Venkateswarlu, K., & Megharaj, M. (2020). Cocoa-laden cadmium threatens human health and cacao economy: A critical view. *Science of the Total Environment, 720*, 137645.

Maeda, E., Murata, K., & Kumazawa, Y. (2019). Associations of environmental exposures to methylmercury and selenium with female infertility: A case control study. *Environmental Research, 168*, 357–363.

Mahaffey, K. R., Joseph, D., Annest, L., Robert, J. M. S., Robert, S., & Murphy, M. S. P. H. (1982). National estimates of blood lead levels: United States, 1976-1980; association with selected demographic and socioeconomic factors. *The New England Journal of Medicine, 307*(10), 537–579.

Mahvi, A. H., & Diels, L. (2001). Biological removal of cadmium by Alcaligenes eutrophus CH34. *International journal of Environmental Science and Technology, 1*(3), 199–204.

Maier, D., Blaha, L., Giesy, J. P., Henneberg, A., & Köhler, H. R. (2014). Biological plausibility as a tool to associate analytical data for micropollutants and effect potentials in wastewater, surface water, and sediments with effects in fishes. *Water Research, 22*(9), 6670–6678.

Maitra, S. (2016). Study of genetic determinants of nickel and cadmium resistance in bacteria-a review. *International Journal of Current Microbiology and Applied Sciences, 5*(11), 459–471.

Maja-Lena, B., Richard, B., Ingemar, R., Ove, E., Jerzy, B., & Kjell, B. (1999). The medieval metal industry was the cradle of modern large-scale atmospheric lead pollution in Northern Europe. *Environmental Science & Technology, 33*(24), 4391–4395.

Mandal, B., & Suzuki, K. (2002). Arsenic round the world: A review. *Talanta, 58*, 201–235.

Martin, S., & Griswold, W. (2009). Human health effects of heavy metals. *Environmental Science and Technology Briefs, 15*, 1–6.

Masindi, V., & Muedi, K. L. (2018). Environmental contamination by heavy metals. In *Heavy metals*. InTech.

Mason, R. P. (2009). Mercury emissions from natural processes and their importance in the global mercury cycle. In *Mercury fate and transport in the global atmosphere* (pp. 173–191). Springer Science and Business Media LLC.

Mathe, I., Benedek, T., Tancsics, A., Palatinszky, M., Lanyi, S., & Marialigeti, K. (2012). Diversity, activity, antibiotic and heavy metal resistance of bacteria from petroleum hydrocarbon contaminated soils located in Harghita Country (Romania). *International Biodeterioration & Biodegradation, 73*, 41–49.

Mathew, B. B., Tiwari, A., & Jatawa, S. K. (2011). Free radicals and antioxidants: A review. *Journal of Pharmacy Research, 4*(12), 4340–4343.

Matsuo, T. (2003). Japanese experiences of environmental management. *Water Science and Technology, 47*, 7–14.

Maynaud, G., Brunel, B., Yashiro, E., Mergeay, M., Cleyet-Marel, J. C., & Le Quere, A. (2014). CadA of Mesorhizobium metallidurans isolated from a zinc-rich mining soil is a P(IB-2)-type ATPase involved in cadmium and zinc resistance. *Research in Microbiology, 165*, 175–189.

Messner, B., & Bernhard, D. (2010). Cadmium and cardiovascular diseases: Cell biology, pathophysiology, and epidemiological relevance. *BioMetals, 23*(5), 811–822.

Mire, C. E., Tourjee, J. A., O'Brien, W. F., Ramanujachary, K. V., & Hecht, G. B. (2004). Lead precipitation by Vibrio harveyi: Evidence for novel quorum-sensing interactions. *Applied and Environmental Microbiology, 70*, 855–864.

Mohamed, R. M., & Abo-Amer, A. E. (2012). Isolation and characterization of heavy-metal resistant microbes from roadside soil and phylloplane. *Journal of Basic Microbiology, 52*, 53–65.

Morais, S., Costa, F. G., & Pereira, M. L. (2012). Heavy metals and human health. In J. Oosthuizen (Ed.), *Environmental health – emerging issues and practice* (pp. 227–246). InTech.

Morel, F., Kraepiel, A. M. L., & Amyot, M. (1998). The chemical cycle and bioaccumulation of mercury. *Annual Review of Ecological Systems, 29*, 543–566.

Moreno, F. N., Anderson, C. W. N., Stewart, R. B., & Robinson, B. H. (2008). Phytofiltration of mercury- contaminated water: Volatilisation and plant-accumulation aspects. *Environmental and Experimental Botany, 62*, 78–85.

Morillo Pérez, J. A., García-Ribera, R., Quesada, T., Aguilera, M., Ramos-Cormenzana, A., & Monteoliva-Sánchez, M. (2008). Biosorption of heavy metals by the exopolysaccharide produced by Paenibacillus jamilae. *World Journal of Microbiology and Biotechnology, 24*, 2699–2704.

Muller, D., Médigue, C., Koechler, S., Barbe, V., Barakat, M., & Talla, E. (2007). A tale of two oxidation states?: Bacterial colonization of arsenic-rich environments. *PLoS Genetics, 3*, e53.

Mundey, M. K., Roy, M., Roy, S., Awasthi, M. K., & Sharma, R. (2013). Antioxidant potential of Ocimum sanctum in arsenic induced nervous tissue damage. *Brazilian Association of Veterinary Pathology, 6*, 95–101.

Najeeb, U., Ahmad, W., Zia, M. H., Malik, Z., & Zhou, W. (2014). Enhancing the lead phytostabilization in wetland plant Juncus eff usus L. through somaclonal manipulation and EDTA enrichment. *Arabian Journal of Chemistry, 10*, 3310–3317.

Nath, S., Deb, B., & Sharma, I. (2012). Isolation and characterization of cadmium and lead resistant bacteria. *Global Advanced Research Journal of Microbiology, 1*, 194–198.

Naujokas, M. F., Anderson, B., Ahsan, H., Vasken Aposhian, H., Graziano, J. H., Thompson, C., & Suk, W. A. (2013). The broad scope of health effects from chronic arsenic exposure: Update on a worldwide public health problem. *Environmental Health Perspectives, 121*, 295–302.

Nawrot, T. S., Staessen, J. A., & Roels, H. A. (2010). Cadmium exposure in the population: From health risks to strategies of prevention. *BioMetals, 23*(5), 769–782.

Nies, D. H., Nies, A., Chu, L., & Silver, S. (1989). Expression and nucleotide sequence of a plasmid determined divalent cation efflux system from Alcaligenes eutrophus. *Proceedings of the National Academy of Sciences of the United States of America, 86*, 7351–7355.

Nithya, C., Gnanalakshmi, B., & Pandian, S. K. (2011). Assessment and characterization of heavy metal resistance in Palk Bay sediment bacteria. *Marine Environmental Research, 71*, 283–294.

Nnaji, J. C., Uzairu, A., Harrison, G. F. S., & Balarabe, M. L. (2007). Evaluation of cadmium, chromium, copper, lead and zinc concentrations in the fish head/viscera of Oreochromis niloticus and Syndontis schall of River Galma, Zaria, Nigeria. *Electronic Journal of Environmental, Agriculture and Food Chemistry, 6*, 2420–2426.

Nordberg, M. (1978). Studies on metallothionein and cadmium. *Environmental Research, 15*, 381–404.

Norouzi, E., Bahramifar, N., & Ghasempouri, S. M. (2012). Effect of teeth amalgam on mercury levels in the colostrums human milk in Lenjan. *Environmental Monitoring and Assessment, 184*, 375–380.

Nriagu, J. O., & Pacyna, J. (1988). Quantitative assessment of worldwide contamination of air, water and soil by trace metals. *Nature, 333*, 134–139.

Obi, E., Okafor, C., Igwebe, A., Ebenebe, J., Johnson Afonne, O., Ifediata, F., Orisakwe, O. E., Nriagu, J. O., & Basu, N. (2015). Elevated prenatal methylmercury exposure in Nigeria: evidence from maternal and cord blood. *Chemosphere, 119*, 485–489.

Ogbomida, E. T., Nakayama, S., Bortey-Sam, N., Oroszlany, B., Tongo, I., Enuneku, A. A., Ogbeide, O., Ainerua, M. O., Fasipe, I. P., Ezemonye, L. I., Mizukawa, H., & Ikenaka, Y. (2018). Accumulation patterns and risk assessment of metals and metalloid in muscle and ofcial of free-range chickens, cattle and goat in Benin City, Nigeria. *Ecotoxicology and Environmental Safety, 151*, 98–108.

Okolo, N. V., Olowolafe, E. A., Akawu, I., & Okoduwa, S. I. R. (2016). Effects of industrial effluents on soil resources in Challawa industrial area, Kano, Nigeria. *Journal of Global Ecology and Environment, 5*, 1–10.

Outten, F. W., Outten, C. E., & Halloran, T. O. (2000). Metalloregulatory systems at the interface between bacterial metal homeostasis and resistance. In G. Storz & R. Hengge-Aronis (Eds.), *Bacterial stress responses* (pp. 145–157). ASM Press.

Oves, M., Saghir Khan, M., & Zaidi, A. (2013). Biosorption of heavy metals by *Bacillus thuringiensis* strain OSM29 originating from industrial effluent contaminated north Indian soil. *Saudi Journal of Biological Sciences, 20*, 121–129.

Öz, M., Inanan, B. E., & Dikel, S. (2018). Effect of boric acid in rainbow trout (Oncorhynchus mykiss) growth performance. *Journal of Applied Animal Research, 46*, 990–993.

Pacyna, E. G., Pacyna, J. M, Steenhuisen, F., Wilson, S. (2006). Global anthropogenic mercury emission inventory for 2000. *Atmospheric Environment, 40*, 4048–4063.

Pais, I., & Jones, J. B. (1997). *The handbook of trace elements*. Saint Lucie Press.

Panagiotaras, D., & Nikolopoulos, D. (2015). Arsenic occurrence and fate in the environment. A geochemical perspective. *Journal of Earth Science and Climatic Change, 6*(4), 1–9.

Panwichian, S., Kantachote, D., Wittayaweerasa, B., & Mallavarapu, M. (2011). Removal of heavy metals by exopolymeric substances produced by resistant purple non sulphur bacteria isolated from contaminated shrimp ponds. *Electronic Journal of Biotechnology, 14*, 3458.

Paperi, R., Micheletti, E., & De Philippis, R. (2006). Optimization of copper sorbing-desorbing cycles with confined cultures of the exopolysaccharide-producing cyanobacterium Cyanospira capsulata. *Journal of Applied Microbiology, 101*, 1351–1356.

Patra, M., Bhowmik, N., Bandopadhyay, B., & Sharma, A. (2004). Comparison of mercury lead and arsenic with respect to genotoxic effects on plant systems and the development of genetic tolerance. *Environmental and Experimental Botany, 52*, 199–223.

Paul, S., Bera, D., & Chattopadhyay, P. (2006). Biosorption of Pb(II) by Bacillus cereus M116 immobilized in calcium alginate gel. *Journal of Hazardous Substance Research, 5*, 2.

Pepi, M., Focardi, S., Tarabelli, A., Volterrani, M., & Focardi, S. E. (2013). Bacterial strains resistant to inorganic and organic forms of mercury isolated from polluted sediments of the Orbetello Lagoon, Italy, and their possible use in bioremediation processes. *E3S Web of Conferences, 1*, 31002.

Pepi, M., Gaggi, C., Bernardini, E., & Focardi, S. (2011). Mercury-resistant bacterial strains *Pseudomonas* and *Psychrobacter* spp. isolated from sediments of Orbetello Lagoon (Italy) and their possible use in bioremediation processes. *International Biodeterioration & Biodegradation, 65*, 85–91.

Perez, M. P. J. A., Garcıa-Ribera, R., Quesada, T., Aguilera, M., Ramos-Cormenzana, A., & Monteoliva-Sanchez, M. (2008). Biosorption of heavy metals by the exopolysaccharide produced by Paenibacillus jamilae. *World Journal of Microbiology and Biotechnology, 24*, 2699–2704.

Pirrone, N., Costa, P., Pacyna, J., & Ferrara, R. (2001). Mercury emissions to the atmosphere from natural and anthropogenic sources in the Mediterranean region. *Atmospheric Environment, 35*, 2997–3006.

Polya, D. A., & Middleton, D. R. (2017). Arsenic in drinking water: Sources & human exposure. In P. Bhattacharya, D. A. Polya, & D. Draganovic (Eds.), *Best practice guide on the control of arsenic in drinking water* (1st ed.). International Water Association Publishing.

Pompeani, D. P., Abbott, M. B., Steinman, B. A., & Bain, D. J. (2013). Lake sediments record prehistoric lead pollution related to early copper production in North America. *Environmental Science & Technology, 47*(11), 5545–5552.

Pongratz, R., & Heumann, K. G. (1999). Production of methylated mercury, lead and cadmium by marine bacteria as a significant natural source for atmospheric heavy metals in polar regions. *Chemosphere, 39*, 89–102.

Porto, J. I. R., Araujo, S. O. C., & Feldberg, E. (2005). Mutagenic effects of mercury pollution as revealed by micronucleus test on three Amazonian fish species. *Environmental Research, 97*(3), 287–292.

Radwan, M. A., & Salama, A. K. (2006). Market basket survey for some heavy metals in Egyptian fruits and vegetables. *Food and Chemical Toxicology, 44*, 1273–1278.

Rahman, M. A., Hasegawa, H., & Lim, R. P. (2012). Bioaccumulation, biotransformation and trophic transfer of arsenic in the aquatic food chain. *Environmental Research, 116*, 118–135.

Rahman, M. A., Hogan, B., Duncan, E., Doyle, C., Krassoi, R., Rahman, M. M., & Hassler, C. (2014). Toxicity or arsenic species to three freshwater organisms and biotranformation or inorganic arsenic by freshwater phytoplankton (Chlorella sp. CE-35). *Ecotoxicology and Environmental Safety, 106*, 126–135.

Rajeshkumar, S., & Li, X. (2018). Bioaccumulation of heavy metals in fish species from the Meiliang Bay, Taihu Lake, China. *Toxicology Reports, 5*, 288–295.

Ravenscroft, P., Brammer, H., & Richard, K. (2009). *Arsenic pollution—a global synthesis*. Wiley-Blackwell.

Rehman, M. Z. U., Zafar, M., Waris, A. A., Rizwan, M., Ali, S., Sabir, M., Usman, M., Ayub, M. A., & Ahmad, Z. (2020). Residual effects of frequently available organic amendments on cadmium bioavailability and accumulation in wheat. *Chemosphere, 244*, 125548.

Reuben, A., Caspi, A., & Belsky, D. W. (2017). Association of childhood blood lead levels with cognitive function and socioeconomic status at age 38 years and with IQ change and socioeconomic mobility between childhood and adulthood. *JAMA, 317*(12), 1244–1251.

Richmond, R. (2015). Coral reefs: present problems and future concerns resulting from anthropogenic disturbance. *Integrative and Comparative Biology, 33*(6), 524–536.

Salehizadeh, H., & Shojaosadati, S. A. (2003). Removal of metal ions from aqueous solution by polysaccharide produced from Bacillus firmus. *Water Research, 37*, 231–4235.

Samina, W., Shams, T., & Masood, A. (2010). Isolation and characterization of a *Pseudomonas fluorescens* strain tolerant to major Indian water pollutants. *Journal of Bioremediation & Biodegradation, 1*, 1–5.

Sanchez-Pena, L. C., Petrosyan, P., Morales, M., Gonzalez, N. B., Gutierrez-Ospina, G., Del Razo, L. M., & Gonsebatt, M. E. (2010). Arsenic species, AS3MT amount, and AS3MT gene expression in different brain regions of mouse exposed to arsenite. *Environmental Research, 110*, 428–434.

Sawut, R., Kasim, N., Maihemuti, B., Hu, L., Abliz, A., Abdujappar, A., & Kurban, M. (2018). Pollution characteristics and health risk assessment of heavy metals in the vegetable bases of northwest China. *Science of the Total Environment, 642*, 864–878.

Sfakianakis, D. G., Renieri, E., Kentouri, M., & Tsatsakis, A. M. (2015). Effect of heavy metals on fish larvae deformities: A review. *Environmental Research, 137*, 246–255.

Shaolin, C., & David, B. W. (1997). Construction and characterization of genetically engineered for bioremediation of Hg2þ contaminated environments. *Applied and Environmental Microbiology, 63*, 2442–2445.

Sheehan, M. C., Burke, T. A., Navas-Acien, A., Breysse, P. N., McGready, J., & Fox, M. A. (2014). Global methylmercury exposure from seafood consumption and risk of developmental neurotoxicity: A systematic review. *Bulletin of the World Health Organization, 92*, 254–269.

Shi, Z., Carey, M., Meharg, C., Williams, P. N., Signes-Pastor, A. J., Triwardhani, E. A., Pandiangan, F. I., Campbell, K., Elliott, C., & Marwa, E. M. (2020). Rice grain cadmium concentrations in the global supply-chain Expo Health (pp. 1–8).

Shi, Z., Taylor, A. W., Riley, M., Byles, J., Liu, J., & Noakes, M. (2018). Association between dietary patterns, cadmium intake and chronic kidney disease among adults. *Clinical Nutrition, 37*(1), 276–284.

Shivakumar, C. K., Thippeswamy, B., Tejaswikumar, M. V., & Prashanthakumara, S. M. (2014). Bioaccumulation of heavy metals and its effect on organs of edible fishes located in Bhadra River, Karnataka. *International Journal of Research in Fisheries and Aquaculture, 4*, 90–98.

Siddiquee, S., Rovina, K., Azad, S. A., Naher, L., Suryani, S., & Chaikaew, P. (2015). Heavy metalcontaminants removal from wastewater using the potential filamentous fungi biomass: A review. *Journal of Microbial and Biochemical Technology, 7*(6), 384–395.

Silbernagel, S. M., Carpenter, D. O., Gilbert, S. G., Gochfeld, M., & Groth, E. (2011). Recognizing and preventing overexposure to methylmercury from fish and seafood consumption: Information for physicians. *Journal of Toxicology, 2011*, 1–7.

Silverberg, B. A., Wong, P. T. S., & Chau, Y. K. (1977). Effect of tetramethyl lead on freshwater green algae. *Archives of Environmental Contamination and Toxicology, 5*, 305–313.

Singh, S., Kang, S. H., Mulchandani, A., & Chen, W. (2008). Bioremediation: Environmental clean-up through pathway engineering. *Current Opinion in Biotechnology, 19*, 437–444.

Sivaperumal, P., Sankar, T. V., & Viswanathan Nair, P. G. (2007). Heavy metal concentrations in fish, shellfish and fish products from internal markets of India vis-a-vis international standards. *Food Chemistry, 102*, 612–620.

Smedley, P., & Kinniburgh, D. (2002). A review of the source, behaviour and distribution of arsenic in natural waters. *Applied Geochemistry, 17*, 517–568.

Soliman, N. F., Younis, A. M., & Elkady, E. M. (2019). An insight into fractionation, toxicity, mobility and source apportionment of metals in sediments from El Temsah Lake, Suez Canal. *Chemosphere, 222*, 165–174.

Sri Kumaran, N., Sundaramanicam, A., & Bragadeeswaran, S. (2011). Adsorption studies on heavy metals by isolated cyanobacterial strain (Nostoc sp.) from Uppanar estuarine water, Southeast coast of India. *Journal of Applied Sciences Research, 7*(11), 1609–1615.

Srivastav, A. K., Rai, R., Suzuki, N., Mishra, D., & Srivastav, S. K. (2013). Effects of lead on the plasma electrolytes of a freshwater fish. Heteropneustes fossilis. *International Aquatic Research, 5*, 4.

Stanbrough, R., Chuaboonmee, S., Palombo, E. A., Malherbe, F., & Bhave, M. (2013). Heavy metal phytoremediation potential of a heavy metal resistant soil bacterial isolate, *Achromobacter* sp. strain AO22. *APCBEE Procedia, 5*, 502–507.

Styblo, M., Del Razo, L. M., Vega, L., Germolec, D. R., Lecluyse, E. L., Hamilton, G. A., Reed, W., Wang, C., Cullen, W. R., & Thomas, D. J. (2000). Comparative toxicity of trivalent and pentavalent inorganic and methylated arsenicals in rat and human cells. *Archives of Toxicology, 74*, 289–299.

Sultana, M., Sanyal, S. K., & Hossain, M. A. (2015). Arsenic pollution in the environment. *Bioremediation, 2015*, 92–119.

Takeuchi, M., Kawahata, H., Gupta, L. P., Kita, N., Morishita, Y., Ono, Y., & Komai, T. (2007). Arseni resistance and removal by marine and non-marine bacteria. *Journal of Biotechnology, 127*, 434–442.

Tan, Z., Seshadri, S., & Beiser, A. (2003). Plasma total cholesterol level as a risk factor for Alzheimer disease: The Framingham Study. *Archives of Internal Medicine, 163*, 1053–1057.

Tang, W., Shan, B., Zhang, H., Zhang, W., Zhao, Y., & Ding, Y. (2014). Heavy metal contamination in the surface sediments of representative limnetic ecosystems in Eastern China. *Scientific Reports, 4*, 7152.

Thompson, J., & Bannigan, J. (2008). Cadmium: Toxic effects on the reproductive system and the embryo. *Reproductive Toxicology, 25*, 304–315.

Thophon, S., Kruatrachue, M., Upatham, E. S., Pokethitiyook, P., Sahaphong, S., & Jaritkhuan, S. (2003). Histopathological alterations of white seabass, Lates calcalifer, in acute and subchronic cadmium exposure. *Environmental Pollution, 121*, 307–320.

Thornton, I. (1992). Sources and pathways of cadmium in the environment. *IARC Scientific Publications, 118*, 149–162.

Timbrell, J. A. (2008). Biochemical mechanisms of toxicity: Specific examples. In *Principles of biochemical toxicology* (4th ed.). Informa Health Care.

Tinkov, A. A., Filippini, T., Ajsuvakovae, O. P., Skalnaya, M. G., Aasethf, J., Bjørklundh Gatiatulinai, E. R., Popova, E. V., Nemereshinai, O. N., & Huangk, P. T. (2018). Cadmium and atherosclerosis: A review of toxicological mechanisms and a meta-analysis of epidemiologic studies. *Environmental Research, 162*, 240–260.

Tsibakhashvili, N., Mosulishvili, L., Kirkesali, E., & Murusidze, I. (2010). NAA for studying detoxification of Cr and Hg by Arthrobacter globiformis 151B. *Journal of Radioanalytical and Nuclear Chemistry, 286*, 533–537.

Usman, A. R. A., Alkredaa, R. S., & Al-Wabel, M. (2013). Heavy metal contamination in sediments and mangroves from the coast of Red Sea: Avicennia marina as potential metal bioaccumulator. *Ecotoxicology and Environmental Safety, 97*, 263–270.

Vahidnia, A., Van der Voet, G., & De Wolff, F. (2007). Arsenic neurotoxicity—a review. *Human & Experimental Toxicology, 26*, 823–832.

Vargas-García, M. D. C., López, M. J., Suárez-Estrella, F., & Moreno, J. (2012). Compost as a source of microbial isolates for the bioremediation of heavy metals: in vitro selection. *Science of the Total Environment, 431*, 62–67.

Vasanthi, N., Muthukumaravel, K., Sathick, O., Sugumaran, J. (2019). Toxic effect of mercury on the freshwater fish *Oreochromis mossambicus*. *Research Journal of Life Sciences Bioinformatics Pharmaceutical and Chemical Sciences, 5*(3), 376. https://doi.org/10.26479/2019.0503.30

Vega, L., Styblo, M., Patterson, R., Cullen, W., & Wang, C. (2001). Differential effects of trivalent and pentavalent arsenicals on cell proliferation and cytokine secretion in normal human epidermal keratinocytes. *Toxicology and Applied Pharmacology, 172*, 225–232.

Veiga, M. M., Maxson, P. A., & Hylander, L. D. (2006). Origin and consumption of mercury in small- scale gold mining. *Journal of Cleaner Production, 14*(3–4), 436–447.

Vijayaraghavan, K., & Yun, Y. S. (2008). Bacterial biosorbents and biosorption. *Biotechnology Advances, 26*, 266–291.

Vipra, A., Desai, S. N., Junjappa, R. P., Roy, P., Poonacha, N., Ravinder, P., Sriram, B., & Padmanabhan, S. (2013). Determining the minimum inhibitory concentration of bacteriophages: potential advantages. *Advances in Microbiology, 3*, 181–190.

Vogel, C., & Fisher, N. S. (2010). Metal accumulation by heterotrophic marine bacterioplankton. *Limnology and Oceanography, 55*(2), 519–528.

Von Canstein, H., Li, Y., & Wagner-Dobler, I. (2001). Long-term performance of bioreactors cleaning mercury-contaminated waste water and their response to temperature and mercury stress and mechanical perturbation. *Biotechnology and Bioengineering, 74*, 212–219.

Wagner-Dobler, I., Lunsdorf, H., Lubbenhusen, T., von Canstein, H. F., & Li, Y. (2000). Structure and species composition of mercury-reducing biofilms. *Applied and Environmental Microbiology, 66*, 4559–4563.

Weidenhamer, J. D., & Clement, M. L. (2007). Widespread lead contamination of imported low-cost jewelry in the US. *Chemosphere, 67*(5), 961–965.

Wieczorek, J., Baran, A., Urbanski, K., Mazurek, R., & Klimowicz-Pawlas, A. (2018). Assessment of the pollution and ecological risk of lead and cadmium in soils. *Environmental Geochemistry and Health, 40*, 2325–2342.

Wu, X., Cobbina, S. J., Mao, G., Xu, H., Zhang, Z., & Yang, L. (2016). A review of toxicity and mechanisms of individual and mixtures of heavy metals in the environment. *Environmental Science and Pollution Research International, 23*(9), 8244–8259.

Yang, K., Zhu, L., Zhao, Y., Wei, Z., Chen, X., Yao, C., Meng, Q., & Zhao, R. A. (2019). Novel method for removing heavy metals from composting system: The combination of functional bacteria and adsorbent materials. *Bioresource Technology, 293*, 122095.

Ye, S., Zeng, G., Wu, H., Liang, J., Zhang, C., Dai, J., & Yu, J. (2019). The effects of activated biochar addition on remediation efficiency of co-composting with contaminated wetland soil. *Resources, Conservation and Recycling, 140*, 278–285.

Ye, S., Zeng, G., Wu, H., Zhang, C., Liang, J., Dai, J., & Cheng, M. (2017). Co-occurrence and interactions of pollutants, and their impacts on soil remediation-a review. *Critical Reviews in Environmental Science and Technology, 47*(16), 1528–1553.

Zamil, S. S., Ahmad, S., Choi, M. H., Park, J. Y., & Yoon, S. C. (2009). Correlating metal ionic characteristics with biosorption capacity of Staphylococcus saprophyticus BMSZ 711 using QICAR model. *Bioresource Technology, 100*, 1895–1902.

Zeitoun, M. M., & Mehana, E. (2014). Impact of water pollution with heavy metals on fish health: Overview and updates. *Global Veterinary, 12*, 219–231.

Zeng, X., Tang, J., Liu, X., & Jiang, P. (2012). Response of P. aeruginosa E1 gene expression to cadmium stress. *Current Microbiology, 65*, 799–804.

Zhang, J., Mu, X., Xu, W., Martin, F. L., Alamdar, A., Liu, L., Tian, M., Huang, Q., & Shen, H. (2014). Exposure to arsenic via drinking water induces 5-hydroxymethylcytosine alteration in rat. *Science of the Total Environment, 497–498*, 618–625.

Zhang, W., Chen, L., & Liu, D. (2012). Characterization of a marine-isolated mercury resistant *Pseudomonas putida* strain SP1 and its potential application in marine mercury reduction. *Applied Microbiology and Biotechnology, 93*, 13051314.

Zhao, H., Xia, B., Fan, C., Zhao, P., & Shen, S. (2012). Human health risk from soil heavy metal contamination under different land uses near Dabaoshan Mine, Southern China. *Science of the Total Environment, 41*, 45–54.

Zheng, N. A., Wang, S., & Dong, W. U. (2019). The toxicological effects of mercury exposure in marine fish. *Bulletin of Environmental Contamination and Toxicology, 102*(5), 714–720.

Zietz, B. P., Lass, J., Dunkelberg, H., & Suchenwirth, R. (2009). Lead pollution of drinking water in lower Saxonomy from corrosion of pipe materials. *Gesundheitswesen, 71*(5), 265–274.

Zoghi, A., Darani, K. K., & Sohrabvandi, S. (2014). *Mini Reviews in Medicinal Chemistry, 14*, 84–98.

Generalities of the Coagulation-Flocculation Process: A Perspective on Biocoagulants

Caroline Lissette Loor-Moreira, Kevin Jhon Fernández-Andrade, Gabriela S. Cedeño-Solórzano, Gema M. Manzaba-Salazar, Yunet Gómez-Salcedo, Joan Manuel Rodríguez-Díaz, and Ricardo J. Baquerizo-Crespo

1 Introduction

Liquid effluent discharges from industrial activities present high levels of polluting substances, which are harmful to ecosystems. The bodies of water receiving the effluents, which are mostly surface waters, show part of this contamination with the opacity of their waters as a result of sediments and interactions with natural organisms (Zhou et al., 2020). The interactions between suspended particles, dissolved

C. L. Loor-Moreira
Programa de Posgrado en Ingeniería Química, Instituto de Posgrado, Universidad Técnica de Manabí, Portoviejo, Ecuador
e-mail: cloor3333@utm.edu.ec

K. J. Fernández-Andrade
Programa de Posgrado en Ingeniería Química, Instituto de Posgrado, Universidad Técnica de Manabí, Portoviejo, Ecuador

Departamento de Investigación y Desarrollo, Empresa Purificadora Aqua Heredia, Aquaher S. A., Rocafuerte, Ecuador

G. S. Cedeño-Solórzano
Departamento Producción, Empresa Purificadora Aqua Heredia, Aquaher S. A., Rocafuerte, Ecuador

G. M. Manzaba-Salazar · Y. Gómez-Salcedo · R. J. Baquerizo-Crespo (✉)
Departamento de Procesos Químicos, Facultad de Ciencias Matemáticas Físicas y Químicas, Universidad Técnica de Manabí, Portoviejo, Ecuador
e-mail: yunet.gomez@utm.edu.ec; ricardo.baquerizo@utm.edu.ec

J. M. Rodríguez-Díaz
Departamento de Procesos Químicos, Facultad de Ciencias Matemáticas Físicas y Químicas, Universidad Técnica de Manabí, Portoviejo, Ecuador

Programa de Pós-graduação em Engenharia Química, Universidade Federal da Paraíba, João Pessoa, Brazil
e-mail: joan.rodriguez@utm.edu.ec

N. R. Maddela, L. C. García (eds.), *Innovations in Biotechnology for a Sustainable Future*, https://doi.org/10.1007/978-3-030-80108-3_16

inorganic chemical species, as well as organic matter generate suspended sediments, which cause the opacity of the waters (Kitchener et al., 2017; Tinterri et al., 2020).

Turbidity is a relative indicator of the opacity of the bodies of waters, which measures the sediment concentration or total solids. The effects on ecosystems caused by turbidity are: (1) the decrease in the passage of sunlight, which hinders photosynthetic processes on the underwater surface (Tinterri et al., 2020), (2) decreased dissolved oxygen (Miljojkovic et al., 2019), (3) retention of organic matter (Soler et al., 2020), among others.

One of the characteristics that stands out of the turbidity of the water is that the suspended particles have negative charges, this generates interparticle interactions that reduce their sedimentation rate (Park & Seo, 2011). These particles with equal charges form a stable colloidal system, this presents attractive and repulsive forces which maintain the suspension.

To remove turbidity, water treatment plants include clarification stages to reduce suspended sediments. These stages use coagulation-flocculation systems, these are low cost, simple and highly efficient operations (Zahrim et al., 2017). The coagulating agents can be synthetic or biocoagulants (Baquerizo-Crespo et al., 2021). Synthetic coagulants are highly efficient, but present limitations such as: (1) the toxicity of the effluents due to trace metals and (2) the generation of large volumes of sewage sludge that are not biodegradable.

The most widely used synthetic coagulants are salts and aluminized polymers; these present high percentages of turbidity removal and are low costs and easy to obtain (Gandiwa et al., 2020). However, the chemical stability is low, since in aqueous media aluminum is soluble, the dissolution of which is harmful to health and to post-clarification treatment processes. Other compounds of wide application for this type of treatment are synthetic cationic polymers that can form large flocs very quickly. This effect occurs because most of them are of plastic origin, which when destabilizing the system, their agglomeration occurs easily. Among the most widely used polymers are aluminum polychlorides, polyamides, ferric polychlorides, polydiallyldimethylammonium, among others (Manda et al., 2016). Despite their efficiency for the removal of turbidity, these polymers are not degradable, so they remain in the residual sludge, which leads to contamination problems and final disposal becomes complex.

Biocoagulants are polymers of plant, animal, or microbial origin, this guarantees the biodegradability of the sludge (Saleem & Bachmann, 2019) and less toxicity (Miyashiro et al., 2021). Biocoagulants from plant extracts are widely accepted due to their abundance and ease of obtaining. Plant extracts with biocoagulant properties are characterized by having large polymeric chains of high molecular weight, these can be proteins, polysaccharides, or polyphenols. Some of these extracts are already in use due to their high yields such as *Cactus Opuntia*, *Moringa Oleifera* (Gandiwa et al., 2020), *Alyssum* (Fard, Hamidi, Yetilmezsoy, et al., 2021), and *Lallemantia* (Fard, Hamidi, Alavi, et al., 2021).

Biocoagulants have proven to be highly efficient for treating surface water as well as wastewater; however, the implementation of this technology on an industrial scale faces challenges in optimizing the process. Factors such as pH, temperature, as well as the polymer composition of the biocoagulant, among others, influence the kinetics of the biocoagulation process. This chapter presents an overview of colloidal systems, turbidity removal mechanisms, and factors that influence water treatment from biocoagulants. In addition, it identifies different mathematical models for the kinetic evaluation of the process.

2 Colloidal Systems

Surface and wastewater are media for the dispersion for a variety of substances such as particulate materials, as well as silt, bacteria, plankton, viruses, biopolymers, and macromolecules. These particles, known as colloids, maintain their dispersed state due to the superiority of the forces associated with the solid-liquid interface (Bratby, 2016). Oyegbile et al. (2016) mention that colloidal particles in suspension (0.001–10 μm) have low settling rates due to interactions with other particles of equal surface charges.

Park and Seo (2011) describe the dispersion of particles by means of the theory of Derjaguin, Landau, Verwey, and Overbeek (DLVO), which states that the main forces that influence the colloidal charge in a solution are: electrostatic repulsion and the van der Waals force. As a result of the combination between the van der Waals force and the electrical repulsion potential, the interaction between two particles is obtained.

A colloid particle has negative charges, which causes attractive interactions with positively charged particles, this generates an electric potential. A colloidal particle has three potentials: (1) the Nernst potential that exists on the surface of the colloid, (2) the Zeta potential (Z) that corresponds to the potential measured on the surface that encloses the fixed layer of ions adhered to the particle, and (3) the Stern potential that exists on the inner surface of the double layer (Shammas, 2005). A colloidal particle has a double electrical layer, which surrounds a particle in dispersed phase (Fig. 1).

Colloidal particles present a layer of ions (surface charge), generally negatively charged, which are adsorbed on the surface of the particle. The Stern layer presents counterions with opposite charges attracted to the surface charge by the electrostatic force. The diffuse layer is the dispersion medium adjacent to the particle, it contains free ions with a greater amount of counterions. The ions in this layer are affected by the electrostatic force of the charged particle (Park & Seo, 2011).

In the electric double layer, the electric potential has a maximum value in the Stern layer, which causes an attractive movement between particles induced by the inclusion of an electrolyte (positive or negative) to destabilize the colloidal medium. This attraction movement causes the distance between particles to decrease and therefore the molecular interaction energy increases. On the contrary, with increasing distance from the surface the potential decreases and reaches zero at the boundary of the electric double layer. A layer of the surrounding liquid remains attached to the particle when it begins to move in the dispersion medium. This layer has a limit called the sliding plane, and the value of the electric potential at this limit

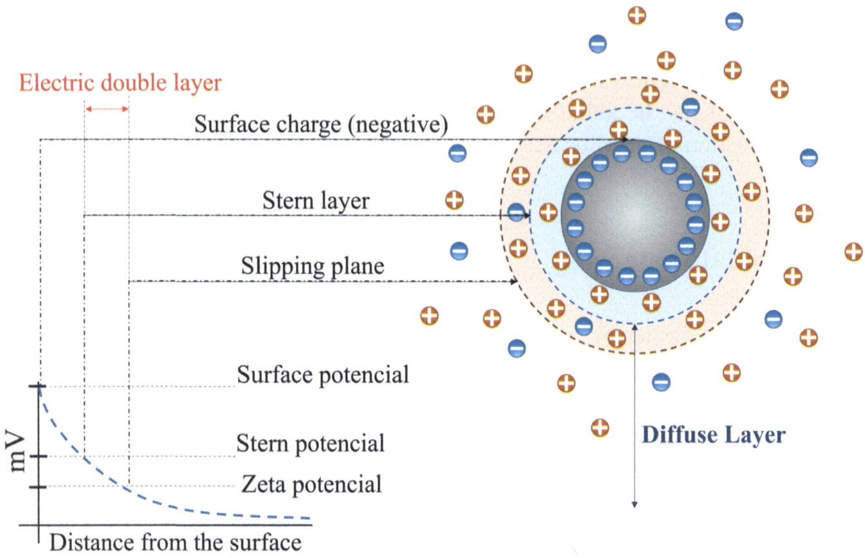

Fig. 1 Double electric layer of a colloidal particle

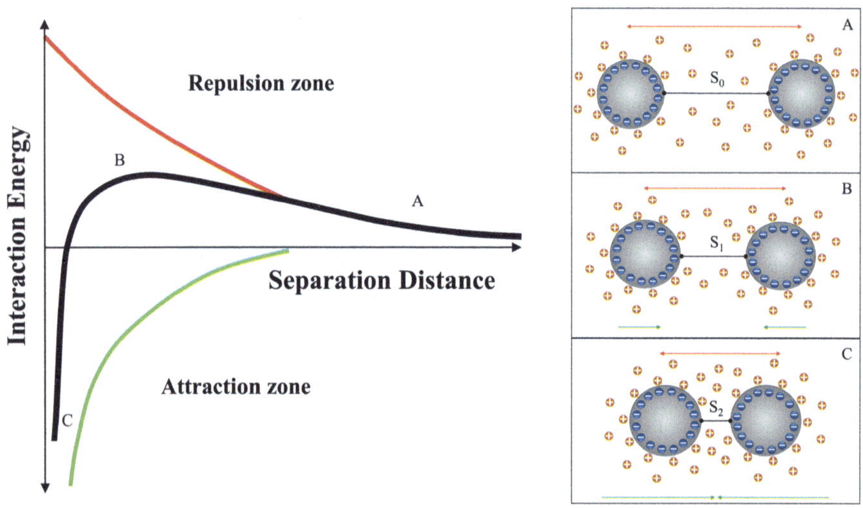

Fig. 2 Separation between colloidal particles and their repulsion and attraction interactions

is known as the Zeta potential, which is a relevant parameter for the interaction of colloids (Oyegbile et al., 2016). The lower the value of the Zeta potential, the greater the intermolecular attractive force. In particular, the dependence of the electric potentials with the distance between particles allows to establish the stability of the system and under what circumstances destabilization can be induced (Oyegbile et al., 2016). The curves in Fig. 2 indicate the strong relationship between the distance and the energy of interactions; however, the thickness of the double layer and the state of valences are also considered. The greater the thickness of the double layer, the greater the energy of molecular interaction at short distances, but that decreases as the particles move away from each other. All these relationships converge in the behavior of Brownian motion, which is one of the main forces that govern the mechanism of flocculation and subsequent sedimentation, as a function of the frequency of collisions (Oyegbile et al., 2016).

The energy of the interaction between particles is the sum of the forces of repulsion and attraction (Fig. 2). The separation of the counterions of a colloidal particle is impossible due to the stability existing between them. This makes the potential Z the only potential that can be estimated directly. For this reason, the potential Z is fundamental since its value determines the electrostatic forces between the particles, although it must be considered that it varies according to the nature of the components in the solution (López et al., 2014). At greater distances between colloids, the repulsive forces decrease, while the attractive forces do not exist (Fig. 2A). The coexistence between particles and the repulsive forces that act, generates the erratic movement of the colloids, this in turn encourages the attraction forces to come closer together. As the colloids reduce their distance, the interaction energy reaches a maximum point, because the attractive forces increase (Fig. 2B). When the repulsive forces exceed those of attraction, the Z potential decreases, without reaching an intimate contact between particles (Fig. 2C).

3 Fundamentals of Coagulation-Flocculation

Coagulation-flocculation systems use chemical agents known as coagulants. These substances have charges, generally positive, to destabilize the colloidal matter dispersed in the water (Teh & Wu, 2014). The neutralization of the colloids forms aggregates of greater size and density, known as flocs, which allow the sedimentation of these particles (Bratby, 2016).

Arboleda (1972) explained the coagulation-flocculation process with a physical model. This physical model refers to the formation of an adhered layer and a diffuse layer around the colloid. The formation of the first layer occurs when the coagulant carried by the dispersing medium reaches the surface of the colloid to neutralize the charge. The second occurs when the counterions separate from the surface of the colloid due to the agitation of the liquid, this prevents the colloids from forming a compact double layer. Thus, the colloids that form the fixed layer are retained by electrostatic and Van der Waals forces; in addition, they resist thermal agitation, which does not happen with the diffuse layer of ions (Pieterse & Cloot, 1997). This physical model of coagulation has limitations regarding the use of biocoagulants, as well as in the treatment of wastewater, because electrolytes can undergo various interactions of addition, electrostatic attraction and repulsion, the hypotheses point towards mechanisms of a chemical type.

The literature explains the coagulation-flocculation process from the perspective of three chemical models. The first model refers to the adsorption of opposite charges contributed by the coagulants, which generates the neutralization of the colloid, this process occurs on the colloidal surface (Saleem & Bachmann, 2019). A second model refers to the destabilization of colloids, the formation of bridges between particles. For this process it is necessary for the polymer to be formed by chemical groups that interact with the colloidal particles, this allows the formation of chemical bridges between the particles and increases their size. These bridges can break under certain conditions (Mer & Healy, 1963). A third model refers to destabilized particles attached to formed flocs that settle, and gradually increase the density of the particle (Duan & Gregory, 2003).

3.1 Phases of Coagulation

According to Baquerizo-Crespo et al. (2021), the coagulation-flocculation process occurs in two phases: (1) perkinetic coagulation and (2) orthokinetic flocculation. Both the perikinetic phase and the orthokinetic phase depend on the motion of the particles. Particles can be displaced due to (1) Brownian motion or (2) by agitation of the liquid, the main cause of collisions. In a way, all the particles are subject to flocculation due to both mechanisms. The greater or lesser influence of one or the other depends on the size of these.

Perkinetic coagulation begins with a rapid shaking regime (between 100 and 200 rpm), and its objective is to disperse the coagulating agent in the raw water to facilitate the destabilization of the colloidal particles (Baptista et al., 2017). Orthokinetic flocculation occurs under conditions of slow agitation, in which destabilized particles collide and due to interactions between them, join together to form larger particles (Baquerizo-Crespo et al., 2021). Criteria for the slow agitation regime vary. Baptista et al. (2017) and Hargreaves et al. (2018) indicated agitation speeds between 15 and 30 rpm, while Kalaitzidou et al. (2020) set an average agitation speed of 50 rpm. Bu et al. (2019) recommended agitation speeds between 40 rpm.

The optimization of the perikinetic coagulation process, as well as orthokinetic flocculation requires the study of various factors to estimate the dose of the coagulant for treatment. The factors include: (1) dose of the coagulant-flocculant, (2) composition of the water, (3) pH, (4) temperature as well as (5) mixing conditions.

3.1.1 Coagulant-Flocculant Dosage

The turbidity reduction effect of biocoagulants is due to the presence of various substances such as carbohydrates and long-chain proteins that are capable of complexing and sedimentation easily. The selection of the coagulating agent has a direct influence on the efficiency of the coagulation process (Zhu et al., 2011). The following situations may arise: (1) the amount of coagulant is insufficient to achieve destabilization of the colloids, (2) enough coagulant has been added to cause destabilization, (3) excess coagulant can cause charge inversion and re-stabilization of the particles (Shammas, 2005).

The selection of the optimal dose to apply is determined by jar test tests and depends on the concentration of colloidal particles and the pH value. Choudhary et al. (2019) reported the removal of 98% of the synthetic turbidity in waters with 500 NTU, with a dosage of extract of *Opuntia ficus-indica* of 1500 mg L^{-1}. On the other hand, Vignesh et al. (2020), applied doses of 200 mg L^{-1} of coagulant based on *Moringa oleifera* to remove 83% of the turbidity in synthetic waters of 250 NTU of turbidity. These biocoagulants (*Opuntia ficus-indica* and *Moringa oleifera*) reported concentrations of divalent minerals in the protein structures that favored the destabilization of colloidal particles. According to Kakoi et al. (2017), achieved a 99% reduction in turbidity, and more than 95% of COD, lead, and chromium from waters with 2750 NTU, with a dosage of 1200 mg L^{-1} of biocoagulant based on *Maerua Decumbent*. In addition, he stated that the efficiency of biocoagulants is linked to the presence of hydroxyl and carboxyl groups in the extracts. Sun et al. (2019) used a flocculant based on carboxylated chitosan and modified with dimethyldiallylammonium chloride and 3-chloro-2-chloropropyltrimethylammonium chloride, which had a yield of 91% of turbidity removal with the use of only 4 mg L^{-1} of flocculant. This occurred due to the presence of chlorinated and ammonium compounds that allowed the destabilization of the system quickly and evenly. Daverey et al. (2019) used only 0.6 mL L^{-1} extracts of *Musa paradisiaca* peel and *Dolichos lablab* seeds, with a removal of turbidity that reached 98.14% at a pH of 9. Although sometimes the dosages can be low, in the study carried out by Muniz et al. (2020) with the application of 2 g L^{-1} of ripe *Abelmoschus esculentus*, turbidity removal reached 91%. Mohamed Noor et al. (2021) added 60 mL of *Moringa oleifera* extract to remove 94% of the turbidity from the water.

3.1.2 Composition of the Water

The composition of the water depends on the source of origin, as well as the factors of the environment with which it interacts. One of the main factors that influences the composition of the water from natural springs is the contamination of the soil, due to the filtration processes of the water prior to its arrival at the water source (Vogt et al., 2001). Another factor that influences is the change of the climatic season that generates changes in turbidity levels from 400 to 600 NTU in dry seasons, up to 13,000 NTU in rainy seasons. In the case of bodies of water whose fluidity is limited, as in lakes and ponds, the degradation of organic matter produces the accumulation of phosphates and nitrogen ions. Other important factors are the chemical oxygen demand (COD) and the biological oxygen demand (BOD), which depend on runoff contaminated with chemical compounds or by the microbial flora of the water source. Surface water bodies with high concentrations of organic matter show coloration and a wide range of pathogenic and non-pathogenic microorganisms. Dissolved salts also vary depending on the origin and nature of the water, since they can be from a well or from a river and have different concentrations with each other and with other types of water. The presence of liquid effluents changes the composition of these bodies of water by adding emerging compounds such as drugs, dyes, pesticides, among others.

The presence of dissolved salts such as calcium and sodium chlorides promote colloidal destabilization and facilitate the coagulation-flocculation process. According to Megersa et al. (2019), the presence of salts of NaCl and NH_4Cl, promoted removal yields of high turbidity of 95%, with the application of extracts of *Moringa stenopetala* (Baker f.) Cufod and *Maerua subcordata* in waters with synthetic turbidity. Likewise, Carvalho et al. (2016) used *Moringa oleifera* extracts and compared the effects of NaCl and $CaCl_2$ in the removal of turbidity caused by *Microcystis aeruginosa* cells, they achieved 75% reduction. This confirms that biocoagulants based on plant extracts also have antimicrobial activity. The presence of salts during the coagulation process promotes: modifications in the optimal pH range, variations in the time required for flocculation and in the amount of biocoagulant required (Gregory, 2013). Rossini et al. (1999) indicate that the excessive presence of dissolved ions in the water could affect the physical-chemical balance of the process.

3.1.3 Alkalinity and pH

The pH influences the balance of two competing forces: (1) between the hydrolysis products of metal and H^+, for the interaction with organic ligands, and (2) between hydroxide ions and organic anions for the interaction with products of metallic hydrolysis (Pallier et al., 2010; Tatsi et al., 2003). The use of biocoagulants must consider an optimal pH range to achieve maximum removal of turbidity. In general, biocoagulants act at slightly acidic pH and close to neutral, however, there are some whose pH is lower (Table 1).

The pH of the waters establishes the charges of the particles that generate turbidity (Saxena et al., 2019), likewise the efficiency of a biocoagulant varies depending on its solubility at a certain pH (Nharingo et al., 2015). The wrong selection of a biocoagulant implies operating expenses in the change of pH of the medium (Ni et al., 2020; Tian & Zhao, 2021).

3.1.4 Water Temperature

This factor affects the movement of fluids, their interactions, as well as the degree of hydrolysis of coagulants and the mechanisms of adsorption and precipitation (Kang & Cleasby, 1995). Coagulation rate is directly related to process temperature (Fitzpatrick et al., 2004). Mamchenko et al. (2011) suggest operating temperatures above 5 °C and below 20 °C, regardless of the biocoagulant and pH. Dolejs (1992)

Table 1 Optimal pH for various types of waters and biocoagulants

Type of water	Biocoagulant	pH	Removal	Reference
Surface water	Chitin extracted from shrimp waste	6	95% turbidity 85% organic matter	(Frantz et al., 2020)
Textile wastewater	*Moringa oleífera* extract	6.5	96 % reactive black 5 dye	(Miyashiro et al., 2021)
Paint-contaminated water	*Mucuna* seed extract	2	89% pigment particle	(Ezemagu et al., 2020)
River water	*Plantago ovata* extract	< 8	95% turbidity	(Ramavandi, 2014)
Colored wastewater	*Moringa oleífera* powder	6-8	90% turbidity and color	(Badrus, 2018)
Wastewater	Rice starch	2.3	98% turbidity	(Usefi & Asadi-Ghalhari, 2019)
Bilge water	*Alyssum* mucilage	7.05	92.30% COD 99.92% turbidity 99% tensoactives	(Fard, Hamidi, Alavi, et al., 2021)
Synthetic water (100 NTU)	Gummy exudate from *Cedrela odorata*	6.3	>80% turbidity	(Mejías et al., 2010)

mentions that temperatures close to 15 °C promote the balanced formation of flocs, because the diffusion speed is slow enough to allow the agglomeration of destabilized particles and in the amount necessary for the formation of flocs of similar sizes throughout the system. The formation of flocs of adequate size allows a greater recovery of sludge and therefore a greater recovery of clarified water. If the temperature is low (<5 °C), the rate of coagulation and flocculation decreases, which causes the formation of excessively large flocs and other excessively small ones that fail to settle. On the contrary, when the temperature is too high (>20 °C), coagulation and flocculation are so fast that the flocs do not reach the adequate weight to settle, so it is necessary to add a greater amount of biocoagulant (Huang et al., 2013; Zhang et al., 2018).

3.1.5 Mixing Conditions

On a laboratory scale, agitation of the medium in three stages is an important aspect to achieve a proper process. A first rapid mixing stage (>100 rpm) of short duration (1–2 min) allows the dispersion of the biocoagulant and the destabilization of the particles, the movement of the medium is chaotic. During the addition of the coagulant, the degree of agitation determines whether the coagulation will be complete. In addition, the turbulence must be uniform to ensure the dispersion of the biocoagulant and the neutralization of charges (Liang et al., 2009).

Subsequently, the second stage corresponds to low speed stirring (<50 rpm) for a prolonged period (between 10 and 15 min). This stage promotes the interaction between destabilized particles with the flocs formed, so in this stage the intensity of the mixture and the flocculation time are factors that influence the physical and chemical properties of the flocs. Cho et al. (2006) indicate that increasing the flocculation time results in a decrease in the size of the flocs. The flocculation time is related to the agglomeration rate, which will depend on the nature of the fluid flow (Bridgeman et al., 2008, 2010). The higher the agitation intensity, the faster the agglomeration speed of the particles (Klimpel & Hogg, 1986), therefore, as the flocs grow in size, the hydrodynamic shear forces that are induced by the velocity gradient also increase (Bouyer et al., 2005). If the flocs exceed their maximum size, the shear forces acquire an intensity that breaks them into smaller particles. Floc resistance is affected by several factors such as size, shape, compaction, nature of the microparticles, the number and shape of the ligaments that hold the particles together (Elmaleh & Jabbouri, 1991). In practice, the variation of the flow modifies the residence times and velocity gradients in the flocculator, so the optimization of this parameter ensures the reduction of turbidity (Manning et al., 2010; Rossini et al., 1999). The increases in stirring speed (from 60 to 150 rpm) negatively influence the removal of turbidity in surface waters, with biocoagulants based on *Moringa oleifera* (from 98.67 to 96.05%) and *Caesalpinia spinosa* (from 94.67 to 92.14%) (Baquerizo-Crespo et al., 2020, 2021).

Finally, reducing the stirring speed to 0 rpm for indefinite times allows free sedimentation of the flocs (Chatsungnoen & Chisti, 2016). In this last stage, the destabilized particles and flocs settle, separating clean water from accumulated impurities (Hriberšek et al., 2011). Sedimentation is considered the last treatment step before discharge to the receiving water, and should produce a supernatant with low suspended solids concentrations that meet the standards (Wilén & Balmér, 1999).

3.2 *Kinetic Aspects of Coagulation-Flocculation*

The kinetic study of the coagulation-flocculation treatment presents different perspectives, related to: (1) floc growth, (2) the coagulation-flocculation process, and (3) sedimentation. Each perspective presents elements that allow establishing the coagulation-flocculation mechanisms, the influence of the factors, and their impact on the performance of the process. It is also a tool for the design of clarification systems (Kang & Cleasby, 1995).

3.2.1 Coagulation-Flocculation Modeling

The classic approach to the coagulation-flocculation process considers a binary coalescence model, known as the Smoluchowski model (Eq. (1)) in honor of its author. This approach considers the union of two particles to form a larger particle (Filbet & Laurençot, 2004; Kang & Redner, 1984). Among the aspects considered by this model, the relationship between the mass of a particle with a collision cross-section and its diffusion in the medium stands out.

$$\frac{dc_k(t)}{dt} = \frac{1}{2}\sum_{i+j=k} K_{ij}c_i(t)c_j(t) - c_k(t)\sum_{j>0} K_{jk}c_j(t) \tag{1}$$

where $c(t)$ represents the concentration of particles of sizes i, j, k, in a time t, i, j, k are subscripts for the colliding particles (i, j) for the formation of a larger particle (k), K_{ij} is rate constant for increasing the concentration of particles of size k, K_{jk} is rate constant for decreasing the concentration of particles of size k.

The model presents a mass balance structure in an unsteady state, to establish the change in the concentration of particles with mass k. The model explains the variation of $c_k(t)$ from the union of particles of sizes i, j and from the consumption of these particles to form others of greater size. An analysis of the expressions reveals that the binding process refers to a second-order constant, so that increases in concentration exponentially raise the concentration. The reduction term refers to a first-order constant, so the reduction of the k size particles is linear. In addition, the model considers the relationship between the mass of a collision cross-sectional particle and its diffusion in the medium; however, it discards convective effects. Another important aspect to highlight is the value of $\frac{1}{2}$, this refers to the fact that the collision of two particles only generates a new particle.

Another approach is that of the Brownian motion of the particles (Eq. (2)). This approach considers the decrease in the concentration of the initial size particles by means of non-specific order kinetics (Obiora-Okafo et al., 2019).

$$\frac{dC}{dt} = -kC^{\alpha} \tag{2}$$

where C is the concentration of particles with the initial size at time t, α is the order of coagulation-flocculation reaction, k is the αth coagulation-flocculation order rate.

This approach relates the kinetic constant of the process with the properties of the medium. Obiora-Okafo et al. (2019) indicate an expression for the estimation of the value of k (Eq. (3)).

$$k = \frac{8\ K_B T}{3\mu} \tag{3}$$

where K_B is the Boltzmann's constant, T is the temperature, μ is the viscosity of the medium.

Parker et al. (1971) mention an approach to the coagulation-flocculation process in which he considers two terms (Eq. (4)). The first term is related to the rate of erosion of flocs due to shear forces with the fluid medium. The second term is related to the reduction of primary particles. The proposed model is:

$$\frac{dn}{dt} = k_B X G^m - k_A X n G \tag{4}$$

where k_B is the floc breakup rate constant, m is the breakup rate exponent, x represents suspended solids, G is the velocity gradient, k_A is aggregation rate coefficient, n is primary particle concentration at time t, t is the time.

The various kinetic coagulation-flocculation approaches provide information on the stability of the dispersion and the interactions, of the particles, which depend on the number and efficiency of the collisions (Oyegbile et al., 2016). According to Gregory (2013), collisions between particles occur at a speed which is controlled by differential sedimentation diffusion transport mechanisms and involves a process with second-order kinetics. However, the applications of these models to the analysis of biocoagulants are limited.

3.2.2 Sedimentation Kinetics

The sedimentation of the flocs is a process that depends on the size and density of these particles. Floc densities are generally between 2.5 and 4.0 kg L^{-1}, with sizes ranging from 0.3 to 2 mm (Kramer et al., 2019).

Kynch (1952) analyzes the sedimentation kinetic with a model based on a microscopic mass balance (Eq. (5)). This model proposes that the variation of the concentration at a point varies according to the speed with which a particle descends.

$$\frac{\partial m}{\partial t} + \frac{\partial \left(V_s\, m\right)}{\partial z} = 0 \tag{5}$$

where m is the concentration of particles at time t in a specific position, V_s is the sedimentation velocity.

This approach considers that the movement of the particles is unidirectional, caused by the force of gravity. Under this premise, the study of sedimentation kinetics requires the observation of the displacement of the flocs in batch sedimentation. These tests measure the advancement of the interfaces formed between the: (1) the supernate and suspension, and (2) the suspension and the sediments. The observation of these zones requires transparent graduated cylinders, generally made of glass (Fig. 3).

The sedimentation curve measures the height of the two interfaces that form versus time. These curves refer to the sedimentation characteristics, defined by the sedimentation rate, the solid/liquid volume fraction, the floc size, the induction period, the soil formation time, the solid content profiles, the permeability, and the sediment compressibility (Kang et al., 2019).

For effective coagulation-flocculation processes, the clarified liquid (supernate) in the upper part does not present particles in suspension. Below there will be an area with flocculent particles that settle (suspension), this will have greater turbidity. The sediments accumulate in the lower part and this generates an increase in

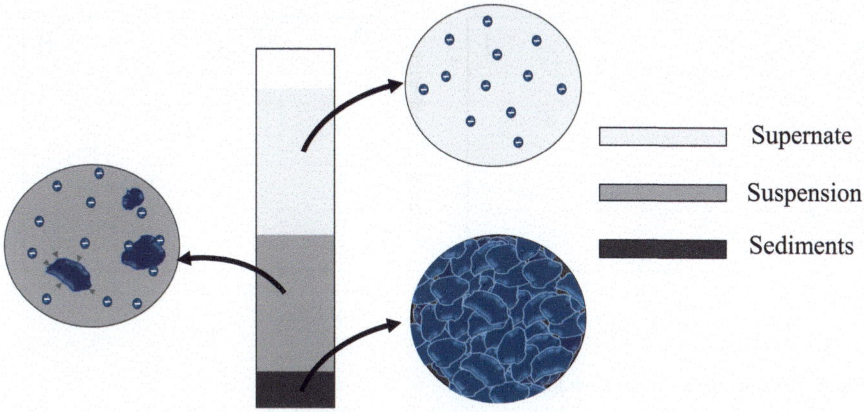

Fig. 3 Batch sedimentation zones

volume. However, not all areas are always visible during sedimentation tests. The literature presents various approaches (Eqs. (6)–(17)) to the estimation of sedimentation velocity, by means of non-linear expressions (Cho et al., 1993; Manning et al., 2010).

$$v_s = v_{s\ \max} e^{-n_s X} \tag{6}$$

$$v_s = v_{s\ \max} X^{-n_s} \tag{7}$$

$$v_s = v_{s\ \max} (1 - n_s X)^{4.65} \tag{8}$$

$$v_s = v_{s\ \max} \frac{(1 - n_s X)^3}{X} \tag{9}$$

$$v_s = v_{s\ \max} \frac{(1 - n_s X)^4}{X} \tag{10}$$

$$v_s = v_{s\ \max} \frac{e^{-n_s X}}{X} \tag{11}$$

$$v_s = v_{s\ \max} \frac{(1 - n_{s,1} X)^4}{X} e^{-n_{s,2} X} \tag{12}$$

$$v_s = v_{s\ \max} X(1 - X) \tag{13}$$

$$v_s = v_{s\ \max} (1 - n_s X)^2 e^{-4.19X} \tag{14}$$

$$v_s = v_{s\ \max} (1 - n_s X)^2 \exp\left(\frac{-n_{s,2} X}{1 - n_{s,3} X}\right) \tag{15}$$

$$v_s = v_{s\ \max} \left(1 - n_{s,1} X + n_{s,2} X^2 + n_{s,3} X^3 + n_{s,4} X^4\right) \tag{16}$$

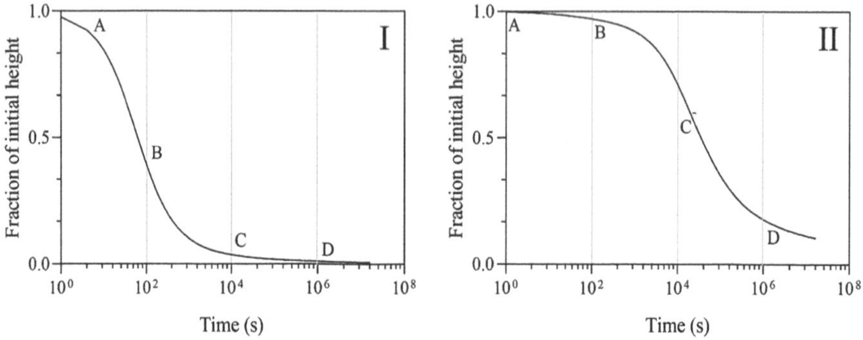

Fig. 4 S-model for sedimentation kinetics

$$v_s = v_{s,1}(1 - n_{s,1}X)^{n_{s,2}} + v_{s,2} \tag{17}$$

where n_s is an empirical factor for the fit of the model.

Kang et al. (2019) propose a 5-parameter S-model (Eq. (18)) for the sedimentation of the flocculated zone. This model considers the height of the suspension zone and the sedimentation zone.

$$\frac{H(t)}{h_0} = \frac{1}{\left\{\ln\left[\exp(1) + \left(\frac{t}{t_h}\right)^{n_s}\right]\right\}^{m_c}} \left\{1 - \left[\frac{\ln\left(1 + \frac{t}{t_s}\right)}{\ln\left(1 + \frac{10^{45}}{t_c}\right)}\right]\right\} \tag{18}$$

where $H(t)$ is the position of the interface at time t, h_0 is the initial height of the suspension, t_h is a parameter related to the induction period of the hindered sedimentation, n_s represents the maximum settling speed of the sedimentation blanket, m_c is a parameter related to the consolidation phase, t_s is a parameter related to the hindered settling, t_c is a parameter related to the final sediment height.

This model considers the extent of the period of hindered sedimentation (Fig. 4A), visualized as a plain in the upper part of the kinetic curve. Followed by a period of constant rate of sedimentation (downward sloping) (Fig. 4B) that culminates when the zone of suspended particles is reduced and allows the zone of clarified liquid and sediment to merge (Fig. 4C). Another process that is considered is the compaction of the mud (Fig. 4D).

The graphs of the model (Kang et al., 2019) allow evaluation of solid concentrations. Thus, for systems with high concentrations of solids (Fig. 4 I), the model presents low times for hindered settling, as well as a zone of compaction of sludge with a low height. For lower concentrations of solids (Fig. 4 II) the hindered settling zone is extended, likewise the zone of compression of sludge indicates that to reach the total sedimentation of the solids extremely long time is required.

Bakiri and Nacef (2020) conceptualized the variation of the height of the interfaces in sedimentation as a two-stage process (hindered settling and compression). This model proposes a simple approach, in which the evaluation of the height takes the non-linear expressions evaluated by Cho et al. (1993), with a change in the independent variable of suspension concentration by the sedimentation time. Due to the characteristics of non-linear expressions, the change of variable is feasible.

$$H(t) = \begin{cases} k_1 e^{-n_1 t}; \quad 0 \leq t \leq t_l & \text{(hindered settling)} \\ k_2 \left(\dfrac{e^{n_2 (t + t_l)}}{\frac{t}{t_l}} + 1 \right); \quad t_l \leq t & \text{(compression)} \end{cases} \tag{19}$$

where $H(t)$: is the height of the interface at time t, k_1, k_2 are kinetic constants, n_1, n_2 are shape factors, t_l is the time limit between the hindered settling and compression.

Despite the extensive literature describing the kinetics of the coagulation-flocculation process, there are few references to the application of these models to biocoagulants. A particular case is that of Padhiyar et al. (2020), whom evaluated the sedimentation kinetics with three wastewaters (dye rich, starch rich, mixed wastewater), with the use of *Moringa oleifera* powder as a biocoagulant, with particle sizes of 600 and 300 μm. In this case, the evaluation was according to kinetic models of chemical reaction of zero, first, and second order. Second-order kinetics stood out in the results. The values were diverse, thus for the dye rich waters, the constants presented equal values (0.0002 NTU^{-1} min^{-1}), while for the starch rich waters the particle sizes of 300 μm favored the kinetics ($k = 0.0001$ NTU^{-1} min^{-1}), contrary to mixed wastewaters in which the 600 μm size presented the highest value of k in the study (0.0008 NTU^{-1} min^{-1}).

4 Final Considerations

The literature presents evidence of the use of biocoagulants in the treatment of surface and wastewater, with advantages over chemical coagulants such as:

- Low volume of sludge.
- Biodegradable sludge.
- Does not affect the ecosystem.
- Sustainability of the process.

However, there are still gaps in the mechanisms of action, as well as in the kinetic aspects in the use of biocoagulants. Some aspects to consider in the future development of biocoagulants are:

- Effects of the temperature of the medium in the process of colloidal destabilization, formation of flocs, and sedimentation.

- Analysis of the agitation regime in the development of flocs.
- Evaluation of the process scaling from the kinetic criteria.
- Analysis of biocoagulant product sludge and its post-treatment alternatives.

References

Arboleda, J. (1972). *Teoría, diseño y control de los procesos de clarificación del agua*. CEPIS (Centro Panamericano de Ingeniería Sanitaria y Ciencias del Ambiente).

Badrus, Z. (2018). Potential of natural flocculant in coagulation-flocculation wastewater treatment process. *E3S Web Conference, 73*, 05006. https://doi.org/10.1051/e3sconf/20187305006

Bakiri, Z., & Nacef, S. (2020). Development of an improved model for settling velocity and evaluation of the settleability characteristics. *Water Environment Research, 92*, 1089–1098. https://doi.org/10.1002/wer.1306

Baptista, A. T. A., Silva, M. O., Gomes, R. G., Bergamasco, R., Vieira, M. F., & Vieira, A. M. S. (2017). Protein fractionation of seeds of *Moringa oleifera* lam and its application in superficial water treatment. *Sep Purif Technol, 180*, 114–124. https://doi.org/10.1016/j.seppur.2017.02.040

Baquerizo-Crespo, R. J., Macías-Alcívar, J. A., Zhingre-Farfán, J. M., Gómez-Salcedo, Y., Córdova, A., & Zambrano-Arcentales, M. A. (2020). Evaluation of the effect of *Moringa oleifera* and *Caesalpinia spinosa* mixtures on surface water turbidity. Afinidad 77.

Baquerizo-Crespo, R. J., Nuñez, Y., Albite, J., Macías-Alcívar, J. A., Cedeño-Zambrano, N., Dueñas-Rivadeneira, A. A., et al. (2021). Biocoagulants as an Alternative for Water Treatment. In: Maddela NR, García Cruzatty LC, Chakraborty S (eds) Advances in the Domain of Environmental Biotechnology: Microbiological Developments in Industries, Wastewater Treatment and Agriculture. Springer, Singapore, pp 313–334.

Bouyer, D., Coufort, C., Liné, A., & Do-Quang, Z. (2005). Experimental analysis of floc size distributions in a 1-L jar under different hydrodynamics and physicochemical conditions. *Journal of Colloid and Interface Science, 292*, 413–428. https://doi.org/10.1016/j.jcis.2005.06.011

Bratby, J. (2016). *Coagulation and flocculation in water and wastewater treatment*. IWA Publishing.

Bridgeman, J., Jefferson, B., & Parsons, S. (2008). Assessing floc strength using CFD to improve organics removal. *Chemical Engineering Research and Design, 86*, 941–950. https://doi.org/10.1016/j.cherd.2008.02.007

Bridgeman, J., Jefferson, B., & Parsons, S. A. (2010). The development and application of CFD models for water treatment flocculators. *Advances in Engineering Software, 41*, 99–109. https://doi.org/10.1016/j.advengsoft.2008.12.007

Bu, F., Gao, B., Shen, X., Wang, W., & Yue, Q. (2019). The combination of coagulation and ozonation as a pre-treatment of ultrafiltration in water treatment. *Chemosphere, 231*, 349–356. https://doi.org/10.1016/j.chemosphere.2019.05.154

Carvalho, M. S., Alves, B. R. R., Silva, M. F., Bergamasco, R., Coral, L. A., & Bassetti, F. J. (2016). $CaCl_2$ applied to the extraction of *Moringa oleifera* seeds and the use for *Microcystis aeruginosa* removal. *Chem Eng J, 304*, 469–475. https://doi.org/10.1016/j.cej.2016.06.101

Chatsungnoen, T., & Chisti, Y. (2016). Continuous flocculation-sedimentation for harvesting Nannochloropsis salina biomass. *Journal of Biotechnology, 222*, 94–103. https://doi.org/10.1016/j.jbiotec.2016.02.020

Cho, M.-H., Lee, C.-H., & Lee, S. (2006). Effect of flocculation conditions on membrane permeability in coagulation–microfiltration. *Desalination, 191*, 386–396. https://doi.org/10.1016/j.desal.2005.08.017

Cho, S. H., Colin, F., Sardin, M., & Prost, C. (1993). Settling velocity model of activated sludge. *Water Research, 27*, 1237–1242. https://doi.org/10.1016/0043-1354(93)90016-B

Choudhary, M., Ray, M. B., & Neogi, S. (2019). Evaluation of the potential application of cactus (*Opuntia ficus-indica*) as a bio-coagulant for pre-treatment of oil sands process-affected water. *Separation and Purification Technology, 209*, 714–724. https://doi.org/10.1016/j.seppur.2018.09.033

Daverey, A., Tiwari, N., & Dutta, K. (2019). Utilization of extracts of *Musa paradisiaca* (banana) peels and *Dolichos lablab* (Indian bean) seeds as low-cost natural coagulants for turbidity removal from water. *Environmental Science and Pollution Research, 26*, 34177–34183. https://doi.org/10.1007/s11356-018-3850-9

Dolejs, P. (1992). The effects of temperature, pH and rapid mixing gradient on the formation of particles in treatment of humic water. In R. Klute & H. Hahn (Eds.), *Chemical water and wastewater treatment II* (pp. 65–77). Springer.

Duan, J., & Gregory, J. (2003). Coagulation by hydrolysing metal salts. *Advances in Colloid and Interface Science, 100–102*, 475–502. https://doi.org/10.1016/S0001-8686(02)00067-2

Elmaleh, S., & Jabbouri, A. (1991). Flocculation energy requirement. *Water Research, 25*, 939–943. https://doi.org/10.1016/0043-1354(91)90141-C

Ezemagu, I. G., Ejimofor, M. I., & Menkiti, M. C. (2020). Turbidimetric study for the decontamination of paint effluent (PE) using mucuna seed coagulant (MSC): Statistical design and coag-flocculation modelling. *Environmental Advances, 2*, 100023. https://doi.org/10.1016/j.envadv.2020.100023

Fard, M. B., Hamidi, D., Alavi, J., Jamshidian, R., Pendashteh, A., & Mirbagheri, S. A., (2021a). Saline oily wastewater treatment using *Lallemantia* mucilage as a natural coagulant: Kinetic study, process optimization, and modeling. *Ind Crops Prod, 163*, 113326. https://doi.org/10.1016/j.indcrop.2021.113326

Fard, M. B., Hamidi, D., Yetilmezsoy, K., Alavi, J., & Hosseinpour, F., (2021b). Utilization of *Alyssum* mucilage as a natural coagulant in oily-saline wastewater treatment. *J Water Process Eng, 40*, 101763. https://doi.org/10.1016/j.jwpe.2020.101763

Filbet, F., & Laurençot, P. (2004). Numerical simulation of the smoluchowski coagulation equation. *SIAM Journal on Scientific Computing, 25*, 2004–2028. https://doi.org/10.1137/S1064827503429132

Fitzpatrick, C. S. B., Fradin, E., & Gregory, J. (2004). Temperature effects on flocculation, using different coagulants. *Water Science and Technology, 50*, 171–175. https://doi.org/10.2166/wst.2004.0710

Frantz, T. S., Farias, B. S. de, Leite, V. R. M., Kessler, F., Cadaval Jr, T. R. S., & Pinto, L. A. de A. (2020). Preparation of new biocoagulants by shrimp waste and its application in coagulation-flocculation processes. J Clean Prod 269:122397. https://doi.org/10.1016/j.jclepro.2020.122397

Gandiwa, B. I., Moyo, L. B., Ncube, S., Mamvura, T. A., Mguni, L. L., & Hlabangana, N. (2020). Optimisation of using a blend of plant based natural and synthetic coagulants for water treatment: (Moringa Oleifera-Cactus Opuntia-alum blend). South Afr J Chem Eng 34:158–164. https://doi.org/10.1016/j.sajce.2020.07.005

Gregory, J. (2013). Flocculation fundamentals. In T. Tadros (Ed.), *Encyclopedia of colloid and interface science* (pp. 459–491). Springer.

Hargreaves, A. J., Vale, P., Whelan, J., Alibardi, L., Constantino, C., Dotro, G., Cartmell, E. et al. (2018). Impacts of coagulation-flocculation treatment on the size distribution and bioavailability of trace metals (Cu, Pb, Ni, Zn) in municipal wastewater. Water Res 128:120–128. https://doi.org/10.1016/j.watres.2017.10.050

Hriberšek, M., Žajdela, B., Hribernik, A., & Zadravec, M. (2011). Experimental and numerical investigations of sedimentation of porous wastewater sludge flocs. *Water Research, 45*, 1729–1735. https://doi.org/10.1016/j.watres.2010.11.019

Huang, X., Gao, B., Yue, Q., Wang, Y., Li, Q., Zhao, S., & Sun, S. (2013). Effect of dosing sequence and raw water pH on coagulation performance and flocs properties using dual-coagulation of polyaluminum chloride and compound bioflocculant in low temperature surface water treatment. *Chem Eng J, 229*, 477–483. https://doi.org/10.1016/j.cej.2013.06.029

Kakoi, B., Kaluli, J. W., Ndiba, P., & Thiong'o, G. (2017). Optimization of Maerua Decumbent bio-coagulant in paint industry wastewater treatment with response surface methodology. *Journal of Cleaner Production, 164*, 1124–1134. https://doi.org/10.1016/j.jclepro.2017.06.240

Kalaitzidou, K., Zouboulis, A., & Mitrakas, M. (2020). Cost evaluation for Se (IV) removal, by applying common drinking water treatment processes: Coagulation/precipitation or adsorption. *Journal of Environmental Chemical Engineering, 8*, 104209. https://doi.org/10.1016/j.jece.2020.104209

Kang, K., & Redner, S. (1984). Fluctuation effects in Smoluchowski reaction kinetics. *Physical Review A, 30*, 2833–2836. https://doi.org/10.1103/PhysRevA.30.2833

Kang, L.-S., & Cleasby, J. L. (1995). Temperature effects on flocculation kinetics using Fe (III) coagulant. *Journal of Environmental Engineering, 121*, 893–901. https://doi.org/10.1061/(ASCE)0733-9372(1995)121:12(893)

Kang, X., Xia, Z., Wang, J., & Yang, W. (2019). A novel approach to model the batch sedimentation and estimate the settling velocity, solid volume fraction, and floc size of kaolinite in concentrated solutions. *Colloids and Surfaces A: Physicochemical and Engineering Aspects, 579*, 123647. https://doi.org/10.1016/j.colsurfa.2019.123647

Kitchener, B. G., Wainwright, J., & Parsons, A. J. (2017). A review of the principles of turbidity measurement. *Progress in Physical Geography: Earth and Environment, 41*, 620–642. https://doi.org/10.1177/0309133317726540

Klimpel, R. C., & Hogg, R. (1986). Effects of flocculation conditions on agglomerate structure. *Journal of Colloid and Interface Science, 113*, 121–131. https://doi.org/10.1016/0021-9797(86)90212-2

Kramer, O. J. I., de Moel, P. J., Baars, E. T., van Vugt, W. H., Padding, J. T., & van der Hoek, J. P. (2019). Improvement of the Richardson-Zaki liquid-solid fluidisation model on the basis of hydraulics. *Powder Technol, 343*, 465–478. https://doi.org/10.1016/j.powtec.2018.11.018

Kynch, G. J. (1952). A theory of sedimentation. *Transactions of the Faraday Society, 48*, 166–176. https://doi.org/10.1039/TF9524800166

Liang, Z., Wang, Y., Zhou, Y., Liu, H., & Wu, Z. (2009). Variables affecting melanoidins removal from molasses wastewater by coagulation/flocculation. *Sep Purif Technol, 68*, 382–389. https://doi.org/10.1016/j.seppur.2009.06.011

López, E. A., Oropeza-Guzman, M. T., Jurado-Baizaval, J. L., & Ochoa-Terán, A. (2014). Coagulation–flocculation mechanisms in wastewater treatment plants through zeta potential measurements. *Journal of Hazardous Materials, 279*, 1–10. https://doi.org/10.1016/j.jhazmat.2014.06.025

Mamchenko, A. V., Gerasimenko, N. G., & Pakhar, T. A. (2011). The impact of temperature on the efficiency of the coagulation process of titanyl sulfate and aluminum sulfate. *Journal of Water Chemistry and Technology, 33*, 315–322. https://doi.org/10.3103/S1063455X11050079

Manda, I. K. M., Chidya, R. C. G., Saka, J. D. K., & Biswick, T. T. (2016). Comparative assessment of water treatment using polymeric and inorganic coagulants. *Physics and Chemistry of the Earth Parts ABC, 93*, 119–129. https://doi.org/10.1016/j.pce.2015.09.008

Manning, A. J., Langston, W. J., & Jonas, P. J. C. (2010). A review of sediment dynamics in the Severn estuary: Influence of flocculation. *Marine Pollution Bulletin, 61*, 37–51. https://doi.org/10.1016/j.marpolbul.2009.12.012

Megersa, M., Gach, W., Beyene, A., Ambelu, A., & Triest, L. (2019). Effect of salt solutions on coagulation performance of *Moringa stenopetala* and *Maerua subcordata* for turbid water treatment. *Sep Purif Technol, 221*, 319–324. https://doi.org/10.1016/j.seppur.2019.04.013

Mejías, D. G., Chávez Delgado, M., Masyrubi, M., Chacín Ramos, E., & Fernández Acosta, N. (2010). Uso potencial del exudado gomoso de Cedrela odorata como agente coagulante para el tratamiento de las aguas destinadas a consumo humano. *Rev For Venez, 54*, 147–154

Mer, V. K. L., & Healy, T. W. (1963). The role of filtration in investigating flocculation and redispersion of colloidal dispersions. *The Journal of Physical Chemistry, 67*, 2417–2420. https://doi.org/10.1021/j100805a038

Miljojkovic, D., Trepsic, I., & Milovancevic, M. (2019). Assessment of physical and chemical indicators on water turbidity. *Physica A: Statistical Mechanics and its Applications, 527*, 121171. https://doi.org/10.1016/j.physa.2019.121171

Miyashiro, C. S., Mateus, G. A. P., dos Santos, T. R. T., Paludo, M. P., Bergamasco, R., & Fagundes-Klen, M. R. (2021). Synthesis and performance evaluation of a magnetic biocoagulant in the removal of reactive black 5 dye in aqueous medium. *Mater Sci Eng C, 119*, 111523. https://doi.org/10.1016/j.msec.2020.111523

Mohamed Noor, M. H., Lee, W. J., Mohd Azli, M. F. Z., Ngadi, N., & Mohamed, M. (2021). *Moringa oleifera* extract as green coagulant for POME Treatment: Preliminary studies and sludge evaluation. *Mater Today Proc*. https://doi.org/10.1016/j.matpr.2021.02.241

Muniz, G. L., Borges, A. C., & da Silva, T. C. F. (2020). Performance of natural coagulants obtained from agro-industrial wastes in dairy wastewater treatment using dissolved air flotation. *Journal of Water Process Engineering, 37*, 101453. https://doi.org/10.1016/j.jwpe.2020.101453

Nharingo, T., Zivurawa, M. T., & Guyo, U. (2015). Exploring the use of cactus Opuntia ficus indica in the biocoagulation–flocculation of Pb (II) ions from wastewaters. *International journal of Environmental Science and Technology, 12*, 3791–3802. https://doi.org/10.1007/s13762-015-0815-0

Ni, C., Wang, J., Guan, Y., Jiang, B., Meng, X., Luo, S., Guo, S., et al. (2020). Self-powered peroxi-coagulation for the efficient removal of p-arsanilic acid: pH-dependent shift in the contributions of peroxidation and electrocoagulation. *Chem Eng J, 391*, 123495. https://doi.org/10.1016/j.cej.2019.123495

Obiora-Okafo, I. A., Onukwuli, O. D., & Ezugwu, C. N. (2019). Application of kinetics and mathematical modelling for the study of colour removal from aqueous solution using natural organic polymer. *Desalination and Water Treatment, 165*, 362–373. https://doi.org/10.5004/dwt.2019.24507

Oyegbile, B., Ay, P., & Narra, S. (2016). Flocculation kinetics and hydrodynamic interactions in natural and engineered flow systems: A review. *Environmental Engineering Research, 21*, 1–14. https://doi.org/10.4491/eer.2015.086

Padhiyar, H., Thanki, A., Kumar Singh, N., Pandey, S., Yadav, M., & Chand Yadav, T. (2020). Parametric and kinetic investigations on segregated and mixed textile effluent streams using Moringa oleifera seed powders of different sizes. *J Water Process Eng, 34*, 101159. https://doi.org/10.1016/j.jwpe.2020.101159

Pallier, V., Feuillade-Cathalifaud, G., Serpaud, B., & Bollinger, J.-C. (2010). Effect of organic matter on arsenic removal during coagulation/flocculation treatment. *Journal of Colloid and Interface Science, 342*, 26–32. https://doi.org/10.1016/j.jcis.2009.09.068

Park, S.-J., & Seo, M.-K. (2011). Chapter 1 - Intermolecular force. In S.-J. Park & M.-K. Seo (Eds.), *Interface science and technology* (pp. 1–57). Elsevier.

Parker, D. S., Kaufman, W. J., & Jenkins, D. (1971). Physical conditioning of activated sludge Floc. *Journal - Water Pollution Control Federation, 43*, 1817–1833.

Pieterse, A. J. H., & Cloot, A. (1997). Algal cells and coagulation, flocculation and sedimentation processes. *Water Science and Technology, 36*, 111–118. https://doi.org/10.1016/S0273-1223(97)00427-7

Ramavandi, B. (2014). Treatment of water turbidity and bacteria by using a coagulant extracted from *Plantago ovata. Water Resources and Industry, 6*, 36–50. https://doi.org/10.1016/j.wri.2014.07.001

Rossini, M., Garrido, J. G., & Galluzzo, M. (1999). Optimization of the coagulation–flocculation treatment: influence of rapid mix parameters. *Water Research, 33*, 1817–1826. https://doi.org/10.1016/S0043-1354(98)00367-4

Saleem, M., & Bachmann, R. T. (2019). A contemporary review on plant-based coagulants for applications in water treatment. *Journal of Industrial and Engineering Chemistry, 72*, 281–297. https://doi.org/10.1016/j.jiec.2018.12.029

Saxena, K., Brighu, U., & Choudhary, A. (2019). Coagulation of humic acid and kaolin at alkaline pH: Complex mechanisms and effect of fluctuating organics and turbidity. *Journal of Water Process Engineering, 31*, 100875. https://doi.org/10.1016/j.jwpe.2019.100875

Shammas, N. K. (2005). Coagulation and flocculation. In L. K. Wang, Y.-T. Hung, & N. K. Shammas (Eds.), *Physicochemical treatment processes* (pp. 103–139). Humana Press.

Soler, M., Serra, T., Folkard, A., & Colomer, J. (2020). Hydrodynamics and sediment deposition in turbidity currents: Comparing continuous and patchy vegetation canopies, and the effects of water depth. *Journal of Hydrology, 2020*, 125750. https://doi.org/10.1016/j.jhydrol.2020.125750

Sun, Y., Sun, W., Shah, K. J., Chiang, P.-C., & Zheng, H. (2019). Characterization and flocculation evaluation of a novel carboxylated chitosan modified flocculant by UV initiated polymerization. *Carbohydr Polym, 208*, 213–220. https://doi.org/10.1016/j.carbpol.2018.12.064

Tatsi, A. A., Zouboulis, A. I., Matis, K. A., & Samaras, P. (2003). Coagulation–flocculation pretreatment of sanitary landfill leachates. *Chemosphere, 53*, 737–744. https://doi.org/10.1016/S0045-6535(03)00513-7

Teh, C. Y., & Wu, T. Y. (2014). The potential use of natural coagulants and flocculants in the treatment of urban waters. *Chemical Engineering Transactions, 39*, 1603–1608. https://doi.org/10.3303/CET1439268

Tian, C., & Zhao, Y.-X. (2021). Dosage and pH dependence of coagulation with polytitanium salts for the treatment of *Microcystis aeruginosa*-laden and *Microcystis wesenbergii*-laden surface water: The influence of basicity. *Journal of Water Process Engineering, 39*, 101726. https://doi.org/10.1016/j.jwpe.2020.101726

Tinterri, R., Civa, A., Laporta, M., & Piazza, A. (2020). Chapter 17 - Turbidites and turbidity currents. In: Scarselli N, Adam J, Chiarella D, et al. (eds) Regional Geology and Tectonics (Second Edition). Elsevier, pp 441–479.

Usefi, S., & Asadi-Ghalhari, M. (2019). Modeling and optimization of the coagulation–flocculation process in turbidity removal from aqueous solutions using rice starch. *Pollution, 5*, 623–636. https://doi.org/10.22059/poll.2019.271649.552

Vignesh, A., Manigundan, K., Santhoshkumar, J., Shanmugasundaram, T., Gopikrishnan, V., Radhakrishnan, M., et al. (2020). Microbial degradation, spectral analysis and toxicological assessment of malachite green by *Streptomyces chrestomyceticus* S20. *Bioprocess Biosyst Eng, 43*, 1457–1468. https://doi.org/10.1007/s00449-020-02339-z

Vogt, R. D., Seip, H. M., Orefellen, H., Skotte, G., Irgens, C., & Tyszka, J. (2001). Trends in Soil Water Composition at a Heavily Polluted Site – Effects of Decreased S-Deposition and Variations in Precipitation. *Water Air Soil Pollut, 130*, 1445–1450. https://doi.org/10.1023/A:1013960930000

Wilén, B.-M., & Balmér, P. (1999). The effect of dissolved oxygen concentration on the structure, size and size distribution of activated sludge flocs. *Water Research, 33*, 391–400. https://doi.org/10.1016/S0043-1354(98)00208-5

Zahrim, A. Y., Dexter, Z. D., Joseph, C. G., & Hilal, N. (2017). Effective coagulation-flocculation treatment of highly polluted palm oil mill biogas plant wastewater using dual coagulants: Decolourisation, kinetics and phytotoxicity studies. *Journal of Water Process Engineering, 16*, 258–269. https://doi.org/10.1016/j.jwpe.2017.02.005

Zhang, Z., Jing, R., He, S., Qian, J., Zhang, K., Ma, G., et al. (2018). Coagulation of low temperature and low turbidity water: Adjusting basicity of polyaluminum chloride (PAC) and using chitosan as coagulant aid. *Sep Purif Technol, 206*, 131–139. https://doi.org/10.1016/j.seppur.2018.05.051

Zhou, B., Shang, M., Feng, L., et al. (2020). Long-term remote tracking the dynamics of surface water turbidity using a density peaks-based classification: A case study in the Three Gorges Reservoir, China. *Ecological Indicators, 116*, 106539. https://doi.org/10.1016/j.ecolind.2020.106539

Zhu, G., Zheng, H., Zhang, Z., Tshukudu, T., Zhang, P., & Xiang, X. (2011). Characterization and coagulation–flocculation behavior of polymeric aluminum ferric sulfate (PAFS). *Chem Eng J, 178*, 50–59. https://doi.org/10.1016/j.cej.2011.10.008

Part V
Medical Biotechnology

Drug Resistance Mechanism in *Staphylococcus aureus*

Anjaneyulu Musini, Priyanka Kandula, and Archana Giri

1 Introduction

Staphylococcus species are largely distributed in nature and comprised of huge populations (Harris et al., 2002). They are natural commensals of both human and animal skin and mucous membranes and retrieved from nature in an omnipresent manner (Huijbers et al., 2015). The real probability we are facing in twenty-first century is that even small wounds and common infections may lead to death of an individual. Among the *Staphylococcus* species, *S. aureus* is the most pathogenic with numerous infections varying from mild to life-threatening infections (Archer & Crossly, 1997). World Health Organization (WHO) has listed methicillin-resistant *Staphylococcus* as one of the significant threats worldwide (2014) owing to its resistance to multi-drugs. Different infections caused by *S. aureus* include pneumonia (lungs inflammations), mastitis (mammary gland infection), skin infections (impetigo, cellulitis and staphylococcal scalded skin syndrome), bone infection (osteomyelitis) and infective endocarditis. *S. aureus* may also cause food poisoning, resulting from the production of enterotoxins.

1.1 Background

It was first found by specialist Alexander Ogston in 1880 in Aberdeen, Scotland, in patients who are experiencing ulcerated injuries. They belong to *Staphylococcus* genus, Firmicutes: it shows positive on gram staining, coordinated in a series of

A. Musini (✉) · P. Kandula · A. Giri
Centre for Biotechnology, Institute of Science and Technology, Jawaharlal Nehru Technological University, Hyderabad, India
e-mail: manjaneyulu@jntuh.ac.in; archanagiriin@jntuh.ac.in

N. R. Maddela, L. C. García (eds.), *Innovations in Biotechnology for a Sustainable Future*, https://doi.org/10.1007/978-3-030-80108-3_17

grapes when seen under a magnifying lens, anaerobic or vigorous, ideally grows at 37 °C and at a pH of 7.4. Most of these are hemolytic, developing a clear haemolytic ring on blood agar plates around the colonies. *S. aureus* does not produce spores or flagella, but has a capsule that can develop golden yellow pigment and decompose mannitol. It is the most prevalent organism found on dry surfaces for a more extended period of time and can endure high temperatures (Guo et al., 2020).

1.2 Risk Groups

Age is a strong factor, for *S. aureus* bacteremia (SAB) occurrence which is having a high infection rate. Previous studies have shown that extreme infection occurs during the first years of life, decreases during young adulthood, and moderately increases advancing with age. Compared to women, the male is more influenced by this. The reason behind this is not known well, incidence of SAB significantly increased in HIV-infected patients. It is a consequence of high drug injections used by the HIV people and the low amount of CD4 count in them. Patients with hemodialysis are at great risk, too (Steven et al., 2015).

1.3 Epidemiology

In many of its description after a century of research, *S. aureus* is an alarming pathogen for human beings. Despite constant improvement in patient care, infections attributed *S. aureus* are still associated with substantial mortality and morbidity to both in hospitals and in society. Clinical and molecular epidemiology of infections caused by *S. aureus* through the advent of CA-MRSA (community-acquired MRSA) infections have improved enormously throughout the last 20 years. These changes are quickly identified, thanks to the uncommon combination in CA-MRSA with low rates of non-β-lactam antimicrobials. These clones come out and spread in patients who are lacking with the classic epidemiologic risk factors for MRSA infections and interaction with healthcare indicates that the CA-MRSA epidemic may emerge from unexpected regions and populations. This worrying surveillance of epidemiological studies on *S. aureus* infection and the molecular depiction of *S. aureus* strains has highlighted the renewed importance of it. This observation has been largely conducted in North America and Europe for the basic identification of novel CA-MRSA clones and livestock-associated MRSA (Rasigade et al., 2014).

Overall, since the turn of the millennium worldwide, the landscape of infections with *Staphylococcus aureus* has grown at a rapid rate. However, in non-western countries until recently most of these changes had gone largely undetected and probably because research efforts are mainly focused on the CA-MRSA catastrophic rise in North America and on the similar epidemic threat in Europe. Owing to the increasing connectivity, a global view of *S. aureus* dynamics is necessary to anticipate the rapid spreading of newly emerged clones. The *S. aureus* of Africa, India and the Far East requires reconnaissance efforts to be continued and more resources to be provided for controlling and investigating infections (Lowy, 1998).

1.4 Genetic Mobile Components

S. aureus has number of genetic mobile components which may encode antimicrobial resistance for different antibiotics (Wang et al., 2018). *S. aureus* is a commensal pathogen due to its rapid adaptive nature to all the pressures exerted by the human host. Versatile hereditary components are believed to be an essential element in the adaption cycle by transferring the hereditary material within and between the forms of microscopic organisms. Most importantly, genetic mobile components encoding for virulence factors show resistance to majority of antibiotics, as well as the gene which demonstrates resistance to all the β-lactam antibiotics in MRSA (Malachowa & DeLeo, 2010).

Versatile hereditary components were first revealed within the genome of maize during the 1940s (Mcclintock, 1951). They serve as an important element for the transfer between eukaryotes and prokaryotes of genetic material. They are small fragments that code for the virulence gene and resistance gene. Besides, they code for the enzyme that mediates their transition into the host and their incorporation. The transfer of MGEs between cells known as horizontal gene transfer has been observed through intracellular and extracellular mobility. It may take place in prokaryotic-to-prokaryotic cells or prokaryotic-to-eukaryotic cells or eukaryotic-to-eukaryotic cells also. Mobile genetic elements include components of the plasmids, insertional sequences, phages, transposons, chromosomal cassettes and pathogenicity islands. The genetic material transfer takes place via the vertical process (Malachowa & DeLeo, 2010).

1.5 **Staphylococcus aureus** *Genome*

The *S. aureus* genome includes the core genome, accessory components and foreign genes. The core genome responsible for building up the backbone of the genes has the main role in metabolic functions. This genome is strongly conserved among the isolates. An accessory part contains mobile genetic components constituting 25% of the total genome of *S. aureus* which includes chromosomal cassette, transposons, pathogenicity islands and genomic islands. Virulence genes are carried by MGEs (mobile genetic elements) (Kirmusaoglu, 2007) (Fig. 1).

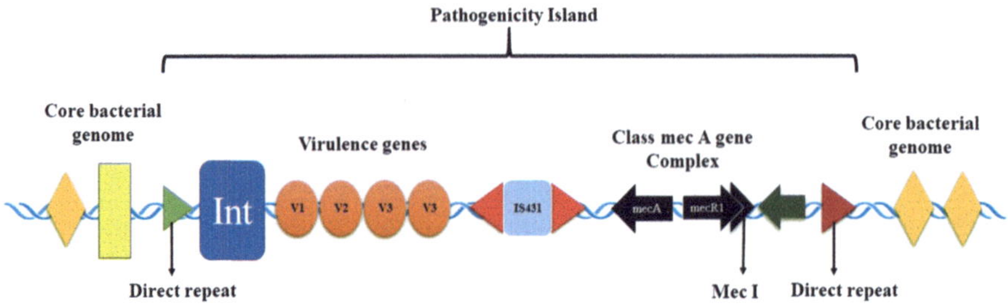

Fig. 1 Genome structure of *Staphylococcus aureus*

2 Bacterial Pathogenesis

S. aureus pathogenesis of infection involves five stages, they are colonization, local infections, sepsis, metastatic infection and toxinosis (Cluff & Reynolds, 1965). In healthy people, roughly around 30% of *S. aureus* colonization happens in the anterior nares of nose, vagina and the perianal area. It is transferred in mucous membrane asymptomatically from days to months. Colonization takes place before the infection occurs. Once infection happens, it may spread locally or enter the bloodstream. The existence of various virulence factors make this organism a pathogen. *S. aureus* has different virulent mechanisms hence this organism is used for the study of pathogenetic infections (Archer & Crossly, 1997).

One of the most significant stages of pathogenesis in *S. aureus* infection is the antiphagocytic polysaccharide capsule (Vlaeminck et al., 2020). Presence of the different virulence factors provides the potential of an organism to cause different infections by releasing enzymes, cell surface proteins, adhesion proteins and evading factors. Some toxins are also responsible for causing diseases, different factors work together to cause pathogenesis. Bacteria are protected by the extracellular polysaccharide capsule which safeguards the bacteria from the opsonization process to live in host cells (Stark, 2013).

Virulence factors which are released by *S. aureus* are very much complex; they not only lyse the cells of the host but also stimulate the process of invasion and disrupt the host cells. Many of the variables of virulence can control both adaptive and innate responses of immune system (Thammavongsa et al., 2015). Surface proteins present on the surface of *S. aureus* aid in adhesion. Many *Staphylococcus* strains express fibrin or fibrinogen binding proteins, which help for their attachment to the blood clots. Osteomyelitis and septic arthritis infection are caused when some strains of *S. aureus* bind to collagen (Fig. 2).

One of the virulence factors is coagulase—an extracellular protein that ties to the prothrombin to frame a composite called staphylothrombin in the cells of the host. Fibrinogen is converted to fibrin by activating protease in the thrombin. Coagulase protects bacteria from phagocytosis and immune system. Protein A also acts as a surface protein of *S. aureus*, this links with the IgG molecule in the wrong direction.

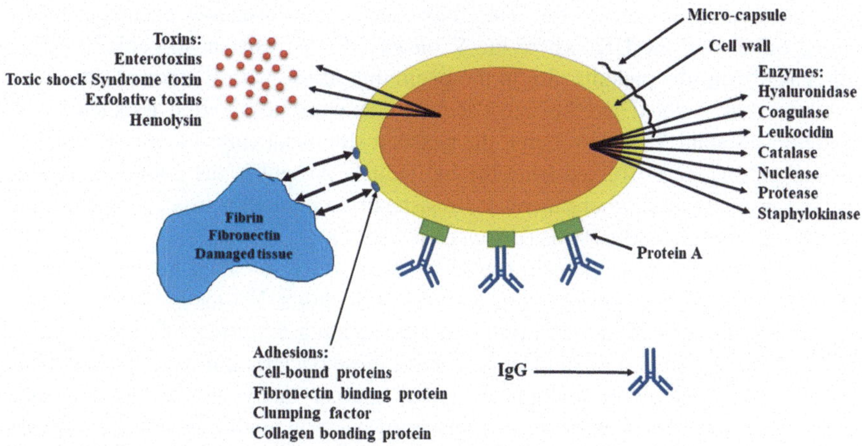

Fig. 2 Virulence factors of *Staphylococcus aureus*

The protein A binding to the IgG molecules occurs in the serum. Binding in incorrect orientation disrupts phagocytosis and opsonization process.

Proteases, lipases and DNase supply nutrients to the bacterial growth in the host cells. α, β, δ, γ toxins act as membrane damaging toxins, among those most powerful one is the α toxin. Platelets and monocytes in human are most sensitive to α toxin. Production of α toxin initiates septic shock. Leukocidin is the main factor responsible for necrotizing infections of the skin. It works together with others to destruct the membranes.

2.1 Plasmids Encode Antibiotic Resistance

Plasmids are autoreplicating DNA molecules. *Staphylococcus* generally contains one or more plasmids per cell, and they differ in gene content. *Staphylococci* plasmids are further separated into three groups, they are: (1) small multiply plasmids carrying a single immune determinant, (2) larger (15–30 kb) low copy (4/6 cells) plasmids that carry many resistant determinants, (3) conjugative multi-resistant plasmids. Larger plasmids undertake DNA replication mechanisms although smaller ones replicate by rolling circular mechanism. *S. aureus* plasmids transfer intercellularly via transduction or conjugation. *S. aureus* remains a free circularized DNA or linearized DNA and merges into the chromosome (Malachowa & DeLeo, 2010).

β-lactams are wide class of antibiotics because of their high effectiveness and low toxicity. They prevent cell wall biosynthesis in bacteria by irreversible binding of penicillin-binding proteins to the transpeptidase domain (Carboneau et al., 2010). A strain of *S. aureus* had progressed antibiotic resistance against penicillin by producing an enzyme called penicillinase, which hydrolyses penicillin. This process was first observed in the late 1940s. They have seen many instances occurring in the United States and the UK, where the *S. aureus* strains resistant to penicillin were segregated from the patients. So, in the beginning, those who showed resistance to *S. aureus* were isolated and they have acquired the word nosocomial associated with penicillin-resistant *S. aureus*. Later, the resistant strains lacking risk factors, like the hospital strains were isolated from the individuals present in the community. This resulted in a situation where increased resistance to penicillin was found from the late 1940s until the early 1960s (Bitrus et al., 2018).

Later a homologue to penicillin was released which is called methicillin as an important drug for the prevention of *S. aureus* infections. Yet methicillin resistance was observed over a duration of one year after treating *S. aureus* infections (Bitrus et al., 2018). We should be in place to control these bacteria quite easily, but this is not the case because of endurance shown against many antimicrobial agents. *S. aureus* is considered as the most transferred bacteria among healthcare workers due to improper cleaning of hands. Recent studies have shown that by simply enhancing the handwashing acquiescence in hospitals not only *S. aureus* infections, but many are reduced (Zoabi et al., 2011).

Inactivation of penicillin by penicillin-resistant strains was first revealed by Kirby (1944). Penicillinase is now produced by more than 90% of *S. aureus*. Penicillin resistance mainly occurs by spreading resistant strains. The transposable element contains a gene for β-lactamase, and it is present on the larger plasmid (Lowy, 2003).

Penicillin resistance in *S. aureus* is mediated by the Blaz gene which codes for β-lactamase. This primary extracellular enzyme is produced when *S. aureus* is exposed to β-lactam antibiotics and hydrolyses the β-lactam ring, thus inactivating the β-lactam. Mainly two regulatory genes control the Blaz, they are antirepressor blaR1 and the repressor blaI. Numerous studies have shown that a sequential split of regulatory proteins BlaR1 and BlaI is required for the signalling pathway of β-lactamase. Upon exposure to β-lactams, cleavage of BlaR1, a transmembrane sensor-transducer gets activated. Zhang et al. (2001) explained that a protein which is cleaved acts as a protease directly or indirectly splits the repressor BlaI and permits enzyme synthesis by blaZ (Lowy, 2003).

2.2 *Antimicrobial Resistance Development in* Staphylococcus aureus

S. aureus provides a greater and tougher model for understanding the complexity; they adopt during antibiotic pressure. These microorganisms can respond fast to the new provocations shown by all the new antibiotics. The evolution of resistance in this pathogen occurs by altering the target drug site, inactivation of antimicrobial agent enzymatically, and efflux pump (Bitrus et al., 2018).

S. aureus cell wall is very thick compared to other Gram-positive bacteria. Increased thickness of the cell wall is a mechanism used by the *Staphylococcus aureus* for resistance against other antibiotics that attack wall of the cell, mainly vancomycin (Cazares-Domínguez et al., 2015). The wall of the cell is a polymer that is comprised of peptidoglycan and composed of disaccharides structures NAG and NAM units. NAM has peptide stems; these peptide stems of neighbouring peptidoglycan are cross-linking points. The peptide stems are different for every organism. *S. aureus* has pentapeptide (Ala-Glu-Lyc(Gly)-Ala-Ala). The peptidoglycan backbone synthesis is a catalytic reaction of transglycosylases and transpeptidases. Transpeptidases are nothing but penicillin-binding proteins (PBP), they perform cross-linking through the wall of bacterial cells. The activity of PBP is more important for bacterial sustainability. The antibiotics like B-lactams also target these PBPs. There are four native PBPs that exist in *S. aureus*. PBP1, PBP2, PBP3, PBP4 (Tipper & Strominger, 1965). Fifth PBP is identified in methicillin-resistant *S. aureus* and named PBP2a. This PBP2a is encoded by the gene *mecA*. The normal PBPs are efficiently inhibited by B-lactam antibiotics, due to the structural similarity but this PBP2a is not well inhibited. So, when normal PBPs activity is inhibited, it will continue the cross-linking of peptidoglycans. *S. aureus* produces Pc1Beta-lactamase, the product of *blaz* gene which resists the beta-lactam antibiotics (Fishovitz et al., 2014).

B-lactamase hydrolyses the B-lactam ring and makes it inactive. Methicillin-resistant *S. aureus* acquired gene mecA which encodes PBP2a. The genes for fem (factor essential for expression of resistance to methicillin) and aux (Auxiliary) factor also responsible for methicillin resistance have been identified (Tomasz & De Lencastre, 1994). The gene mecA has been acquired by *S. aureus* from an unknown source and its autologous has been identified in *S. sciuri*, *S. vitulinus* and *S. fleurettii* (Fishovitz et al., 2014).

2.3 The Action of Antibiotics and Mechanism

Serine acylation takes place when serine active site strikes the carbonyl ring of β-lactam. β-lactam antibiotics bind to the transpeptidase or carboxypeptidase of the penicillin-binding protein domains. β-lactam derived acyl-enzyme complex is stable compared to the temporary acyl-enzyme complex obtained from the D-Ala-D-Ala which is incapable of transpeptidation and goes through extremely slow hydrolysis for the regeneration of free serine. Irreversible acylation is observed. Hence, this irreversible acylation ceases the activity of transpeptidase, and carboxypeptidase engaged in the cell wall formation (Walsh & Timothy, 2016).

Transpeptidase PBP2 in the *S. aureus* enzyme comprises of the trans glycosylase domain and the transpeptidase domain. The transglycosylase and transpeptidase domains are separated well in the *S. aureus* containing bifunctional proteins PBP2. The β-lactam antibiotic targets the trans glycosylase region which transfers the disaccharide pentapeptide of peptidoglycan from membrane-bound lipid II to the growing polysaccharide chains. The transpeptidase (TP) region cross-links the glycine cross bridge to the 4th D-alanine of the adjacent chain (Peacock & Paterson, 2015).

β-lactam antibiotics due to their structural analogy to D-Ala4-D-Ala5 binds to the serine residues present in the active site of TP in PBP2, as a result the β-lactam bond is broken, and penicilloyl-o-serine intermediate is formed. Penicilloyl-o-serine intermediate is extremely stable than the intermediate formed by the peptidoglycan acyl-enzyme. It takes about 1–4 h for the addition of a water molecule to penicilloyl-o-serine to regenerate the serine active site and release penicilloic acid as the product. Blocking of the TP (Transpeptidase) enzyme active site, the peptidoglycan biosynthesis is stopped. The β-lactam antibiotics induce the toxic malfunctioning of cell wall biosynthesis apparatus which results in the cycle of synthesis followed by autolysis.

The β-lactamase of *S. aureus* is a typical β-lactamase (BlaZ) which is same as the acyl-enzyme intermediate as transpeptidase of PBP2 and is resistant to penicillin. In this process the catalytic difference between *BlaZ* and TP is the kinetics of deacylation involved. In the presence of *BlaZ,* the addition of water is fast which results in regeneration of the serine active site and release of open ring penicilloic acid which is a hydrolytic degradation product without inhibitory activity (Foster, 2017).

2.3.1 *Staphylococcus aureus* Resistance to Penicillin and Mechanism

Resistance to penicillin by *S. aureus* is mediated by *blaZ* gene which codes for B-lactamase. The *blaZ* structural gene which is accountable for β-lactamase is transferred by the elements like transposon Tn552. It is situated on a large plasmid or it is integrated into the chromosome of bacteria (Plata et al., 2009). *BlaI* repressor and the *BlaR* sensor are responsible for the expression of the enzyme. Lack of β-lactam antibiotics stops the transcription of *blaZ* because the *BlaI* which is bound to the promoter-operator region suppresses *blaz* and *blaI-blaRI* operon. In the presence of β-lactam antibiotics, the binding of β-lactams to *BlaRI* takes place, then separation of zinc metalloprotease domain of *blaRI* occurs and splits the *blaI*. *BlaZ* gene is transcribed in the presence of β-lactam to β-lactamase and allows *S. aureus* to grow continuously. This primary extracellular enzyme, which is produced when staphylococci are exposed to β-lactam antibiotics, hydrolyses the β-lactam ring, thus inactivating the β-lactam (Kirmusaoglu, 2007) (Fig. 3).

2.3.2 *Staphylococcus aureus* Resistant to Methicillin

Infections of MRSA are different types based on where they are acquired. For instance, healthcare acquired MRSA has been documented in 1960s, particularly in hospitals and other healthcare facilities (HA-MRSA). Patients with the weakened immune system, who have undergone recent surgery or those with medical devices implanted are at higher risk for HA-MRSA and the prevalence of this infection has

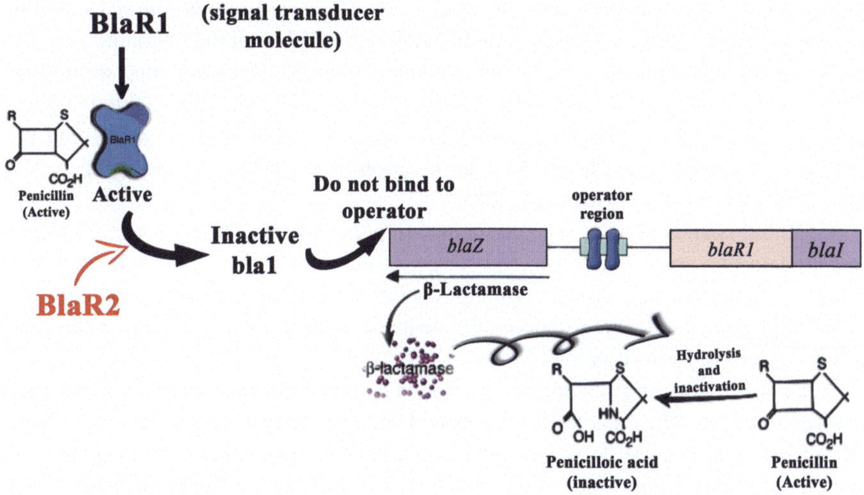

Fig. 3 Penicillin resistance mechanisms in *S. aureus*

been increasing steadily (Moosdeen et al., 1986). A second and more alarming type of infection with MRSA has emerged in the 1990s, which is called community-acquired MRSA. The point of origin for this infection can be more complex to remember, but they are thought to be common in the childcare setting, athletic settings, term care facilities, or sharing of objects. The infections appear as a simple infection in the beginning but can progress into a life-threatening illness. The extensive spread of MRSA is seen in community-acquired infections.

MRSA strain can develop many genetic changes by making it highly resistant to different antibiotics. This organism can develop drug resistance very rapidly (Aziz & Hassan, 2019). The incidence of mecA gene in the chromosome helps in the transfer of bacterial cells to other cells. Studies of Ito et al. (2001) revealed that the incidence of *mecA* gene in all the DNA strands of MRSA acts as antibiotic resistance. Two genes with recombinase ccrA and ccrB regulate the *mecA* gene. Previous experiments proved that by removing *mecA* gene from MRSA it lost its resistance mechanism. *S. aureus* shows resistance to methicillin only if *mecA* gene is present along with ccrA and ccrB. They both help in the expression of *mecA* gene. Protein production is inhibited by *mecA* gene and it makes it difficult for the antibiotics to bind (Dever & Dermody, 1991).

The chromosome of *S. aureus* contains *Orfx* genes in which the large DNA fragments called Staphylococcus cassette chromosome is inserted. It may encode resistance to antibiotics or virulence determinants. *S. aureus* obtains resistance to methicillin by the introduction of a staphylococcal cassette chromosome, which carries the *mecA* gene, inside the chromosome (Malachowa & DeLeo, 2010).

The *Scc* element is exactly extracted from the N315 chromosome and integrated into the *S. aureus* chromosome, containing a distinctive set of recombinase genes, CcrA and CcrB. The distribution of *Scc mec* in MRSA was first observed in Japan during the 1990s. The studies of Ito et al. (2001) reported two additional types of genetic elements that carry *mecA* gene in MRSA isolated from many countries in the world. Structural analysis of the new genetic element disclosed the sharing of chromosomal integration site with *Scc mec*, preserved terminal inverted repeats and direct repeats, the genetic organization is conserved among the *mecA* and the presence of Ccr genes. It helps for identification of the elements belonging to *Scc mec* family of *S. aureus* (Ito et al., 2001).

Genomic islands and scc mec elements are present in the staphylococcal species and contain two crucial elements, the ccr genes complex and the *mec* gene complex (mec). Ccr gene domain contains *orfs* and genes for ccr, whereas *mec* gene contains *mecA* gene, genes for regulation and sequences responsible for upstream and downstream insertion of *mecA*.

It has led to the classification of *Scc mec* because of the existence of various mec and ccr allotypes. They are Type I *Scc mec*, Type II *Scc mec*, Type III Scc mec, Type IV Scc mec and Type V. Type I *Scc mec* contains class B *mec* and *Ccr* type I; Type II scc mec contains class A mec and ccr type II, Type III Scc mec contains *mecA* class and type III ccr, Type IV Scc mec contains *mecB* class and type II *ccr*, Type IV contains *mec* C2 class and type 5 ccr (Piriyaporn et al., 2006). Type V *Scc mec* element does not include any antibiotic resistance genes except β-lactam resistance which is transferred by a novel genetic complex (Ito et al., 2004).

MecA Gene, Resistance Mechanism

mecA gene codes for an altered penicillin-binding protein, PBP-2a, which is not reticent by the existing beta-lactam antibiotics and helps in continuous cell wall biosynthesis. Most of the researchers in this field suggest that all the MRSA strains contain a *mecA* gene. Hence, we can consider this as a cornerstone that is responsible for the MRSA phenomenon.

Mec operon is present in MRSA strains and absent in MSSA strains. Studies revealed that the *mecA* gene present in MRSA and *mecA* of *Staphylococcus sciuri* disclosed 88% similarity. But the difference is *mecA* of *Staphylococcus sciuri* showed sensitivity to methicillin. This reinforces that MRSA is an offspring of ancestral strains in the developmental process (Chambers, 1997).

Transcription of *mecA* is stopped when there are no β-lactam antibiotics because the *mecI* that is bound to the promotor-operator region suppresses *mecA* along with mecI-mecRI operon. MecRI acts as a β-lactam sensing signal transducer. When β-lactams are present, they bind to MecRI and cleave the MecI which is already bound to the operator region. Transcription of *mecA* occurs, the transcribed *mecA* produces PBP2a which shows less affinity to β-lactams. Because of the low affinity of PBP2a to β-lactams, the cell wall biosynthesis occurs continuously and shows resistance (Kirmusaoglu, 2007) (Fig. 4).

The basic function of PBP2 may be replaced by the PBP2A which acts as a surrogate transpeptidase in methicillin resistance *S. aureus* strains. Hence, PBP2 is no more important for the MRSA to grow. But they face the problem when they are exposed to β-lactam antibiotics. In this incidence, the activity of transglycosylase of

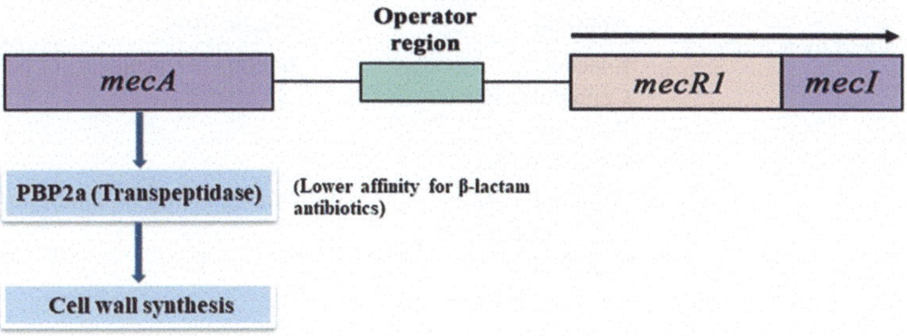

Fig. 4 Methicillin resistance mechanism in *S. aureus*

PBP2 plays an important role in the cell wall biosynthesis and development of bacteria (Leski & Tomasz, 2005).

In a previous literature, Anne et al. (1995) explained that *S. aureus* resistance towards methicillin is an intrinsic resistance unlike the plasmid-borne in penicillin. This intrinsic resistance is observed in many other pathogenic bacteria which is related to the modification of PBP; it may be either in the amount or their affinity towards the β-lactams. Hence, the immediate attention turned out for understanding the mechanism of PBPs in methicillin-resistant *S. aureus.*

The quantity of amino acid, amino sugar and teichoic acid of MRSA and *MSSA* showed no noticeable variations in the wall composition (Wyke et al., 1982). The biosynthesis of peptidoglycan has been researched by Smith and Wilkinson (1981) in vitro on *S. aureus* which is highly resistant to methicillin. They described two different types of peptidoglycan synthesis which shows different susceptibilities to methicillin. Inhibition of 'cell wall thickening' occurs when the methicillin is added in low concentration, whereas high concentration of methicillin showed slight effect on their growth and synthesis of cell surface.

Wyke et al. (1982) observed the growth and viability of the cells when methicillin was applied to *S. aureus* which was in exponential phase. They observed that turbidity increased gradually with no change in the rate for over one generation and remained constant and similar was with the high MRSA strains. Hence methicillin acts as bacteriostatic and not bactericidal.

Past examinations have established that components encoded on β-lactam plasmids are engaged with the acceptance of PBP2a and it has been stated that those components are the β-lactamase administrative qualities. These examinations gave the primary direct proof that *blaI* and *bla*RI are the plasmid determined qualities engaged with the managed expression of *mecA* (Wyke et al., 1982).

2.3.3 Kinetics Reaction Mechanism

Mechanism of the kinetics for the reaction of PBP2a with all the β-lactams is identical to that of all the normal PBPs. Dissociation constant K_d is formed when

Table 1 Different groups of antibiotics and their mode of action

Antibiotics	Mechanism
β-lactams Cephalosporins Penicillin Glycopeptides	Inhibits the synthesis of cell wall
Aminoglycosides Chloramphenicol Macrolides Oxazolidinones Streptogramins	Protein synthesis inhibition
Quinolones Fluoroquinolones	Nucleic acid synthesis inhibition
Sulphonamides Trimethoprim	Obstruct metabolic pathway

β-lactam reversibly binds with the active site of the enzyme (serine). Dissociation constant K_d is converted to the acyl-enzyme complex (k_2) by the process known as acylation. Acylation inhibits the cross-linking behaviour of the transpeptidase resulting in deacylation (k_3). This is the only process to free the active site present.

$$\underset{(K_d)}{\mathrm{PBP} + \beta - \mathrm{lactam} \leftrightarrow} \quad \underset{(k_2)}{\mathrm{PBP}.\beta - \mathrm{lactam} \rightarrow} \quad \underset{(k_3)}{\mathrm{PBP} + \beta - \mathrm{lactam}}$$

Slow hydrolysis occurs releasing the β-lactam open ring as a product. It has been observed that the deacylation process takes a longer time in normal PBPs than cell viability in the presence of β-lactam antibiotics. The rate of acylation occurs fast in normal PBPs compared to that of PBP2a. It shows 3–4 orders of smaller magnitude than the acylation rate, which is observed in normal PBPs. On analysis through circular dichroism spectroscopy, it exhibits a modification of conformation upon acylation of the PBP2a. The slow rate of deacylation has been observed in PBP2a (Peacock & Peterson, 2015).

There are wide variety of antimicrobial drugs available that have been used to treat *S. aureus* infection. Antimicrobial resistance is often based on the mechanism of the activity of antimicrobial drugs (Reygaert, 2018). The antimicrobial agents based on their mechanism of action are listed below (Table 1).

2.3.4 Drug Resistance to Different Antibiotics

Hospital-acquired and community infections with *S. aureus* have improved and strengthened antibiotic resistance of *S. aureus* isolated strains from 2009 to 2014 from various clinical samples at Yuzuncu Yil University, Dursun odabas medical center, microbiology laboratory and their antibiotic susceptibility test results were compared and analysed. The isolates were described using traditional methods in

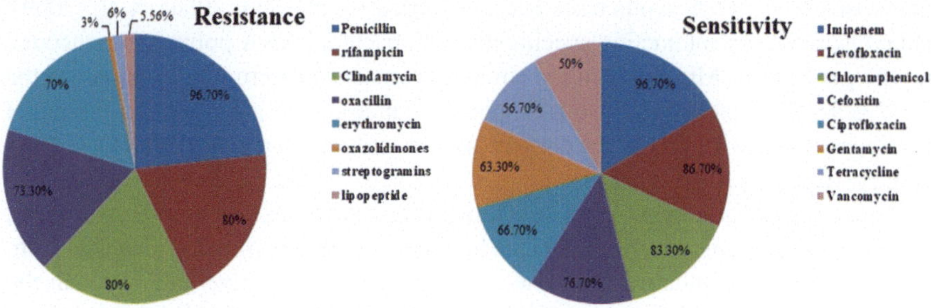

Fig. 5 List of antibiotics showing resistance/sensitivity against *S. aureus*

compliance with the clinical and laboratory standards of the institute and performed using phoenix susceptibility tests (Ragbetli et al., 2016).

Ragbetli et al. (2016) reported the study of the susceptibility test for the years of 2009 to 2014, all *S. aureus* isolates have been described as susceptible to daptomycin, vancomycin, linezolid and levofloxacin. Nitrofurantoin, quinupristin-dalfopristin and trimethoprim sulphamethoxazole have resistance rates, as 0.3%, 2.4% and 6.1%, respectively. Clindamycin, gentamicin, rifampicin, erythromycin and penicillin have resistance rates as 11%, 14%, 14%, 18% and 100%, respectively. In the year 2009, the highest percentage of methicillin resistance was about 30% and then decreased in subsequent years such as 20%, 19%, 16%, 13% and 21% (Gursoy et al., 2009). Resistance of *S. aureus* to antibiotics varied considerably with the highest resistance recorded to ampicillin and penicillin, *S. aureus* revealed varying susceptibility to different antibiotics, highest susceptibility to imipenem and levofloxacin (Fig. 5) (Akanbi et al., 2017).

Despite finding new antibiotics against bacteria, they are adapting and making favourable to them especially because those bugs can evolve into a superstructure known as biofilm, which makes them resistant and survive with that in the habitat. Eighty percent of the nosocomial infections are linked with biofilm formation which is mostly observed in the *S. aureus* species. It is one of the main species in this domain (Reffuveille et al., 2017) Treatment of *Staphylococcus aureus* becomes difficult due to the methicillin resistance and biofilm formation. It was identified that the presence of the mecA gene increases the biofilm formation by disabling the agr system (Kirmusaoglu, 2007).

Oxazolidinones (linezolid and tedizolid) bind to the bacterial 23S rRNA at the ribosomal peptide-transferase centre, interrupting transitional RNA positioning. Point mutations in the genes encoding the 23S rRNA are most common, with majority of mutations occurring in the central loop of domain V of the 23S rRNA (Munita et al., 2015). Most bacteria have multiple copies of the 23S rRNA gene (*S. aureus* has four to seven), and the accumulation of mutations determines the degree of linezolid resistance, i.e., the mutant gene dosage effect (Pillai et al., 2002).

The *ileS-2* gene from *S. aureus* has been found to be responsible for conferring mupirocin resistance (Nunes et al., 1999). Subsequently, high-level resistance to mupirocin was found to be conferred by the *mupA* and *mup B* genes, which encode novel isoleucyl-tRNA synthetases and are carried by plasmids (Plata et al., 2009). Rifampicin acts by interacting specifically with bacterial RNA polymerase encoded by the gene *rpoB*. Mutations in the rifampin resistance-determining regions of the *rpoB* gene of *S. aureus* leads to rifampin resistance due to alterations in the target leading to reduced affinity of the enzyme for the antibiotic (Damon et al., 1998; Zhou et al., 2012).

The *erm* gene product confers clindamycin resistance on *S. aureus* (Levin et al., 2005). Inducible clindamycin resistance has been reported in *S. aureus* isolated from clinical samples. Clindamycin is one of the essential alternative antibiotics in the therapy of *S. aureus* infections. Clinical failure of clindamycin therapy has been reported due to numerous mechanisms that confer resistance to macrolides, lincosamides and Streptogramin B (MLSB) antibiotics (Fiebelkorn et al., 2003; Sasirekha et al., 2014). Resistance to these antibiotics can occur by efflux mechanism encoded by the msrA gene and ribosomal target modification encoded by the erm gene (MLSB resistance). Macrolides, which include erythromycin, are one of the antibiotic groups used against infections caused by *S. aureus*. The repercussion of their utmost use is the high incident of mechanisms responsible for resistance to these drugs. Numerous mechanisms may be involved in *S. aureus* resistance to macrolides. Three genes msr (A), mph(C) and erm(Y) confer resistance to macrolide antibiotics on *S. aureus* (Matsuoka et al., 2003). Another known cause of resistance in *S. aureus* is the synthesis and activity of macrolide-inactivating enzymes called phosphorylases (Piątkowska et al., 2012).

2.4 Biofilm and Antibiotic Resistance

Biofilm forming bacteria show resistance to antibiotics, chemical disinfectants, phagocytosis and the immune defensive system of the human body. Communication of biofilm containing bacteria occurs using certain molecules which in turn help inactivation of certain genes that are responsible for the biofilm structure formation and producing some virulence factors. This event is called quorum sensing. The extent of quorum sensing relies upon the total quantity of bacteria present (Hoiby et al., 2011).

Gram-positive species use oligopeptide molecules which act as signalling molecules during cell to cell communication process. For the treatment of bacterial biofilm infections, quorum sensing system is the future target for many researchers. Research conducted by Yen Chen et al. (2016) showed that baicalein has antibacterial activity and inhibits the formation of biofilm in *S. aureus* by inhibiting the quorum sensing (Singh et al., 2017).

Formation of Biofilm

Biofilm production is a complex process involving four stages. It begins with the initial attachment of the microorganisms to a substrate surface and permanent binding is induced with colonization following the exponential growth process (Sharma et al., 2019).

Biofilm formation was being thoroughly studied in last 20 years and the method of attachment and initial biofilm formation was deeply understood. Close examination of the cell surface and substratum is necessary for a better understanding of the attachment. The substratum may be rough or smooth. Characteristics of substratum play an important role in the attachment of microbes. The substratum which is rough and hydrophobic helps in the development of biofilm very quickly. Not only the characteristics of substratum but the cell surface features also play significant role. Cell surface characteristics like flagella and pili enhance the rate of attachment to the substratum (Donlan, 2001).

Biofilm formation has been observed on different metals and minerals. They are found in water, underground surfaces, and on the ground surface also. Their presence is widely distributed in nature. They are able to expand on tissues of plants and animals. Biofilm formation has been observed on implanted devices of medical apparatus such as catheters, pacemakers, endotracheal tubes, intrauterine devices and mechanical heart valves. They cause primary and secondary diseases in the human body. The National Institute of Health announced that 80% of the microbial infections in the body are due to biofilm formation (Donlan, 2002).

The capability of an organism to acquire resistance temporarily or permanently during multiplication would intend to stop or inhibit the growth of other microorganisms belonging to the same strain only under some conditions (Donlan, 2002).

The resistance to various antibiotics is very well understood for the last 20 years, but resistance to disinfectants and antiseptics is not understood. Regulation and interaction of the gene expression acts as a main mechanism adopted by bacteria within the biofilm. The formation of biofilm has increased resistance to several antibiotics leading to many diseases worldwide. Biofilm acquires various mechanisms that afford resistance to different antibiotics (Singh et al., 2017).

The Reason Behind Biofilm Formation

Biofilm formation increases bacteria's resistance to various environmental factors. The capacity of surface attachment increases thus bacteria escapes from washing off by the water flow or blood stream (Rabin et al., 2015). It offers more resistance to antibiotics and confers drug resistance. Bacterial mobility decreases and the cell density increases, which increases the exchange of genetic materials through conjugation. The best environmental conditions for the plasmids are given by an increase in the cell density. Horizontal transfer of genes takes place during biofilm formation.

Antibiotic Resistance in Biofilm Cultures

The population of biofilm contributes to different chronic infections because of the resistance to antibiotics. The resistance mechanism shown by the biofilm population is different from the planktonic organisms. Resistance to antibiotics occurs due to several mechanisms in biofilm communities like partial penetration of antibiotics into the biofilm and alteration of the chemical pathway within the biofilm. This leads to the outcome of multicellular existence in biofilms showing antibiotic resistance within the biofilm population as well as failure to the therapeutic approach (Sharma et al., 2019).

The production of biofilm is very well organized sequence of events by intracellular and intercellular signalling throughout its formation. Upregulation and downregulation of a set of genes have been observed for the bacterial attachment to the substrate layer.

Antibiotic resistance is due to the multicellular nature of biofilm. By destroying any steps of the multicellular biofilm structure, the effectiveness of antibiotics and the host immune system may be increased (Sharma et al., 2019).

2.5 Quorum Sensing

Cell to cell communication in the bacterial biofilm occurs by a process called quorum sensing (Sharma et al., 2019). Gram-positive bacteria use oligopeptide molecules for signalling and intraspecific communication occurs between bacteria (Chen et al., 2016). Quorum sensing helps in regulating bacterial behaviour during different circumstances.

Quorum sensing system governs many cellular processes which involve the regulation of luminescence, spore formation, the appearance of biofilm, production of toxins and resistance to drugs (Zhao et al., 2020).

Their expression of gene is coordinated by responding and producing autoinducers which act as inter- and intracellular signalling molecules. In gram-negative bacteria, quorum sensing gene expression is mediated by small peptides and autoinducers. The amount of autoinducers in the extracellular media relies on the density of the cells present. As the cell's density increases, the concentration of the autoinducers also increases. When the signal concentration of molecules reaches a certain level, it passes through the phosphorylation/dephosphorylation cascade by binding to the receptor protein for the movement of oligopeptide molecules to the promotor region for the initiation of transcription and post-translational modifications (Zhao et al., 2020). In turn, it stimulates or suppresses target gene expression in bacterial biofilm (Miller, 2001).

N-acyl homoserine lactones serve as a signalling molecule in gram-positive bacteria, whereas in Gram-positive bacteria antimicrobial peptides are produced as a virulence response (Kleerebezem et al., 1997). Quorum sensing is also observed in fungi that produce biofilm. It has been observed in the species of *Aspergillus* and *Candida albicans* (Ramage et al., 2002).

The accessory regulators of gene (agr) locus in *S. aureus* are considered as a quorum sensing signalling mechanism in *S. aureus*. It is correlated with infections including endocarditis and osteomyelitis. However, the exact role of the *agr* system differs on the infection model type used (Jessica et al., 2014).

The existence of an accessory gene regulatory system (agr) reduces the cell surface protein expression and relates to the increased expression of many secreted virulence factors in *S. aureus* mainly during the transfer process from exponential to stationary phase. Behaviour in biofilm and dispersal are dependent on the expression of *agr* system. It regulates many of the products which are involved in the development of biofilms (Yarwood et al., 2003).

The locus of Agr is 3.5 kb in size and comprises of two divergent transcriptional units, RNAII and RNA-III and their transcription is done by the P2 and P3 promoters. The RNA II locus comprises of four genes, *argB*, *argD*, *argC and argA*. The transcript argD encodes for a precursor polypeptide for the extracellular quorum signal of *Agr* called the autoinducing peptide (AIP) (Ji et al., 1997). AIP is 7–9 amino acids long containing a thiolactone ring between the centrally located cysteine and the C terminal. The transmembrane endopeptidase encodes the ArgB gene whose function is to introduce the thiolactone alteration, C-terminal cleavage and AIP export. *Agr A* along with *AgrC* gene encodes a system of two-component signal transduction involving a sensor for histidine kinase. The *Agr C* gets activated upon phosphorylation by AIP. This activation results in the binding of *AgrA* to the promoter of P2region of RNAII and the P3 promoter region of RNA II. Variation in the *AgrB*, *AgrC* and *AgrD* gene sequence produces AIPs with varied signalling specificities, allowing them to self-activate or cross inhibit *Agr* groups of nonself. It has been observed that SrrAB and SarA regulatory molecules and environmental factors such as glucose concentration and pH can activate *Agr* (Le & Otto, 2015).

RNA-III is an intracellular Agr system effector molecule that regulates Agr targets. It also acts as an mRNA containing the hld genes for delta toxin. RNA-III blocks translation by antisense base pairing with 5′UTR and forming an RNA duplex. RNA-III blocks the translation of repressor of toxin (Rot) proteins belonging to the staphylococcal accessory regulator (Sar) transcriptional regulator family (McNamara et al., 2000). Blocking of transcription takes place by binding Rot to the promotor region of many toxins and exoproteins. Furthermore, expression of gene analysis genome-wide shows that besides controlling of virulence determinants, Agr control also contains a sequence of metabolic targets. Such general physiological changes can help bacteria to adjust to the modifications occurred during infections. Recent works show that RNA III normally acts by antisense base-pairing mechanism by regulating many target genes through controlling repressor protein gene called as Rot. It belongs to a SarA family (Queck et al., 2008).

Developing to the presence of high antibiotic resistance in bacterial biofilm, appropriate therapy for biofilm-associated bacterial infections is impaired. Traditional antibiotic treatment seems unable to completely kill the bacterial cells, contributing to the development of biofilm. Therefore, to tackle this process alternative strategies and new antibiofilm forming agents are being found in previous studies (Sharma et al., 2019). There are different approaches to prevent the infections caused by bacterial biofilm. They are, use of some antimicrobial peptides which have antibiofilm activity (Francolini & Donelli, 2010). Bacteriophage's treatment to fight biofilm-caused infections (Shahid et al., 2019). Some small molecules having an antibiofilm activity that reduce virulence (Van Tilburg et al., 2015). Enzymes that dissolve or disperse biofilm. Use of plant extracts, honey and essential oils as a natural antibiofilm strategy (Shahid et al., 2019). Nanotechnology-based strategies, use of pohotodynamic therapy, and CRISPR/cas technology for controlling and managing biofilms (Sharma et al., 2019). Novel therapeutic strategies for MRSA treatment are quorum sensing inhibition, lectin inhibition, iron chelation, phage therapy and nanotechnology (Guo et al., 2020; Li et al., 2017).

3 Conclusion

Antibiotic resistance is a deadly threat to humankind globally. A probable line of action in the fight against drug resistant bacteria depends upon analysing the molecular mechanism of various networks associated with drug resistance and their coordination. A combination of new strategies along with drug repurposing or in combination with antimicrobials may work for combating drug resistance in bacteria. Discovery of antibiotics gave all the researchers a positive hope to combat *S. aureus*; however, this did not last for a longer period. Because the bacteria started to show resistance against them by acquiring some genes from other microorganisms by horizontal gene transfer or mutation. Almost all the strains of *S. aureus* are resistant to β-lactam drugs. Some drugs are effective against *S. aureus,* however, there is still a big question mark. They are multi-drug resistant. Compared with other bacteria *S. aureus* has the quick ability to become resistant on exposure with any antibiotic for a longer period. Treating *S. aureus* is the most difficult problem because of MRSA and biofilm formation. The past few decades have witnessed progress in understanding of biofilm formation and related virulence properties. Hopefully, the inhibition of quorum sensing during biofilm formation might be the cornerstone for developing a new drug to control the infections which are related to biofilm. This chapter highlights several recent developments, which may provide wide-ranging advances to alleviate antibiotic resistance. The novel approaches may pave the way for further advances in drug discovery to combat drug resistance. The discovery of alternative control strategies is the need of the hour.

References

Akanbi, O. E., Njom, H. K., Fri, J., Otigbu, A. C., & Clarke, A. M. (2017). Antimicrobial susceptibility of *Staphylococcus aureus* isolated from recreational waters and beach sand in Eastern cape province of south Africa. *International Journal of Environmental Research and Public Health, 14*(9), 1001. https://doi.org/10.3390/ijerph14091001

Anne S., & Reisman R. E. (1995). Risk of administering cephalosporin antibiotics to patients with histories of penicillin allergy. *Ann Allergy Asthma Immunol. Feb;74*(2), 167–70.

Archer, G. L., & Crossly, K. B. (1997). *The Staphylococci in human disease*. Elsevier.

Aziz, Z. S., & Hassan, M. A. (2019). Phenotypic and molecular study of *mecA* gene in MRSA isolated from clinical isolates in Misan Province. *Indian Journal of Public Health Research and Development, 10*, 553–558. https://doi.org/10.5958/0976-5506.2019.00350.4

Bitrus, A. A., Peter, O. M., Abbas, M. A., & Goni, M. D. (2018). *Staphylococcus aureus*: A review of antimicrobial resistance mechanisms. *Research Reviews, 4*(2), 43–54.

Carboneau, C., Benge, E., Jaco, M. T., & Robinson, M. (2010). A lean six sigma team increases hand hygiene compliance and reduces hospital-acquired MRSA infections by 51%. *Journal for Healthcare Quality, 32*, 61–70.

Cazares-Domínguez, V., Cruz-Cordova, A., Ochoa, S. A., Escalona, G., Arellano-Galindo, J., & Rodríguez-Leviz. (2015). Vancomycin tolerant, methicillin-resistant *Staphylococcus aureus* the effects of vancomycin on cell wall thickening. *PLoS One, 10*(3), e0118791. https://doi.org/10.1371/journal.pone.0118791

Chambers, H. F. (1997). Methicillin resistance in Staphylococci molecular and biochemical basis and clinical implications. *Clinical Microbiology Reviews, 10*(4), 781–791.

Chen, Y., Liu, T., Wang, K., Hou, C., Cai, S., Huang, Y., Du, Z., Huang, H., Kong, J., & Chen, Y. (2016). Baicalein inhibits *Staphylococcus aureus* biofilm formation and the quorum sensing system in vitro. *PLoS One, 11*, e153468. https://doi.org/10.1371/journal.pone.0153468

Cluff, L. E., & Reynolds, R. J. (1965). Management of Staphylococcal infections. *The American Journal of Medicine, 39*, 812–825.

Damon, H. A., Soussy, C. J., & Courvalin, P. (1998). Characterization of mutations in the *rpoB* gene that confer rifampin resistance in *Staphylococcus aureus*. *Antimicrobial Agents and Chemotherapy, 42*(10), 2590–2594.

Dever, L. A., & Dermody, T. S. (1991). Mechanisms of bacterial resistance to antibiotics. *Archives of Internal Medicine, 151*(5), 886–895. https://doi.org/10.1001/archinte.1991.00400050040010

Donlan, R. M. (2001). Biofilm formation: A clinically relevant microbiological process. *Clinical Infectious Diseases, 33*, 1387–1392. https://doi.org/10.1086/322972

Donlan, R. M. (2002). Microbial life on surfaces. *Emerge Infectious Diseases, 8*(9), 881–890. https://doi.org/10.3201/eid0809.020063

Fiebelkorn, K. R., Crawford, S. A., McElmeel, M. L., & Jorgensen, J. H. (2003). Practical disc diffusion method for detection of inducible clindamycin resistance in Staphylococcus aureus and coagulase-negative Staphylococci. *Journal of Clinical Microbiology, 41*, 4740–4744.

Fishovitz, J., Hermoso, J. A., Chan, M., & Mobashery, S. (2014). Penicillin- binding protein 2a of methicillin-resistant *Staphylococcus aureus*. *IUBMB Life, 66*(8), 572–577. https://doi.org/10.1002/iub.1289

Foster, T. J. (2017). Antibiotic resistance in *Staphylococcus aureus*. Current status and future prospects. *FEMS Microbiology Reviews, 41*, 430–449.

Francolini, I., & Donelli, G. (2010). Prevention and control of biofilm-based medical-device-related infections. *FEMS Immunology and Medical Microbiology, 59*(3), 227–238. https://doi.org/10.1111/j.1574-695X.2010.00665x

Guo, Y., Song, G., Sun, M., & Wang, J. (2020). Prevalence and therapies of antibiotic resistance in Staphylococcus aureus. *Frontiers in Cellular and Infection Microbiology, 10*, 107. https://doi.org/10.3389/fcimb.2020.00107

Gursoy, N. C., Ersoy, Y., Gunal, S., & Kuzucu, C. (2009). Antibiotic resistance in Staphylococcus aureus strains isolated from blood cultures. *Ankem Dergisi, 23*(1), 26–29.

Harris, L. G., Foster, S. J., & Richards, R. G. (2002). An Introduction to *staphylococcus aureus* and techniques for identification and quantifying *Staphylococcus aureus* adhesins about adhesion to biomaterials: Review. *European Cells & Materials, 4*, 4. https://doi.org/10.22203/eCM.v004a04

Hoiby, N., Ciofu, O., Johansen, H. K., Song, Z. J., Moser, C., Jensen, P. O., Molin, S., Givskov, M., & Tolker-Neilsen, T. (2011). The clinical impact of bacterial biofilms. *International Journal of Oral Science, 3*(2), 55–65. https://doi.org/10.4248/IJOS11026

Huijbers, P. M., Blaak, H., de Jong, M. C., Graat, E. A., Vandenbroucke-Grauls C. M., & de Roda Husman A. M. (2015). Role of the Environment in the Transmission of Antimicrobial Resistance to Humans: A Review. https://doi.org/10.1021/acs.est.5b02566

Ito, T., Katayama, Y., Asada, K., Mori, N., & Tsutsumimoto, K. (2001). Structural comparison of three types of *Staphylococcal* cassette chromosome mec integrated in the chromosome in methicillin-resistant *Staphylococcus aureus*. *Antimicrobial Agents and Chemotherapy, 45*, 1323–1336. https://doi.org/10.1128/AAC.45.5.1323-1336.2001

Ito, T., Ma, X. X., Takeuchi, F., Okuma, K., Yuzawa, H., & Hiramatsu, K. (2004). Novel type V *Staphylococcal* cassette chromosome mec driven by a novel cassette chromosome recombinase, ccrC. *Antimicrobial Agents and Chemotheraphy, 7*, 2637–2651. https://doi.org/10.1128/AAC.48.7.2637-2651

Jessica, L., Alexander, R., Roy, J., & Carver, L. A. (2014). *Staphylococcus aureus* biofilms: Recent developments in biofilm dispersal. *Frontiers in Cellular and Infection Microbiology, 2014*, 178. https://doi.org/10.3389/fcimb.2014.00178

Ji, G., Beavis, R., & Novick, R. P. (1997). Bacterial interference caused by autoinducing peptide variants. *Science, 276*, 2027–2030. https://doi.org/10.1126/science.276.5321.2027

Kirby, W. M. (1944). Extraction of a highly potent penicillin inactivator from penicillin resistant *staphylococci. Science, 99*, 452–453.

Kirmusaoglu, S. (2007). *MRSA and MSSA: The mechanism of methicillin resistance and the influence of methicillin resistance on biofilm phenotype of Staphylococcus aureus*. IntechOpen.

Kleerebezem, M., Quadri, L. E., & Kuipers, O. P. (1997). Quorum sensing by peptide pheromones and two- component signal-transduction signals in Gram-positive bacteria. *Molecular Biology, 24*(5), 895–904. https://doi.org/10.1046/j.1365-2958.1997.4251782.x

Le, K. Y. & Otto, M. (2015) Quorum-sensing regulation in staphylococci—an overview. Front. Microbiol. 6:1174. https://doi.org/10.3389/fmicb.2015.01174

Leski, T. A., & Tomasz, A. (2005). Role of penicillin binding protein 2 (PBP2) in the antibiotic susceptibility and cell wall cross-inking of evidence for the cooperative functioning of PBP2, PBP4, and PBP2A. *Journal of Bacteriology, 187*(5), 1815–1824. https://doi.org/10.1128/JB.187.5.1815-1824

Levin, T. P., Suh, B., Axelrod, P., Truant, A. L., & Fekete, T. (2005). Potential clindamycin resistance in clindamycin-susceptible, erythromycin- resistant Staphylococcus aureus: report of a clinical failure. *Antimicrobial Agents and Chemotherapy, 49*, 1222–1224.

Li, M., Zou, P., Tyner, K., & Lee, S. (2017). Physiologically based pharmacokinetic (PBPK) modeling of pharmaceutical nanoparticles. *The AAPS Journal, 19*, 26–42. https://doi.org/10.1208/s12248-016-0010-3

Lowy, F. D. (1998). *Staphylococcus aureus* infections. *The New England Journal of Medicine, 339*, 520–532.

Lowy, F. D. (2003). Antimicrobial resistance: the example of *staphylococcus aureus. The Journal of Clinical Investigation, 111*, 1265–1273. https://doi.org/10.1172/JCI200318535

Malachowa, N., & DeLeo, F. R. (2010). Mobile genetic elements of *staphylococcus aureus. Life Sciences, 67*, 3057–3071. https://doi.org/10.1007/s00018-010-0389-4

Matsuoka, M., Inoue, M., Endou, K., & Nakajima, Y. (2003). Characteristic expression of three genes, *msr(A), mph(C)* and *erm(Y)*, that confer resistance to macrolide antibiotics on *Staphylococcus aureus. FEMS Microbiology Letters, 220*, 287–293.

Mcclintock, B. (1951). Chromosome organisation and genic expression. *Cold Spring Harbor Symposia on Quantitative Biology, 16*, 13–47. https://doi.org/10.1101/SQB.1951.016.01.004

McNamara, P. J., Monroe, K. C., Khalili, S., & Proctor, R. A. (2000). Identification, cloning and initial characterization of rot, a locus encoding a regulator of virulence factor expression in *Staphylococcus aureus. Journal of Bacteriology, 182*, 3197–3202. https://doi.org/10.1128/JB.182.11.3197-3203

Miller, M. B. (2001). Quorum sensing in bacteria. *Annual Review of Microbiology, 55*(1), 165–199.

Moosdeen, F., & Williams, J. D. (1986). Antibiotic resistance in: epidemiology, mechanisms and therapeutic possibilities. Rev Infect Dis 8(Suppl 5): S555–S561.

Munita, J. M., Bayer, A. S., & Arias, C. A. (2015). Evolving resistance among Gram-positive pathogens. *Clin Infect Dis*, 61. https://doi.org/10.1093/cid/civ523

Nunes, L. D. C., Santos, K. R. N. D., Mondino, P. J. J., De Freire Bastos, M. D. C. D. F., & Giambiagi-Demarval, M. (1999). Detection of iles-2 gene encoding mupirocin resistance in methicillin- resistance Staphylococcus aureus by multiplex PCR. https://doi.org/10.1016/S0732-8893(99)00021-8

Peacock, S. J., & Paterson, G. K. (2015). Mechanism of MRSA resistance. *Annual Review of Biochemistry, 84*, 577–601. https://doi.org/10.1146/annurev-biochem-060614-034516

Peacock, S. J., & Peterson, G. K. (2015). Mechanisms of methicillin resistance in Staphylococcus aureus. *Annual Review of Biochemistry, 2015*, 34516. https://doi.org/10.1146/annurev-biochem-060614-034516

Piątkowska, E., Tkowski, J., & Mordarsk, A. P. (2012). The strongest resistance of *staphylococcus aureus* to erythromycin is caused by decreasing uptake of the antibiotic into the cells. *Cellular & Molecular Biology Letters, 17*(4), 633–645. https://doi.org/10.2478/s11658-012-0034-3

Piriyaporn, C., Ito, T., Ma, X. X., Kondo, Y., Trakulsomboon, S., Tiensasitorn, C., Jamklang, M., Chavalit, T., Song, J.-H., & Hiramatsu, K. (2006). Staphylococcal Cassette Chromosome mec (SCC mec) typing of Methicillin- Resistant Staphylococcus aureus strains Isolated in 11 Asian countries: a proposal for a New Nomenclature for SCC mec Elements. Antimicrobial agents and Chemotherapy, March 2006, p. 1001–1012. https://doi.org/10.1128/AAC.50.3.1001-1012. 2006.

Pillai, S. K., Sakoulas, G., Wennersten, C., Eliopoulos, G. M., Moellering, R. C. Jr, Ferraro, M. J., & Gold, H. S. (2002). Linezolid resistance in Staphylococcus aureus: characterization and stability of resistant phenotype. *J Infect Dis, 186*(11), 1603–7. https://doi.org/10.1086/345368.

Plata, K., Rosato, A. E., & Wegrzyn, G. (2009). *Staphylococcus aureus* as an infective agent: overview of biochemistry and molecular genetics of its pathogenicity. *Acta Biochemica Polnica, 56*(4), 597–612.

Queck, S. Y., Jameson, L. M., Villaruz, A. E., Bach, T. H., Burhan, A., Khan, B. A., Sturdevant, D. E., Ricklefs, S. M., Li, M., & Ott, M. (2008). RNAIII- independent target gene control by the agr quorum-sensing system: Insight into the evolution of virulence regulation in Staphylococcus aureus. *Molecular Cell, 32*, 150–158.

Rabin, Z., Opoku-Temeng, D., Bonsu & Sintim. (2015). *Future Med. Chem., 7*(4), 493–512.

Ragbetli, C., Parlak, M., Bayram, Y., Guducuoglu, H., & Ceylan, N. (2016). Evaluation of antimicrobial resistance in *Staphylococcus aureus* isolates by years. *Interdisciplinary Perspectives on Infectious Diseases, 2016*, 9171395. https://doi.org/10.1155/2016/9171395

Ramage, G., Savile, S. P., Wickes, B. L., & Lopez, R. J. L. (2002). Inhibition of Candida albicans biofilm formation by farnesol, a quorum sensing molecule. *Applied and Environmental Microbiology, 68*, 5459–5463.

Rasigade, J. P., Dumitrescu, O., & Lina, G. (2014). New epidemiology of *staphylococcus aureus* infections. *Clinical Microbiology and Infection, 20*, 587–588.

Reffuveille, F., Josse, J., Valle, Q., Mongaret, C., & Gangloff, S. C. (2017). *Staphylococcus aureus biofilms and their impact on the medical field*. IntechOpen.

Reygaert, W. C. (2018). An overview of the antimicrobial resistance mechanisms of bacteria. *AIMS Microbiology, 4*(3), 482–501. https://doi.org/10.3934/microbiol.2018.3.482

Sasirekha, B., Usha, M. S., Amruta, J. A., Ankit, S., Brinda, N., & Divya, R. (2014). Incidence of constitutive and inducible clindamycin resistance among hospital-associated Staphylococcus aureus. *3 Biotech, 4*, 85–89. https://doi.org/10.1007/s13205-013-0133-5

Shahid, A., Rasool, M., Akhter, N., Aslam, B., Hassan, A., & Sana, S. (2019). *Innovative strategies for the control of biofilm formation in clinical settings*. IntechOpen.

Sharma, D., Misba, L., & Khan, A. U. (2019). Antibiotics versus biofilm: an emerging battleground in microbial communities. *Antimicrobial Resistance and Infection Control, 8*, 76. https://doi.org/10.1186/s13756-019-0533-3

Singh, S., Kumar, S., & Indrajit. (2017). Understanding the mechanism of bacterial biofilms resistance to antimicrobial agents. *The Open Microbiology Journal, 11*, 53. https://doi.org/10.2174/1874285801711010053

Smith, P. F., & Wilkinson, B. J. (1981). Synthesis of methicillin resistant *Staphylococcus aureus*. *Journal of Bacteriology, 148*, 610–617.

Stark, L. (2013). *Staphylococcus aureus-aspects of pathogenesis and molecular epidemiology*. Elsevier.

Steven, Y. C. T., Joshua, S., Eichenberger, E., Thomas, L., & Vance, G. (2015). *Staphylococcus aureus* infections: Epidemiology, pathophysiology, clinical manifestations and management. *Clinical Microbiology Reviews, 28*(3), 603–661. https://doi.org/10.1128/CMR.00134-14

Thammavongsa, V., Kim, H. K., Missiakas, D., & Schneewind, O. (2015). *Staphylococcal* manipulation of host immune responses. *Nature Reviews. Microbiology, 13*(9), 529–543.

Tipper, D. J., & Strominger, J. L. (1965). Mechanism of action of penicillins: A proposal based on their structural similarity to acyl-d-amyl-d-alanine. *Proceedings of the National Academy of Sciences of the United States of America, 54*, 1133–1141.

Tomasz, A., & De Lencastre, H. (1994). Reassessment of the number of auxiliary genes essential for expression of high-level methicillin resistance in *staphylococcus aureus*. *Antimicrobial Agents and Chemotherapy, 38*, 2590–2598.

Van Tilburg, B. E., Lewenza, S., & Reckseidler, Z. S. (2015). Current research approaches to target biofilm infections. *Journal of Postdoctoral Research, 3*(6), 36–49.

Vlaeminck, J., Raafat, D., Surmann, K., Timbermont, L., Normann, N., Sellman, B., Willem, J. B., & Wamel, M.-k. S. (2020). Exploring virulence factors and alternative therapies against *Staphylococcus aureus*. *Toxins, 12*(11), 721. https://doi.org/10.3390/toxins12110721

Walsh, C. T., & Timothy, A. W. (2016). *Antibiotics: Challenges, mechanisms opportunities*. ASM Press.

Wang, W., Lin, X., Jiang, T., Peng, Z., Xu, J., Yi, L., Li, F., Fanning, S., & Baloch, Z. (2018). Prevalence and Characterization of Staphylococcus aureus Cultured from Raw Milk Taken From Dairy Cows With Mastitis in Beijing, China. Front. Microbiol. https://doi.org/10.3389/fmicb.2018.01123

Wyke, A. W., Ward, J. B., & Hayes, M. V. (1982). Synthesis of peptidoglycan in vivo in methicillin-resistant *Staphylococcus aureus*. *European Journal of Biochemistry, 127*, 553–558.

Yarwood, J. M., Patrick, M., & Schlievert, P. M. (2003). Quorum sensing in Staphylococcus infections. *The Clinical Investigator, 112*(11), 1620–1625. https://doi.org/10.1172/JCI20442

Zhang, H. Z., Hackbarth, C. J., Chansky, K. M., & Chambers, H. F. (2001). A proteolytic transmembrane signalling pathway and resistance to β-lactams in Staphylococci. *Science, 291*, 1962–1965.

Zhao, X., Yu, Z., & Ding, T. (2020). Quorum sensing regulation of antimicrobial resistance in bacteria. *Microorganisms, 8*(3), 425. https://doi.org/10.3390/microorganisms8030425

Zhou, W., Shan, W., Ma, X., Chang, W., Zhou, X., Lu, H., & Dai, Y. (2012). Molecular characterization of rifampicin-resistant Staphylococcus aureus isolates in a Chinese teaching hospital from Anhui, China. *BMC Microbiology, 12*, 240.

Zoabi, M., Keness, Y., Titler, N., & Bisharat, N. (2011). Compliance of hospital staff with guidelines for the active surveillance of methicillin-resistance *Staphylococcus aureus* (MRSA) and its impact on rates of nosocomial MRSA bacteremia. *The Israel Medical Association Journal, 2011*(13), 740–744.

Anticancer Secondary Metabolites Found in Native Ecuadorian Plant Species *Uncaria tomentosa* DC. (Rubiaceae), *Croton lechleri* Müll. Arg. (Euphorbiaceae), and *Equisetum giganteum* L. (Equisetaceae)

Michelle Sánchez García and Carla Quilumbango Grijalva

1 Introduction

Ecuador is a country that is located northeast of the South American continent, it limits the north with Colombia, the south and east with Peru, and the west with the Pacific Ocean. It is one of the smallest countries in South America with an area of approximately 283,560 km^2, or 1.5% of the total surface of the continent (Sierra et al., 2002). Despite this, Ecuador is recognized as one of the countries with the highest biodiversity worldwide (Bailon-Moscoso et al., 2015a). In terms of vegetation, diversity is extraordinary, containing an immense number of native and endemic species (Malagón et al., 2003). The diversity of species in Ecuador is due to many factors such as the variety of climates and its distinguished regional separation, mainly due to the presence of the Andes Mountains (Sierra et al., 2002). It has four geographic regions: Amazonia, Sierra, Costa, and Insular, in which the composition of the soil, temperature, humidity, latitude, and altitude change, thus giving a great variety of ecosystems and therefore an immense diversity of animal and plant species.

For a long time, indigenous cultures around the world, such as the Ecuadorian population nowadays, maintain their ancestral traditions regarding the use of native and endemic plants as natural remedies to combat different diseases (Seca & Pinto, 2018; Tene et al., 2007). This knowledge in ethnobotany is orally transmitted from generation to generation (Tene et al., 2007). There is a wide variety of native plants known in Ecuador thanks to their great diversity with great medicinal potential, specifically with an anticancer effect (Bailon-Moscoso et al., 2015a). There are reports that around 3000 plants worldwide have these anticancer properties and

M. Sánchez García (✉) · C. Quilumbango Grijalva
School of Biological Sciences and Engineering, Yachay Tech University, Urcuquí, Ecuador
e-mail: michelle.sanchez@yachaytech.edu.ec

N. R. Maddela, L. C. García (eds.), *Innovations in Biotechnology for a Sustainable Future*, https://doi.org/10.1007/978-3-030-80108-3_18

many of them can be found in Ecuador. These plants are of great interest since cancer nowadays is the disease with the highest mortality and incidence rates worldwide (Sung et al., 2021). According to the GLOBOCAN 2020, there are proximally 19.3 million new cancer case diagnosed in 2020 and around 10.0 million cancer deaths were reported (Sung et al., 2021). Moreover, it is estimated 28.4 million new cancer cases for 2040. It represents an increase of 47% in comparison with 2020 (Sung et al., 2021). In Ecuador in 2020 there was 29,273 new cancer cases, and 15,123 deaths in which breast, prostate, colorectal, stomach, and thyroid cancers predominate (World Health Organization, 2020). This is an alarming number, for this reason the search for a treatment that helps fight this disease is of utmost importance (Seca & Pinto, 2018).

The compounds that behave as the most important anticancer agents and are found in these plants are called secondary metabolites (Demain & Fang, 2000). In this chapter, the general characteristics of secondary metabolites will be described and the effect of this metabolites found in different plants of the world finally focuses on three Ecuadorian species, *Uncaria tomentosa* DC., *Croton lechleri* Müll. Arg., and *Equisetum giganteum* L. against different types of cancer.

2 Secondary Metabolites

Secondary metabolites are organic molecules produced within the plant, which are not necessary for its development, growth, and reproduction (Seca & Pinto, 2018). They have unique structures and are found in specific cells and organs where they accumulate in vacuoles. The production of secondary metabolites depends mainly on the adaptations and interactions of plants with their environment (Baikar & Malpathak, 2010). Furthermore, it has been proved that some secondary metabolites have anticancer activity. Observe Table 1 about plants around the world with secondary metabolites that were used for studies to demonstrate their anticancer properties. Secondary metabolites are classified according to the route by which they are synthesized, so they have three main groups: alkaloids, terpenoids, and phenols (Seca & Pinto, 2018).

2.1 Alkaloids

Alkaloids are considered among the most active secondary metabolites and widely distributed in the plant kingdom, mainly in angiosperms (Coqueiro & Verpoorte, 2019; Wink, 2015). Their structures consist of one to several nitrogen atoms in a ring structure (true alkaloids) or in a side chain (pseudoalkaloids). In addition to carbon, hydrogen, and nitrogen, this group may also contain oxygen, sulfur, and rarely elements like chlorine, bromine, and phosphorus (Kabera, 2018). The alkaloids are synthesized by various organisms, such as bacteria, fungi, animals and plants,

Table 1 Plant species around the world with secondary metabolites and anticancer properties

Species	Distribution[a]	Compounds used for cytotoxicity study	Type of cancer	Line cell	IC_{50} values	References
Ormenis eriolepis Coss	Moroccan	Dichloromethane Fraction (Oe-DF) Hexanic Extract (Oe-HE)	T cell leukemia Mantle Cell Lymphoma Glioblastoma Caucasian prostate adenocarcinoma	Jurkat Jeko-1 LN229 PC-3	11.63 ± 5.37 μg/ml 13.33 ± 1.67 μg/ml 41.67 ± 1.98 μg/ml 19.31 ± 4.88 μg/ml	(Belayachi et al., 2017)
Podophyllum peltatum L.	Canada, United States	Podophyllotoxin	Esophageal squamous cell carcinoma (ESCC)	KYSE 30 KYSE 70 KYSE 410 KYSE 450 KYSE 510	0.27 μM 0.30 μM 0.17 μM 0.30 μM 0.30 μM	(Yoon et al., 2020)
Camptotheca acuminata Decne	China	Camptothecin derivative, CPT211	Breast cáncer	MCF-7 MDA-MB-231	191.2 nM 478.4 nM	(Chiu et al., 2020)
Linum narbonense L.	Portugal, Switzerland, Corsica	Arylnaphthalene lignan Justicidin B	Chronic myeloid leukemia Hodgkin lymphoma Human urinary bladder carcinoma	K-562 LAMA-84 HD-MY-Z EJ cells	0.597 ± 0.019 μg/ml 0.029 ± 0.05 μg/ml 0.620 ± 0.022 μg/ml 0.812 ± 0.031 μg/ml	(Ionkova et al., 2013)
Cotula cinerea Delile	Algeria, Morocco, Red Sea region, Sinai, Qattara Depression, and Mali	Hexane	Human embryonal rhabdomyosarcoma Monkey kidney cancer	RD Vero	57.21 ± 3.43 μg/mL 142.27 ± 11.33 μg/mL	(Guaouguaou et al., 2018; Lakhdar, 2018)
Cotula cinerea Delile	Algeria, Morocco, Red Sea region, Sinai, Qattara Depression, and Mali	Ethyl acetate	Human embryonal rhabdomyosarcoma Monkey kidney cancer	RD Vero	187.52±6.27 μg/mL 212.83±9.02 μg/mL	(Guaouguaou et al., 2018; Lakhdar, 2018)

(continued)

Table 1 (continued)

Species	Distribution[a]	Compounds used for cytotoxicity study	Type of cancer	Line cell	IC_{50} values	References
Cotula cinerea Delile	Algeria, Morocco, Red Sea region, Sinai, Qattara Depression, and Mali	n-butanol	Human embryonal rhabdomyosarcoma Monkey kidney cancer	RD Vero	>500 μg/mL 447.38 ± 6.52 μg/mL	(Guaouguaou et al., 2018; Lakhdar, 2018)
Cotula cinerea Delile	Algeria, Morocco, Red Sea region, Sinai, Qattara Depression, and Mali	Essential oil	Human embryonal rhabdomyosarcoma Monkey kidney cancer	RD Vero	173.05 ± 4.46 μg/mL 72.72 ± 2.18 μg/mL	(Guaouguaou et al., 2018; Lakhdar, 2018)
Berberis vulgaris L.	Canada, United States, most of Europe	Ethanol extract Ethyl acetate extract	Human breast adenocarcinoma	MCF-7	3.54 μg/mL 596.71 μg/mL	(El Khalki et al., 2018)
Solanum aculeastrum Dunal	Tropical to South Africa	Solamargine	Human neuroblas-toma Breast cancer	SH-SY5Y SK-Br3	13.54 μg/mL 16.12 μg/mL	(Burger et al., 2018)
Leonotis leonurus (L.) R.Br.	South Africa	Marrubiin	Human colorectal adenocarcinoma	HT-29	50 μg/mL	(Kee et al., 2008)
Hypoxis hemerocallidea Fisch. & C.A. Mey.	South Africa	Hypoxoside	Cervical cancer Colorectal Adenocarcinoma Breast adenocarcinoma	HeLa HT-29 MCF-7	26.46 μg/mL 82.80 μg/mL 45.62 μg/mL	(Boukes, 2010)
Nelumbo nucifera Gaertn.	Australia, Bhutan, Burma, China, Guyana, India, Japan, Nepal, Pakistan, Philippines, Russian, South Korea, Thailand, United States	Neferine	Hepatocellular car-cinoma (HCC)	HepG2	10 μM	(Poornima et al., 2013)

Podophyllum peltatum L.	Canada, United States	Glucoside derivatives of podophyllotoxin	Leukemia hepatoma Lung cancer Breast cancer Colon cancer	HL-60 SMMC-7721 A-549 MCF-7 SW480	11.37 ± 0.52 μM 8.41 ± 0.48 μM 10.74 ± 0.37 μM 9.18 ± 0.49 μM 3.27 ± 0.21 μM	(Zi et al., 2019)
Zingiber officinale Roscoe	Australia, Belize, Bhutan, Bolivia, Burma, Cambodia, China, Colombia, Costa Rica, Ecuador, Gabon, Guatemala, Honduras, India, Japan, Laos, Madagascar, Mexico, Nicaragua, Panama, Philippines, Thailand, Vietnam	[10]-gingerol	Triple negative breast cancer (TNBC)	4T1Br4 MDA-MB-231BrM b	29.9 ± 4.8 μM 65.2 ± 5.1 μM	(Martin et al., 2017)
Inula viscosa (L.) Aiton	Mediterranean Basin	Hexane extract	Acute T cell leukemia Non-Hodgkin lymphomas Glioblastoma Colorectal adenocarcinoma Prostate adenocarcinoma	Jurkat Jeko-1 LN229 SW620 SW480 PC-3	7.52 ± 4.57 μg/mL <1 μg/mL 5.9 ± 3.57 μg/mL 8.40 ± 3.31 μg/mL 10.36 ± 6.59 μg/mL 6.63 ± 1.42 μg/mL	(Belayachi, 2013; Talib et al., 2012)
Azima tetracantha Lam.	Madagascar, South Africa	Friedelin	Breast cancer	MCF-7	1.2 μM	(Subash-Babu et al., 2017)
Aristolochia longa L.	Europe	Hexanic extract	Rhabdomyosarcoma Kidney adenocarcinoma of hamster	RD BSR	30 ± 0.70 μg/mL 18 ± 3.53 μg/mL	(Aneb et al., 2016)

(continued)

Table 1 (continued)

Species	Distribution[a]	Compounds used for cytotoxicity study	Type of cancer	Line cell	IC_{50} values	References
Aristolochia longa L.	Europe	Dichloromethanic extract	Rhabdomyosarcoma Kidney adenocarcinoma of hamster	RD BSR	15 ± 1.06 μg/mL 60 ± 2.47 μg/mL	(Aneb et al., 2016)
Carica Papaya L.	America	Lyophilized papaya leaf juice extract (LJP)	Prostate cancer (PCa)	RWPE-1 BPH-1 PC-3 LNCaP	0.22 mg/mL 0.79 mg/mL 0.95 mg/mL 0.96 mg/mL	(Pandey et al., 2017)
Pistacia lentiscus L.	Spain Algeria, Egypt, Libya, Morocco, Tunisia, Cyprus, Israel, Jordan, Lebanon, Syria, Turkey, Albania, Greece, Italy, France, and Portugal	*P. lentiscus* essential oils obtained from fruits (PLFEO)	Rhabdomyosarcoma	RD	26.43 ± 2.18 μg/mL	(Bouyahya et al., 2019)
Trollius chinensis Bunge	China	Orientin Vitexin	Esophageal cancer	EC-109	5–8 μM	(An et al., 2015)
Cajanus cajan (L.) Huth	America, Asia, Africa	Cajanstilbenoids A	Hepatocarcinoma Breast adenocarcinoma Lung cancer	Hep2 MCF-7 A549	2.50 ± 0.2 μM 2.56 ± 0.3 μM 2.14 ± 0.2 μM	(Zhang et al., 2018)
Cajanus cajan (L.) Huth	America, Asia, Africa	Cajanstilbenoids B	Hepatocarcinoma Human breast adenocarcinoma Human lung cancer	Hep2 MCF-7 A549	5.99 ± 0.3 μM 22.63 ± 0.4 μM 6.18 ± 0.3 μM	(Zhang et al., 2018)
Celtis Africana Burm. f.	Angola, Madagascar, Mozambique, Nigeria, South Africa, Zambia, Zimbabwe	Ethyl acetate extract	Mouse lymphoma	L5178Y	8.3 μg/mL	(Perveen et al., 2015)

Elephantopus mollis Kunth	America, Fiji, Gabon, Micronesia Federated S tates, Tanzania, Uganda	Methanol extract	Lung cancer Mesothelioma Hepatocarcinoma Colorectal adeno-carcinoma Breast adenocarcinoma	A549 SPC212 HepG2 DLD-1 MCF-7	24.38 ± 1.86 μg/mL 4.05 ± 0.69 μg/mL 3.74 ± 0.07 μg/mL 3.29 ± 0.04 μg/mL 3.97 ± 0.48 μg/mL	(Kuete et al., 2017)
Enantia chlorantha Oliv.	Gabon	Methanol extract	Lung cancer Mesothelioma Hepatocarcinoma Colorectal adeno-carcinoma Breast adenocarcinoma	A549 SPC212 HepG2 DLD-1 MCF-7	>40 μg/mL 25.16 ± 1.30 μg/mL 17.32 ± 0.13 μg/mL >40 μg/mL >40 μg/mL	(Kuete et al., 2017)
Kalanchoe crenata (Andrews) Haw.	Belize, Brazil, Ecuador, El Salvador, Guatemala, Guyana, Honduras, Madagascar, Sierra Leone, South Africa, United States	Methanol extract	Lung cancer Mesothelioma Hepatocarcinoma Colorectal Adenocarcinoma Breast adenocarcinoma	A549 SPC212 HepG2 DLD-1 MCF-7	8.23 ± 0.15 μg/mL 2.33 ± 0.23 μg/mL 28.96 ± 3.51 μg/mL 23.87 ± 1.69 μg/mL 19.31 ± 0.79 μg/mL	(Kuete et al., 2017)
Lophira alata Banks ex C.F. Gaertn.	Gabon	Methanol extract	Lung cancer Mesothelioma Hepatocarcinoma Colorectal adeno-carcinoma Breast adenocarcinoma	A549 SPC212 HepG2 DLD-1 MCF-7	>40 μg/mL >40 μg/mL 32.98 ± 4.20 μg/mL >40 μg/mL >40 μg/mL	(Kuete et al., 2017)

(continued)

Table 1 (continued)

Species	Distribution[a]	Compounds used for cytotoxicity study	Type of cancer	Line cell	IC_{50} values	References
Millettia macrophylla Benth.	Austral Africa	Methanol extract	Lung cancer Mesothelioma Hepatocarcinoma Colorectal adeno-carcinoma Breast adenocarcinoma	A549 SPC212 HepG2 DLD-1 MCF-7	>40 μg/mL 7.54 ± 0.26 μg/mL 2.01 ± 0.04 μg/mL 31.02 ± 2.86 μg/mL 25.99 ± 1.68 μg/mL	(Kuete et al., 2017)
Phragmanthera capitata (Spreng.) Balle	Gabon	Methanol extract	Lung cancer Mesothelioma Hepatocarcinoma Colorectal adeno-carcinoma Breast adenocarcinoma	A549 SPC212 HepG2 DLD-1 MCF-7	>40 μg/mL >40 μg/mL >40 μg/mL >40 μg/mL >40 μg/mL	(Kuete et al., 2017)
Withania adpressa Cors.	North Africa, Mediterranean basin, and India	Withanolide F	Multiple myeloma cancer stem cells Plasmacytoma	MM-CSCs RPMI 8226	5.3 ± 0.9 μM 0.1 ± 0.01 μM	(Ben Bakrim et al., 2018)
Withania adpressa Cors.	North Africa, Mediterranean basin, and India	Withafein A	Multiple myeloma cancer stem cells Plasmacytoma	MM-CSCs RPMI 8226	0.33 ± 0.02 μM 0.17 ± 0.03 μM	(Ben Bakrim et al., 2018)
Cedrus deodara Cors.	Afghanistan, Bolivia, China, India, Nepal, and Pakistan	*Cedrus deodara* total lignans CTL	Lung carcinoma	A549	39.82 ± 1074 *μ*g/mL	(Shi et al., 2019)
Dysoxylum nutans Cors.	India, China, Malaysia, Indonesia, Australia, and New Zealand	3α-Hydroxystigmast-5 (6), 22-Dien-7-en	Breast adenocarcinoma	MCF-7	20.13 ± 0.06 μM	(Mayanti et al., 2020)

Peganum Harmala L.	Afghanistan, China, Kazakhstan, Kyrgyzstan, Mongolia, Pakistan, Russian Federation, South Africa, Tajikistan, United States, Uzbekistan	Harmine	Breast Lung Colon Cervic Colon	HBL-100 A549 HT-29 HEeLAa HCT-116	32.56 ± 0.65 μM 106-71 ± 2.49 μM 45.55 ± 0.87 μM 61.81 ± 0.65 μM 33.97 ± 0.21 μM	(Ayoob et al., 2017)
Allium wallichii Kunth	Bhutan, Burma, China, India, Nepal	Crude extract of *Allium wallichii*	Prostate cancer Breast cancer Cervical cancer	PC3 MCF-7 HeLa	69.69 μg/mL 55.29 μg/mL 46.51 μg/mL	(Bhandari et al., 2017)
Mitracarpus frigidus (Willd. ex Roem. & Schult.) K. Shum	Bolivia, Brazil, Guyana, Peru, Venezuela	Psychorobrin (14)	Breast cancer Human promyelocytic leukemia T lymphocyte	MCF-7 HL60 Jurkat	1.1 μM 4.5 μM 5.6 μM	(Fabri et al., 2012)
Sphagneticola trilobata (L.) J.F. Pruski	Australia, Belize, Bolivia, Brazil, Caribbean, Colombia, Costa Rica, Ecuador, French Guiana, Guatemala, Guinea, Guyana, Honduras, Malaysia, Mexico, Micronesia Federated States, Nicaragua, Panama, Peru, Philippines, United States, Venezuela	Ethyl acetate extract of *S. trilobata*	Breast cancer	MCF-7	58.143 μg/mL	(Mardina et al., 2020)

[a]The distribution information was obtained by accessing the database of the web pages: Tropicos.org (2021) and GBIF.org (2021)

especially by the latter. Unlike the other secondary metabolites, alkaloids have a great structural diversity and are very heterogeneous, so they do not have a very precise classification (Coqueiro & Verpoorte, 2019). Most are toxic to other organisms and are obtained using the acid-base extraction technique (Kabera, 2018). They have pharmacological effects and a great use as medicines. Some alkaloids are being used in cancer therapies such as chemotherapeutics. Similarly, there are alkaloids that act as neuroreceptors, or modulate the transduction of neuronal signals. Others interfere with DNA, telomeres, telomerase, or apoptosis-inducing protein biosynthesis (Tiwari, 2015; Wink, 2015).

2.2 *Terpenoids*

Also known as isoprenoids, they are the most numerous and structurally diverse natural products, with more than 23,000 known structures. They are derived from polymeric isoprene and are synthesized from the mevalonic acid pathway (Kabera, 2018). They are classified according to the number and structural organization of the isoprene units (Tiwari, 2015; Wink, 2015). For this reason, isoprene can be considered as the basic component of terpenoids, it is 2-methylbuta-1,3-diene (C_5H_8). The most basic class of terpenoids is hemiterpenoids which consists of only one isoprene unit (Ludwiczuk et al., 2017). Monoterpenoids are major components of essential oils and are known for their aromatic properties. These are the main components of spruce (38%), pine (30%), Angelica species (73%), rose (54%), among more species. Flower and seed essential oils have more specialized monoterpenoids. Diterpenoids are found mainly in the Lamiaceae family and have antimicrobial and antiviral properties. And in the same way, each of the other divisions can be found in specific plant families (Ludwiczuk et al., 2017; Wink, 2015). Terpenoids have great pharmacological activity and are used in the treatment of various diseases. According to researchers they have anticancer activity and their mechanism of action is the prevention of tumor cell proliferation through necrosis or apoptosis induction (Kabera, 2018; Ludwiczuk et al., 2017; Wink, 2015).

2.3 *Phenols*

Phenolic compounds are secondary metabolites that are universally distributed in all plants (Ghasemzadeh & Ghasemzadeh, 2014; Kabera, 2018). They have various structures, but all have hydroxylated aromatic rings. Most phenolic compounds polymerize into larger molecules, such as proanthocyanidins and lignins. Likewise, they can be found in food plants such as glycosides or esters (Kabera, 2018; Tiwari, 2015). In tea, coffee, berries, and fruits, phenolic compounds could be found up to a total of 103 mg/100 g of fresh weight (Kabera, 2018). These play an important role in the development and reproduction of plants. They also are produced in response to different environmental factors (light, drought, cold, etc.) and to defend the plant from pathogens or herbivores. Phenolic compounds provide odor, taste, and antioxidant, anti-inflammatory, anticancer properties among many more found in different species of the plant kingdom (Ghasemzadeh & Ghasemzadeh, 2014; Kabera, 2018).

3 Plant Families and their Native Ecuadorian Plant Species that have Secondary Metabolites for Anticancer Activity: Three Case Studies

3.1 *Rubiaceae*

Rubiaceae is a member of the Gentianales and represents one of the four most species-rich families of angiosperms (flowering plants), with 13,143 species classified in 611 genera (Davis et al., 2009), more than 40 tribes, and 3 subfamilies (Bremer & Eriksson, 2009; Goevarts et al., 2006). Rubiaceae have a cosmopolitan distribution, so they are widespread and are located in all major regions of the world, including the Antarctic continent, but the diversity of species and biomass is found in tropical and subtropical areas. Species belonging to this family occupy many types of habitat in different biogeographic regions (Bremer & Eriksson, 2009; Davis et al., 2009). The diversity of the family is enormous, with life forms ranging from small grasses to large trees, with flowers adapted to different types of pollinators, with fruits that have many types of dispersal mechanisms, and with a wide variety of different chemicals that accumulate in the plants (Bremer & Eriksson, 2009). And according to Ulloa (2006), in Ecuador we have 658 species distributed in 88 genera, positioning itself as the third richest family in species, after Orchidaceae and Asteraceae, in the country.

To the present, several species of the Rubiaceae family have been reported that have a wide range of medicinal uses and pharmacological activities such as antioxidant, antifungal, antidiabetic, antibacterial, antidiarrheal, anti-inflammatory, anticancer, and more (Ekalu, 2021). All this is due to the bioactive secondary metabolites that have been detected in plants, thus allowing the development of new drugs for the treatment of diseases such as cancer (Singh et al., 2020). Because of this, various plants of this family have been studied for possible cancer treatments, some examples are: (1) *Mitracarpus* species where bioactive secondary metabolites with anticancer activity have been detected such as psychorubin, ursolic acid, rutin, kaempferol-3-O-rutinoside, kaempferol, stigmasterol, and quercetin (Ekalu, 2021); (2) *Gardenia jasminoides* Elli with the bioactive compound genipin, extracted from desiccative ripe fruits, as a chemopreventive agent to prevent cancer (Chen et al., 2020); (3) *Hedyotis corymbosa* from which the secondary metabolite asperuloside is isolated (Manzione et al., 2020); (4) *Mitragyna speciosa* Korth, recognized as a rich source of alkaloids, exhibits the bioactive compound mitragynine with cytotoxic activity against breast cancer cells (Firmansyah et al., 2021); and (5) *Uncaria* species, which are an important source of bioactive indole alkaloids (Liang et al., 2020; Qin et al., 2021); in the last 20 years around 100 new secondary metabolites have been elucidated in which they include alkaloids, triterpenes, and flavonoids that provide anticancer biological functions among many more (Liang et al., 2020). *Uncaria tomentosa* native to Ecuador is the most studied species among the numerous species of genus *Uncaria* because it contains bioactive compounds with anticancer properties (Martins & Nunez, 2015) and will be described in more detail below.

3.1.1 *Uncaria tomentosa* DC

It belongs to the family Rubiaceae (Bacher et al., 2006; Bailon-Moscoso et al., 2015a; De la Torre et al., 2008; Farias et al., 2013; Rinner et al., 2009), is a native liana to Ecuador and is distributed in Asia (Bailon-Moscoso et al., 2015a) and in different countries in South America (Amazon of Brazil, Colombia, Peru, Bolivia, Venezuela, and Guyana) and Central America (El Salvador, Guatemala, Belize, Hondura, El Salvador, Nicaragua, Costa Rica, and Panama) (Farias et al., 2013). It is known by its spanish common name "uña de gato" and translated as cat's claw (Bailon-Moscoso et al., 2015a; De la Torre et al., 2008). In Ecuador, it is distributed in the provinces of Napo, Orellana, Sucumbíos, and Zamora-Chinchipe at altitudes of 0 to 900 meters above sea level (masl) (Bailon-Moscoso et al., 2015a). Throughout history, it has been used as traditional medicine by indigenous peoples of Ecuador, Peru, and Bolivia (Bacher et al., 2006; Bailon-Moscoso et al., 2015a) and the method of administration of this plant is by taking the extract of the bark or root boiled with water or by maceration in alcohol (Pilarski et al., 2013; Rinner et al., 2009). The extracts of this plant are used for treatments of inflammation, rheumatism, arthritis, infections, and cancer, due to its antiproliferative activity, which induces apoptosis in cancer cells (Bacher et al., 2006; Bailon-Moscoso et al., 2015a; Farias et al., 2013; Rinner et al., 2009).

Uncaria tomentosa contains several active secondary metabolites with pharmacological activity, as well as sterols, tannins, procyanidins, flavonoids, triterpene derivatives (polyhydroxylated triterpenes, oleanolic and ursolic acid), quinovic acid glycosides and oxindole alkaloids (including speciophylline, mitraphylline, uncarine pteropodine, isomitraphylline, and isopteropodine) (García Prado et al., 2007; Pilarski et al., 2013).

The mechanism of *U. tomentosa* cytotoxicity is involved in the inhibition of proliferation and inactivation of NF-$_{k}$B and the synthesis and release of pro-inflammatory cytokines such as TNF-α (Fig. 1) (Pilarski et al., 2013). In normal cells, NF-$_{k}$B proteins regulate cell proliferation and cell survival. However, in tumor cells there is a failure in the NF-$_{k}$B pathway. More studies demonstrated that NF-$_{k}$B inhibition of TNF secretion in monocyte-like THP-1 cells exposed by *U. tomentosa* extracts is associated with the deactivation of the c-Jun, JunB, p65, RelB, and p50 subunits (Allen-Hall et al., 2010). Therefore, the development of anticancer therapies that include the inhibition of NF-$_{k}$B that can stop the proliferation of cancer

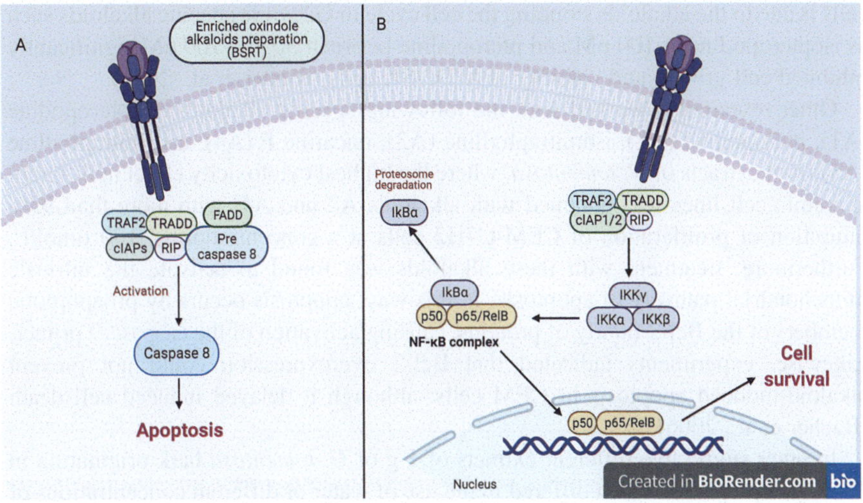

Fig. 1 Apoptosis induced by mechanism BSRT + TNF-α vs cell survival by TNF-α/NF-$_k$B pathway. (**a**) Tumor cell pretreated with BSRT inhibits the NF-$_k$B complex and promote the action of TNF-α to activate proapoptotic activity. (**b**) Tumor cell without BSRT has a misregulation of NF-$_k$B complex that causes the tumor cell survival (Pilarski et al., 2013)

cells, produce cell death, or improve the response of antitumor drugs, is one of the greatest challenges in today's pharmaceutical industry.

In a study from 2013 researchers used 10 g of the bark of *U. tomentosa* originating in Peru to obtain an enriched preparation of oxindole total alkaloids named BSRT. According to the high-performance liquid chromatography (HPLC) analysis, more than 50% of pentacyclic oxindole alkaloids (POA) in dry weight of BSRT was obtained with predominance of pteropodine, speciophylline, and isopteropodine. Besides, POA show greatest pharmacological activity, as opposed tetracyclic oxindole alkaloids (TOA) which were found in very low concentrations (García Prado et al., 2007; Pilarski et al., 2013). Cytotoxicity tests indicate that the BSRT concentration required to inhibit the 50% growth (IC_{50}) of promyelocytic leukemia HL-60 cells is 60 μg/mL. And analysis by flow cytometry confirms that the alkaloid enriched extract induces apoptosis in promyelocytic leukemia HL-60 cells and the highest result was obtained with the BSRT treatment combined with TNF-∞ with a 39.9% apoptotic increase (Pilarski et al., 2013).

Medullary thyroid carcinoma (MTC) is a calcitonin producing tumor of the parafollicular C-cells (Rinner et al., 2009). MTC is insensitive to radiation therapy as well as chemotherapy. Furthermore, it has been shown that resistance to chemotherapy is due to the expression of the mdrl gene that causes resistance to multiple drugs. Also, studies indicate that different MTC cell lines have an increase in the anti-apoptotic protein Bcl-2, which allows tumor cells to survive (Yang et al., 1991). Rinner and colleagues reported that extracts from *U. tomentosa* affected the growth, viability, and apoptosis of MTC-SK cells. The antiproliferative effect in MTC-SK cells is due to the alkaloids stopping the cell cycle in G2 while specific alkaloids such as isopteropodine at 100 μM and pteropodine between 50 and 100 μM significantly inhibited cell growth and viability of MTC-SK cells (Rinner et al., 2009).

Other researchers worked with the following specific alkaloids: isopteropodine (A1), pteropodine (A2), isomitraphylline (A3), uncarine F (A4), and mitraphylline (A5) from extracts of *U. tomentosa,* where the highest cytotoxicity effect in different leukemia cell lines was obtained with alkaloids A2 and A4, with more than 95% inhibition of proliferation of CEM-C7H2 cells at a concentration of 100 μmol/L. Furthermore, treatment with these alkaloids was found to activate the intrinsic mitochondrial pathway of apoptosis. In this way, apoptosis occurs by proapoptotic members of the Bcl-2 family of proteins, causing activation of the caspase-9 primer. Likewise, experiments indicated that Bcl-2 overexpression could not prevent alkaloid-induced apoptosis in CEM cells, although it delayed induced cell death (Bacher et al., 2006).

In other study, five different extracts of 1 g of *U. tomentosa* bark originating in Peru were prepared, which differed in the use of water or different concentrations of ethanol with or without boiling (B/W_{37}, B/W_b, $B/50E_{37}$, B/E_b, $B/96E_{37}$). They also obtained a bark preparation rich in alkaloids with another extraction technique (BSRT), using 10 g of the same raw material as the rest. The highest alkaloid content was obtained with the B/SRT preparation where just over 50% of the dry weight was pure oxindole alkaloids. Following $B/96E_{37}$, B/E_b, $B/50E_{37}$, B/W_b, and B/W_{37} with a 3581, 3408, 1897, 509, 430 mg/100 g of pure oxindole alkaloids content, respectively. Cytotoxicity test results indicated that the highest rate of antiproliferative activity was by $B/96E_{37}$ with growth inhibition in Lewis lung carcinoma (LL/2) (IC_{50} = 25.06 μg/ml), cervical carcinoma (KB) (IC_{50} = 49.06 μg/ml), and colon adenocarcinoma (SW707) (IC_{50} = 49.06 μg/ml). Preparation B/SRT obtained an inhibition of proliferation in cervical carcinoma (KB) (IC_{50} = 23.57 μg/ml), breast carcinoma (MCF-7) (IC_{50} = 29.86 μg/ml), and lung carcinoma (A-549) (IC_{50} = 40.03 μg/ml). However, the authors conclude that the idea that the carcinogenic activity of *U. tomentosa* is only by the quantity of alkaloids is wrong because the BSRT preparation did not have the highest antiproliferative activity among the rest of the extracts. Therefore, it is concluded that there are other phytochemicals responsible for the pharmacological activity that *U. tomentosa* possesses (Pilarski et al., 2010).

3.2 *Euphorbiaceae*

The Euphorbiaceae family is one of the largest of the flowering plants, having 300 genera and approximately 10,000 species (Aleksandrov et al., 2019). It is widely distributed in the Indo-Malayan region, tropical America, tropical Africa, Mediterranean Basin, Middle East, South Africa, and southern USA (Mahbubur Rahman & Iffat Ara Gulshana, 2014). This family contains plants of all kinds such as large woody trees, climbing lianas to simple weeds. They can survive different hot and dry conditions such as tropical climates or even hot and humid such as the rainforest (Mwine & van Damme, 2011). The Euphorbiaceae family has been used frequently with medicinal uses, thanks to its active components such as alkanes, triterpenes, phytosterols, tannins, polyphenols, and flavonoids (Aleksandrov et al., 2019). Some plants have been tested as cancer treatments thanks to these secondary metabolites, some examples are: *Euphorbia hirta* L., *Euphorbia tirucalli* L., *Euphorbia helioscopia* L. (Aleksandrov et al., 2019), *Croton caudatus* (Shantabi et al., 2020), *Croton oblongifolius* (Sajon, 2019). In Ecuador, there are 244 registered species, of which 46 are endemic (Cerón et al., 2011). These species are mainly found in foothill forests and inter-Andean vegetation, littoral forests, and Galapagos. The genus Croton has the highest number of endemic species in the country (Cerón et al., 2011). Next, the anticancer effect of *Croton lechleri* will be described in detail which is native to Ecuador.

3.2.1 *Croton lechleri* Müll. Arg.

The genus Croton of the Euphorbiaceae family that contains around 1300 species is generally found in tropical and subtropical areas of the world (De Lima et al., 2018). Specifically, *Croton lechleri* a small-size Amazonian tree is distributed on the slopes of the eastern Andes of Peru, Colombia, Ecuador, Bolivia, Venezuela, Brazil, and Mexico (Alonso-Castro et al., 2012; Rossi et al., 2011). In Ecuador, *Croton lechleri* is native and it is found at elevations of 0 to 2000 masl, in the Andean and Amazon region of the country in the provinces of Loja Esmeraldas, Carchi, Morona-Santiago, Napo, Pastaza, Sucumbíos, Tungurahua, and Zamora-Chinchipe (Bailon-Moscoso et al., 2015b). It is better known by its Spanish name "Sangre de Drago" due to the red thick latex that is obtained from it (Alonso-Castro et al., 2012). In Latin America, the red latex of this species has been used as a traditional medicine by indigenous Amazonians for some purposes like heal wounds, to treat gastrointestinal diseases and also as a treatment for cancer due to its cytotoxic and antiproliferative activity (Alonso-Castro et al., 2012; Bailon-Moscoso et al., 2015b; Rossi et al., 2013).

Polyphenolic compounds are the main components of the sap of the *C. lechleri* in Ecuador since it contains 90% dry weight of proanthocyanidin oligomers (Gonzales & Valerio, 2008). Among them catechin-(4α-8)-epigallocatechin, gallocatechin-(4-α-8)-epicatechin, gallocatechin-(4α-6)-epigallocatechin, catechin-(4α-8)-gallocatechin-(4α-8)-gallocatechin and gallocatechin-(4α-8)-gallocatechin-(4α-8)-epi-gallocatechin mixtures of proanthocyanidins, flavon-3-ols (Gonzales & Valerio, 2008). Other elements present in *Croton lechleri* are steroids like β-sitosterol and β-sitostenine, aromatic compounds including 1,3,5-trimethoxybenzene, 2,4,6-trimethoxyphenol, 3,4-dimethoxyphenol, 3,4-dimethoxy benzylalcohol, and 4-hydroxyphenethylalcohol; diterpenoids and a very important compound taspine alkaloid (Gonzales & Valerio, 2008).

In recent years, there has been a growth in interest in the *C. lechleri* sap and its taspine alkaloid. Since proanthocyanidins and other compounds in this plant are known to protect against various pathologies like cancer, taspine is the distinctive

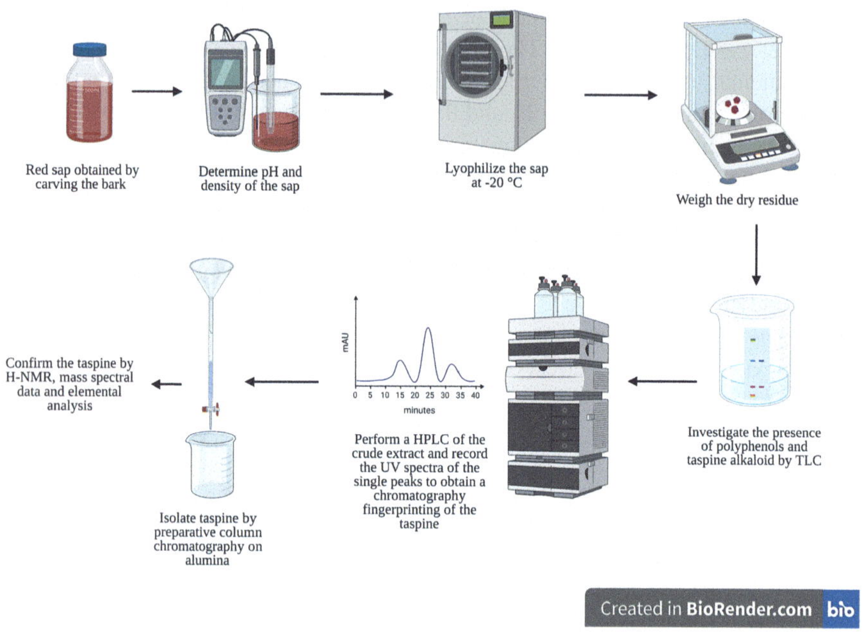

Fig. 2 Representation of the extraction, isolation, and determination of the taspine, secondary metabolite of the *Croton lechleri* (Froldi et al., 2009)

alkaloid of the *C. lechleri* sap and is an active compound with wound healing properties and anticancer activity.

In a study published in 2012 in the Journal of Ethnopharmacology, *Croton lechleri* was evaluated in cell lines such as melanoma (SK-23) and colorectal carcinoma (LoVo and HT29) to determine its anticancer activity (Montopoli et al., 2012). The study was based on cell viability and cell cycle of melanoma and colon cancer cell lines in culture by applying the sap of *C. lechleri* and isolated taspine (Montopoli et al., 2012). This is because the incidence rate of cutaneous malignant melanoma and colorectal cancer has been increasing and the known treatments are not sufficient. For example, malignant melanoma is the most serious skin cancer and does not respond to chemotherapy or radiotherapy, therefore, in vitro studies have been carried out with the use of *C. lechleri* and taspine on epithelial cancer as its traditional use, which is by topical application (Montopoli et al., 2012).

To carry out the research about melanoma and colorectal carcinoma, the red sap of *C. lechleri* was collected from trees in the province of Napo, Ecuador (Montopoli et al., 2012). The sap was lyophilized and stored at 20 °C, and taspine was isolated according to the previously reported (Fig. 2) (Froldi et al., 2009). In this study, the activity of *C. lechleri* sap and taspine in cell proliferation was measured using the MTT test. For this, melanoma and colorectal carcinoma cells were treated for 24 h with the sap in concentrations from 0.01 μg/ml to 100 μg/ml and as a result it was obtained that at 0.5 μg/ml, the sap decreased cell proliferation in 67.57 ± 3.4% in melanoma cells, and there were no significant results in colorectal carcinoma cells. While inhibition colorectal carcinoma cell proliferation started with 10 μg/ml of sap. The case of taspine was evaluated from 1 ng/ml to 1 μg/ml, showing significant inhibition activity at 0.1 μg/ml in the melanoma and colorectal carcinoma cell lines (Montopoli et al., 2012).

To assess the sensitivity of cells during the cell cycle, the melanoma cell line was exposed to *C. lechleri* sap and taspine for 24 h. Sap at 1.0 μg/ml had no influence on the cell cycle, but at 10.0 μg/ml the cells increased in the G1/G0 phase from $17.77 \pm 3.5\%$, while the S and G2/M phases decreased from $7.97 \pm 3.1\%$ and $11.77 \neq 2.8\%$, respectively. When the sap was used at 50.0 μg/mL, we observed a sharp decrease in the G1/G0 and S phases with a surprising increase in subG0 cells, which proves the cytotoxic action of the *C. lechleri* sap. In the case of the alkaloid taspine (0.1 and 0.5 μg/mL), there were no changes in the cell cycle (Montopoli et al., 2012). With this result there is a probe of the anticancer activity of the secondary metabolites present in *C. lechleri.*

3.3 *Equisetaceae*

The Equisetaceae family is one of the oldest found among vascular plants (Christenhusz et al., 2019). It has fossil remains dating from the Carboniferous (Christenhusz et al., 2019). It is known as the horsetail family, because it contains its only surviving genus Equisetum, which has approximately 15–30 species (Xu & Deng, 2017). This family is widely distributed around the world, mainly in the northern hemisphere, South America, Africa, and Asia (Christenhusz et al., 2019). However, it is absent in Australasia and Antarctic (Boeing et al., 2021). They are generally small plants that rarely measure up to one meter in height (Mello & Budel, 2013), rhizomatous perennial plants that have a characteristic jointed-looking stem, containing small leaves fused at a node (Hauke, 1990). They can be adapted to high temperatures, tropical and cold regions (Boeing et al., 2021). In the same way, they can live in both terrestrial and aquatic environments (Xu & Deng, 2017). The genus Equisetum is known to contain a wide variety of secondary metabolites such as saponins, flavonoids, tannis, and alkaloids (Mello & Budel, 2013). By containing these compounds, the genus has several medicinal properties such as antitumor, antimicrobial, antioxidant, anti-inflammatory, and diuretic (Boeing et al., 2021). Speaking of anticancer properties, some species have been shown to have it, such as *Equisetum arvense* (Bhat et al., 2020) and *Equisetum ramosissimum* (Li et al., 2016). In Ecuador, this family is widely distributed in the provinces of Bolívar, Carchi, Chimborazo, Cotopaxi, El Oro, Galapagos, Imbabura, Loja, Morona-Santiago, Napo, Pastaza, Pichincha, Sucumbíos, Tungurahua, Azuay, and Zamora-Chinchipe. The *Equisetum giganteum* native to Ecuador contains these anticancer properties that will be discussed in more detail below.

3.3.1 *Equisetum giganteum* L.

The Equisetaceae family is currently made up of 15 species, *Equisetum giganteum* belonging to it and is better known by its spanish name "cola de caballo" (Ricco et al., 2011). It is a plant endemic to South America and Central America (Alavarce et al., 2015). In Ecuador, *E. giganteum* is native and is found at altitudes ranging from 0 to 3000 masl, in the Coastal, Sierra, and Amazonian regions. Specifically, it is found in the provinces of Azuay, Chimborazo, Cotopaxi, Imbabura, Loja, Morona-Santiago, Napo, Pastaza, Pichincha, Tungurahua, and Zamora-Chinchipe. It is used in traditional medicine as a diuretic and hemostatic in urinary disorders and inflammatory conditions, among other applications such as cancer treatment (Alavarce et al., 2015).

The composition of *E. giganteum* has shown a great abundance of phenolic compounds derived from caffeic and ferulic acids and flavonoid heterosides derived from quercetin and kaempferol, in addition to styrylpyrones (Francescato et al., 2013). Phenolic compounds are among the most studied secondary metabolites due to their beneficial effect for good health. Several in vitro and in vivo studies have reported the important role of phenolic compounds in the fight against cancer and among them are ferulic acid, caffeic acid, and quercetin present in *E. giganteum*, among others. These secondary metabolites have mechanisms of antitumor activity, although they are not fully understood (Jabeur, 2016).

In a 2017 study by the Journal Food & Function, cytotoxicity of *E. giganteum* was evaluated using three tumor cell lines, HeLa (cervical carcinoma), HepG2 (hepatocellular carcinoma), and MCF-7 (breast adenocarcinoma) (Jabeur et al., 2017). It was shown that a low concentration of the *E. giganteum* extract can give 50% inhibition of growth in human tumor cell lines. In general, total phenolic compounds, phenolic acids, and flavonoids are highly related to antioxidant, anti-inflammatory, and antitumor activities, presenting correlation factors for the human tumor cell lines HeLa (cervical carcinoma), HepG2 (hepatocellular carcinoma), and MCF-7 (breast adenocarcinoma) (Inés Jabeur et al., 2017). *E. giganteum* gave a total flavonoid content of 21.7 ± 0.4 mg/g and phenolic acids 4.98 ± 0.03 mg/g, giving a very interesting biological potential, thus confirming that it is an option for the treatment of these types of cancer (Inés Jabeur et al., 2017).

4 Conclusions

It is clear that Ecuador has a great diversity of plant species, resulting in turn in a wide variety of compounds, that we may not know in its entirety, with potential pharmacological use. As presented in this review, *Uncaria tomentosa*, *Croton lechleri,* and *Equisetum giganteum* have a considerable content of secondary metabolites, among which oxindole alkaloids, taspine alkaloids, proanthocyanidin oligomers, flavonoids, and phenolic acids are the most prominent. Furthermore, in vivo and in vitro tests indicate that the secondary metabolites present in these three plants present a great pharmacological activity for the treatment of different cancer cell lines. The mechanism of action is summarized in that they possess antiproliferative activity and are also capable of inducing apoptosis in different tumor cell lines. In addition, it is important to emphasize the importance of starting to do studies with samples of species from Ecuador because the content of metabolites can vary according to the ecological interactions between the plant and its environment. Furthermore, it is necessary to start doing more in vivo clinical trials in order to understand in more specific detail the mechanisms of action that are involved in the inhibition of cancer cell proliferation and induced apoptosis.

References

Alavarce, R. A. S., Saldanha, L. L., Almeida, N. L. M., Porto, V. C., Dokkedal, A. L., & Lara, V. S. (2015). The beneficial effect of equisetum giganteum L. against Candida biofilm formation: New approaches to denture stomatitis. *Evidence-based Complementary and Alternative Medicine, 2015*, 939625. https://doi.org/10.1155/2015/939625

Aleksandrov, M., Maksimova, V., & Koleva Gudeva, L. (2019). Review of the anticancer and cytotoxic activity of some species from genus euphorbia. *Agriculturae Conspectus Scientificus, 84*(1), 1–5.

Allen-Hall, L., Arnason, J. T., Cano, P., & Lafrenie, R. M. (2010). Uncaria tomentosa acts as a potent TNF-α inhibitor through NF-κB. *Journal of Ethnopharmacology, 127*(3), 685–693. https://doi.org/10.1016/j.jep.2009.12.004

Alonso-Castro, A. J., Ortiz-Sánchez, E., Domínguez, F., López-Toledo, G., Chávez, M., Ortiz-Tello, A. D. J., & García-Carrancá, A. (2012). Antitumor effect of Croton lechleri Mull. Arg. (Euphorbiaceae). *Journal of Ethnopharmacology, 140*(2), 438–442. https://doi.org/10.1016/j.jep.2012.01.009

An, F., Wang, S., Tian, Q., & Zhu, D. (2015). Effects of orientin and vitexin from Trollius chinensis on the growth and apoptosis of esophageal cancer EC-109 cells. *Oncology Letters, 10*(4), 2627–2633. https://doi.org/10.3892/ol.2015.3618

Aneb, M., Talbaoui, A., Bouyahya, A., Boury, H., Amzazi, S., Benjouad, A., Dakka, N., & Bakri, Y. (2016). In vitro cytotoxic effects and antibacterial activity of moroccan medicinal plants Aristolochia longa and Lavandula multifida. *European Journal of Medicinal Plants, 16*(2), 1–13. https://doi.org/10.9734/ejmp/2016/28534

Ayoob, I., Hazari, Y. M., Lone, S. H., Shakeel-u-Rehman, K., Fazili, K. M., & Bhat, K. A. (2017). Phytochemical and cytotoxic evaluation of peganum harmala: Structure activity relationship studies of harmine. *ChemistrySelect, 2*(10), 2965–2968. https://doi.org/10.1002/slct.201700232

Bacher, N., Tiefenthaler, M., Sturm, S., Stuppner, H., Ausserlechner, M. J., Kofler, R., & Konwalinka, G. (2006). Oxindole alkaloids from Uncaria tomentosa induce apoptosis in proliferating, G0/G1-arrested and bcl-2-expressing acute lymphoblastic leukaemia cells. *British Journal of Haematology, 132*(5), 615–622. https://doi.org/10.1111/j.1365-2141.2005.05907.x

Baikar, S., & Malpathak, N. (2010). Secondary metabolites as DNA topoisomerase inhibitors: A new era towards designing of anticancer drugs. *Pharmacognosy Reviews, 4*(7), 12–26. https://doi.org/10.4103/0973-7847.65320

Bailon-Moscoso, N., Romero-Benavides, J. C., Tinitana-Imaicela, F., & Ostrosky-Wegman, P. (2015a). Medicinal plants of Ecuador: A review of plants with anticancer potential and their chemical composition. *Medicinal Chemistry Research, 24*(6), 2283–2296. https://doi.org/10.1007/s00044-015-1335-7

Bailon-Moscoso, N., Romero-Benavides, J. C., Tinitana-Imaicela, F., & Ostrosky-Wegman, P. (2015b). Medicinal plants of Ecuador: A review of plants with anticancer potential and their chemical composition. *Medicinal Chemistry Research, 24*(6), 2283–2296. https://doi.org/10.1007/s00044-015-1335-7

Belayachi, L. (2013). Screening of North African medicinal plant extracts for cytotoxic activity against tumor cell lines. *European Journal of Medicinal Plants, 3*(3), 310–332. https://doi.org/10.9734/ejmp/2013/3403

Belayachi, L., Aceves-Luquero, C., Merghoub, N., de Mattos, S. F., Amzazi, S., Villalonga, P., & Bakri, Y. (2017). Induction of cell cycle arrest and apoptosis by ormenis eriolepis a morrocan endemic plant in various human cancer cell lines. *African Journal of Traditional, Complementary, and Alternative Medicines, 14*(2), 356–373. https://doi.org/10.21010/ajtcam.v14i2.37

Ben Bakrim, W., El Bouzidi, L., Nuzillard, J. M., Cretton, S., Saraux, N., Monteillier, A., Christen, P., Cuendet, M., & Bekkouche, K. (2018). Bioactive metabolites from the leaves of withania adpressa. *Pharmaceutical Biology, 56*(1), 505–510. https://doi.org/10.1080/13880209.2018.1499781

Bhandari, J., Muhammad, B. T., Thapa, P., & Shrestha, B. G. (2017). Study of phytochemical, anti-microbial, anti-oxidant, and anti-cancer properties of Allium wallichii. *BMC Complementary and Alternative Medicine, 17*(1), 1–9. https://doi.org/10.1186/s12906-017-1622-6

Bhat, A. A., Ahamad, B., Rehman, M. U., & Ahmad, P. (2020). Impact of ethanolic extract of Equisetum arvense (EA1) on pancreatic carcinoma AsPC-1 cells. *Saudi Journal of Biological Sciences, 27*(5), 1260–1264. https://doi.org/10.1016/j.sjbs.2020.01.029

Boeing, T., Tafarelo Moreno, K. G., Gasparotto Junior, A., Mota da Silva, L., & de Souza, P. (2021). Phytochemistry and pharmacology of the genus equisetum (Equisetaceae): A narrative review of the species with therapeutic potential for kidney diseases. *Evidence-based Complementary and Alternative Medicine, 2021*, 6658434. https://doi.org/10.1155/2021/6658434

Boukes, G. J. (2010). *The in vitro biological activities of three Hypoxis species and their active compounds. January.*

Bouyahya, A., Chadon Assemian, I. C., Mouzount, H., Bourais, I., Et-Touys, A., Fellah, H., Benjouad, A., Dakka, N., & Bakri, Y. (2019). Could volatile compounds from leaves and fruits of Pistacia lentiscus constitute a novel source of anticancer, antioxidant, antiparasitic and antibacterial drugs? *Industrial Crops and Products, 128*, 62–69. https://doi.org/10.1016/j.indcrop.2018.11.001

Bremer, B., & Eriksson, T. (2009). Time tree of rubiaceae: Phylogeny and dating the family, subfamilies, and tribes. *International Journal of Plant Sciences, 170*(6), 766–793. https://doi.org/10.1086/599077

Burger, T., Mokoka, T., Fouché, G., Steenkamp, P., Steenkamp, V., & Cordier, W. (2018). Solamargine, a bioactive steroidal alkaloid isolated from Solanum aculeastrum induces non-selective cytotoxicity and P-glycoprotein inhibition. *BMC Complementary and Alternative Medicine, 18*(1), 1–11. https://doi.org/10.1186/s12906-018-2208-7

Cerón, C., Riina, R., Santiana, J., & Tye, A. (2011). Euphorbiaceae. In *Libro rojo de las plantas endémicas del Ecuador* (2nd ed., pp. 317–325). Pontificia Universidad Católica del Ecuador.

Chen, L., Li, M., Yang, Z., Tao, W., Wang, P., Tian, X., Li, X., & Wang, W. (2020). Gardenia jasminoides Ellis: Ethnopharmacology, phytochemistry, and pharmacological and industrial applications of an important traditional Chinese medicine. *Journal of Ethnopharmacology, 257*(March), 112829. https://doi.org/10.1016/j.jep.2020.112829

Chiu, C. F., Lin, Y. Q., Park, J. M., Chen, Y. C., Hung, S. W., Chiu, C. C., & Chang, C. F. (2020). The novel camptothecin derivative, CPT211, induces cell cycle arrest and apoptosis in models of human breast cancer. *Biomedicine and Pharmacotherapy, 128*(February), 110309. https://doi.org/10.1016/j.biopha.2020.110309

Christenhusz, M. J. M., Bangiolo, L., Chase, M. W., Fay, M. F., Husby, C., Witkus, M., & Viruel, J. (2019). Phylogenetics, classification and typification of extant horsetails (Equisetum,

Equisetaceae). *Botanical Journal of the Linnean Society, 189*(4), 311–352. https://doi.org/10.1093/botlinnean/boz002

Coqueiro, A., & Verpoorte, R. (2019). Alkaloids. In *Encyclopedia of Analytical Science* (pp. 77–84). Elsevier. https://doi.org/10.1016/B978-0-12-409547-2.11675-0

Davis, A. P., Govaerts, R., Bridson, D. M., Ruhsam, M., Moat, J., & Brummitt, N. A. (2009). A Global assessment of distribution, diversity, endemism, and taxonomic effort in the rubiaceae. *Annals of the Missouri Botanical Garden, 96*(1), 68–78. https://doi.org/10.3417/2006205

De la Torre, L., Navarrete, H., Muriel, P., Macía, M. J., & Balslev, H. (2008). Enciclopedia de las Plantas Útiles del Ecuador. in *Herbario QCA de la Escuela de Ciencias Biológicas de la Pontificia Universidad Católica del Ecuador & Herbario AAU del Departamento de Ciencias Biológicas de la Universidad de Aarhus.*

De Lima, E. J. S. P., Alves, R. G., D'Elia, G. M. A., Da Anunciação, T. A., Silva, V. R., Soares, M. B. P., Cardozo, N. M. D., Costa, E. V., Da Silva, F. M. A., Koolen, H. H. F., & Bezerra, D. P. (2018). Antitumor effect of the essential oil from the leaves of Croton matourensis Aubl. (Euphorbiaceae). *Molecules, 23*(11), 1–12. https://doi.org/10.3390/molecules23112974

Demain, A. L., & Fang, A. (2000). The natural functions of secondary metabolites. *Advances in Biochemical Engineering/Biotechnology, 69*, 1–39. https://doi.org/10.1007/3-540-44964-7_1

Ekalu, A. (2021). Medicinal uses, phytochemistry, and pharmacological activities of Mitracarpus species (Rubiaceae): A review. *Scientific African, 11*, e00692. https://doi.org/10.1016/j.sciaf.2020.e00692

El Khalki, L., Tilaoui, M., Jaafari, A., Ait Mouse, H., & Zyad, A. (2018). Studies on the dual cytotoxicity and antioxidant properties of Berberis vulgaris extracts and its main constituent berberine. *Advances in Pharmacological Sciences, 2018*, 3018498. https://doi.org/10.1155/2018/3018498

Fabri, R. L., Grazul, R. M., De Carvalho, L. O., Coimbra, E. S., Cardoso, G. M. M., De Souza-Fagundes, E. M., Da Silva, A. D., & Scio, E. (2012). Antitumor, antibiotic and antileishmanial properties of the pyranonaphthoquinone psychorubrin from Mitracarpus frigidus. *Anais da Academia Brasileira de Ciências, 84*(4), 1081–1089. https://doi.org/10.1590/S0001-37652012005000064

Farias, J. G., Frescura, V. D. S., Tedesco, S. B., Farias, I. L. G., Barzotto, F., Dalla Possa, J. S., Schetinger, M. R. C., & Nicoloso, F. T. (2013). Uncaria tomentosa reduces lipid peroxidation and DNA-damage from chemotherapy. *Latin American Journal of Pharmacy, 32*(3), 340–345.

Firmansyah, A., Sundalian, M., & Taufiq, M. (2021). Kratom (Mitragyna speciosa korth) for a new medicinal: A review of pharmacological and compound analysis. *Biointerface Research in Applied Chemistry, 11*(2), 9704–9718. https://doi.org/10.33263/BRIAC112.97049718

Francescato, L. N., Debenedetti, S. L., Schwanz, T. G., Bassani, V. L., & Henriques, A. T. (2013). Identification of phenolic compounds in Equisetum giganteum by LC-ESI-MS/MS and a new approach to total flavonoid quantification. *Talanta, 105*, 192–203. https://doi.org/10.1016/j.talanta.2012.11.072

Froldi, G., Zagotto, G., Filippini, R., Montopoli, M., Dorigo, P., & Caparrotta, L. (2009). Activity of sap from Croton lechleri on rat vascular and gastric smooth muscles. *Phytomedicine, 16*(8), 768–775. https://doi.org/10.1016/j.phymed.2009.02.003

García Prado, E., García Gimenez, M. D., De la Puerta Vázquez, R., Espartero Sánchez, J. L., & Sáenz Rodríguez, M. T. (2007). Antiproliferative effects of mitraphylline, a pentacyclic oxindole alkaloid of Uncaria tomentosa on human glioma and neuroblastoma cell lines. *Phytomedicine, 14*(4), 280–284. https://doi.org/10.1016/j.phymed.2006.12.023

GBIF.org. (2021). *GBIF Home Page*. Retrieved March 27, 2021, from https://www.gbif.org

Ghasemzadeh, A., & Ghasemzadeh, N. (2014). Flavonoids and phenolic acids: Role and biochemical activity in plants and human. https://doi.org/10.5897/JMPR11.1404

Goevarts, R., Ruhsam, M., Andersson, L., Robbrecht, E., Bridson, D., Davis, A., Schanzer, I., & Sonke, B.. (2006). World checklist of Rubiaceae. *Royal Botanic Gardens, Kew.*

Gonzales, G., & Valerio, L. (2008). Medicinal plants from Peru: A review of plants as potential agents against cancer. *Anti-Cancer Agents in Medicinal Chemistry, 6*(5), 429–444. https://doi.org/10.2174/187152006778226486

Guaouguaou, F. E., Bebaha, M. A. A., Taghzouti, K., Bouyahya, A., Bakri, Y., Dakka, N., & Es-Safi, N. E. (2018). Cytotoxicological investigation of the essential oil and the extracts of cotula cinerea and Salvia verbenaca from Morocco. *BioMed Research International, 2018*, 7463961. https://doi.org/10.1155/2018/7163961

Hauke, R. L. (1990). Equisetaceae. In *Pteridophytes and gymnosperms* (pp. 46–48). Springer. https://doi.org/10.1007/978-3-662-02604-5_12

Ionkova, I., Sasheva, P., Ionkov, T., & Momekov, G. (2013). Linum narbonense: A new valuable tool for biotechnological production of a potent anticancer lignan Justicidine B. *Pharmacognosy Magazine, 9*(33), 39. https://doi.org/10.4103/0973-1296.108138

Jabeur, I. (2016). The broad spectrum of bioactive properties of phenolic extracts: a prospective study in three different plants. In مجلة الحاسب للعلوم المحاسبية و المراجعة (Vol. 23, Issue 45). Retrieved from http://hdl.handle.net/10198/13100

Jabeur, I., Martins, N., Barros, L., Calhelha, R. C., Vaz, J., Achour, L., Santos-Buelga, C., & Ferreira, I. C. F. R. (2017). Contribution of the phenolic composition to the antioxidant, anti-inflammatory and antitumor potential of Equisetum giganteum L. and Tilia platyphyllos Scop. *Food & Function, 8*(3), 975–984. https://doi.org/10.1039/c6fo01778a

Kabera, J. (2018). *Plant secondary metabolites: Biosynthesis, classification, function and pharmacological classification, function and pharmacological properties. January.*

Kee, N. L. A., Mnonopi, N., Davids, H., Naudé, R. J., & Frost, C. L. (2008). Antithrombotic/anticoagulant and anticancer activities of selected medicinal plants from South Africa. *African Journal of Biotechnology, 7*(3), 217–223. https://doi.org/10.4314/ajb.v7i3.58372

Kuete, V., Fokou, F. W., Karaosmanoğlu, O., Beng, V. P., & Sivas, H. (2017). Cytotoxicity of the methanol extracts of Elephantopus mollis, Kalanchoe crenata and 4 other Cameroonian medicinal plants towards human carcinoma cells. *BMC Complementary and Alternative Medicine, 17* (1), 1–9. https://doi.org/10.1186/s12906-017-1793-1

Lakhdar, M. (2018). Traditional uses, phytochemistry and biological activities of Cotula cinerea Del: A review. *Tropical Journal of Pharmaceutical Research, 17*(2), 365–373. https://doi.org/10.4314/tjpr.v17i2.24

Li, P. H., Chiu, Y. P., Shih, C. C., Wen, Z. H., Ibeto, L. K., Huang, S. H., Chiu, C. C., Ma, D. L., Leung, C. H., Chang, Y. N., & Wang, H. M. D. (2016). Biofunctional activities of Equisetum ramosissimum extract: Protective effects against oxidation, melanoma, and melanogenesis. *Oxidative Medicine and Cellular Longevity, 2016*, 2853543. https://doi.org/10.1155/2016/2853543

Liang, J. H., Wang, C., Huo, X. K., Tian, X. G., Zhao, W. Y., Wang, X., Sun, C. P., & Ma, X. C. (2020). The genus Uncaria: A review on phytochemical metabolites and biological aspects. *Fitoterapia, 147*, 104772. https://doi.org/10.1016/j.fitote.2020.104772

Ludwiczuk, A., Skalicka-Woźniak, K., & Georgiev, M. I. (2017). Terpenoids. In *Pharmacognosy: Fundamentals, applications and strategy*. https://doi.org/10.1016/B978-0-12-802104-0.00011-1

Mahbubur Rahman, H. M., & Iffat Ara Gulshana, M. (2014). Taxonomy and medicinal uses on Amaranthaceae family of Rajshahi, Bangladesh. *Applied Ecology and Environmental Sciences, 2*(2), 54–59. https://doi.org/10.12691/aees-2-2-3

Malagón, O., Vila, R., Iglesias, J., Zaragoza, T., & Cañigueral, S. (2003). Composition of the essential oils of four medicinal plants from Ecuador. *Flavour and Fragrance Journal, 18*(6), 527–531. https://doi.org/10.1002/ffj.1262

Manzione, M. G., Martorell, M., Sharopov, F., Bhat, N. G., Kumar, N. V. A., Fokou, P. V. T., & Pezzani, R. (2020). Phytochemical and pharmacological properties of asperuloside, a systematic review. *European Journal of Pharmacology, 883*, 173344. https://doi.org/10.1016/j.ejphar.2020.173344

Mardina, V., Ilyas, S., Harmawan, T., Halimatussakdiah, H., & Tanjung, M. (2020). Antioxidant and cytotoxic activities of the ethyl acetate extract of Sphagneticola trilobata (L.) J.F. Pruski on MCF-7 breast cancer cell. *Journal of Advanced Pharmaceutical Technology & Research, 11*(3), 123–127. https://doi.org/10.4103/japtr.JAPTR3120

Martin, A. C. B. M., Fuzer, A. M., Becceneri, A. B., da Silva, J. A., Tomasin, R., Denoyer, D., Kim, S.-H., McIntyre, K. A., Pearson, H. B., Yeo, B., Nagpal, A., Ling, X., Selistre-de-Araújo, H. S., Vieira, P. C., Cominetti, M. R., & Pouliot, N. (2017). [10]-Gingerol induces apoptosis and inhibits metastatic dissemination of triple negative breast cancer in vivo. *Oncotarget, 8*(42), 72260–72271. https://doi.org/10.18632/oncotarget.20139

Martins, D., & Nunez, C. V. (2015). Secondary metabolites from Rubiaceae species. *Molecules, 20* (7), 13422–13495. https://doi.org/10.3390/molecules200713422

Mayanti, T., Amir, N. I., Katja, D. G., Fajriah, S., Darmawan, A., Supratman, U., Awang, K., & Shiono, Y. (2020). Steroids from the stem bark of Dysoxylum nutans (Meliaceae) and their cytotoxic effect against MCF-7 breast cancer cell lines. *Jurnal Kimia Valensi, 6*(2), 133–139. https://doi.org/10.15408/jkv.v6i2.15976

Mello, M., & Budel, J. M. (2013). Equisetum L. (Equisetaceae): Uma Revisão. *Cadernos Da Escola de Saúde, 1*(9), 1–15.

Montopoli, M., Bertin, R., Chen, Z., Bolcato, J., Caparrotta, L., & Froldi, G. (2012). Croton lechleri sap and isolated alkaloid taspine exhibit inhibition against human melanoma SK23 and colon cancer HT29 cell lines. *Journal of Ethnopharmacology, 144*(3), 747–753. https://doi.org/10.1016/j.jep.2012.10.032

Mwine, J. T., & van Damme, P. (2011). Why do euphorbiaceae tick as medicinal plants? a review of euphorbiaceae family and its medicinal features. *Journal of Medicinal Plant Research, 5*(5), 652–662.

Pandey, S., Walpole, C., Cabot, P. J., Shaw, P. N., Batra, J., & Hewavitharana, A. K. (2017). Selective anti-proliferative activities of Carica papaya leaf juice extracts against prostate cancer. *Biomedicine and Pharmacotherapy, 89*, 515–523. https://doi.org/10.1016/j.biopha.2017.02.050

Perveen, S., Fawzy, G., Khan, A., Proksch, P., Al-Taweel, A., & El-Shafae, A. (2015). Cytotoxic glucosphingolipid from Celtis Africana. *Pharmacognosy Magazine, 11*(42), 1. https://doi.org/10.4103/0973-1296.157662

Pilarski, R., Filip, B., Wietrzyk, J., Kuraś, M., & Gulewicz, K. (2010). Anticancer activity of the Uncaria tomentosa (Willd.) DC. preparations with different oxindole alkaloid composition. *Phytomedicine, 17*(14), 1133–1139. https://doi.org/10.1016/j.phymed.2010.04.013

Pilarski, R., Gurrola-Díaz, C. M., García-López, P. M., Soldevila, G., Olejnik, A., Grajek, W., & Gulewicz, K. (2013). Enhanced proapoptotic response of the promyelocytic leukemia HL-60 cells treated with an Uncaria tomentosa alkaloid preparation. *Journal of Herbal Medicine, 3*(4), 149–156. https://doi.org/10.1016/j.hermed.2013.04.002

Poornima, P., Quency, R. S., & Padma, V. V. (2013). Neferine induces reactive oxygen species mediated intrinsic pathway of apoptosis in HepG2 cells. *Food Chemistry, 136*(2), 659–667. https://doi.org/10.1016/j.foodchem.2012.07.112

Qin, N., Lu, X., Liu, Y., Qiao, Y., Qu, W., Feng, F., & Sun, H. (2021). Recent research progress of Uncaria spp. based on alkaloids: phytochemistry, pharmacology and structural chemistry. *European Journal of Medicinal Chemistry, 210*, 112960. https://doi.org/10.1016/j.ejmech.2020.112960

Ricco, R. A., Agudelo, I., Garcés, M., Evelson, P., Wagner, M. L., & Gurni, A. A. (2011). Polifenoles y actividad antioxidante en Equisetum giganteum L. (Equisetaceae). *Boletin Latinoamericano y del Caribe de Plantas Medicinales y Aromaticas, 10*(4), 325–332.

Rinner, B., Li, Z. X., Haas, H., Siegl, V., Sturm, S., Stuppner, H., & Pfragner, R. (2009). Antiproliferative and pro-apoptotic effects of Uncaria tomentosa in human medullary thyroid carcinoma cells. *Anticancer Research, 29*(11), 4519–4528.

Rossi, D., Guerrini, A., Maietti, S., Bruni, R., Paganetto, G., Poli, F., Scalvenzi, L., Radice, M., Saro, K., & Sacchetti, G. (2011). Chemical fingerprinting and bioactivity of Amazonian Ecuador Croton lechleri Müll. Arg. (Euphorbiaceae) stem bark essential oil: A new functional food ingredient? *Food Chemistry, 126*(3), 837–848. https://doi.org/10.1016/j.foodchem.2010.11.042

Rossi, D., Guerrini, A., Paganetto, G., Bernacchia, G., Conforti, F., Statti, G., Maietti, S., Poppi, I., Tacchini, M., & Sacchetti, G. (2013). Croton lechleri Müll. Arg. (Euphorbiaceae) stem bark essential oil as possible mutagen-protective food ingredient against heterocyclic amines from cooked food. *Food Chemistry, 139*(1–4), 439–447. https://doi.org/10.1016/j.foodchem.2013.01.076

Sajon, S. R. (2019). In-silico study of mapk inhibition based lead identification from the isolated compounds of Croton Oblongifolius roxb for the treatment of hepatocellular carcinoma. *Pharmacology, 1*, 301–318.

Seca, A. M. L., & Pinto, D. C. G. A. (2018). Plant secondary metabolites as anticancer agents: Successes in clinical trials and therapeutic application. *International Journal of Molecular Sciences, 19*(1), 263. https://doi.org/10.3390/ijms19010263

Shantabi, L., Jagetia, G. C., Moirangthem, D. S., & Nongalleima, K. (2020). Anticancer activity of an ehnomedicinal plant Croton caudatus Geiseler, Kam sabut in cultured HeLa cells. *Biocatalysis and Agricultural Biotechnology, 23*, 101500. https://doi.org/10.1016/j.bcab.2020.101500

Shi, X., Du, R., Zhang, J., Lei, Y., & Guo, H. (2019). Evaluation of the anti-cancer potential of Cedrus deodara total lignans by inducing apoptosis of A549 cells. *BMC Complementary and Alternative Medicine, 19*(1), 1–7. https://doi.org/10.1186/s12906-019-2682-6

Sierra, R., Campos, F., & Chamberlin, J. (2002). Assessing biodiversity conservation priorities: Ecosystem risk and representativeness in continental Ecuador. *Landscape and Urban Planning, 59*(2), 95–110. https://doi.org/10.1016/S0169-2046(02)00006-3

Singh, G., Passari, A. K., Momin, M. D., Ravi, S., Singh, B. P., & Kumar, N. S. (2020). Ethnobotanical survey of medicinal plants used in the management of cancer and diabetes. *Journal of Traditional Chinese Medicine, 40*(6), 1007–1017. https://doi.org/10.19852/j.cnki.jtcm.2020.06.012

Subash-Babu, P., Li, D. K., & Alshatwi, A. A. (2017). In vitro cytotoxic potential of friedelin in human MCF-7 breast cancer cell: Regulate early expression of Cdkn2a and pRb1, neutralize mdm2-p53 amalgamation and functional stabilization of p53. *Experimental and Toxicologic Pathology, 69*(8), 630–636. https://doi.org/10.1016/j.etp.2017.05.011

Sung, H., Ferlay, J., Siegel, R. L., Laversanne, M., Soerjomataram, I., Jemal, A., & Bray, F. (2021). Global cancer statistics 2020: GLOBOCAN estimates of incidence and mortality worldwide for 36 cancers in 185 countries. *CA: a Cancer Journal for Clinicians, 2020*, 1–41. https://doi.org/10.3322/caac.21660

Talib, W. H., Abu Zarga, M. H., & Mahasneh, A. M. (2012). Antiproliferative, antimicrobial and apoptosis inducing effects of compounds isolated from Inula viscosa. *Molecules, 17*(3), 3291–3303. https://doi.org/10.3390/molecules17033291

Tene, V., Malagón, O., Finzi, P. V., Vidari, G., Armijos, C., & Zaragoza, T. (2007). An ethnobotanical survey of medicinal plants used in Loja and Zamora-Chinchipe, Ecuador. *Journal of Ethnopharmacology, 111*(1), 63–81. https://doi.org/10.1016/j.jep.2006.10.032

Tiwari, R. (2015). *Plant secondary metabolites: A review. October.*

Tropicos.org. (2021). *Missouri botanical garden*. Available March 27, 2021, from https://www.tropicos.org

Ulloa C. (2006). Riqueza de Plantas Vasculares. *Revista Botánica.*

Wink, M. (2015). Modes of action of herbal medicines and plant secondary metabolites. *Medicine, 2015*, 251–286. https://doi.org/10.3390/medicines2030251

World Health Organization. (2020). GLOBOCAN 2020: Estimated cancer incidence, morality and prevalence Ecuador in 2020.

Xu, Z., & Deng, M. (2017). Equisetaceae. In *Identification and control of common weeds: volume 2* (Vol. 2, pp. 3–6). Springer.

Yang, K. P., Liang, Y. F., & Samaan, N. A. (1991). Intrinsic drug resistance in a human medullary thyroid carcinoma cell line: association with overexpression of mdrl gene and low proliferation fraction. *Anticancer Research, 11*(3), 1065–1068.

Yoon, G., Lee, M. H., Kwak, A. W., Oh, H. N., Cho, S. S., Choi, J. S., Liu, K., Chae, J. I., & Shim, J. H. (2020). Podophyllotoxin isolated from Podophyllum peltatum induces G2/M Phase Arrest and mitochondrial-mediated apoptosis in esophageal squamous cell carcinoma cells. *Forests, 11* (1), 10008. https://doi.org/10.3390/F11010008

Zhang, N., Shen, X., Jiang, X., Cai, J., Shen, X., Hu, Y., & Qiu, S. X. (2018). Two new cytotoxic stilbenoid dimers isolated from Cajanus cajan. *Journal of Natural Medicines, 72*(1), 304–309. https://doi.org/10.1007/s11418-017-1138-x

Zi, C. T., Yang, L., Kong, Q. H., Li, H. M., Yang, X. Z., Ding, Z. T., Jiang, Z. H., Hu, J. M., & Zhou, J. (2019). Glucoside derivatives of podophyllotoxin: Synthesis, physicochemical properties, and cytotoxicity. *Drug Design, Development and Therapy, 13*, 3683–3692. https://doi.org/10.2147/DDDT.S215895

The Carao (*Cassia grandis* L.): Its Potential Usage in Pharmacological, Nutritional, and Medicinal Applications

Jhunior Marcía-Fuentes, Ricardo Santos-Aleman, Isabel Borrás-Linares, and Jesús Lozano Sánchez

1 Introduction

Worldwide, more than 500 species of *Cassia* are reported (Sadiq et al., 2012; Korlam et al., 2016). *Cassia grandis* (*C. grandis*) belongs to the *fabaceae* family, and it is commonly known by different names such as carague, cañandonga, and carao in Spanish (Lagarto & Guerra, 2005; Carvalho, 2006; Ramos et al., 2014). Its natural inhabitant is low elevation open field forest, and it is found in South American countries, such as Brazil, Colombia, Mexico, and Cuba. However, it is considered endemic to Central America. It is planted for its ornamental value, and it is traditionally used as a sweetener and medicine (called Jarabe or Miel) by indigenous groups (Fig. 1) (Lafourcade et al., 2014; Bomfim et al., 2018).

In Honduras, it has been reported in the departments of Olancho, Choluteca, Francisco Morazán, Gracias a Dios, and Comayagua (House & Lagos-Witte, 1989; Marcía Fuentes et al., 2020a; Tropicos.org, 2021).

The *C. grandis* is a tree that grows up to 30 m in height with a diameter that can reach 100 cm. It has a cylindrical stem with sympodial growth, and its bark is smooth and brownish-gray with a thickness of 20–30 mm. Its leaves are composed of 15–20 pairs of petioles having a rounded base and a green color that could be 2–5 cm long

J. Marcía-Fuentes (✉)
Technology Sciences Faculty, Universidad Nacional de Agricultura, Catacamas, Honduras

R. Santos-Aleman
School of Nutrition and Food Sciences, Louisiana State University, Baton Rouge, LA, USA

I. Borrás-Linares
Functional Food Research and Development Centre (CIDAF), Granada, Spain

J. L. Sánchez
Functional Food Research and Development Centre (CIDAF), Granada, Spain

Department of Food Science and Nutrition, University of Granada, Granada, Spain

N. R. Maddela, L. C. García (eds.), *Innovations in Biotechnology for a Sustainable Future*, https://doi.org/10.1007/978-3-030-80108-3_19

Fig. 1 Carao fruit (*C. grandis*)

Fig. 2 *C. grandis* tree

and 1–1.5 cm wide. The inflorescences have 15 or more flowers of intense pink color. The seeds are between 2 and 4 cm long and 1.5–2.5 cm wide, and the pod that contains them can be up to 75 cm in length (Fig. 2) (Lafourcade et al., 2014; Ramos et al., 2014; Marcía Fuentes et al., 2020b).

By its seeds, it has vegetative reproduction, and there are no notable reports regarding its pest and plant pathology. It can grow in dry soils with moderate to low humidity and slightly acidic pH conditions. It is considered a half-hardy perennial tree since during the year it drops some of its leaves, especially during the winter. The fruit with woody characteristics contains between 60 and 80 seeds (Marcía Fuentes et al., 2020b), and the immature and mature pods are green and black, respectively, remaining on the tree for approximately 1 year. This tree belonging to the *Cassia* genus group has the characteristic that its seeds are divided transversely and a rounded, flattened, tan-colored seed is found in each partition, which is surrounded by a dark brown, viscous, bittersweet pulp with a potent odor (ICRAF, 2014).

Regarding its usage, *C. grandis* mainly has been used for ornamental purposes throughout the world, and it has also been utilized for the revegetation of areas that have been periodically flooded. In agriculture, it has been applied for animal feeding especially for cattle nutrition (ICRAF, 2014; PIER, 2014).

C. grandis tree is utilized as firewood and charcoal, and its wood is used in carpentry principally towards building beams to be used in construction operations. Regarding its seed, chewing gums can be obtained from the fruit and it can be applied as a binding agent in pharmaceutical and food manufacturing (ICRAF, 2014).

Also, the fruit can be used in many medicinal applications (ICRAF, 2014; Fuentes et al., 2021). Its functionality can be associated with its composition in bioactive compounds, which can potentially be implemented in pharmacological, nutritional, and medical applications.

1.1 Cassia grandis*: Its Pharmacological, Nutritional, and Medicinal Significance*

Medicinal plants are generally known for their several health benefits that they provide in the human organism, and one of them is their supplement of phytochemicals and antioxidants in significant levels in the body (Rojas et al., 2015; Rao & Chatterjee, 2016; Fernández et al., 2021; Marcía Fuentes et al., 2020a). Currently, the pharmaceutical industry made more than 50% of modern clinical drugs from natural products due to their high solubility and absorption not causing adverse health effects (Usman et al., 2014; Kumar et al., 2016).

In India, Kotipalli et al. (2017) conducted studies to determine the physicochemical characteristics and phytoconstituents present in *C. grandis* extracts. The results indicated considerable amounts of phenols, flavonoids, and alkaloids highlighting the functional interest of this plant as a great antioxidant source. Besides, the presence of total ash was about 3.7% reflecting significant levels of minerals (Kotipalli et al., 2017).

In Egypt, Hegazi and Hashim (2016) determined the bioactive molecules and the antioxidant activity from the *C. grandis* leaf. These results showed the presence of a new type of phenolic compound named Grandisina, and was also found meaningful amounts of Quercetin, Flavonol, and Kaempferol. The antioxidant capacity of this plant's leaf was also examined using the DPPH method (2.2-diphenyl-1-picrylhydrazyl), obtaining an IC50 value greater than 60 g/mL (Hegazi & Hashim, 2016).

In Brazil, Albuquerque et al. (2017) reported some observations on the immobilization of bioactive compounds in *Cassia grandis* galactomannan-based films. The results indicated that galactomannan films produce immobilization of lactoferrin, peptides, and phytosterols, so for their gelling capacity, they could be used in the food biotechnology industry for the development of new products (Albuquerque et al., 2017).

In Honduras, Marcía Fuentes et al. (2020b) fortified egg powder with *C. grandis* flour. The results showed that the use of this flour can improve sensory attributes especially acceptance. Also, it could be used as a potential functional food aimed at people with special feeding regimens such as high-performance and malnourished athletes (Marcía Fuentes et al., 2020b). Other research carried out by Marcía Fuentes et al. (2020a) determined that in the pulp, bark, and seed there is high amounts of carotenoids, precursors of vitamin A, which promotes the formation and preservation of teeth, bone tissues, mucous membranes, and skin. It also contains minerals such as calcium, magnesium, potassium, sodium, copper, zinc, manganese, phosphorus, sulfur, and iron in high levels, which are relevant in the treatment of ferropenic anemia. In addition, the proximal analysis showed a high concentration of proteins especially in the seed that could be up to 10.11% (Marcía Fuentes et al., 2020a).

Some studies endorse the nutraceutical potential of *C. grandis*. Romero et al. (2018) and Prada et al. (2020) reveal that *C. grandis* seed can potentially be used as an antidiabetic agent because of its trypsin inhibitory effect. In vivo studies reveal that the pulp of the fruit showed a reduction in blood glucose levels (Lodha et al., 2010; Prada et al., 2019, 2020).

Its nanodispersion exerts a hypoglycemic effect with significant inhibition of alpha-glucosidase and pancreatic lipase (Prada et al., 2019). Besides, the fruit has demonstrated its antianemic potential from in vivo studies due to its high inorganic iron content and its high bioavailability (Tillán et al., 2004; Lafourcade et al., 2014; Prada et al., 2016).

Other research determined that the species *Cassia* contains high levels of iron and other micronutrients and macronutrients with potential pharmaceutical interest, as well as high antioxidant activity (Deshpande & Bhalsing, 2011; Kabila et al., 2017). Fuentes et al. (2021) identified and quantified the phenolic molecules of the seed, referring to the family of chemical structures such as hydroxybenzoic acids, flavonoids (flavonols, flavanols, flavanones, and proanthocyanidins), and flavone derivatives, promoting this fruit as a potential functional and nutraceutical food.

1.2 Bioactive Compounds and Its Properties

Bioactive compounds are minor components of food that affect the human organism as a biological system, organ, tissues, or cells since they are applied in medical and pharmacological applications (Kris-Etherton et al., 2002). Cárdenas et al. (2015) state that the consumption of these phytochemicals such as phytosterols, fatty acids, carotenoids, peptides, non-pro-vitamin A, and polyphenols can treat and prevent diseases such as cancer, diabetes, neurodegenerative disorders, and cardiovascular illnesses.

The origin of the development of these diseases may be due, among many causes, to the disintegration of the cell membrane induced by reactive oxygen species (ROS), which lead to damage to membrane proteins and it can even generate DNA mutation (Ravishankar et al., 2013; Xiao et al., 2014).

Due to their antioxidant capacity and the high levels of some vitamins such as vitamin A (retinol), many studies have focused on the potential effects of certain bioactive compounds against the development of different types of cancer which have shown inhibition of cell proliferation in kidney malignant cells (Wegert et al., 2011) and it can reduce the growth of endometrial cancerous cells (Chen et al., 2011). Likewise, vitamin D in its form of D2 (ergocalciferol) and D3 (cholecalciferol) has positive effects on prostate carcinogenic cells (Alimirah et al., 2011) and ovarian cancer precursors (Shen et al., 2011). Also, some minerals such as zinc act favorably against colon cancerous cells (Cohen et al., 2012), and polyphenols such as isoflavones have demonstrated their efficacy against prostate malignant cells (Li et al., 2012) and proanthocyanidins upon head and neck carcinogenic cells (Sun et al., 2012).

In addition, a diet rich in foods with functional properties improves the immune system, prevents obesity, and assists in preserving various organs such as the brain, heart, and pancreas (Barberá Mateos, 2008).

1.2.1 Classification of Bioactive Compounds

Bioactive compounds in food can be classified according to their chemical structure (Olmedilla & Granado, 2007):

(a) Inorganic bioactive compounds:

Minerals such as calcium, selenium, zinc, potassium, or copper.

(b) Organic bioactive compounds:

Isoprenoids (terpenoids) like carotenoids, saponins, tocotrienols, tocopherols, and terpenes.

Carbohydrates and derivatives of the like kind as ascorbic acid, oligosaccharides, and polysaccharides such as starch.

Proteins, amino acids, and derivatives naming isothiocyanates, capsaicinoids, allyl-S compounds, indoles, and folates.

Lipids including unsaturated fatty acids (an example of which is omega 3) and phytosterols.

Phenolic compounds such as phenolic acids, tannins, lignans, flavonoids, among others.

1.2.2 Phenolic Compounds

Phenolic compounds are secondary metabolites that are generated via the phenylpropanoid metabolization within the shikimic acid of plants and pentose phosphate. These compounds contain hydroxylated aromatic rings, which are made up of a functional hydroxyl group bonded to the carbon chain in a benzene molecule, and they are known for their antioxidant properties (Peñarrieta et al., 2014).

They can contribute to the sensory attributes of food products especially in beverages of plant origin giving astringency flavors. Their intake is linked with various beneficial effects considering them as important antioxidants in the diet, and they can commonly be found in vegetables, cereals, fruits, and roots (Gómez, 2010; Peñarrieta et al., 2014).

Gómez (2010) classifies phenolic compounds based on different factors, according to their basic chemical structure:

Simple Phenols
In this group, these compounds are characterized by having at least one hydroxyl group linked to an aromatic ring. Phenols, benzoquinones, and phenolic acids are the most common ones, and also can include benzoic, phenylacetic, and cinnamic acids. Additionally, it covers acetophenones, phenylpropenes, coumarins, isocoumarins among which hold chromones and naphthoquinones.

Polyphenols
These compounds are known for having multiple phenol units. Polyphenols include xanthones, stilbenes, anthraquinones, flavonoids, lignans, neolignans, tannins, and lignans. The most abundant in the diet are phenolic acids and flavonoids (Gómez, 2010).

Another common classification is to differentiate them between flavonoids and non-flavonoids (Gonçalves et al., 2013; Valencia-Avilés et al., 2017):

Flavonoids are the most plentiful group of phytochemicals which include flavonols, flavones, flavan-3-ols, isoflavones, flavanones, dihydroflavonols, anthocyanidins, and chalcones.

The non-flavonoid phenolic components in food include hydroxybenzoic acids, hydroxycinnamic acids, volatile polyphenols, stilbenes and lignans, and coumarins.

They can also be classified by the degree of solubility. The water-soluble compounds incorporate phenolic acids, phenylpropanoids, flavonoids, and quinones, while those insoluble in water include tannins, lignins, and hydroxycinnamic acids, which commonly are bind to the cell wall of plant cells (Haminiuk et al., 2012).

1.2.3 Bioactive Properties

The bioactive properties that have been associated with phenolic compounds are very diverse including anticancerous, anti-inflammatory, antihypertensive, and estrogenic characteristics. In addition, they can exert potentially beneficial effects against cardiovascular diseases (Muñoz Jáuregui & Ramos Escudero, 2007) and can provide important advantages regarding inhibition of antiallergic, antimicrobial, and antineoplastic activity (Finley et al., 2011; Zapata et al., 2013).

Not only phenolic compounds may have many health benefits, but also they are mostly known for their antioxidant properties, which lie in that they can sequester free radicals, donate hydrogen molecules, scavenge superoxide molecules, and chelate transition metals (Muñoz Jáuregui & Ramos Escudero, 2007).

Concerning anticancerous properties, the consumption of foods with a high amount of flavonoids such as quercetin, kaempferol, and luteolin reduces the risk of contracting various types of cancer, although the mechanisms of these protective effects are not yet fully understood and are being studied (Muñoz Jáuregui & Ramos Escudero, 2007). In addition, recent studies show that resveratrol, antifungal stilbenes produced by plants mainly in response to the stress of fungal infections (Atanacković et al., 2012), has properties to potentially prevent degenerative diseases and breast cancer (Gresele et al., 2011; Pandey et al., 2011).

1.3 Extraction and Characterization Techniques of Phenolic Compounds

1.3.1 General Extraction and Characterization Techniques of Phenolic Compounds

In the last decade, different methodologies have been applied to evaluate bioactive compounds in extraction systems. Advanced extraction systems that have been implemented include novel methods such as ultrasound, supercritical fluids, pressurized fluids, or microwave-assisted extraction (Nasti et al., 2018; Leyva-Jiménez et al., 2019; Fuentes et al., 2021). The advantages of these new techniques compared to conventional extraction methods are that they increase extraction performance, and have more efficient, speed up the process, the amount of solvent required, and more eco-friendly.

However, recent studies on the extraction of bioactive molecules from the bean fruit determined that the pressurized fluid method is the most appropriate for obtaining high recovery yields for this type of molecule (Fuentes et al., 2021). Therefore, in this chapter, the methodology of use for this technique will be discussed.

1.3.1.1 Pressurized Fluid Extraction (PFE)

Pressurized fluid extraction (PLE) is a very useful technique in the case of plants, facilitating and improving the extraction of polar compounds in this type of matrices (Mendiola et al., 2007; Fuentes et al., 2021). It is based on extraction, generally solid material, in which the liquid solvent is at a temperature above its boiling point. However, it managed to keep it in a liquid state due to high pressures. This combination of high temperatures and pressures increases the speed and efficiency of extraction, improves performance, and reduces extraction time.

Specifically, increasing the extraction temperature improves the solubility and transfer of the compounds of interest to the extraction solvent. It also reduces the viscosity and surface tension of the solvent allowing it to more easily reach all areas of the matrix and thus improving the extraction performance (Fanali et al., 2018).

The extraction solvent generally used for this type of polar compound is water and its hydroalcoholic mixtures, mainly ethanol. Ethanol has the advantage of being economically affordable compared to other solvents used in conventional methods. In addition, it is not hazardous, and it is included in the list of GRAS solvents (generally recognized as safe). It does not pollute the environment, so its use means that the application process itself can be considered a green technology (Zhang et al., 2019).

The apparatus system of a pressurized fluid extractor consists of a solvent tank, a high-pressure pump, extraction cells, an oven, and valves. The solvent flows through the cell and carries out the extraction at the selected temperature and is deposited in a collection vial. After extraction, the system is purged with nitrogen gas. Figures 3 and 4 show an outline of this type of extractor.

1.4 Characterization of Phenolic Compounds

High-performance liquid chromatography (HPLC) is a separation technique, and the determination of analytes is performed after the separation of the different compounds of the sample. The stationary phase is fixed to a small diameter column, while the mobile phase is passed by pressure. Both phases are immiscible, so sample components that have a greater affinity for the stationary phase move slowly with the flow of the mobile phase while components with less affinity for the stationary phase move quickly. As a result of the different mobility, the sample components are separated, and the different compounds can be collected in fractions for quantitative or qualitative analysis. It is important to note that this separation process results in a dilution of the analytes.

As shown in Fig. 5, the basic components of an HPLC are mobile phase tank pumps, a sample injection system, the column that contains the stationary phase, the detector, and a data processing system that generates the chromatogram.

In the case of gradient elution, the mobile phase is a mixture of different solvents of various polarity whose ratio composition changes throughout the

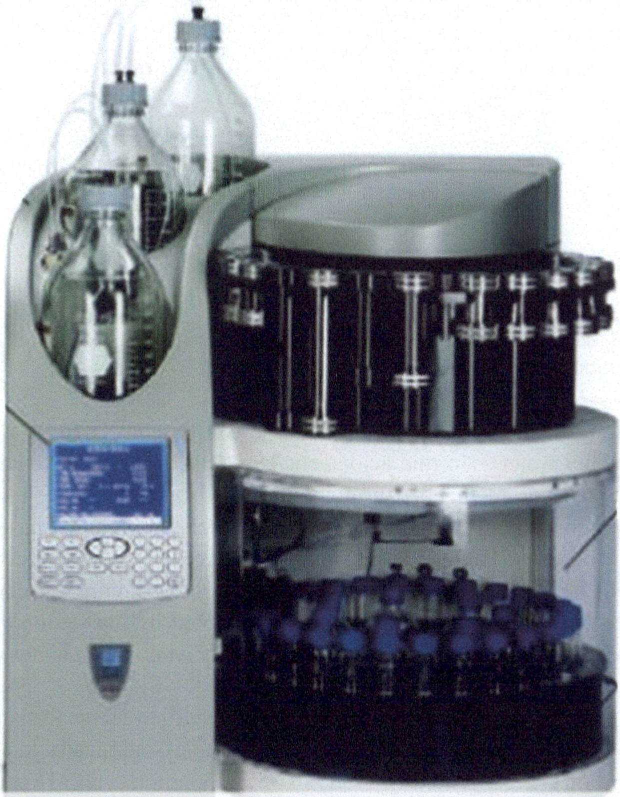

Fig. 3 Extraction system PLE. (https://www.ujaen.es/servicios/scai/recursos/pf07-sistema-de-extraccion-con-liquidos-presurizados)

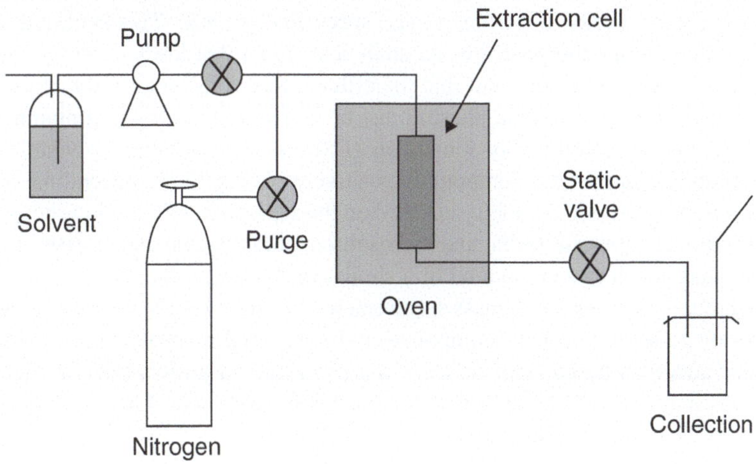

Fig. 4 Parts of an extractor PLE. (https://www.directindustry.es/prod/dionex/product-28284-1124221.html)

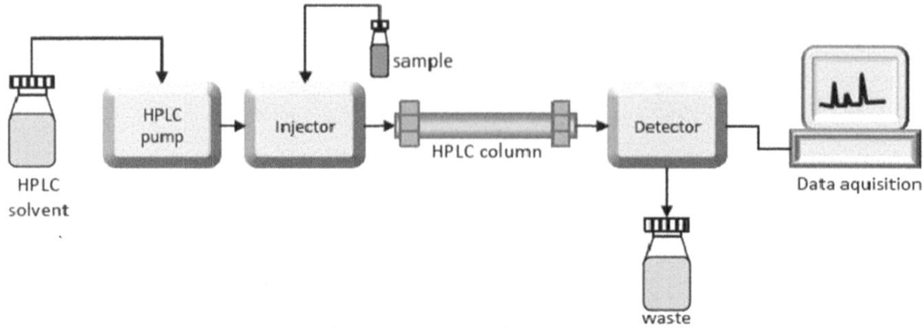

Fig. 5 HPLC equipment. (http://laboratoryinfo.com/wp-content/uploads/2015/07/High-performance-liquid-chromatography-hplc.jpg)

chromatographic elution where a mobile phase can be composed of a more polar eluent A (usually water + formic acid/acetic/etc.) or a more apolar eluent B (commonly acetonitrile or methanol) throughout the process. The gradients used for phenolic compounds start with a small percentage of the organic phase (generally acetonitrile or methanol) and end with high percentages of these solvents. The acids used must not be too strong to prevent damage to the column, and they must be adjusted to the pH within the range of values accepted by the stationary phase, as well as allowing a good ionization of the compounds when the analytes detection is carried out by mass spectrometry detector.

The stationary phases can be solid and/or liquid, and, in the analysis of phenolic compounds, the most used are those that have fillers of chemically modified silica with hydrocarbon chains, generally n-octyl (C-8) or n-octadecyl (C-18), with particle sizes ranging between 1.8 and 5 μm. Taking into account the polarities of the mobile phase and the stationary phase, the chromatographic separations of these compounds are generally carried out in reverse or reverse phase.

Finally, the identification and quantification of the phenolic compounds separated by HPLC can be carried out using a mass spectrometer (MS). This is placed at the outlet of the column and generates an analytical signal that leads to the chromatogram. It is important to point out that the effluent that comes out of the chromatographic column (liquid mobile phase) must be converted into a gas containing the charged analytes. Electrospray ionization (ESI) is very suitable to achieve this purpose, and it ionizes the compounds positively or negatively depending on the selected mode. Mass spectrometry is based on the separation of these ions according to their mass/charge ratio (m/z), and their subsequent detection can be registered in software and match to a compound of a similar mass/charge ratio.

The basic components of a mass spectrometer are an ionization chamber in which the ions are generated from the compounds to be analyzed, an analyzer that separates and generates ions based on their mass/charge ratio, and an ion detector that produces an amplified electrical signal for each ion generated. Thanks to a signal processor, the chromatogram can be obtained (Fig. 6).

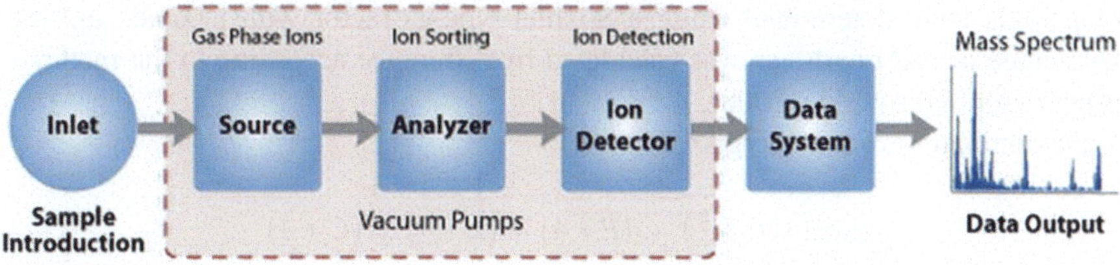

Fig. 6 Components of mass spectrometer. (https://www.intechopen.com/books/spectroscopic-analyses-developments-and-applications/application-of-mass-spectroscopy-in-pharmaceutical-and-biomedical-analysis)

1.5 Proximal, Mineralogical, and Bioactive Molecules of Different Parts of Carao (C. grandis)*: Pulps, Shell, Seeds, and Whole Fruits*

This section includes a review of the results previously reported in the literature by Marcía Fuentes et al. (2020a) and Fuentes et al. (2021) about the determination of the proximal, mineralogical, and bioactive molecules characterization of the carao fruit (*C. grandis*). It has to be taken into account that a careful selection of articles has been followed. With the search criteria established, not too many studies could be included in this section. However, the selected works included the main objectives of this chapter, and the quality of the selection provides great reliability in the results presented.

1.5.1 Samples

The investigations of Marcía Fuentes et al. (2020a) and Fuentes et al. (2021) were carried out using fresh samples of fruit (*C. grandis*) from Guapinol Biological Reserve, Marcovia, Department of Choluteca (Honduras). The seeds were manually separated from the fruit. After that, they were dried in an air circulation oven and grounded for phytochemical characterization and extraction of bioactive molecules (Marcía Fuentes et al., 2020a; Fuentes et al., 2021). In both works, the authors used analytically graded reagents.

1.6 Proximal Analysis of Pulps, Shell, Seeds, and Whole Fruits

The proximal analysis and total calories (Eq. 1) in the different parts of the *C. grandis* fruit were moisture content analyzed in an oven at 105 °C for 3 h and total ash content determined in a muffle oven at 600 °C for 30 min. Proteins were determined by the Kjeldahl distillation method with previous sulfuric digestion. The total lipids were determined using a Soxhlet-type extractor with hexane, and the percentage of carbohydrates was calculated by difference according to the methodology described by IAL (2008).

Determination of the energy value

$$\left(\text{kcal}\, 100\, \text{g}^{-1}\right) = (P * 4) + (L * 9) + (C * 4) \tag{1}$$

where P = protein value (%), L = lipid value (%), C = carbohydrate value (%), 4 = kcal conversion factor determined in the calorimetric bomb for proteins and carbohydrates, and 9 = conversion factor kcal determined in the lipid bomb calorimeter.

1.7 Mineral Analysis of Pulps, Shell, Seeds, and Whole Fruits

For the determination of minerals, the samples were first subjected to perchloric nitric digestion (3: 1), determining the following elements: Ca ($\lambda = 422.70$ nm), Mg ($\lambda = 285.21$ nm), Fe ($\lambda = 248.33$ nm), Zn ($\lambda = 213.80$ nm), Mn ($\lambda = 279.48$ nm), Cu ($\lambda = 324.75$ nm) by Atomic Flame Absorption Spectrometry (FAAS), Na and K by EAS Atomic Emission Spectrometry and UV visible Molecular Spectrophotometry, phosphorus ($\lambda = 660$ nm), and sulfur ($\lambda = 420$ nm) according to the methodology described by EMBRAPA (2009) and Montero et al. (2020).

1.8 Total Phenolic Content and Antioxidant Activity of Pulps, Shell, Seeds, and Whole Fruits

Total phenolic compounds were determined using the Folin Ciocalteu method with the formation of a blue complex using gallic acid as a reference standard, and absorbance readings were performed in a UV visible spectrophotometer at 765 nm according to the methodology described by Wolfe et al. (2003). To determine the antioxidant activity, the 1,1-diphenyl-2-picrylhydrogen radical (DDPH) method and the iron reduction technique were used. In the first method, the absorbance reading was carried out at 515 nm (Miranda & Fraga, 2006); the calibration was carried out from dilutions of a 60 mM DPPH solution in methanol. The second method to determine the antioxidant activity was the method based on the reduction of Fe^{3+} to Fe^{2+} according to the methodology proposed by Sanchez-Moreno et al. (1998), with the readings in UV visible Molecular Absorption Spectrophotometry at 690 nm.

1.9 Total Carotenoids of Pulps, Shell, Seeds, and Whole Fruits

For the quantification of total carotenoids, 1 g of sample was added to 18 ml of acetone. The recording was performed on a molecular absorption spectrophotometer at 470 nm, 661 nm, and 664 nm, respectively, and was calculated using formula 2 described by Lichtenthaler and Buschmann (2001).

Determination of carotenoids

$$\text{Carotenoids C}\left(\text{mg mL}^{-1}\right) = (1000\,\text{A470} - 1.90\,\text{Ca} - 63.14\,\text{Cb})/214 \quad (2)$$

where:

Ca (mg mL^{-1}) = 11.24 A661–2.04 A644.
Cb (mg mL^{-1}) = 20.13 A664–4.19 A661.

1.10 Extraction and Characterization of Phenolic Compounds from Carao Seeds

Fuentes et al. (2021) developed a new PLE green extraction method for recovering phenolic compounds from seeds. To achieve this goal, these authors used an Accelerated Solvent Extraction system. Extractions were performed in static cycles using different extraction solvents ratios (water and ethanol), process temperatures, and extraction time. All extractions were carried out at constant pressure (11 MPa).

For each extraction, seed sample was mixed with sand in order to increase the contact between sample and extraction solvents mixture. All runs were developed in ASE™ 350 extractor. This equipment used the Chromeleon 7 console to set each of the extraction conditions.

1.11 Experimental Design Applied for Recovering Phenolic Compounds from Carao Seeds

With the purpose of optimizing the phenolic compounds extraction, these authors applied a response surface methodology (RSM) (Fuentes et al., 2021). The RSM involved three aspects: design, model, and optimization. The authors selected a central composite design 2^3 (CCD). This model is broadly used due to its high flexibility. Independent variables are the variables set by researchers during the experiments for the evaluation of their effects on the response variable. The three factors established as independent variables were the extraction temperature (range 40–180 °C), percentage of ethanol (range 15–85%), and static cycle times

Table 1 Composite central model 2^3

Experimental runs	Independent factor 1: temperature (°C)	Independent factor 1: %ethanol	Independent factor 1: static cycle time (min)
E1	40	15	20
E2	40	85	5
E3	110	5	12.5
E4	110	50	22
E5	40	15	5
E6	20	50	12.5
E7	110	50	3
E8	110	50	12.5
E9	110	50	12.5
E10	40	85	20
E11	180	85	5
E12	180	85	20
E13	110	95	12.5
E14	200	50	12.5

Values of independent factors. Adapted from Fuentes et al., (2021)

(5–20 min). For each variable, it was run at two levels, maximum and minimum. The axial points located below and above the maximum and minimum values allowed the determination of the curvature of the response surface for each factor.

The experimental design generated by the software was based on a total of 14 experiments that were performed in a random order (Table 1). This optimization was focused on maximizing, first, a target value of a single response variable, yield or phenolic compounds, and second, optimization of several responses (multiple response optimization).

1.11.1 Characterization of Phenolic Compounds from Carao Seeds

Fuentes et al. (2021) analyzed the Carao seed extracts by reverse phase liquid chromatography which allowed to improve separation between very similar compounds. As it has previously been described reversed-phase partition chromatography uses relatively non-polar stationary phase and a polar mobile phase. In this study, chromatographic separation was carried out on a C18 1.8 μm 150 × 4.6 mm analytical column. The mobile phase used was water with 0.1% formic acid as eluent A and acetonitrile as eluent B.

The detection of the compounds was carried out by mass spectrometry in negative ionization mode considering a mass range of 50–1000 *m*/*z*. Authors identified and quantified analytes present in the Carao seed extracts thanks to the analytical platform. Indeed, coupling liquid chromatography with mass spectrometry further offers a potent analytical alternative, which can be applied in characterizing food products. The mass spectrometer was externally calibrated before identification of compounds. The mass data of the molecular ions were processed through DataAnalysis 4.0 software (Bruker Daltonics), which provided a list of possible compounds using the Generate-Molecular Formula Editor. The latter uses a CHNO algorithm providing standard 202 functionalities such as minimum/maximum elemental range, electron 203 configuration, and ring-plus double bonds equivalent.

In addition, these authors also used available commercial standards for both identification and quantitation purposes: gallic acid, catechin, epicatechin, epigallocatechin-gallate, quercetin-3-glucoside, and kaempferol-3-rutinoside.

1.12 Bioactive Molecules Content of Carao (C. grandis)

1.12.1 Proximal Analysis

Table 2 shows the values of the nutritional composition and the total energy for the different parts of the carao fruit (Marcía Fuentes et al., 2020a).

The highest moisture content values for the different parts of the carao fruit are found in the pulp, with 26.72%, and the part that showed the least amount of moisture content was the rind with only 8.19%. The ash content in this fruit showed that the seeds have the highest mineral value with 3.78%. Among the parameters that contribute to the caloric content of the fruit are lipids, carbohydrates, and proteins; the amount of lipids is very low, reaching 1.17% for the seeds. The amount of protein is higher for the seeds with 10.11%. Carbohydrates, including fibers, are the main component of carao, and the peel has the highest percentage of carbohydrates with 88.01%.

Given the high percentage of carbohydrates found in legumes especially in *C. grandis*, this can be used as an alternative food source. The carbohydrate content in this fruit is higher than that found in some tropical legumes such as *Inga*, whose percentage reaches 27.62% (Mendoza et al., 2016). As for the caloric content, the part of the fruit that has an important contribution is the shell with 358.26 kcal 100 g^{-1}. The daily energy recommendations for legumes are around 2000 kcal 100 g^{-1} following the specifications of the European Economic Community 90/496/EEC of 24 September 1990.

Table 2 Proximal composition and caloric content of *Cassia grandis* (Carao)

Fruit part	Moisture (%)	Ash (%)	Lipid (%)	Carbohydrates (%)	Proteins (%)	Calories (kcal 100 g^{-1})
Pulp	26.72a	2.80b	0.21b	61.93c	8.34b	282.97c
Cortex	8.19c	2.42c	0.14c	88.01a	1.24c	358.26a
Seed	9.58b	3.74a	1.17a	75.40b	10.11a	352.57b
Complete fruit	17.31	3.14	0.74	71.4	7.41	321.9

[*]Means with different letters in the same column indicate statistical differences ($P \leq 0.05$) with Tukey's test

1.13 *Mineral Analysis of the* C. Grandis *Fruit*

Table 3 shows the values of the mineral composition for the different parts of the *C. grandis* fruit (Marcía Fuentes et al., 2020a).

Table 2 shows the values of the analysis of the mineral for the different parts, as well as for the whole fruit. Among macro minerals, magnesium stands out, being most concentrated in the seed with levels of 18.27 ± 0.14 mg 100 g^{-1}. This element is of great importance to the human body as it is involved in numerous metabolic processes (Wolf & Cittadini, 2003). Intake recommendations for this element according to DRI (2011) are 420 mg per day for men and 320 mg per day for women. Another essential element within macro minerals is calcium which, like magnesium, is at higher concentrations in the seed with 7.31 ± 0.21 mg 100 g^{-1}. This element is essential for the preservation of bones and teeth (De Franco & Martini, 2014) having intake recommendations of 1000 mg per day for both sexes (DRI, 2011). Sodium and potassium are also two important elements for maintaining electrolyte balance in cells (Cuppari & Bazanelli, 2010). In the seed, potassium is found in higher amounts than sodium, reaching values of 8.23 ± 0.18 mg 100 g^{-1}, and, on the other hand, sodium is in higher concentrations in the pulp with 2.56 ± 0.13 100 mg g^{-1}.

Daily intake recommendations for this element in adulthood are 8 mg per day for men, 18 mg per day for women aged 19–50 years, and 8 mg per day for women above 50 years (DRI, 2011). For manganese, it was found in high concentrations in seeds with 0.51 ± 0.12 mg 100 g^{-1}. Manganese is the second micronutrient after the iron of interest to plants (Malavolta, 2006), but at the same time plays an opposing role towards iron in the body, since, in the diet, excess manganese can cause less absorption of iron, which also causes anemia and can affect the central nervous system (Roels et al., 1997). Zinc has different physiological functions in the cell, such as liver mobilization of vitamin A, sexual maturation, fertility and reproduction,

Table 3 Mineral composition, in the different parts of the *C. grandis* fruit

Concentration (mg 100 g^{-1})	Seed	Pulp	Cortex	Complete fruit
Ca	7.31 ± 0.21a	5.67 ± 0.12b	4.67 ± 0.17c	6.21 ± 0.12
Mg	18.27 ± 0.14a	14.31 ± 0.12b	11.21 ± 0.07c	15.46 ± 0.07
K	8.23 ± 0.18a	3.45 ± 0.07b	2.43 ± 0.14c	4.31 ± 0.08
Na	0.85 ± 0.07c	2.56 ± 0.13a	1.31 ± 0.07b	1.47 ± 0.31
Fe	1.71 ± 0.23a	1.54 ± 0.12b	0.81 ± 0.07c	1.14 ± 0.21
Cu	0.21 ± 0.08b	0.14 ± 0.03c	0.71 ± 0.12a	0.44 ± 0.11
Zn	0.46 ± 0.09a	0.34 ± 0.11b	0.21 ± 0.07c	0.27 ± 0.13
Mn	0.51 ± 0.12a	0.25 ± 0.07b	0.21 ± 0.07c	0.38 ± 0.08
P	0.47 ± 0.07a	0.21 ± 0.02b	0.14 ± 0.07c	0.26 ± 0.08
S	0.08 ± 0.01c	0.17 ± 0.04a	0.11 ± 0.03b	0.04 ± 0.01

*Means with different letters in the same column indicate statistical differences ($P \leq 0.05$) with Tukey's test

and phagocytic, cellular, and humoral immune functions (Roels et al., 1997). The concentration in *C. grandis* seeds is 0.46 ± 0.09 mg 100 g^{-1}. Copper is another essential nutrient not synthesized by the body, found in fruits at concentrations between 0.02 and 0.66 mg 100 g^{-1} according to Amancio (2017).

Copper concentrations found in the *C. grandis* were very low. The bark was the one with the highest concentration with 0.71 ± 0.12 mg 100 g^{-1}. Two other elements analyzed in this fruit were phosphorus and sulfur. The highest concentration of phosphorus was found in the seed with 0.47 ± 0.07 mg 100 g^{-1} which appears on ATP energy metabolism involved in carbohydrate metabolism and is present at the same time in the synthesis of phosphate sugars, nucleic acids, and coenzymes (Epstein & Bloom, 2006). Sulfur was found in low concentrations in the fruit with 0.17 ± 0.04 mg 100 g^{-1}, being an element that is also part of the structure of proteins, and it is found in the body in concentrations of up to 140 g (Lisboa, 2015).

1.14 Phenolic Compounds, Antioxidant Activity, and Total Carotenoids

Table 4 shows the values of phenolic compounds, antioxidant activity, and total carotenoids in the different parts of *C. grandis* fruit (Marcía Fuentes et al., 2020a; Maldonado et al., 2020; Montero et al., 2020).

The total phenolic compounds determined in the different parts of the fruit varied between 2.3 ± 0.1 mg EAG 100 g^{-1} (shell) and 11.1 ± 0.3 mg EAG 100 g^{-1} (seeds). Compared to other legumes such as *Vicia faba*, these values are within those determined by Valente et al. (2018), reaching values of 13 ± 0.1 mg EAG 100 g^{-1}. Antioxidant activity was carried out using two methods (the DPPH technique and the iron reduction method). The seed is the one with the highest antioxidant activity with values of 7.31 ± 0.11 g^{-1} by the DPPH method and 0.34 ± 0.04 mg g^{-1} by the iron reduction method. Godevac et al. (2008) studied the antioxidant activity of nine

Table 4 Phenolic compounds, antioxidant activity, and total carotenoids in different parts of the face

Fruit part	Total phenolic compounds (mg EAG 100 g^{-1})	Antioxidant activity DPPH (μg g^{-1})	 Iron reduction (mg g^{-1})	Totals carotenoids (μg mL^{-1})
Pulp	5.6 ± 0.2b	6.07 ± 0.02b	0.21 ± 0.01b	4.12 ± 0.11[a]
Cortex	2.3 ± 0.1c	5.12 ± 0.04c	0.18 ± 0.02c	2.21 ± 0.07c
Seed	11.1 ± 0.3a	7.31 ± 0.11a	0.41 ± 0.02a	3.76 ± 0.03b
Whole fruit	6.3 ± 0.1	6.48 ± 0.07	0.34 ± 0.04	2.56 ± 0.04

*Means with different letters in the same column indicate statistical differences ($P \leq 0.05$) with Tukey's test

species of *Fabaceae*, obtaining values higher than those obtained for carao fruit. Other researchers, such as Pinela et al. (2011), studied antioxidant activity in the *Genisteae* also belonging to the *Fabaceae* family, obtaining values from 0.15 to 0.50 mg mL^{-1}, being lower than those reported in Table 4. In addition, total carotenoids in the carao fruit give this fruit additional biotechnological attributes as they are precursors of vitamin A and possess antioxidant, anti-inflammatory, and anti-tumor properties (Rehman et al., 2020). The concentrations of this group of substances in *C. grandis* ranged from 2.21 ± to 0.07 g mL^{-1} for the bark (lowest values) and 4.12 ± 0.11 g mL^{-1} for the pulp where the most carotenoids are concentrated.

1.15 Identification of Phenolic Compounds and Other Polar Compounds Present in Carao Seeds

Table 5 shows the main phenolic compounds reported in literature for Carao seeds in agreement with Fuentes et al. (2021).

Two family groups were mostly identified (hydroxybenzoic acids and flavonoids). With respect to hydroxybenzoic acids two chemical compounds were reported: galloyl glucoside and its derivative. Authors determined that both of these phenolic compounds were previously described in plants belonging to the same group (*Magnoliophyta*) and class (*magnoliopsida*) as *Cassia grandis* (*Sclerocarya birrea,* Jiménez-Sánchez et al., 2015).

However, the most representative chemical group was flavonoids. These authors supported these results according to the scientific literature. Indeed, different flavonoids (i.e., flavonols, flavanols, flavanones) have been widely described in other *Cassia* families (Sobeh et al., 2018).

Concerning flavanols, the presence of astilbin, quercetin-3-glucoside, quercetin-rhamnoside, kaempferol-rhamnoside and their isomers were characterized according to the MS data, commercial standards as well as scientific literature (*Cassia bakeriana,* Da Costa Silva et al., 2019; *Cassia abbreviate*, Sobeh et al., 2018).

The flavanol subclass included theaflavin (Sobeh et al., 2018), catechin and (epi)-catechin and (epi)-afzelechin or its isomers (Sobeh et al., 2018). In addition, they also characterized diflavanoids, its isomer, or chemical compounds derived from flavanols (Ucar et al., 2013).

Regarding flavanones, these authors carried out a tentative identification of the following compounds: pinocembrin-7-neohesperidoside, and pinocembrin-7-rutinoside (Xu et al., 2011; Wang et al., 2011; Demirkiran et al., 2011; Meng et al., 2016). These compounds have also been reported on the species *Litchi chinensis, Euphorbia decipiens*, and *Ziziphora clinopodioides* that belong to the same division and class as *Cassia grandis* (Xu et al., 2011; Wang et al., 2011; Demirkiran et al., 2011; Meng et al., 2016).

Table 5 Identification of bioactive phenolic compounds from *Cassia grandis* seed

Phenolic compounds	Chemical structure
Hydroxybenzoic acids	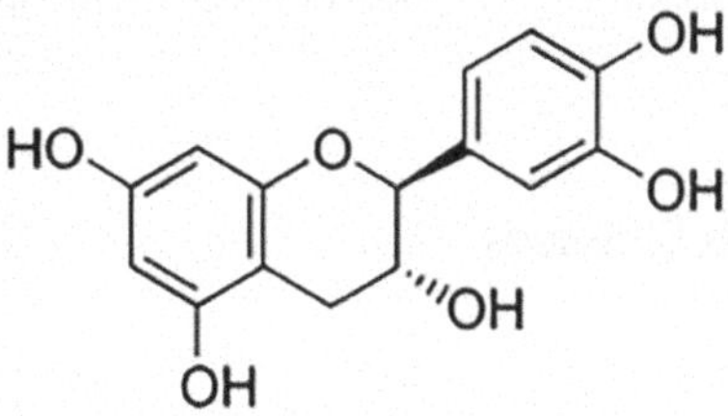
Galloyl glucoside	
Galloyl glucoside derivative	
Flavanols	
Theaflavin derivative	
Catechin	
(Epi)-catechin	
(Epi)-afzelechin isomers	Catechin
Diflavanoid isomers	
Flavonols	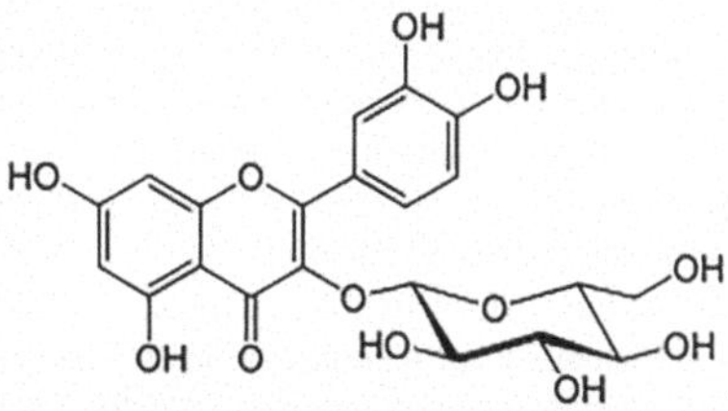
Astilbin	
Quercetin-3-glucoside	
Quercetin-rhamnoside	
Kaempferol-rhamnoside isomers	
Flavanones	
Pinocembrin 7-neohesperidoside	
Pinocembrin 7-rutinoside	Quercetin-3-glucoside
Flavones	
Hexametoxyflavone	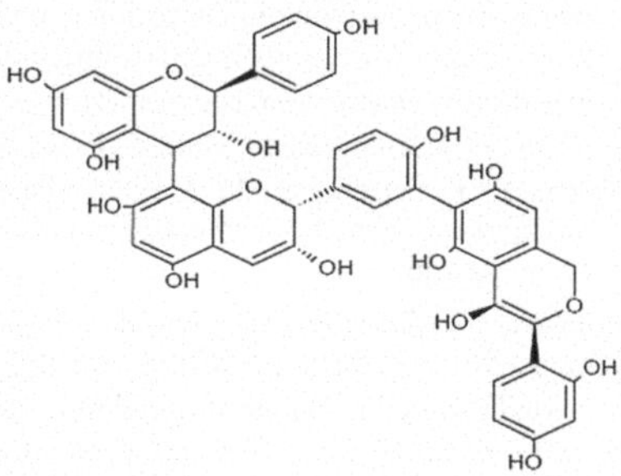
Proanthocyanidins/prodelphinidins	
(Epi)-galocatechin-(epi)-catechin isomers	
Procyanidin derivatives	
(Epi)-catechin-(epi)-catechin	
Cassanidin A	
(Epi)-guibourtini dol-(epi)-catechin isomers	
Catechin-guibourtinidol-cassiaflavan	
	Cassanidin A

Finally, it is important to remark that proanthocyanidins have also been identified in Carao seed extracts (Table 5).

2 Conclusion

Carao (*C. grandis*) is a plant used in popular medicine mainly with indigenous groups in South America to treat different diseases since ancient times. So this chapter summarizes its chemical composition and describes its potential nutritional, pharmacological, and medicinal applications. Due to its proximal, mineral, and bioactive compounds content, the carao fruit is considered a potential functional and nutraceutical food, which can be used as an active ingredient for the fortification and enrichment of foods in people with special diets.

References

Albuquerque, P. B. S., Cerqueira, M. A., Vicente, A., Teixeira, J. A., & Carneiro da Cunha, M. G. (2017). Immobilization of bioactive compounds in *Cassia grandis* galactomannan-based films: influence on physicochemical properties. *International Journal of Biological Macromolecules, 96*, 727–735. https://doi.org/10.1016/j.ijbiomac.2016.12.081

Alimirah, F., Peng, X., Murillo, G., & Mehta, R. G. (2011). Functional significance of vitamin D receptor Fokl polymorphism in human breast cancer cells. *PLoS One, 6*(1), e16024. https://doi.org/10.1371/journal.pone.0016024

Amancio, O. M. S. (2017). *Funções Plenamente reconhecidas de nutrientes cobre*. Série de Publicações ILSI Brasil.

Atanacković, M., Petrović, A., Jović, S., Bukarica, L. G., Bursać, M., & Cvejić, J. (2012). Influence of winemaking techniques on the resveratrol content, total phenolic content and antioxidant potential of red wines. *Food Chemistry, 131*(2), 513–518. https://doi.org/10.1016/j.foodchem.2011.09.015

Barberá Mateos, J. M. (2008). La imagen de la salud. En: *Alimentos funcionales. Aproximación a una nueva alimentación*. Dirección general de salud pública y alimentación (pp. 10–29). http://www.madrid.org/cs/Satellite?blobcol=urldata&blobheader=application%2Fpdf&blobheadername1=Content-Disposition&blobheadervalue1=filename%3Dt065&blobkey=id&blobtable=MungoBlobs&blobwhere=1220428576848&ssbinary=true

Bomfim, G. I., Ferreira, R. A., & Silva-Mann, R. (2018). Genetic variability in natural populations of *Cassia grandis* L. f. *Floresta e Ambiente, 25*(4), e20160309. https://doi.org/10.1590/2179-8087.160309

Cárdenas, G., Arrazola, G., & Villalba, M. (2015). Frutas tropicales: Fuente de compuestos bioactivos naturales en la industria de alimentos. *Ingenium, 17*(33), 29–40. file:///C:/Users/lavoz/Downloads/Dialnet-FrutasTropicales-5327083%20(1).pdf

Carvalho, P. E. R. (2006). Cássia Rósea y Cassia grandis. *Colombo: Embrapa Florestas, Circular Técnica, 117*, 1–8. http://ainfo.cnptia.embrapa.br/digital/bitstream/CNPF-2009-09/40890/1/circ-tec117.pdf

Chen, Y. H., Utsunomuya, H., Pavone, M. E., Yin, P., & Bulun, S. E. (2011). Retinoic acid inhibits endometrial cancer cell growth via multiple genomic mechanisms. *Journal of Molecular Endocrinology, 46*(2), 139–153. https://doi.org/10.1530/JME-10-0064

Cohen, L., Azriel-Tamir, H., Arotsker, N., Sekler, I., & Hershfinkel, M. (2012). Zinc sensing receptor signaling, mediated by GPR39, reduces butyrate-induced cell death in HT29 colonocytes via upregulation of clusterin. *PLoS One, 7*(4), e35482. https://doi.org/10.1371/journal.pone.0035482

Cuppari, L., & Bazanelli, A. P. (2010). *Funções Plenamente reconhecidas de nutrientes Potássio*. Série de Publicações ILSI Brasil.

Da Costa Silva, T., Justino, A. B., Prado, D. G., Koch, G. A., Martins, M. M., de Souza Santos, P., Lemos de Morais, S. A., Goulart, L. R., Scalon Cunha, L. C., Ferreira de Sousa, R. M., et al. (2019). Chemical composition, antioxidant activity and inhibitory capacity of α-amylase, α-glucosidase, lipase and non-enzymatic glycation, in vitro, of the leaves of *Cassia bakeriana* Craib. *Industrial Crops and Products, 140*, 111641. https://doi.org/10.1016/j.indcrop.2019.111641

De França, N. A., & Martini, L. A. (2014). *Funções Plenamente reconhecidas de nutrientes Cálcio*. Série de Publicações ILSI Brasil.

Demirkiran, O., Topcu, G., Hussain, J., Uddin Ahmad, V., & Choudhary, M. I. (2011). Structure elucidation of two new unusual monoterpene glycosides from Euphorbia decipiens, by 1D and 2D NMR experiments. *Magnetic Resonance in Chemistry, 49*, 673–677. https://doi.org/10.1002/mrc.2795

Deshpande, A. H., & Bhalsing, R. S. (2011). Phytochemical analysis of *Cassia obtusifolia, Cassia auriculata, Tephrosia purpurea, Helicteres isora* and *Centella asiatica*. *International Journal of Pharma and Biosciences, 2*(3), 363–367. https://ijpbs.net/abstract.php?article=OTI2

DRI, Dietary Reference Intakes. (2011). *Estimated average requirements* (pp. 1–8). Food and Nutrition Board, Institute of Medicine, National Academies.

Empresa Brasileira de Pesquisa Agropecuária, EMBRAPA. (2009). *Manual de análisis químicos de suelos, plantas y fertilizantes*. 2a edición revisada y ampliada, Brasilia, DF, 627 p.

Epstein, E., & Bloom, A. J. (2006). *Nutrição mineral de plantas: princípios e perspectivas* (403p). Editora Planta.

Fanali, C., Leyva-Jiménez, F., Lozano-Sánchez, J., Borrás-Linares, I., Arráez-Román, D., & Segura-Carretero, A. (2018). Comparative study of conventional and pressurized liquid extraction for recovering bioactive compounds from *Lippia citriodora* leaves. *Food Res Int, 109*, 213–222. https://doi.org/10.1016/j.foodres.2018.04.035

Fernández, I. M., Maldonado, S. A. S., Ferraz, V. P., Fuentes, J. A. M., & Filho, A. A. M. (2021). Chemical characterization of seeds of Amazon fruits as nutritional contribution with functional medicinal potential. *African Journal of Pharmacy and Pharmacology, 14*(4), 67–76. https://doi.org/10.5897/AJPP2020.5124

Finley, J. W., Kong, A. N., Hintze, K. J., Jeffery, E. H., Ji, L. L., & Lei, X. G. (2011). Antioxidants in foods: State of the science important to the food industry. *Journal of Agricultural and Food Chemistry, 59*(13), 6837–6846. https://doi.org/10.1021/jf2013875

Fuentes, J. A. M., López-Salas, L., Borrás-Linares, I., Navarro-Alarcón, M., Segura-Carretero, A., & Lozano-Sánchez, J. (2021). Development of an innovative pressurized liquid extraction procedure by response surface methodology to recover bioactive compounds from carao tree seeds. *Food, 10*(2), 398. https://doi.org/10.3390/foods10020398

Godevac, D., Zdunic, G., Sabikin, K., Vajs, V., & Menkovié, N. (2008). Antioxidant activity of nine *Fabaceae* species growing in Serbia and Montenegro. *Fitoterapia, 79*, 185–187. https://doi.org/10.1016/j.fitote.2007.10.001

Gómez, M. (2010). *Desarrollo y evaluación de estrategias analíticas para la caracterización de compuestos bioactivos en alimentos funcionales*. Departamento de Química Analítica de la Facultad de Ciencias, Universidad de Granada. https://dialnet.unirioja.es/servlet/tesis?codigo=63095

Gonçalves, J., Silva, C. L., Castilho, P. C., & Câmara, J. S. (2013). An attractive, sensitive and high-throughput strategy based on microextraction by packed sorbent followed by UHPLC-PDA analysis for quantification of hydroxybenzoic and hydroxycinnamic acids in wines. *Microchemical Journal, 106*, 129–138. https://doi.org/10.1016/j.microc.2012.05.037

Gresele, P., Cerletti, C., Guglielmini, G., Pignatelli, P., de Gaetano, G., & Violi, F. (2011). Effects of resveratrol and other wine polyphenols on vascular function: An update. *The Journal of Nutritional Biochemistry, 22*(3), 201–211. https://doi.org/10.1016/j.jnutbio.2010.07.004

Haminiuk, C. W. I., Maciel, G. M., Plata-Oviedo, M. S. V., & Peralta, R. M. (2012). Phenolic compounds in fruits—An overview. *International Journal of Food Science and Technology, 47* (10), 2023–2044. https://doi.org/10.1111/j.1365-2621.2012.03067.x

Hegazi, N. M., & Hashim, A. N. (2016). Grandisin, 2-methoxy 6,7,2′,6′-tetrahydroxy flavanone 6-O-glucoside, from *Cassia grandis* leaves—Antioxidant and cytotoxic activities. *Pharmacology, 71*(9), 544–547. https://doi.org/10.1691/ph.2016.6634

House P, & Lagos-Witte, S. (1989). *Manual de 50 plantas medicinales de Honduras* (p. 48). CONSH/CIIR/UNAH.

ICRAF, AgroForestryTree Database. (2014). International Centre for Research in Agroforestry. https://www.worldagroforestry.org/output/agroforestree-database

Instituto Adolfo Lutz. (2008). *Métodos físico-químicos para análise de alimentos* (4th ed.). IAL.

Jiménez-Sánchez, C., Lozano-Sánchez, J., Gabaldón-Hernández, J. A., Segura-Carretero, A., & Fernández- Gutiérrez, A. (2015). RP-HPLC–ESI–QTOF/MS2 based strategy for the comprehensive metabolite profiling of *Sclerocarya Birrea* (Marula) bark. *Industrial Crops and Products, 71*, 214–234. https://doi.org/10.1016/j.indcrop.2015.01.068

Kabila, B., Sidhu, M. C., & Ahluwalia, A. S. (2017). Phytochemical profiling of different *Cassia species* (review). *International Journal of Pharmaceutical & Biological Archives, 8*(2), 12–20. https://www.ijpba.info/ijpba/index.php/ijpba/article/view/1517

Korlam, S., Papani, S., Mylisetti, V. P., & Chitoor, M. S. (2016). Preliminary phytochemical screening of fruits of *Cassia spectabilis* D.C. *Journal of Biomedical and Pharmaceutical Research, 5*(4), 41–44. http://ijseas.com/volume2/v2i9/ijseas20160908.pdf

Kotipalli, H., Battu-Ganga, R., & Devarakonda, R. (2017). Qualitative physicochemical, phytochemical analysis and quantitative estimation of total phenols, flavonoids and alkaloids of *Cassia grandis*. *Journal of Global Trends in Pharmaceutical Sciences, 8*(2), 3860–3867. https://www.jgtps.com/admin/uploads/r8BH7U.pdf

Kris-Etherton, P., Hecker, K., Bonanome, A., Coval, S., Binkoski, A., Hilpert, K. et al. (2002). Bioactive compounds in foods: Their role in the prevention of cardiovascular disease and cancer. The American Journal of Medicine, 113(9): 71-88. Doi: https://doi.org/10.1016/s0002-9343(01)00995-0.

Kumar, R. S., Anburaj, G., & Vasantha, S. (2016). Phytochemical investigation of ethanol, methanol, hydro—Alcoholic and aqueous flower extracts of *Cassia* species. *Asian Journal of Innovative Research, 1*(3), 32–35.

Lafourcade, P. A., Rodríguez, A. J., Escalona A. J., & Fuenzalida, C. (2014). State of the art in *Cassia grandis* L. f. (Cañandonga). *Revista Cubana de Plantas Medicinales, 19*(1), 21–28. http://www.revplantasmedicinales.sld.cu/index.php/pla/article/view/147/52

Lagarto, P. A., & Guerra, S. M. (2005). Toxicidad aguda oral a partir de 3 formas farmacéuticas de *Cassia grandis* L. *Revista Cubana de Plantas Medicinales, 5*, 68–70. http://scielo.sld.cu/scielo.php?script=sci_arttext&pid=S1028-47962000000200009

Leyva-Jiménez, F., Lozano-Sánchez, J., Borrás-Linares, I., Arráez-Román, D., & Segura-Carretero, A. (2019). Manufacturing design to improve the attainment of functional ingredients from *Aloysia citriodora* leaves by advanced microwave technology. *Journal of Industrial and Engineering Chemistry, 79*, 52–61, ISSN 1226-086X. https://doi.org/10.1016/j.jiec.2019.04.060

Li, Y., Kong, D., Ahmad, A., Bao, B., & Sarkar, F. H. (2012). Targeting bone remodeling by isoflavone and 3,3′-diindolylmethane in the context of prostate cancer bone metastasis. *PLoS One, 7*, e33011. https://doi.org/10.1371/journal.pone.0033011

Lichtenthaler, H. K., & Buschmann, C. (2001). Chlorophylls and carotenoids: Measurement and characterization by UV-VIS spectroscopy. *Current Protocols in Food Analytical Chemistry, F4*, 3–F4.8. https://doi.org/10.1002/0471142913.faf0403s01

Lisboa, W. (2015). *Ciclo do Enxofre-Bacterias Sulfitogenica*. https://prezi.com/whnr68fmklir/ciclo-do-enxofre-bacterias-sulfitogenica

Lodha, S. R., Joshi, S. V., Vyas, B. A., Upadhye, M. C., Kirve, M. S., & Salunke, S. S. (2010). Assessment of the antidiabetic potential of *Cassia grandis* using an in vivo model. *Journal of Advanced Pharmaceutical Technology & Research, 1*, 330–333. https://doi.org/10.4103/0110-5558.72429

Malavolta, E. (2006). *Manual de nutrição mineral de plantas*. Editora Agronômica Ceres.

Maldonado, S. A. S., Fernandez, I. M., Aleman, R. S., Fuentes, J. A. M., & Ferreira, M. C. C. (2020). Determination of total phenolic compounds, antioxidant activity and nutrients in Brazil nuts (*Bertholletia excelsa* H. B. K.). *Medicinal Plants Research, 14*(8), 373–376. https://doi.org/10.5897/JMPR2020.6953

Marcía Fuentes, J., Montero Fernández, I., Saravia, S., Varela, I., Silva, C., Hernandez, F., et al. (2020a). Physical-chemical evaluation of the *Cassia grandis* L. as fortifying egg powder. *Journal of Agricultural Science, 12*(8), 277–282. https://doi.org/10.5539/jas.v12n8p277

Marcía Fuentes, J., Montero Fernández, I., Zumbado, H., Lozano Sánchez, J., Santos Alemán, R., Navarro Alarcón, M., Bórras Linares, I., & Saravia, S. (2020b). Quantification of bioactive molecules, minerals and bromatological analysis in Carao (*Cassia grandis*). *Journal of Agricultural Science, 12*(3), 88–94. https://doi.org/10.5539/jas.v12n3p88

Mendiola, J., Herrero, M., Cifuentes, A., & Ibañez, E. (2007). Use of compressed fluids for sample preparation: Food applications. *Journal of Chromatography A, 1152*(1–2), 234–246. https://doi.org/10.1016/j.chroma.2007.02.046

Mendoza, N. A. S., Martínez, C. J., Martínez, A. C., Barba, S. M. C., & Ortiz, G. D. (2016). Physical, nutritional and non nutritional characterization of Inga paterno seeds. *Revista Chilena Nutricion, 43*, 400–407. https://doi.org/10.4067/S0717-75182016000400010

Meng, Q., Li, G., Luo, B., Wang, L., Lu, Y., & Liu, W. (2016). Screening and isolation of natural antioxidants from *Ziziphora clinopodioides* Lam. with high performance liquid chromatography coupled to a post-column Ce(IV) reduction capacity assay. *RSC Advances, 6*, 62378–62384. https://doi.org/10.1039/C6RA08588A

Miranda, A. L. P., & Fraga, C. A. M. (2006). Atividade Seqüestradora de Radical Livre Determinação do Potencial Antioxidante de Substâncias Bioativas. In A. Monge & C. R. Ganellin (Eds.), *Practical studies for medicinal chemistry*. IUPAC.

Montero, I. F., Saravia, S. A. M., Santos, R. A., dos Santos, R. C., Marcía, J. A. F., & da Costa, H. N. R. (2020). Nutrients in Amazonian fruit pulps with functional and pharmacological interest. *African Journal of Pharmacy and Pharmacology, 14*(5), 118–127. https://doi.org/10.5897/AJPP2020.5136

Muñoz Jáuregui, A. M., & Ramos Escudero, F. (2007). Componentes fenólicos de la dieta y sus propiedades biomedicinales. *Revista Horizonte Médico, 7*(1), 1–10. ISSN: 1727-558X. https://www.redalyc.org/pdf/3716/371637115003.pdf

Nasti, N., Borrás-Linares, I., Lozano-Sánchez, J., Svac-Gajic, J., & Segura-Carretero, A. (2018). Optimization of the extraction of phytochemicals from black mulberry (*Morus nigra* L.) leaves. *Journal of Industrial and Engineering Chemistry, 68*, 282–292. https://doi.org/10.1016/j.jiec.2018.07.055

Olmedilla, B., & Granado, F. (2007). *Componentes bioactivos. Alimentos funcionales: aproximación a una nueva alimentación. Instituto de Nutrición y Trastornos alimentarios.* Dirección general de Salud Pública y alimentación. http://www.madrid.org/bvirtual/BVCM009703.pdf

Pandey, P. R., Okuda, H., Watabe, M., Pai, S. K., Liu, W., Kobayashi, A., et al. (2011). Resveratrol suppresses growth of cancer stem-like cells by inhibiting fatty acid synthase. *Breast Cancer Research and Treatment, 130*, 387–398. https://doi.org/10.1007/s10549-010-1300-6

Peñarrieta, J. M., Tejeda, L., Mollinedo, P., Vila, J. L., & Bravo, J. A. (2014). Compuestos fenólicos y su presencia en alimentos. *Revista Boliviana de Química, 31*(2), 68–81. https://www.redalyc.org/pdf/4263/426339682006.pdf

PIER, Pacific Islands Ecosystems at Risk. (2014). *Results of the risk assessment: Cassia grandis*. HEAR, University of Hawaii. http://www.hear.org/pier/wra/pacific/cassia_grandis_htmlwra.htm

Pinela, J., Barros, L., Carvalho, A. M., & Ferreira, I. C. F. R. (2011). Influence of the drying method in the antioxidant potential and chemical composition of four shrubby flowering plants from the tribe Genisteae (Fabaceae). *Food and Chemical Toxicology, 49*, 2983–2989. https://doi.org/10.1016/j.fct.2011.07.054

Prada, A. L., Achod, L. D., Keita, H., Carvalho, J. C., & de Sousa, T. P. (2020). Development, pharmacological and toxicological evaluation of a new tablet formulation based on *Cassia grandis* fruit extract. *Sustainable Chemistry and Pharmacy, 16*, 100244. https://doi.org/10.1016/j.scp.2020.100244

Prada, A. L., Bitencourt, A. P., Amado, J. R., Cruz, R. A., Carvalho, J. C., & Fernández, C. P. (2016). Development and characterization of *Cassia grandis* and *Bixa orellana* nanoformulations. *Current Topics in Medicinal Chemistry, 16*(18), 2057–2065. https://doi.org/10.2174/1568026616666160215161103

Prada, A. L. Keita, H., de Sousa, T. P., Lima, E. S., Acho, L., da Silva, M., Carvalho, J., & Amado, J. (2019). *Cassia grandis* Lf nanodispersion is a hypoglycemic product with a potent α-glucosidase and pancreatic lipase inhibitor effect. *Saudi Pharmaceutical Journal, 27*(2), 191–199. https://doi.org/10.1016/j.jsps.2018.10.003

Ramos, E., Paz, J., Ortiz, G., & Núñez, K. (2014). *Determinación del contenido de hierro, saponinas y porfirinas en Cassia grandis L., procedente de Masaya, Chinandega y Jalapa* (p. 46). Universidad Nacional Autónoma de Nicaragua.

Rao, M. R., & Chatterjee, B. (2016). Preliminary phytochemical, antioxidant and antimicrobial activities of different extracts of *Cassia tora* and *Trichodesma indicum*. *International Journal of Pharmacy and Technology, 8*(2), 12578–12597. http://www.ijptonline.com/wp-content/uploads/2016/07/12578-12597.pdf

Ravishankar, D., Rajora, A., Greco, F., & Osborn, E. (2013). Flavonoids as prospective compounds for anti-cancer therapy. *The International Journal of Biochemistry and Cell Biology, 45*(12), 2821–2831. https://doi.org/10.1016/j.biocel.2013.10.004

Rehman, A., Tong, Q., Jafari, S. M., Assadpour, E., Shehzad, Q., Aadil, R. M., et al. (2020). Carotenoid-loaded nanocarriers: A comprehensive review. *Advances in Colloid and Interface Science, 275*, 102048. https://doi.org/10.1016/j.cis.2019.102048

Roels, H. M. G., Meires, M., Delos, I., Ortega, R., Lauwerys, J. P., & Buschet, D. L. (1997). Influence of the route o administration and the chemical form ($MnCl_2$, MnO_2) on the absorption and cerebral distribution of manganese in rats. *Archives of Toxicology, 71*, 223–230. https://doi.org/10.1007/s002040050380

Rojas, S., Lopera, J. S., Uribe, A., Correa, S., Perilla, N., & Marín, J. S. (2015). Consumo de nutracéuticos, una alternativa en la prevención de las enfermedades crónicas no transmisibles. *Revista Biosalud, 14*(2), 91–103. https://doi.org/10.17151/biosa.2015.14.2.9

Romero, M. P., Costa, B., Ferreira, V., & Porto, A. L. (2018). CgTI, a novel thermostable Kunitz trypsin inhibitor purified from *Cassia grandis* seeds: Purification, characterization and termiticidal activity. *International Journal of Biological Macromolecules, 118*, 2296–2306. https://doi.org/10.1016/j.ijbiomac.2018.07.110

Sadiq, I. S., Shuaibu, M., Bello, A. B., Tureta, S. G., Isah, A., & Izuagie, T. (2012). Phytochemistry and antimicrobial activities of *Cassia occidentalis* used for herbal remedies. *Journal of Chemical Engineering, 1*(1):38–41. https://www.researchgate.net/profile/Sadiq_Shina2/publication/284034861_Phytochemistry_and_antimicrobial_activities_of_Cassia_occidentalis_used_for_herbal_remedies/links/57f8a96c08ae91deaa607280/Phytochemistry-and-antimicrobial-activities-of-Cassia-occidentalis-used-for-herbal-remedies.pdf

Sanchez-Moreno, C., Larrauri, J. A., & Saura-Calixto, F. (1998). A procedure to the antiradical efficiency of polyphenolds. *Journal of the Science of Food and Agriculture, 76*, 270–276. https://doi.org/10.1002/(SICI)1097-0010(199802)76:2<270::AID-JSFA945>3.0.CO;2-9

Shen, Z., Zhang, X., Tang, J., Kasiappan, R., Jinwal, U., Li, P., et al. (2011). The coupling of epidermal growth factor receptor down regulation by 1 alpha, 25-dihydroxyvitamin D3 to the hormone-induced cell cycle arrest at the G1-S checkpoint in ovarian cancer cells. *Molecular and Cellular Endocrinology, 338*, 58–67. https://doi.org/10.1016/j.mce.2011.02.023

Sobeh, M., Mahmoud, M. F., Abdelfattah, M. A. O., Cheng, H., El-Shazly, A. M., & Wink, M. (2018). A proanthocyanidin-rich extract from *Cassia abbreviata* exhibits antioxidant and hepatoprotective activities in vivo. *Journal of Ethnopharmacology, 213*, 38–47. https://doi.org/10.1016/j.jep.2017.11.007

Sun, Q., Prasad, R., Rosenthal, E., & Katiyar, S. K. (2012). Grape seed proanthocyanidins inhibit the invasiveness of human HNSCC cells by targeting EGFR and reversing the epithelial-to-mesenchymal transition. *PLoS One, 7*, e31093. https://doi.org/10.1371/journal.pone.0031093

Tillán, C. J., Rodríguez, C. J., Gómez, M. J., Pardo, R. Z., & Agüero, F. S. (2004). Actividad antianémica de la *Cassia grandis* L. *Revista Cubana de Farmacia, 38*(3), 1. http://scielo.sld.cu/scielo.php?script=sci_arttext&pid=S0034-75152004000300009

Tropicos.org. (2021). Jardín Botánico de Missouri; *Cassia grandis* (en línea). Recuperado el 07 de marzo de 2021. Obtenido de http://www.tropicos.org

Ucar, M. B., Ucar, G., Pizzi, A., & Gonultas, O. (2013). Characterization of Pinus brutia bark tannin by MALDI-TOF MS and 13C NMR. *Industrial Crops and Products, 49*, 697–704. https://doi.org/10.1016/j.indcrop.2013.06.010

Usman, W. A., Jada, M. S., & Abdulazeez, M. B. (2014). Crude extract of leaf and stem bark of *Cassia siamea* have in vitro antimicrobial activity. *Open Journal of Biochemistry, 1*(1), 43–48. http://citeseerx.ist.psu.edu/viewdoc/download?doi=10.1.1.908.6037&rep=rep1&type=pdf

Valencia-Avilés, E., Ignacio-Figueroa, I., Sosa-Martínez, E., Bartolomé-Camacho, M. C., Martínez-Flores, H. E., & García-Pérez, M. E. (2017). Polifenoles: propiedades antioxidantes y toxicológicas. *Revista de la Facultad de Ciencias Químicas, 16*, 1–15. ISSN: 1390-1869. https://publicaciones.ucuenca.edu.ec/ojs/index.php/quimica/article/view/1583

Valente, I. M., Maia, M. R. G., Malushi, N., Oliveira, H. M., Papa, L., Rodrigues, J. A., & Cabrita, A. R. J. (2018). Profiling of phenolic compounds and antioxidant properties of European varieties and cultivars of *Vicia faba* L. pods. *Phytochemistry, 152*, 223–229. https://doi.org/10.1016/j.phytochem.2018.05.011

Wang, L., Lou, G., Ma, Z., & Liu, X. (2011). Chemical constituents with antioxidant activities from litchi (*Litchi chinensis* Sonn.) seeds. *Food Chemistry, 126*, 1081–1087. https://doi.org/10.1016/j.foodchem.2010.11.133

Wegert, J., Bausenwein, S., Kneitz, S., Roth, S., Graf, N., Geissinger, E., et al. (2011). Retinoic acid pathway activity in Wilms tumors and characterization of biological responses in vitro. *Molecular Cancer, 10*, 1–12. https://doi.org/10.1186/1476-4598-10-136

Wolf, F. I., & Cittadini, A. (2003). Chemistry and biochemistry of magnesium. *Molecular Aspects of Medicine, 24*, 11–26. https://doi.org/10.1016/s0098-2997(02)00087-0

Wolfe, K., Wu, X., & Liu, R. H. (2003). Antioxidant activity of apple peels. *Journal of Agricultural and Food Chemistry, 51*, 609–614. https://doi.org/10.1021/jf020782a

Xiao, J., Muzashvili, T., & Georgiev, M. (2014). Advances in the biotechnological glycosylation of valuable flavonoids. *Biotechnology Advances, 32*(6), 1145–1156. https://doi.org/10.1016/j.biotechadv.2014.04.006

Xu, X., Xie, H., Hao, J., Jiang, Y., & Wei, X. (2011). Flavonoid glycosides from the seeds of *Litchi chinensis*. *Journal of Agricultural and Food Chemistry, 59*, 1205–1209. https://doi.org/10.1021/jf104387y

Zapata, K., Cortes, F. B., & Rojano, B. A. (2013). Polifenoles y Actividad Antioxidante del Fruto de Guayaba Agria (*Psidium araca*). *Información Tecnológica, 24*(5), 103–112. https://doi.org/10.4067/S0718-07642013000500012

Zhang, H., Birch, J., Ma, Z., Xie, C., Yang, H., Bekhit, A., et al. (2019). Optimization of microwave-assisted extraction of bioactive compounds from New Zealand and Chinese *Asparagus officinalis* L. roots. *Journal of Food Science and Technology, 56*(2), 799–810. https://doi.org/10.1007/s13197-018-3540-0

Challenges and Future Prospects of Biotechnology

S. A. Aransiola, M. O. Victor-Ekwebelem, A. A. Ikhumetse, and O. P. Abioye

1 Introduction

There has been a tremendous progress in the field of biotechnology over the years, and its application cuts across agriculture, marine biotechnology, bioengineering, bio-manufacturing, biomedical engineering, drug discovery, vaccine development and in environmental microbiology (Daniotti & Re, 2021). Despite its numerous benefits, biotechnology remains low due to several socio-economic, ethical, health, or political concerns (Ivase et al., 2019). Therefore, this chapter reviews the challenges and future prospects of biotechnology.

Biotechnology posed a great promising area based on current and future applications of it, to solve medical, environment, energy, agricultural, and military problems. This promise is from direct application of biotechnology and from the growing area of multidisciplinary research that combines biotechnology with other sciences like materials science, physics, chemistry, and engineering just to name a few. However, there is also growing concern over possible dangers from this area of science based on the potential for mishaps from honest scientists and applications purposefully developed to cause harmful or even devastating effects (Munis et al., 2019). The areas of biotechnology with tremendous future opportunities that can impact all include: (a) Drugs and pharmaceuticals development; (b) Medical device and diagnostics; (c) Noninvasive sensor in agriculture; (d) Rapid testing of

S. A. Aransiola (✉)
Bioresources Development Centre, National Biotechnology Development Agency, Ogbomoso, Nigeria

M. O. Victor-Ekwebelem
Department of Biology/Microbiology/Biotechnology, Alex Ekwueme Federal University, Abakaliki, Ebonyi, Nigeria

A. A. Ikhumetse · O. P. Abioye
Department of Microbiology, Federal University of Technology, Minna, Nigeria

N. R. Maddela, L. C. García (eds.), *Innovations in Biotechnology for a Sustainable Future*, https://doi.org/10.1007/978-3-030-80108-3_20

pathological conditions; (e) Quick testing for food adulteration; (f) Precision agriculture and biofortification; and (g) Machine learning and artificial intelligence in biotechnology.

2 Marine Biotechnology in Sustainable Industrial Development

2.1 *Discovery of New Products*

It is estimated that over 90% of marine biodiversity is completely unknown as our knowledge of the environment is restricted. This is due to the technical difficulty of accessing deep seas, thus impeding the collection of samples (Daniotti & Re, 2021). Also, majority of organisms isolated from the marine environment are difficult to analyze and therefore taxonomic classification is complex, resulting in potential errors that could compromise the entire process of drug discovery due to the impossibility of reproducing the isolation event and the subsequent identification of the bioactive compound (Kasanah & Triyanto, 2019).

2.2 *Sustainable Production*

When using marine organisms for the production of high value-added substances and biofuels, the vulnerability of the marine environment must be considered because the quantities supplied directly by marine organisms do not support industrial requirements and the more limited ones of the drug discovery process (Vlachou et al., 2018). The unswerving collection of bioactive compounds of industrial interest is never maintainable, particularly as most of these species are at the verge of extinction and their unwarranted abuse could destroy the intricate balance of the ecosystem (Daniotti & Re, 2021).

2.3 *Extremozymes in Biotechnology Applications*

There is usually low yield or activity of enzymes when they are cultivated on a large bioreactor scale, usually required for commercial applications, and this has been a significant drawback in utilizing extremozymes in biotechnology (Sarmiento et al., 2015). Active alcohol dehydrogenase (HvADH), laccase (LccA) enzymes and bacterioruberin and halocins proteins produced from *Haloferax volcanii* must be done on a large scale to realize their true biotechnological capacity (Haque et al., 2020), but expression and purification of bulk quantities of highly active proteins in a halophilic environment is however demanding as co-purification of unwanted nonspecific proteins with the target protein is often encountered following affinity tagged protein purification (Sarmiento et al., 2015). When using *Haloferax volcanii* as the host, their intrinsic ability to form biofilms is a deadlock on large-scale protein over expression (Chimileski et al., 2014), since these biofilms can interfere with expensive sensors present in the bioreactor during the fermentation process and may also alter the characteristics of expressed proteins. Furthermore, when culturing *Haloferax volcanii,* there is the need to use reactors made of alternative materials rather than the readily available stainless steel as the molar concentration of salt required to grow this organism can quickly corrode stainless-steel bioreactors (Haque et al., 2020).

3 Agricultural Biotechnology

3.1 Plant Biotechnology

New technologies, especially in transgenes and gene editing, will be required as the demands of human beings such as higher contents of vitamins and micronutrients found in cereals, increased shelf lives of vegetables and fruits and reduced allergens increases perpetually (Nguyen & Ly, 2018). There is still a large gap in successfully creating new crop varieties with desired traits for human consumption, even when gene/genome editing technologies have been successfully tried in many research laboratories (Moshelion & Altman, 2015). Also, having an insight into the roles of genes governing complex traits to actively improve agronomic performance or control adaptations to abiotic stresses is a matter of concern, as genetically modified organisms (GMOs), transgenic crops, and recombinant DNA technology are the future trends in plant biotechnology. The complex traits of interest include a crop's ability to grow efficiently in drought, salinity, acidic, or aluminum-containing soils, competition with weeds, flowering time, heterosis, and durable resistance to diseases (Nguyen & Ly, 2018).

There is a rise in energy prices because of the reduction of fossil fuels and this rise requires new processes for the production of renewable energy sources called biofuels (Nguyen & Ly, 2018). One of the favorable materials for biofuel production through enzymatic fermentation and chemical transformations is lignocelluloses (Den et al., 2018), but a major challenge for biotechnology in the degradation of the stable polymer chain into sugar molecules for further fermentation and conversion is the modification or alteration of the properties of the polysaccharide profile in the cell walls of plant materials (Popa, 2018).

3.2 Animal Biotechnology

Semen sexing technology for selecting the sex of embryos relies on the principle of flow cytometric separation of fluorescence labeled sex chromosomes, but the low number of sexed sperms produced and the occurrence of sperms being damaged during the sorting process that reduces the efficiency of fertilization in later steps are the main drawbacks of this technique (Espinosa-Cervantes & Cordova-Izquierdo, 2013). Cloned animals derived from cloning technology on the other hand, often suffer from severe injuries or are not able to reproduce (Nguyen & Ly, 2018). In animal biotechnology, the issues of animal welfare should also be taken into consideration. Depending on one's personal beliefs, some people oppose the use of animals for any purpose, while others have specific concerns about the impacts that genetic engineering and cloning may bring by producing human therapeutic or industrial proteins (Nabavizadeh et al., 2016).

Another drawback in animal biotechnology is in the field of transplantation of living cells, tissues, or organs, where there is always shortage of organs for clinical implantation in patients who need a replacement organ at the end stage of failure (Nguyen & Ly, 2018). Although, tissues or organs from some animals from the order primates or from pigs could serve as candidates for transplantation in humans, but the lifespans of the donor animals are shorter than humans; therefore, the aging of the grafted tissues at a quicker rate is still a challenge in xenotransplantation technology (Hryhorowicz et al., 2017). Only a few temporarily successful cases of xenotransplantation have thus far been published as animal rights activists have also objected to xenotransplantation on ethical grounds.

3.3 Microbial Biotechnology

Research in microbial biotechnology is focused on three main areas in various application fields; agricultural practices, microbial enzymes for industry, and environment treatments. Stubble, straw, and sawdust which are byproducts in agriculture and forestation production contain stubborn polymers like lignin, cellulose, and hemicellulose, which are a challenge for the development of new technology for biodegradation to convert them into biofuels, feeds, and biofertilizers (Kilbane, 2016). Reactions of some enzymes such as lipozyme, lipase, cellulase, amylase, and xylose isomerase in organisms are efficiently performed under physiological conditions, but industrial conditions are far different with high substrate concentrations, sheering forces, high or low temperatures, and organic solvents. In addition, the requirements of regiospecific, chemospecific, and stereospecific reactions are challenging for industrial and pharmaceutical enzymes (Chapman et al., 2018). Therefore, most enzymes found in soil and water microbes are not able to display their desired activities under industrial conditions (Nguyen & Ly, 2018).

3.4 *Health and Medicinal Biotechnology*

Functional genomics is still a big challenge in gene identification, analysis of gene interactions, and the relationships between genotypes and phenotypes in complex diseases (Nguyen & Ly, 2018).

3.5 *Environmental Biotechnology*

Environmental challenges require newer technologies for environmental control, protection, and remediation. The constituents of environmental contaminants are becoming diverse, and as such, require more effective microorganisms for environmental treatment and management (Vujic et al., 2015). Also, genetically modified organisms are seen as biohazard to the environment in some industrial processes (Nguyen & Ly, 2018). In general, the intricate balances between hosts, pests, humans, and the environment should be seen as a challenge for biotechnology in the future.

4 Challenges Faced in Using CCRISPR (Clustered Regularly Interspaced Short Palindromic Repeats)

One of the key challenges in using CRISPR (clustered regularly interspaced short palindromic repeats) is designing the RNA guide. Research has shown that some RNA guides are less efficient than others, some are inactive while others are promiscuous, and in the absence of a good RNA guide, multiple off-target effects can occur. In addition, several potential RNA guides can perform the same editing task, but each has different off-target outcomes. As a result, selecting the right RNA for the task at hand adds to the challenge of achieving proper design. In other words, the specificity and precision of CRISPR is largely conditioned by the type and specificity of the RNA guide (Vogel & Ben-Ouagrham-Gormley, 2018).

5 Future Prospects of Biotechnology in Machine Learning and Artificial Intelligence in Biotechnology

The rapid progress in biotechnology and in information technology has occurred in parallel over the last twenty years. The large data available and the new experimental technologies developed in the recent years make it easier and cheaper to carry out several biotechnology experiments that would have taken years to complete. The large amounts of data mostly derived through omics-technologies that are generated and stored, in the biotechnology research, create an array of new opportunities for researchers, as well as for companies that offer products and services in this area. At this point artificial intelligence (AI) and machine learning (ML) technologies to process, explore, and analyze these large data are new emerging area in biotechnology and therefore, AI and ML are promising areas important in advancing the benefits of biotechnology. Adaptation of AI and ML will allow biotechnology to solve complex societal challenges through its own systematic testing. For example, now electronic health records systems are gradually integrating in health sector so that health-related data can be accessed globally, opening the door to a holistic outlook (Sharma, 2020). Machine learning technology also holds promise for the future of the clinical trial. Biotech companies can quickly analyze data from current trials to predict the effectiveness of treatments down to a molecular level; they can also revisit data from previous trials to see if anything may have been missed, or if there may be new or different uses for an existing drug (Brian, 2020).

6 The Future Prospects of Biotechnology in Medicine, Medical Device, and Diagnostics

Genetic engineering can potentially regulate hereditary diseases. Brain-computer interfaces developed by using new nanomaterials that provide bidirectional neural communication will enable the patients to write words which they are imagine or to control the devices. These noninvasive interfaces may seem utopian, but can be a marvel of genetic engineering in the new era. In addition, studies are being carried out on treatments that may extend life, such as telomere modifications and reversing aging.

In modern medicine, doctors evaluate patients to clinically diagnose, treat, and prevent diseases. Thus, there are also benefits of a treatment regime that is based on the genotype of each individual to include epigenetic factors for the development of individualized medication selection, dose adjustment and individualized therapies to overcome a more traditional trail-and-error approach. For example, in the field of oncology, which plays a role in the prevention, management, and treatment of non-future cancers, there have been enormous strides that can be attributed to the development of immunotherapy, genomic and genetic engineering technology (Munis et al., 2019).

Another new face of modern genetic engineering is CRISPR/Cas9 technology and its therapeutic potential is excellent. As technology develops, the therapeutic potential of CRISPR/Cas9 will continue to increase. Nonetheless, CRISPR/Cas9 has many difficulties in fully developing its potential. Cas9 nickases and mutants that reduce nonspecific DNA binding are designed to alleviate these problems, although this is a problematic solution. The gene cargo distribution system remains the biggest obstacle to the routine use of CRISPR/Cas9, and a multipurpose delivery method has not yet emerged (Lino et al., 2018). Gene therapy is an experimental method to correct a defective gene that is responsible for the emergence of the disease. In addition to advances in delivery and expression technologies, future efforts will focus on new areas of gene therapy practice, such as new resistance genes and chimeric T-cell receptors. The sequence of the human genome is useful in many fields, from molecular medicine to human evolution. With the development of gene therapy technologies, the modelling of genetic diseases also increased (Kulkarni et al., 2018). In recent years, the renowned method of gene therapy has been used to treat a number of diseases on genetic models developed by CRISPR technology. For example, metabolic diseases, cardiovascular diseases, monogenic diseases. Also developed drugs are tested on these genetic models. If the correct gene is determined and the appropriate vectors are selected, then there should not be disease that cannot be treated.

7 Nanotechnology and Medical Biotechnology

Advances in nanotechnology also provide a rich framework for future developments in medical biotechnology. Related to disease prevention is the widespread use of antibiotics and antibacterial coatings to reduce the impact of infection on causing the failure of implanted medical devices. Nonetheless, such approaches have limited effectiveness. Nanotechnology now provides the tools for nano-texturing the surfaces of materials for medical implants, with the aim to mimic the bactericidal properties of some animal, plant and insect species, and their topographical features. For example, the surface nanostructures of cicada, dragonfly and butterfly wings, shark skin, gecko feet, taro and lotus leaves provide self-cleaning and bactericidal properties (Jaggessar et al., 2017). That type of bioinspiration provides great innovations in providing some biological-like characteristics that can be used to guide the surface structuration and synthesis of materials into functional devices and processes.

In biological systems there is a self-assembly y of molecules to create elegant nanostructured systems. A central feature of such biological nanostructured systems is the assembly of phospholipid bilayer membranes that both provide compartments (i.e., biological cells) to rationalize the overall function of complex organisms (e.g., plants, animals) and also to provide an environment in which to stabilize membrane proteins to assist in the sensing and actuating functions of biological cells (Bentley et al., 2018). These components of biological cells provide the basis for transport of ions and molecules between the cell and the surrounding environment inside the body, which supports the ability of cells to participate in physiological control of the body. Extending the concept of bioinspiration to include such elegant nanostructured biological self-assembly then provides an additional dimension to applying nanotechnology to medical biotechnology. That additional dimension is to include nanostructured bio-membranes into hybrid medical devices that can provide the ability for the medical device to communicate ions and molecules with the body. This then leads to the notion that an implanted medical device should integrate and

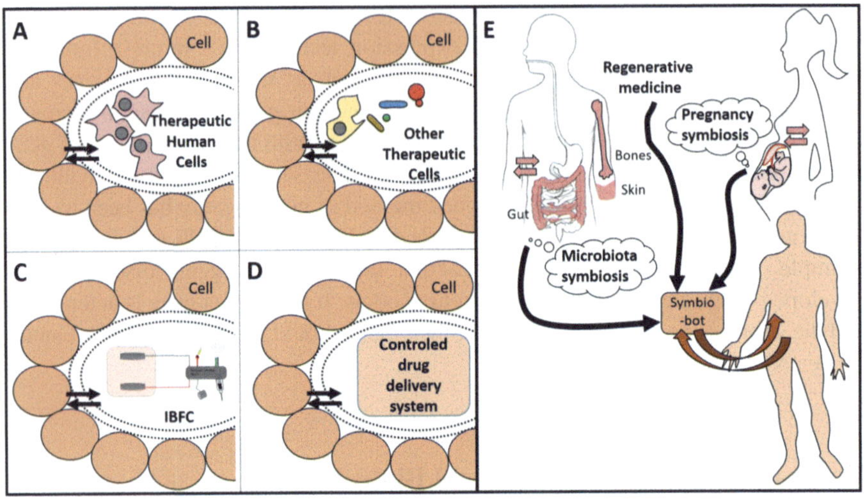

Fig. 1 Examples of symbiotic devices (symbio-bots) (**a–d**) that can be created in a bio-inspired way (**e**). Each device is separated with a smart porous packaging that allows a duplex communication. Therapeutic cells (**a**, **b**) need a porous encapsulation y that avoids an immune reaction and allows protection from both sides. They may be human cells, as MSC or specialized cells such as β-cell from Langerhans islets (**a**), or other eukaryotic or prokaryotic cells (**b**). Panel (**c**) shows an IBFC linked to an electronic medical device. Panel (**d**) shows a generic device delivering a therapeutic molecule. Panel (**e**) Existing symbiosis (i.e., microbiota or pregnancy) are a source of bioinspiration to establish a duplex communication between the body and its implants. Regenerative medicine should embrace this concept of bioinspiration for a better design and integration of implants, especially for future symbio-bots (reproduced with permission from Alcaraz et al. (2018))

become symbiotic with the body (Fig. 1). This extends the definition of a biocompatible system to one that requires stable exchange of materials between the implanted device and the body. Having this novel concept in mind will guide research in a new field between medical implant and regenerative medicine to create actual symbiotic devices (Alcaraz et al., 2018).

8 The Use of Digestible Sensors

Additionally, a treatment method in the future of medicine is the digestible sensors placed in pills. Data from these sensors are transmitted to doctors and family. Thanks to the biotechnology industry, more objects can now be printed using 3D printers. In the near future, printing of medical devices in underdeveloped areas and printing of live tissues, cells, or drugs may be imagined. But everyone is able to print medicines containing patented molecules in their homes, so ethical problems can emerge. Also, 3D printing is not the only way used to create body parts and artificial organs, it can be grown in the laboratory environment using biomaterials. The artificial organ is a device or biological material that is implanted in the body to alter a natural organ or function. The other innovation can be the sensors that can be digested to make a quick diagnosis. These sensors can be swallowed directly for gastrointestinal diseases. Further, the sensors embedded in the tooth can detect jaw movements and speech. More complex microchips that can mimic the whole human body are needed, and this final solution could arrive soon. As the amount of information increases, cognitive computers may be used instead of human in medical decision making (Munis et al., 2019)

9 Conclusion

In summary, despite series of challenges, the future of biotechnology research is strong and very promising. We presume that very soon a day will come, when breakthrough drugs will lead to a world without COVID-19, cancer, or AIDS, and many more life-threatening diseases, a world with sustainable research development that will tackle the need for food, environmental safety, low cost medical devices, energy, and many more societal giant challenges without compromising the world's resources.

References

Alcaraz, J. P., Cinquin, P., & Martin, D. K. (2018). Tackling the concept of symbiotic implantable medical devices with nanobiotechnologies. *Biotechnology Journal, 13*, 1800102.

Bentley, W. E., Martin, D. K., & Gotoh, T. (2018). Biomimetic and bioinspired biotechnology. *Biotechnology Journal, 13*, 1800670.

Brian, E. (2020). *Top biotechnology trends in 2020.* Retrieved from www.northeastern.edu/graduate/blog/emerging-biotechnology-trends/

Chapman, J., Ismail, A. E., & Dinu, C. Z. (2018). Industrial applications of enzymes: Recent advances, techniques, and outlooks. *Catalysts, 8*(6), 238. https://doi.org/10.3390/catal8060238

Chimileski, S., Franklin, M. J., & Papke, R. T. (2014). Biofilms formed by the archaeon *Haloferax volcanii* exhibit cellular differentiation and social motility and facilitate horizontal gene transfer. *BMC Biology, 12*, 65–65. https://doi.org/10.1186/s12915-014-0065-5

Daniotti, S., & Re, I. (2021). Marine biotechnology: Challenges and development market trends for the enhancement of biotic resources in industrial pharmaceutical and food applications. A Statistical analysis of scientific literature and business models. *Marine Drugs, 19*(61), 1–35.

Den, W., Sharma, V. K., Lee, M., Nadadur, G., & Varma, R. S. (2018). Lignocellulosic biomass transformations via greener oxidative pretreatment processes: Access to energy and value-added chemicals. *Frontiers in Chemistry, 6*, 141. https://doi.org/10.3389/fchem.2018.00141

Espinosa-Cervantes, R., & Cordova-Izquierdo, A. (2013). Sexing sperm of domestic animals. *Tropical Animal Health and Production, 45*(1), 1–8.

Haque, R. U., Paradisi, F., & Allers, T. (2020). Haloferax volcanii for biotechnology applications: Challenges, current state and perspectives. *Applied Microbiology and Biotechnology, 104*, 1371–1382.

Hryhorowicz, M., Zeyland, J., Słomski, R., & Lipiński, D. (2017). Genetically modified pigs as organ donors for xenotransplantation. *Molecular Biotechnology, 59*(9-10), 435–444.

Ivase, T. J. P., Ali, J. B., Moveh, S., Dodo, Y. A., Otitolaiye, V. O., & Ogenyi, B. (2019). Current status and challenges of agricultural biotechnology in Nigeria: A concise review. *Journal of Multidisciplinary Engineering Science and Technology, 6*(9), 10656–10662.

Jaggessar, A., Shahali, H., Mathew, A., & Yarlagadda, P. K. D. V. (2017). Bio-mimicking nano and micro-structured surface fabrication for antibacterial properties in medical implants. *Journal of Nanobiotechnology, 15*(1), 64.

Kasanah, N., & Triyanto, T. (2019). Bioactivities of halometabolites from marine actinobacteria. *Biomolecules, 9*, 225.

Kilbane, J. J. (2016). Future Applications of Biotechnology to the Energy Industry. *Frontiers in Microbiology, 7*, 86. https://doi.org/10.3389/fmicb.2016.00086

Kulkarni, K. M., Khot, A. M., Lokapure, S. G., & Jadhav, S. A. (2018). Brief review on gene therapy. *Indo American Journal of Pharmaceutical Sciences, 5*(5), 3288–3299.

Lino, C. A., Harper, J. C., Carney, J. P., & Timlin, J. A. (2018). Delivering CRISPR: a review of the challenges and approaches. *Drug Delivery, 25*(1), 1234–1257.

Moshelion, M., & Altman, A. (2015). Current challenges and future perspectives of plant and agricultural biotechnology. *Trends in Biotechnology, 33*(6), 337–342.

Munis, D., Satya, P., Ratnesh, L., & Donald, K. M. (2019). Future biotechnology. *The Eurotech Journal of Medicine and Biotechnology, 3*(2), 52–56.

Nabavizadeh, S. L., Mehrabani, D., Vahedi, Z., & Manafi, F. (2016). Cloning: A review on bioethics, legal, jurisprudence and regenerative issues in Iran. *World Journal of Plastic Surgery, 5*(3), 213–225.

Nguyen, D. B., & Ly, T. T. (2018). Current research, challenges, and perspectives of biotechnology: An overview. *Vietnam Journal of Agricultural Sciences, 1*(2), 187–199.

Popa, V. I. (2018). Biomass for fuels and biomaterials. Biomass as renewable raw material to obtain bioproducts of high-tech value (pp. 1–37).

Sarmiento, F., Peralta, R., & Blamey, J. M. (2015). Cold and hot extremozymes: Industrial relevance and current trends. *Frontiers in Bioengineering and Biotechnology, 3*, 148–148. https://doi.org/10.3389/fbioe.2015.00148

Sharma, A. K. (2020). *Current progress in Biotechnology research has greater potential to be the basis of subsequent innovation to solve society's big challenges*. Retrieved from www.educationtimes.com/article/careers-science

Vlachou, P., Le-Goff, G., Alonso, C., Álvarez, P., Gallard, J. F., Fokialakis, N., & Ouazzani, J. (2018). Innovative approach to sustainable marine invertebrate chemistry and a scale-up technology for open marine ecosystems. *Marine Drugs, 16*, 152.

Vogel, K. M., & Ben-Ouagrham-Gormley, S. (2018). Anticipating emerging biotechnology threats; A case study of CRISPR. *Politics and the Life Sciences, 37*(2), 203–219.

Vujic, G., Gonzalez-Roof, A., Stanisavljević, N., & Ragossnig, A. M. (2015). Municipal solid waste development phases: Evidence from EU27. *Waste Management & Research, 33*(12), 1112–1120.

Index

N. R. Maddela, L. C. García (eds.), *Innovations in Biotechnology for a Sustainable Future*, https://doi.org/10.1007/978-3-030-80108-3

D

E

The manufacturer's authorised representative in the EU is Springer Nature Customer Service Centre GmbH, Europaplatz 3, 69115 Heidelberg, Germany. If you have any concerns regarding our products, please contact ProductSafety@springernature.com

Printed and bound by CPI Group (UK) Ltd, Croydon, CR0 4YY
30/09/2026
02225506-0012